Wetlands: Ecology, Conservation and Management

Volume 6

Series Editor
C. Max Finlayson,
Albury, New South Wales, Australia

The recognition that wetlands provide many values for people and are important foci for conservation worldwide has led to an increasing amount of research and management activity. This has resulted in an increased demand for high quality publications that outline both the value of wetlands and the many management steps necessary to ensure that they are maintained and even restored. Recent research and management activities in support of conservation and sustainable development provide a strong basis for the book series. The series presents current analyses of the many problems afflicting wetlands as well as assessments of their conservation status. Current research is described by leading academics and scientists from the biological and social sciences. Leading practitioners and managers provide analyses based on their vast experience.

The series provides an avenue for describing and explaining the functioning and processes that support the many wonderful and valuable wetland habitats, such as swamps, lagoons and marshes, and their species, such as waterbirds, plants and fish, as well as the most recent research directions. Proposals cover current research, conservation and management issues from around the world and provide the reader with new and relevant perspectives on wetland issues.

More information about this series at http://www.springer.com/series/7215

C. Max Finlayson • Gurdeep Rastogi
Deepak R. Mishra • Ajit K. Pattnaik
Editors

Ecology, Conservation, and Restoration of Chilika Lagoon, India

Springer

Editors
C. Max Finlayson
Institute for Land Water and Society
Charles Sturt University
Albury, NSW, Australia

IHE Delft
Institute for Water Education
Delft, The Netherlands

Deepak R. Mishra
Center for Geospatial Research,
Department of Geography
University of Georgia
Athens, GA, USA

Gurdeep Rastogi
Chilika Development Authority
Bhubaneswar, Odisha, India

Ajit K. Pattnaik
Chilika Development Authority
Bhubaneswar, Odisha, India

ISSN 1875-1261 ISSN 1875-127X (electronic)
Wetlands: Ecology, Conservation and Management
ISBN 978-3-030-33423-9 ISBN 978-3-030-33424-6 (eBook)
https://doi.org/10.1007/978-3-030-33424-6

This Springer imprint is published by the registered company Springer Nature Switzerland AG.
The registered company address is: Gewerbestrasse 11, 6330 Cham, Switzerland

Foreword

Wetlands are in deep trouble worldwide. Despite the many and hugely important benefits they provide to people, for centuries, wetlands have been drained and otherwise converted to other land uses, and the loss of wetlands continues largely unabated.[1] We also know that deterioration in the state (ecological character) of our remaining wetlands, including designated Ramsar sites (Wetlands of International Importance), is becoming increasingly widespread.[2]

The reasons behind this continuing loss and deterioration of wetlands are complex, and frequently, there are multiple different, but interrelated, pressures leading to this situation.[3] These often involve inter alia conversion for agriculture and urbanization, pollution, upstream water management actions leading to reduced water and sediment flows, the spread of invasive species, overexploitation of natural resources, and changes in climate. The consequence is often a loss of livelihoods for the local communities who depend on wetlands for their health and well-being.

In this very difficult environment for wetlands, is it possible to reverse their continuing degradation and restore their ecological character? Although hugely challenging, from the evidence provided in this important book, the answer is, at least for Chilika Lagoon, a resounding "yes"! It can and has been done.

So, how has this been achieved for Chilika Lagoon, and by whom? At the time of its designation as one of India's first Ramsar sites in 1981, the Lake was already

---

[1]Davidson 2014. How much wetland has the world lost? Long-term and recent trends in global wetland area. *Marine and Freshwater Research* 65(10): 934–941.; Darrah et al. 2019. Improvements to the Wetland Extent Trends (WET) index as a tool for monitoring natural and human-made wetlands. *Ecological Indicators* 99, 294–298.

[2]Davidson et al. 2020. Trends in the ecological character status of wetlands reported to the Ramsar Convention. *Marine and Freshwater Research* 71(1): 127–138.; McInnes et al. in press. Citizen science; global assessment; ecological character; wetland status and trends. *Wetlands*

[3]Ramsar Convention Secretariat 2018. Global Wetland Outlook. State of the world's wetlands and their services to people; McInnes et al. in press. Citizen science; global assessment; ecological character; wetland status and trends. *Wetlands*

recognized as being degraded, and that degradation continued in the following decade. It was national and regional government recognition of this, leading to the establishment of the Chilika Development Authority (CDA) in 1991 which has provided the "authorizing environment" for all the subsequent on-the-ground information gathering and action. Scientific research and modeling, understanding ecosystem service values delivered by the Lagoon, and engaging with multiple sectors affecting the Lagoon and with local communities to better understand their needs and livelihoods have all contributed to informing appropriate actions to turn the state of the Lagoon round to the benefit of multiple stakeholders.

The Chilika Lagoon process also provides an exemplar of how a Ramsar Contracting Party can utilize and benefit from all the available mechanisms under the Ramsar Convention on Wetlands and through these to comply with the Convention's wise use provisions. By placing Chilika Lagoon on the Convention's "Montreux Record," the government of India drew international attention to the plight of the lagoon. Following huge efforts to restore the lagoon, "Ramsar Advisory Mission" (RAM) visited in 2001 to review the management actions undertaken and concluded that in view of the achievements, the Lagoon should be removed from the Montreux Record – this was done in 2002. This was accompanied by an ongoing commitment from the Government of India and the CDA to continue to develop and implement an overall management planning process for the Ramsar site, which has now been developed and implemented.

The 2001 Ramsar Advisory Mission (RAM) also recommended that the Convention should consider using Chilika as an exemplary good practice case study of the application of the various Ramsar guidelines, and the use of the Convention's tools and approaches, to address complex site and catchment management issues. In recognition of their successful restoration and wise use efforts for Chilika Lagoon, the Chilika Development Authority received international recognition for their work through their receipt of the Ramsar Wetland Conservation Award and Evian Special Prize in 2002, on the occasion of the 8th Ramsar Conference of the Contracting Parties (COP8, Valencia, Spain).

This important new book provides a wealth of information about Chilika Lagoon as the exemplary good practice case study of the application of the various Ramsar guidelines, and the use of the Convention's tools and approaches, to address complex site and catchment management issues called for by the 2001 RAM.

This book is absolutely essential reading for all of us around the world striving to manage and restore our degrading wetlands. Read the book and be encouraged and excited by it. Can you achieve restoration and wise use of the world's important wetlands? The Chilika Lagoon story says "yes, you can"! Be inspired by this book – and never again say that it is not possible to turn around the fate of our wetlands, for the great benefit of so many people.

Nick Davidson

Institute for Land, Water & Society,
Charles Sturt University, Bathurst,
Australia

Nick Davidson Environmental, London,
UK

Former Deputy Secretary General,
Ramsar Convention on Wetlands,
Gland, Switzerland
March 2019

Preface

Coastal lagoons have immense ecological, economical, and social values. These are among the most productive and dynamic ecosystems as they are positioned at the interface between rivers and sea. The Chilika Lagoon, the largest brackish water lagoon in Asia, is located along the eastern coast of India in the state of Odisha. The lagoon provides a range of ecosystem services to the coastal communities. Chilika Lagoon has been considered as a wetland of international importance and has been designated as a "Ramsar site." At the same time, the lagoon is exposed to many natural and anthropogenic pressures such as siltation, weed infestation, illegal aquaculture, as well as cyclonic storms. In 1991, the lagoon was included in the "Montreux Record," a list of threatened "Ramsar sites." Realizing the problems of Chilika Lagoon, the Chilika Development Authority (CDA) adapted an ecosystem restoration approach to prevent the degradation of the lagoon. In the year 2000, CDA performed a major hydrological intervention in the form of opening a new channel between the lagoon and the sea which helped to improve the salinity levels, biodiversity, fish catches, and livelihood of dependent communities. Chilika was removed from the "Montreux Record" in 2001, and the restoration effort was recognized with the prestigious "Ramsar Award" to CDA in 2002.

Sustainable management of Chilika Lagoon is crucial not only for maintaining the rich biodiversity and productivity but also for the wise use of common resources by the communities. Management of Chilika Lagoon needs an interdisciplinary approach to effectively use the information available on different aspects of wetland ecology. The recent recognition of wetlands in the context of their ecosystem services has promoted worldwide conservation efforts with research on all aspects of wetland ecology, biodiversity, hydrology, and conservation. This has resulted in an increasing demand for successful case studies to be conducted on model wetlands such as "Chilika" where research proved to be a vital element in sustainable management and conservation of the lagoon.

Recent research and innovative management practices for the conservation of the Chilika Lagoon have provided a strong foundation for this book. A book addressing all major aspects of wetland ecology including conservation and governance issues must be made available to meet the needs of researchers, wetland managers, and students. This book will serve as an invaluable resource to aid research on ongoing studies in Chilika Lagoon. The book also identifies existing knowledge gaps for further research and technological developments in wetland studies. The book will also be of interest to those wetland managers who are working in similar coastal lagoon ecosystems around the world where lessons learned from Chilika Lagoon could be applicable for sustainable management and conservation purposes.

Albury, NSW, Australia — C. Max Finlayson
Bhubaneswar, Odisha, India — Gurdeep Rastogi
Athens, GA, USA — Deepak R. Mishra
Bhubaneswar, Odisha, India — Ajit K. Pattnaik

Acknowledgments

The editors are thankful to Sri Susanta Nanda, IFS, chief executive, Chilika Development Authority, for his constant encouragement and advice. Several chapters in the books are based on the studies which were carried out through the support received from the World Bank-funded Integrated Coastal Zone Management Project (ICZMP) Odisha (Cr. No. 4765-IN) to Chilika Development Authority. These chapters are the following: Spatiotemporal Variation in Physicochemical Parameters of Water in the Chilika Lagoon; Spatiotemporal Assessment of Phytoplankton Communities in the Chilika Lagoon; Fish and Fisheries of Chilika Lake: Post-restoration Scenario; Avifauna of Chilika; Odisha: Assessment of Spatial and Temporal Changes; Biodiversity of Benthic Fauna in Chilika Lagoon; Microbial Ecology of Chilika Lagoon; and Survey, Characterization, Ecology, and Management of Macrophytes in Chilika Lagoon. This book was possible due to the contribution of the researchers and scientists who took active part in writing and submitting their chapters. The editors would like to acknowledge Madhusmita Mohapatra and Stiti Prangya Dash, research fellows at WRTC, CDA, for providing technical help in the compilation of the chapters received by different authors. We acknowledge the following reviewers who provided valuable comments to improve the manuscripts and overall quality of this project.

1. Prof. Nick Davidson, Nick Davidson Environmental & Institute for Land, Water & Society, Charles Sturt University, UK
2. Dr. Mark Everard, University of the West of England (UWE Bristol)
3. Prof. Robert McInnes, RM Wetlands & Environment Ltd., Oxfordshire, SN7 8EQ, UK
4. Prof. Philip A. Meyers, Department of Earth and Environmental Sciences, University of Michigan, USA
5. Dr. P. Chandramohan, Indomer Coastal Hydraulics (P) Ltd., Chennai, India
6. Dr. Federica Semprucci, Dipartimento di Scienze Biomolecolari (DiSB), Sezione di Microscopia, Università degli Studi di Urbino "Carlo Bo," Campus Scientifico, Via Cà Le Suore, Urbino (PU), Italy
7. Dr. Taej Mundkur, Wetlands International, Wageningen, Netherlands

8. Dr. B. C. Jha, Central Inland Fisheries Research Institute (CIFRI), Barrackpore, Kolkata, India
9. Sri. S. K. Mohanty, Chilika Development Authority, Bhubaneswar, Odisha, India
10. Dr. Ritesh Kumar, Wetlands International South Asia, New Delhi, India

About This Book

This book is based on the original contribution of leading scientists and experts who have worked on Chilika Lagoon for several decades. It covers issues pertaining to the management, governance, and restoration of the lagoon ecosystem, ecosystem services, and lagoon-specific research topics on hydrodynamic modeling, catchment modeling, water quality, sediment dynamics, and spatial and temporal trends in biodiversity (fisheries, avifauna, benthic fauna, phytoplankton, microbial communities, and macrophytes). The book is a significant contribution to research in the increasingly important discipline of wetland management and their conservation using Chilika Lagoon as a case study.

List of Lead Authors

1. Prof. C Max Finlayson, Charles Sturt University, Institute for Land, Water & Society, Elizabeth Mitchell Drive, Albury, 2640, Australia, & IHE Delft, Institute for Water Education, NL-2601 DA, Delft, Netherlands
2. Dr. Ritesh Kumar, Wetlands International South Asia, New Delhi, India
3. Prof. Torsten Vennemann, Institute of Earth Surface Dynamics, University of Lausanne, Geopolis 4634, CH-1015 Dorigny, Switzerland
4. Dr. R. S. Kankara, Coastal Processes & Shoreline Management (CPSM), National Centre for Coastal Research (NCCR), Ministry of Earth Sciences, Govt. of India, NIOT Campus, Pallikaranai, Chennai – 600100, India
5. Dr. Priyabrata Santra, Division of Agricultural Engineering and Renewable Energy, ICAR-Central Arid Zone Research Institute, Jodhpur Rajasthan 342003, India
6. Dr. Deepak R. Mishra, Department of Geography, University of Georgia, Athens, 210 Field Street, Athens, GA 30602, USA
7. Dr. Pradipta R. Muduli, Wetland Research and Training Center, Chilika Development Authority, Barkul, Odisha- 752030, India.
8. Prof. Rajiv Sinha, Department of Earth Sciences, Indian Institute of Technology Kanpur 208016 (UP) India
9. Dr. Gurdeep Rastogi, Wetland Research and Training Center, Chilika Development Authority, Barkul, Odisha- 752030, India.
10. Sri. S. K. Mohanty, Chilika Development Authority, Bhubaneswar, Odisha, India
11. Dr. S. Balachandran, Bombay Natural History Society, Hornbill House, S.B. Singh Road, Mumbai – 400001.
12. Prof. Punyasloke Bhadury, Indian Institute of Science Education and Research, Department of Biological Sciences, Kolkata, India
13. Dr. Ajit K. Pattnaik, Chilika Development Authority, Bhubaneswar – 751020, Odisha, India

Contents

About the Editors

C. Max Finlayson is an internationally renowned wetland ecologist with extensive experience internationally in water pollution, agricultural impacts, invasive species, climate change, and human well-being and wetlands. He has participated in global assessments such as those conducted by the Intergovernmental Panel on Climate Change, the Millennium Ecosystem Assessment, and the Global Environment Outlook 4 & 5 (UNEP). Since the early 1990s, he has been a technical adviser to the Ramsar Convention on Wetlands and has written extensively on wetland ecology and management. He has also been actively involved in environmental NGOs and from 2002 to 2007 was president of the governing council of global NGO Wetlands International. He has contributed to over 450 journal articles, reports, guidelines, proceedings, and book chapters on wetland ecology and management and to the development of concepts and methods for wetland inventory, assessment, and monitoring and undertaken many site-based assessments in many countries. He is the Editor-in-Chief of the journal *Marine and Freshwater Research* published by CSIRO Publishing and of the book series *Wetland, Ecology, Conservation and Management* published by Springer.

Gurdeep Rastogi is working as a Senior Scientist at the Wetland Research and Training Centre (WRTC) of the Chilika Development Authority (CDA). He has received his PhD in the area of Environmental Microbiology from the National Centre for Cell Sciences, Pune, in the year 2006. He has completed his postdoctoral research at the South Dakota School of Mines & Technology and the University of California, Davis, USA. He is currently heading the molecular ecological research at WRTC, a nodal lab of CDA which has been entrusted with the responsibility of managing the Chilika Lake by the Odisha State Government. His research area is focussed on understanding the microbial biogeography and functions in coastal systems with special emphasis on bacteria, archaea, and phytoplankton. Over the past 15 years, he has published more than 40 papers in peer-reviewed international journals which have been well cited by the scientific communities, worldwide. He is an active reviewer for many international journal published by Springer and Elsevier.

Deepak R. Mishra is a Professor of Geography at the University of Georgia, Athens. His research interests are mainly in the area of remote sensing and satellite image processing as applied to inland and coastal environments. His specific interests are water quality, wetlands biophysical characteristics including gross primary productivity, carbon sequestration potential, application of big data in environmental informatics, and cyber-physical systems research. He completed his PhD in Natural Resources from the University of Nebraska, Lincoln. He has published numerous papers in various high-impact international journals and book chapters, presented his research in many national and international meetings, edited special issues and books, and received external grants from various federal agencies. He currently serves on the editorial board of *GIScience & Remote Sensing*. He is an associate editor of the *ISPRS Journal of Photogrammetry and Remote Sensing* and a senior associate editor of *Remote Sensing* and coeditor of *Southeastern Geographer*.

Ajit K. Pattnaik as the former chief executive of Chilika Development Authority, headed the most successful project of restoration and sustainable management of Chilika Lake for 16 years. He completed his PhD in the "Characterization of Phytodiversity of Chilika Lake" from Utkal University, Bhubaneswar. Chilika was removed from the Montreux Record in 2002, due to the successful ecological restoration. He has 37 years of varied professional experience in the management and restoration of coastal ecosystems, climate change, and disaster management. He has extensive experience in project formulation, planning, management, wetland restoration, ecotourism, and water and community development projects. Presently, he is a consultant to the World Bank and UNDP and holds the position of the vice president of Wetlands International South Asia. He is a member of several international and national scientific and technical committees and has published over 100 scientific articles. He is also member of the expert advisory panel of journal *Lakes & Reservoirs: Science, Policy and Management for Sustainable Use* published by Wiley.

Chapter 1
Introduction

C. Max Finlayson, Gurdeep Rastogi, Deepak R. Mishra, and Ajit K. Pattnaik

Abstract This book provides an overview of the decades-long work of studying, analyzing, and reversing the environmental pressures that threatened India's Chilika Lagoon, the largest brackish-water lagoon in the region, and the second largest in the world. Following the establishment of the Chilika Development Authority (CDA) steps were taken to gather information and devise a restoration plan that benefits the ecosystems of the lagoon, with sensitivity to the needs and livelihoods of local communities. The restoration plan included a major hydrological intervention to re-establish hydrological and salinity regimes, biodiversity, and fish catches, and help protect the livelihood of lagoon-dependent communities. Expert contributors detail the work of analysis, planning and implementation, including extensive coverage of such topics as: implementing Ramsar wise use guidelines; sedimentologic, chemical, and isotopic impacts; hydrodynamics and salinity; runoff and sediment in watersheds; water quality and continued monitoring; bio-optical models for cyclone impact assessment; geomorphology, land use, and sedimentary environments; spatiotemporal assessment of phytoplankton communities; post-restoration scenario for fish and fisheries; and the status of waterbirds, species diversity and migration patterns.

C. M. Finlayson (✉)
Institute for Land Water and Society, Charles Sturt University, Albury, NSW, Australia

IHE Delft, Institute for Water Education, Delft, The Netherlands
e-mail: mfinlayson@csu.edu.au

G. Rastogi
Wetland Research and Training Centre, Chilika Development Authority, Balugaon, Odisha, India

D. R. Mishra
Center for Geospatial Research, Department of Geography, University of Georgia, Athens, GA, USA
e-mail: dmishra@uga.edu

A. K. Pattnaik
Chilika Development Authority, Bhubaneswar, Odisha, India

C. M. Finlayson et al. (eds.), *Ecology, Conservation, and Restoration of Chilika Lagoon, India*, Wetlands: Ecology, Conservation and Management 6,
https://doi.org/10.1007/978-3-030-33424-6_1

Keywords Chilika Lagoon · Ramsar Site · Wetlands · Sustainable management · Climate change

Located at the land-water interface, coastal wetlands are affected by both land and ocean processes, and function as valuable sources of primary and secondary productivity and biodiversity, which are crucial to support life on our planet. Wetlands are highly productive ecosystems and at the same time, they are highly vulnerable to anthropogenic and natural disturbances (Ramsar Convention 2018). Considering the socioecological importance of wetlands, the international community in 1971 established the Ramsar Convention on Wetlands (Matthews 1993). Under the Convention wetlands include areas that are either permanently or seasonally inundated with water and, depending upon the geomorphology, hydrological regime, and vegetation, they comprise various types such as mangroves, peatlands, marshes, estuaries, rivers, lakes, and flooded forests. Being a transitional ecosystem, they pose many challenges with respect to management, monitoring, and conservation as well as the multitude of stakeholders depending on them for their livelihoods. One such wetland, Chilika Lagoon, the largest brackish water lagoon in Asia, is an extremely important natural asset for the State of Odisha, India. Successful management of Chilika requires not only conserving and preserving the biodiversity of this lagoon and its ecosystem services, but also the livelihoods of the coastal communities (Kumar and Pattnaik 2012). Chilika presents a role model of successful implementation of the Ramsar Convention in which an ecosystem approach has been used for conservation and sustainable management of natural resources.

This book is designed to highlight the theories, past developments, and current state-of-the-art knowledge in management and conservation of the coastal lagoon. Chilika Lagoon has been intensively studied by numerous physical and social scientists for many decades. However, because of the lack of a coherent and comprehensive synthesis of the multi-decadal research on this important environment, the focus of this book is squarely placed on Chilika Lagoon. The book contains 16 chapters covering key topics on geomorphology, ecology, water resources, ecosystem management and restoration pertaining to Chilika Lagoon, as well as making it immensely helpful for the management of similar lagoon ecosystems elsewhere.

The eco-restoration approach which considered ecological, social, and economical inter-connectedness has been described in Chap. 2 "**An overview of the restoration and management of Chilika Lagoon: successful application of the Ramsar wise use guidelines**". The integrated management planning framework for Chilika, and the wise use of natural resources in the context of sustaining the ecological character and ecosystem services of the lagoon have been discussed in detail.

The ecological character of Chilika mostly depends on its hydrological regime which is linked with both natural and anthropogenic factors. Ecological character is therefore an indicator of the overall health of the Chilika ecosystem as it includes all critical components (e.g., bathymetry, hydrology, water quality, biodiversity), processes (e.g., fish recruitment, sedimentation, inlet migration), and ecosystem services (e.g., provisioning, regulating, and cultural). The framework provided for such characterization under the Ramsar Convention has been discussed in the context of Chilika in Chap. 3 "**Ecological characterization of Chilika: defining strategies and management needs for wise use**" and Chap. 4 "**Ecosystem services: implications for managing Chilika**". From a management perspective, the identification of key ecological characters, processes, and services and threats to the ecological character and ecosystem services have been summarized in detail.

Chilika Lagoon is subjected to many anthropogenic stresses such as siltation, weed infestation, and nutrient loading. To help trace and quantify the anthropologic effects on Chilika Lagoon, Chap. 5 "**Sedimentologic, chemical, and isotopic constraints on the anthropogenic influence on Chilika Lake, India**" presented a comprehensive geochemical dataset acquired during both the dry and monsoon season from the lake. The trends in isotope composition (Hydrogen, Oxygen, Carbon, and Nitrogen) in addition to salinity, Dissolved Inorganic Carbon, and POM were presented which revealed that the mixing of freshwater with seawater mainly controlled the geochemical composition of the lagoon ecosystem. Seasonal and sectoral variability was also observed. The data presented on N-isotope composition is also important for the evaluation of the invasive macrophytes that proliferate along the shores and are seen as a potential environmental hazard, but may actually be effective filters for excess nitrate and nutrient fluxes into the lagoon.

The ecology of Chilika Lagoon entirely depends on salinity which is determined by freshwater inputs and tidal flux. The hydrodynamic circulation is dependent on many physical processes such as wind directions, water currents, and position and cross section of the seawater inlets. Chapter 6 "**Modelling of hydrodynamics and salinity characteristics in Chilika Lagoon**" complemented the findings presented in the previous chapters by shedding light on the hydrodynamic circulation which controlled the geochemical and biological properties of the lagoon. A fully integrated time generalized hydrodynamic model with effects from tide, wind, and freshwater sources and sinks was presented. A shift in key forcings from wind and tide in summer to freshwater influx during monsoon which controlled the hydrodynamic and salinity patterns of the lagoon was observed. The study concluded that the hydrological intervention and restoration measures have facilitated better exchange with the sea resulting in an improvement in salinity distribution and ecology of the lagoon. However, shifting of the inlet(s) and siltation in the dredged channels remain as significant concerns.

Chilika has a vast catchment area of approximate 4406 km^2 which contributes a large sediment load to the lagoon through freshwater discharge from the rivers. This sediment load enriches the lagoon with nutrients and organic matter leading to extensive colonisation of macrophytes. Chapter 7 "**Assessment of runoff and sediment yield from selected watersheds in the Western Catchment of the Chilika Lagoon**" presented a hydrological model to estimate runoff and sediment load in the western catchment which drives siltation and affects the overall water quality of the lagoon. A Soil and Water Assessment Tool (SWAT) model for two river basins in the western catchment was calibrated and validated with the results showing that rainfall was the main source of runoff which brought a significant amount of eroded sediment into the lagoon. The study concluded that the sediment load was harmful to the sustainability of the lagoon and needed to be stopped at the source, which is the catchment itself.

Chilika is a turbid water lagoon due to a high amount of suspended sediments in the water column which determines the quantity and quality of the light available to phytoplankton for primary production. Chapter 8 "**Long-term analysis of water quality in Chilika Lagoon and application of bio-optical models for cyclone impact assessment**" examined the long-term water quality of the lagoon in terms of total suspended sediment and chlorophyll-a (a proxy for phytoplankton abundance) using NASA's MODIS satellite data. The study also presented the differential impact of the recent anniversary super cyclones, Phailin and Hudhud on the lagoon. Analysis of a 14-year dataset revealed that the seasonal variability of Total Suspended Solids was dominant in all the three sectors of the lagoon compared to inter-annual variability. The study concluded that many factors including the location of the landfall, intensity, trajectory, and speed of the cyclone played a role in determining the outcome (high turbidity versus high phytoplankton) for the lagoon.

Systematic and comprehensive monitoring of water quality constitutes an important step in assessing the ecological health of Chilika Lagoon. Chapter 9 "**Spatio-temporal variation in physicochemical parameters of water in the Chilika lagoon**" discussed the long-term water quality variability using a large dataset collected between 1999 and 2015 from 30 permanent stations. The chapter presents an overview of seasonal and sectoral variation in physicochemical factors such as salinity, nutrients, dissolved oxygen in relation to major physical processes such as mixing of freshwater with seawater, rainfall patterns, river water discharge, and tidal influx from the Bay of Bengal. The outcomes were also compared with thresholds prescribed by Central Pollution Control Board, New Delhi for water quality guidelines set for the propagation of wildlife and fishery.

The Land Use/Land Cover (LULC) changes in Chilika affect many physical and biological processess in the lagoon through change in salinity, increased nutrient inputs and weed infestation. Remote sensing and GIS are important tools to document changes in geomorphic and anthropogenic processes. Chapter 10 "**Geomorphology, land use/land cover and sedimentary environments of the Chilika basin**" presented the outcomes of geomorphic studies in and around Chilika from 1980 to 2015 using remote sensing data. LULC mapping was carried out to examine the anthropogenic changes surrounding the lagoon which could be playing a role in degrading the water quality. The study concluded that the lagoon is facing a significant problem of siltation mainly due to improper utilization of LULC. Agriculture plantations and barren lands are more vulnerable due to the impact of urbanization, such as engineering construction, settlements, and transport. Changes in the island landforms within the lagoon are mainly due to the hydrodynamic circulation.

The productivity of the Chilika lagoon, thus, the entire trophic food chain relies on the phytoplankton communities, the primary producers of the system. The spatiotemporal distribution of phytoplankton communities provides a vital clue regarding the trophic status of the system and are used as bioindicators for several biological processes such as eutrophication and harmful algal blooms. Chapter 11 "**Spatiotemporal assessment of phytoplankton communities in the Chilika lagoon**" provided a detailed assessment of group-wise inventory of the phytoplankton species composition and new records from Chilika based on surveys carried out between 2000 and 2014. The impact of the very severe cyclonic storm '*Phailin*' on the phytoplankton communities is also elaborated. This chapter also provided an insight on major environmental factors that shape the phytoplankton community in Chilika lagoon. The need to further study the diversity of small-size phytoplankton (nano and picophytoplankton) through DNA sequencing is highlighted for a complete understanding of the phytoplankton communities of the lagoon.

Chilika Lagoon has experienced ecological degradation during the 1990s due to the natural closure of the seawater inlet by siltation. This led to a dramatic decrease in the biotic diversity of the lagoon, including the species used for fisheries. In September 2000, the lagoon was restored through the opening of a new mouth for entry of seawater from the Bay of Bengal. This hydrological intervention resulted in a spectacular enhancement in fishery species diversity and catches during the post-restoration period (2000–2004). Chapter 12 "**Fish and fisheries of Chilika Lake: post-restoration scenario**" provided detailed information on the changes in fish diversity before and after the hydrological intervention. The latest inventory on fish and shellfish fauna diversity, their habitat, and conservation status have been provided. In addition, the biology and ecology of commercially important fishes, management challenges, and recommendations for sustainable management of fishery resources in Chilika Lagoon have been discussed.

The bird diversity of Chilika is well recognized for providing several ecosystem services and is considered a key component of the biota along with fisheries. Assessment of the bird diversity and population status have been systematically monitored by the Bombay Natural History Society (BNHS) since 2000. Chapter 13 "**Avifauna of Chilika, Odisha: assessment of spatial and temporal Changes**" provided an overview of the population status of waterbirds, their migration pattern, and species diversity, based on the monitoring studies carried out between 2000 and 2014. The study has highlighted the importance of the Nalabana Bird Sanctuary in providing ideal feeding, resting, and breeding ground for several exotic bird species. Issues related to habitat management such as invasion by grasses on islands, loss of mudflats, and increased human interference have been highlighted to be considered in wetland management planning.

Benthic macro-and meiofauna play a crucial role in the decomposition of organic matter which is accumulated into sediments. The benthic fauna, thus, plays an important role in recycling the nutrients and drive the nutrient cycling leading to the flow of energy in the trophic food chain in Chilika. The species composition of benthic macro- and meiofauna also provides a bioindicator to track natural and anthropogenic disturbances. The benthic fauna also provides a rich source of food for many species of birds and fish. Chapter 14 "**Biodiversity of benthic fauna in Chilika lagoon**" summarized the latest information available on the benthic fauna based on the monitoring survey carried out between 2014 and 2017. The chapter also highlighted the need for conducting long-term monitoring to understand the impact of fishing, continuous dredging, oil pollution and sewage discharge on benthic communities. The changes induced by these anthropogenic activities would impact the fishery and bird resources of Chilika lagoon and eventually the livelihood of coastal communities.

Microbial ecology of Chilika lagoon, especially with reference to bacterial and archaeal communities, is an understudied area. Despite the fact that microbial communities present in sediments and the water column play a crucial role in the biogeochemical cycling of nutrients, not many studies are available on this subject. Chapter 15 "**Microbial ecology of Chilika lagoon**" summarized the microbial ecological studies available from Chilika and discussed in detail the culture-based and culture-independent approaches. Recent developments in microbial ecology due to high-throughput DNA sequencing and their application in studying the structure and function of microbial communities through metagenomics have been discussed in detail. The role of different biotic and abiotic drivers in structuring the sediment microbial communities have been highlighted.

Macrophytes in a wetland system provide a range of ecosystem services such as sheltering grounds for many faunal communities from their predators, as well as breeding and foraging ground for many ecologically important species of birds, finfish and shellfish. In addition, they play a crucial role in water and sediment biogeochemistry leading to the supply of organic matter into the sediments and water column. Chilika lagoon is a macrophyte dominated system which supports a diverse macrophyte community due to a variety of salinity and nutrient regimes. Chapter 16 "**Survey, characterization, ecology, and management of macrophytes in Chilika lagoon**" described changes in macrophyte diversity due to post-hydrological intervention. Based on the ground survey carried out on Chilika, 748 species of angiosperm were documented. The data on the spread of seagrasses, invasive weeds like *Phragmites karka* along with management recommendations have been discussed. The ecology of macrophytes in relation to water quality parameters have been presented.

The content of this book summarizes the progress that has been made so far by the scientific community studying the lagoon. The methods, models, and analysis synthesized in this book will hopefully address some of the existing challenges in monitoring geomorphic, geochemical, biological properties, water quality, analyzing their interrelationship, and quantifying their impact on other biota such as seagrasses and benthic algae in Chilika Lagoon.

References

Kumar R, Pattnaik AK (2012) Chilika – an integrated management planning framework for conservation and wise use. Wetlands International – South Asia/Chilika Development Authority, New Delhi/Bhubaneswar

Matthews GVT (1993) The Ramsar convention; its history and development. Ramsar Convention Bureau, Gland

Ramsar Convention (2018) Global wetland outlook; state of the world's wetlands and their ecosystem services. Ramsar Convention, Gland

Chapter 2
An Overview of the Restoration and Management of Chilika Lagoon: Successful Application of the Ramsar Wise Use Guidelines

C. Max Finlayson

Abstract Lake Chilika was listed as a Ramsar site in 1981 and after a period of ongoing degradation was placed on a register of sites (The Montreux Record of the Ramsar Convention), that are in need of further management and restoration. Following a committed management effort through the Chilika Development Authority the site was restored and an active wise use program implemented. These steps were sufficient for the site to be removed from the Record in November 2002. The management effort received international recognition and the Lake is now seen as an example of how to apply the guidance provided by the Convention to ensure the maintenance of the ecological character of a Ramsar site. The history of the application of the Convention to Lake Chilika is described here in recognition of the ongoing management efforts, and as an example for other Ramsar site managers.

Keywords Wetland · Lagoon · Ecological character · Ramsar Convention

2.1 Introduction

The Indian Government acceded to the Ramsar Convention on 1 October 1981 with the Convention formally coming into force some 4 months later on 2 February 1982 with the Ministry of Environment and Forests (MOEF) Government of India being the Administrative Authority for national implementation, including for meeting the requirements to nominate at least one wetland as internationally important (known as Ramsar Sites) and to make wise use of all wetlands. MOEF listed six wetlands as internationally important, including Lake Chilika (Fig. 2.1) which is located in

C. M. Finlayson (✉)
Institute for Land Water and Society, Charles Sturt University, Albury, NSW, Australia

IHE Delft, Institute for Water Education, Delft, The Netherlands
e-mail: mfinlayson@csu.edu.au

C. M. Finlayson et al. (eds.), *Ecology, Conservation, and Restoration of Chilika Lagoon, India*, Wetlands: Ecology, Conservation and Management 6,
https://doi.org/10.1007/978-3-030-33424-6_2

Fig. 2.1 Sign commemorating the designation of Lake Chilika as a Ramsar site in late 1981. (Photograph © CM Finlayson)

Odisha State in eastern India and covers 116,500 ha (Fig. 2.2). The importance of the Lake as a Ramsar site was based initially on four of the criteria used by the Convention at the time (Table 2.1). With the addition of further criteria and the collection of further information two more were applied in May 2001. The most recent version of the Ramsar Information Sheet that was used to describe the ecological character of Lake Chilika is lodged with the Secretariat of the Convention and is accessible through the Ramsar Site Information Service (https://rsis.ramsar.org/RISapp/files/RISrep/IN229RIS.pdf). The Information Sheet was provided initially in 1982 and updated in 2001; a further update is now overdue. Further information on the ecology and management of the Lake has been collated and summarised by Kumar and Pattnaik (2012) in support of the development of an integrated management planning framework for the Lake.

The Ramsar Information Sheet highlights the importance of Lake Chilika for its biodiversity and for the economic importance it has for local people. It is a biodiversity hotspot and supports a valuable fishery resource for more than 0.2 million people. The biodiversity includes over a million migratory waterbirds, including shorebirds (waders); more than 400 invertebrate species; and an assemblage of marine, brackish and freshwater species, as well as several rare, endangered and

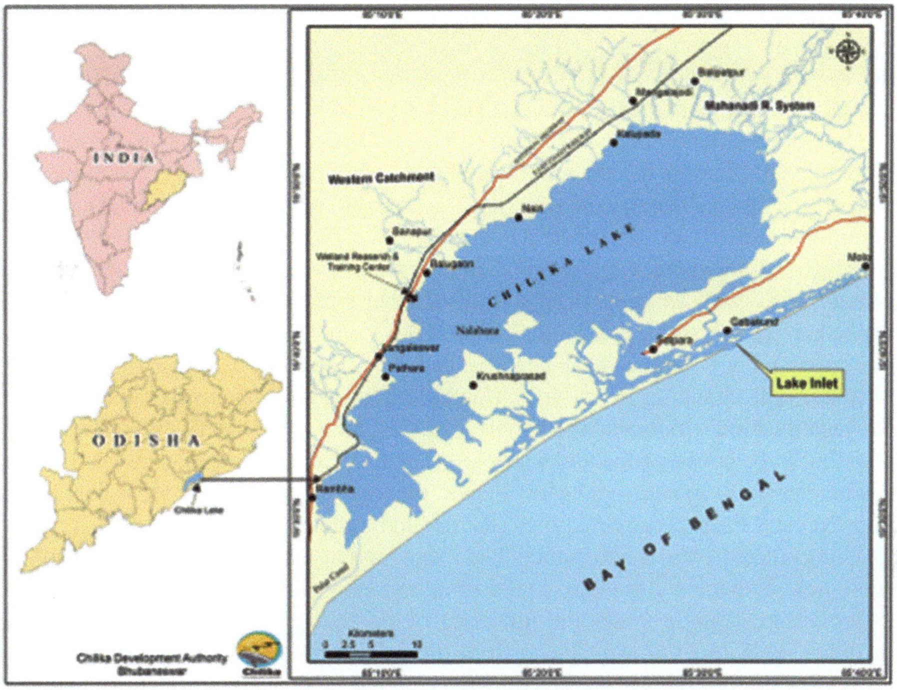

Fig. 2.2 Location of Lake Chilika along the eastern coastline of Odisha State, India (Chilika Development Authority)

Table 2.1 Criteria used to list Lake Chilika as internationally important, initially in 1982 and as updated in 2001

1 February 1982		
Criterion	1	Contains a representative, rare, or unique example of a natural or near-natural wetland type found within the appropriate biogeographic region
	2	Supports vulnerable, endangered, or critically endangered species or threatened ecological communities
	3	Supports populations of plant and/or animal species important for maintaining the biological diversity of a particular biogeographic region
	5	Regularly supports 20,000 or more waterbirds
15 May 2001		
Criterion	7	Supports a significant proportion of indigenous fish subspecies, species or families, life-history stages, species interactions and/or populations that are representative of wetland benefits and/or values and thereby contributes to global biological diversity
	8	Important source of food for fishes, spawning ground, nursery and/or migration path on which fish stocks, either within the wetland or elsewhere, depend

threatened species. Detailed species inventories are available, for the micro- and macrophytic vegetation, as well as for invertebrate and vertebrate animals although the knowledge about the biodiversity contains many gaps (see descriptions and tabulation in Kumar and Pattnaik 2012).

Chilika is the largest coastal lagoon on the east coast of India and considered to be the lifeline of the state of Odisha. It is a highly complex ecosystem and is influenced by a diverse range of social and economic factors within its catchment and also within the coastal zone. Not unexpectedly therefore, the management of Chilika is also complex – all the more so given the dual purpose of ensuring the rich biodiversity is conserved as well as supporting the sustainable livelihoods for the communities dependant on the wetland resources. Given this complexity the management is not prescriptive, but rather has been adaptive and developed in order to enable managers to respond to changing needs and information, especially concerning the linkages between the biodiversity and the people. The long-term objective of the management planning is the conservation and wise use of the Lake, integrating management of the catchments and coastal zones to ensure the ecological security and livelihood improvement for local communities (Kumar and Pattnaik 2012).

Unfortunately, the Lake which had been undergoing adverse changes to its ecological character since the 1950s prior to being nominated as a Ramsar site continued to degrade under the influence of multiple pressures (Fig. 2.3). These included increased sediment loads from the catchment which infilled parts of the lake, particularly in the north-western area, and led to reduced connectivity with the ocean, which in turn, resulted in changes in the water salinity. Invasive weeds, including *Eichhornia crassipes*, *Azolla pinnata*, and *Potamogeton pectinatus*, also established and a process of terrestrialisation was underway with a reduction in the volume and depth of the water. The introduction and expansion of shrimp ponds added further pressure on the ecological character of the lagoon ecology and also ultimately led to significant disruption of the community institutions that had traditionally managed the fisheries in a sustainable manner. These changes have been outlined by Kumar and Pattnaik (2012) and have been the subject of many discussions, including technical workshops, community consultations and political dialogue. Information from these activities has also been used to develop public awareness and encourage knowledge exchange about the Lake and its management.

These changes were leading to many adverse consequences for the biodiversity in the lagoon as well as having a large impact on the livelihoods of the communities, especially those dependent on fishing. In the terms of the Ramsar Convention the Lake was recognised as undergoing an adverse change in ecological character and needed urgent managerial intervention. As the Lake provided an important setting for human wellbeing and livelihoods (*sensu* Horwitz and Finlayson 2011) with a large population of local people being dependent on its resources, the situation had a critical human dimension that if not effectively addressed could have had dire outcomes. This situation formed the background for the inclusion on 16 June 1993

Fig. 2.3 Pressures on the ecological character of Lake Chilika: invasive weeds; extensive fisheries; and increased tourism. (Photographs © CM Finlayson)

of Chilika into the Montreux Record of the Ramsar Convention at the request MOEF. This is a voluntary record for listing sites that have or are undergoing adverse human-induced change in ecological character (Finlayson 1996) where ecological character is defined as "the combination of the ecosystem components, processes and benefits/services that characterize a wetland at a given point in time".

In this paper the managerial responses that have occurred since the Lake was placed on the Montreux Record in June 1993 are assessed, including in particular how the Chilika Development Authority and the Government of India have responded to the recommendations that followed. In this respect the Lake is presented as an example of how the managers have adhered to the requirements under the Convention to manage the Lake and restore its ecological character. This is seen as a major achievement given the size and complexity of the site and the prevailing socio-economic conditions. It also provides a case study for other countries seeking to make full use of the procedures and guidance available through the Convention.

2.2 The Chilika Development Authority

In response to the deterioration of the ecological character of the Lake the Government of Odisha created an institution i.e. the Chilika Development Authority (CDA) in 1991 to lead an urgently needed and complex ecosystem restoration program. Financial support for the CDA came from the State Government and also from the Ministry of Environment and Forests, Government of India. The principal objectives of the CDA are:

(i) to protect the lake ecosystem and its genetic biodiversity;
(ii) to survey, plan and prepare a proposal for integrated resource management in and around the lake;
(iii) to undertake multi-dimensional and multi-disciplinary development activities; and
(iv) to cooperate and collaborate with other institutions for development of the lake.

The establishment of the CDA was an important step and heralded a successful and concerted effort that over the past 25 years has seen the situation in the Lake turned around, with many beneficial changes for people and for the biodiversity that characterises the Lake.

The CDA has initiated a number of major management programmes including:

(i) rectification of some of the land use and land cover problems in the degraded catchments;
(ii) an intensive hydrobiological monitoring effort;
(iii) hydrological intervention based on the outcome of numerical modelling to restore the hydrology and salinity regime;
(iv) the sustainable development of fisheries;
(v) wildlife conservation and the development of ecotourism;
(vi) extensive community participation and development; and
(vii) capacity building at various levels.

In 2000, after detailed investigations, rigorous modelling, and consultation a major hydrological intervention was undertaken with the opening of a new mouth to the Bay of Bengal (Fig. 2.4). This was designed to help restore the salinity regime, facilitate auto-recruitment breeding and migration to enhance the fish catch, reduce the area under invasive species and improve the overall water quality. The ecological recovery that occurred resulted in significant improvements in the livelihoods of the communities dependent on the Lake. In response to these improvements the Government of India requested the Secretariat of the Ramsar Convention to consider removing Chilika from the Montreux Record. This occurred in 2002 following the submission of a Montreux Record Questionnaire to the Ramsar Secretariat (on 30 April 2001) and acceptance of the recommendations of a Ramsar Advisory Mission to the Lake (Finlayson et al. 2001).

The restoration effort was recognized with The Ramsar Wetland Conservation Award and Evian Special Prize being presented to the CDA in 2002. The citation for

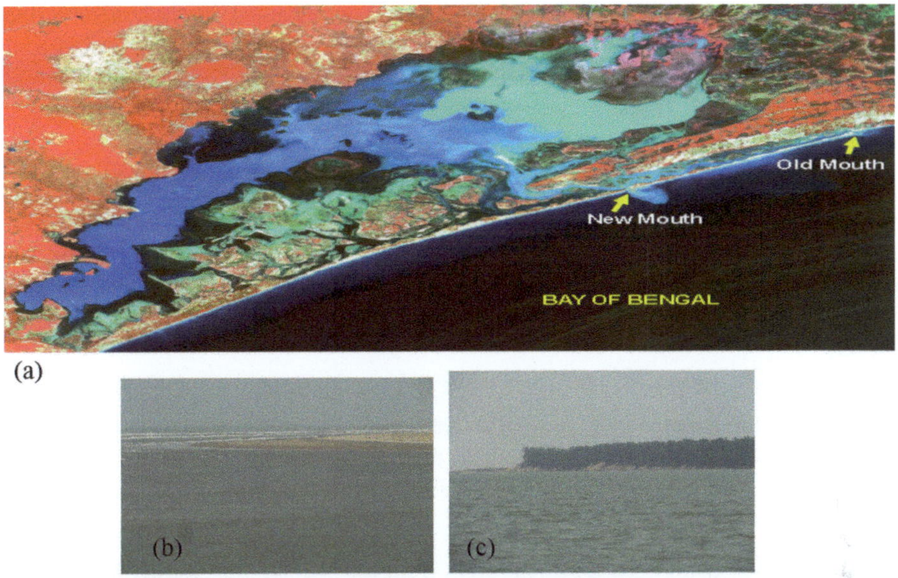

Fig. 2.4 New mouth to connect Lake Chilika to the Bay of Bengal. (**a**) Map from Chilika Development Authority; (**b**) & (**c**) photographs © CM Finlayson)

Fig. 2.5 The Chilika Development Authority was awarded the Ramsar Wetland Conservation Award and Evian Special Prize in 2002 based on the application of the principles of wise use and integrated management, and with a major emphasis on the participation of the local population including for (**a**) fishing and (**b**) prawn culture. (Photographs © CM Finlayson)

this award recognised that "the restoration was carried out based on the principles of wise use and integrated management, and with a major emphasis on the participation of the local population and their shared decision-making, as well as capacity building." (http://www.ramsar.org/activities/award-2-2002; Fig. 2.5).

The success of these measures has been witnessed over the ensuring years with the cutting of the new channel to the Bay of Bengal providing the impetus for a massive change in the ecological character and a sustained improvement in the fisheries in the Lake. The ecological changes that occurred as a consequence of this intervention have been described by Kumar and Pattnaik (2012). The information that formed the basis of these processes is accessible in separate documents at the Ramsar Secretariat and is summarised below.

2.3 Ramsar Advisory Mission

In response to the formal request from MOEF, a Ramsar Advisory Mission visited the Lake in December 2001 to:

(i) review the management actions undertaken;
(ii) assess the reported improvements to the ecological character of the Lake; and
(iii) prepare a report as a basis for considering the removal of the site from the Montreux Record.

The Mission comprised representatives from the Ramsar Secretariat, MOEF, the CDA, Wetlands International South Asia, and the Asia/Oceania representative on the Convention's Scientific and Technical Review Panel, as an independent scientific advisor. The Mission visited Bhubaneswar, the capital of Odisha, and Chilika Lake and met with officials from the Odisha Department of Forests and Environment and representatives of local communities from a number of villages around the lake (Fig. 2.6). The text that follows was largely derived from the report the Advisory Mission submitted to the Convention (Finlayson et al. 2001).

The Advisory Mission was able to determine that the monitoring and assessment projects that had been implemented and coordinated by the CDA had addressed many of the major pressures on the ecological character of the Lake. The primary drivers behind these pressures were the rapid increase in the human population in the catchment with subsequent catchment degradation, along with a low awareness of the ecological processes that maintained the Lake ecosystem and the benefits that had formerly been available to the local people. The resultant problems, including

Fig. 2.6 The Ramsar Advisory Mission to Lake Chilika. (Photographs © CM Finlayson)

increased siltation, weed infestation, hunting of birds, and pollution, posed a major threat to the sustainability of the fisheries as well as to the wildlife and water quality of the lake. The biodiversity and productivity, including that of economically valuable species in the lake, were under severe threat, and exacerbated by the uncontrolled expansion of prawn aquaculture into the lake. Each of these major problems is described in brief below.

Siltation Soil erosion is prevalent in the catchment due to over-grazing, the illicit felling of trees and shrubs, cultivation along the hill slopes, and the clearance of vegetation for agriculture. At that time an estimated 365,000 tonnes of sediment was being added annually to the lake. The resultant siltation led to a reduction in the area of the lake and clogging of the natural connection to the Bay of Bengal. The restriction in the capacity of the inlet that had connected the lake to the sea had led to a reduction in flushing and a decrease in salinity with the loss of some marine and brackish water species from the Lake.

Weed Infestation Increased siltation and decreased salinity promoted the spread of invasive weed species that were tolerant of the fresh-brackish water conditions. These included *Eichhornia crassipes*, *Azolla pinnata*, and *Potamogeton pectinatus* which extended from an area of 20 km^2 in 1973 to nearly 400 km^2 in 1993, mainly in the north-western part of the lake, restricting the free flow of sediments and causing increased siltation and infilling, and the loss of feeding and breeding grounds of many fish of economic importance.

Bird Hunting For some years, many villagers had been poaching birds from the lake as their sole means of livelihood. However, the extent of this activity had become a serious threat to the populations of some species and severely disturbed many others which used the Lake for roosting or feeding.

Pollution Although water pollution from industrial sources was not a major problem, fertilizer and pesticide residues from agricultural fields posed a problem in the northern part of the lake. Similarly, sewage and the waste water from small villages and towns was posing a pollution problem, although not considered a significant threat. Although pollution was not yet seen as a major problem for the Lake, it had the potential to increase.

2.4 Management Actions Undertaken in Lake Chilika

In response to the problems that had been identified the CDA invested in a cooperative and collaborative program with a number of local and national institutions, and sought guidance from a rapidly developing international network. In particular, there was a large degree of cooperation with governmental agencies and institutions for data collection and analysis as well as consultation with local communities in the catchment of the lake. In these activities it was strongly supported by Wetlands International South Asia (http://south-asia.wetlands.org/) in the formulation of action plans, technical documentation, and the dissemination of information through the publication of newsletters and brochures.

Specific management activities were supported by MOEF and through special grants. Major activities focussed on: (i) controlling the silt load delivered by the rivers; (ii) managing invasive weeds; (iii) steps to ensure the sustainable development of fisheries and aquaculture; (iv) the conservation of wildlife, including the restoration of islands and bird habitat; (v) moderation of the lake level; (vi) environmental impact assessments; (vii) construction of infrastructure such as roads to support socio-economic development; (viii) the promotion and regulation of ecotourism; and (ix) capacity building for local institutions and villages. These activities were undertaken in collaboration with other agencies and institutions and with village cooperatives and self-help groups. The mix of advice obtained from technical expertise provided by specialist institutions and local knowledge came to characterise the operating environment that the CDA created and maintained.

The specific management actions undertaken by the CDA in response to the adverse changes in ecological character in the Lake were described in the Montreux Record Questionnaire and are summarised below.

Engineering Works The most significant management actions were the hydrological intervention by way of opening of the mouth to the Bay of Bengal (started in January 2000 and completed on 23 September 2000) and the dredging of a lead channel between the Bay and the lake. Given the scale and importance of these activities they were the subject of much debate with the National Institute of Oceanography being commissioned to assess the impact of the reduction in siltation and the opening of the new outlet.

It was expected that positive changes would include an increase in the desired salinity regime and a reduction in the spread of invasive weeds, and in particular, a decline in the area covered by the floating plant *Eichhornia crassipes* (water hyacinth). The positive responses occurred very soon afterwards with changes in the salinity regime leading to increased fisheries (fish, prawn and crab) yields; data for the years 2000–2002 indicated that since the opening of the channel to the sea and the new lake mouth, fish landings increased from a previous average of 1745 in 1999–2000 to 4982 million tonnes in 2000–2001 with 12,235 million tonnes in 2015–2016 (Chilika Development Authority data).

Monitoring Data collection to characterise the hydrobiological features of the Lake is carried out by CDA in collaboration with Wetlands International South Asia and other partners. The overall objective of the monitoring is to ensure that the flow regime in the Lake is enhanced and the salinity levels are optimised to ensure the maintenance of the Lake's biodiversity (Kumar and Pattnaik 2012).

Community Consultation The management actions being undertaken are coordinated by the CDA in consultation with many local stakeholders and relevant agencies and experts. This has resulted in a powerful and complex network of interested parties who have worked together to take the actions which have led to the positive changes that have been observed and measured in the ecological character of the Lake. The CDA is also very actively involved in many socio-economic activities that provide support for the local communities that depend on the Lake for their livelihoods and well-being.

Bird Hunting The CDA has taken specific actions to address the illegal hunting of birds by local villagers. This has been done by working with local Non-Governmental Organisations and Community-Based Organizations to form a Bird Protection Committee and community based nature tourism as a means of alternate livelihoods. This has included arrangements to provide soft loans and to encourage the villagers to take up other activities. Through such steps the Committee has obtained commitments from local communities to abandon poaching. This has been recognised at an official level and provided a sound basis for better management of the bird populations of the Lake.

Pollution While pollution was not regarded as a major management issue for the lake, the CDA has kept the situation under review and has plans to address pollution issues as required in the future.

2.5 Integrated Management Responses and the Ramsar Convention

The CDA has documented much of the information about these activities and presented many talks about its work at major scientific conferences, such as the Asian Wetland Symposia, World Lake Conferences and INTECOL Wetland Conferences. These have been used to demonstrate the nature of the integrated approach that has been used to manage the lake and ensure its ecological character is restored and maintained in a way that enables the sustainable use of ecosystem services. In this respect alone it can be regarded as an excellent example of the whole ecosystem approach to management advocated by the Convention on Biological Diversity and in line with the Ramsar Convention's Wise Use concept (see Finlayson et al. 2011).

Through the above-mentioned management actions and by making good use of the information supplied by relevant authorities, the responses at Chilika Lake were seen as sufficient for the Lake to be removed from the Montreux Record. This was accompanied by an ongoing commitment from the Government of India and the CDA to continue to develop and implement an overall management planning process for the Ramsar site.

The Ramsar Advisory Mission also recommended that the Convention should consider using Lake Chilika as an exemplary good-practice case study of the application of the various Ramsar guidelines, and the use of the Convention's tools and approaches, to address complex site and catchment management issues (Finlayson et al. 2001). It was anticipated that with the development and implementation of a management plan, the lake could serve as an example for other Contracting Parties to the Convention interested in the suite of wise use measures that were being applied, including, for example, participatory planning, awareness and education, monitoring and integrated management. The abovementioned management activities and monitoring provide ample evidence that Lake Chilika has been managed in accordance with the Ramsar wise use guidance (see Ramsar Convention Secretariat 2010a). In addition, this example could assist the Convention to develop further guidance in support of the whole ecosystem approach to wetland management and provide an example of adaptive management practices for wetlands.

Opportunities for wide dissemination of materials through the Convention's media, describing the experience and approach of Chilika Lake management, and opportunities for exchange visits and other mechanisms for sharing of expertise, have been explored and widely used to promote the Lake and its management to local people and the international wetland community.

In response to the Ramsar Advisory Mission the CDA has successfully adapted those parts of the Ramsar guidelines on management planning (Ramsar Convention Secretariat 2010b) that could usefully support the further management of the Lake. In this respect there has been widespread recognition that the management approaches should support clearly articulated and widely agreed goals and objectives, with the latter being specifically related to actions and interventions to maintain or restore the ecological character of the lake. Importantly, this planning approach has further encouraged participatory planning and consultation with key stakeholders within the lake and its immediate environs, as well as further afield in the river basin and adjacent coastal zone. In this respect the Ramsar Guidelines on integrating wetland conservation and wise use into river basin management and those for integrated coastal zone management (Ramsar Convention Secretariat 2010c, d) have been successfully applied.

The participatory management practices that were outlined and demonstrated to the Ramsar Advisory Mission have been continued and extended as far as practicable in accordance with the Convention's guidelines on local community and indigenous people's participation in the management of wetlands (Ramsar Convention Secretariat 2010e) and used to assist in further developing this successful aspect of the management procedures for the lake. These have been coupled with community education and awareness programs that also represent many features of the guidelines provided by the Convention (Ramsar Secretariat 2010f). Completion of the Lake's wetland interpretation and training centre has provided a remarkably successful base for capacity building and developing awareness amongst local stakeholders and technical experts (Fig. 2.7) of the multiple values of the lake, including the development of appropriate guidance for the expanding ecotourism in the Lake. In particular those tourist activities which may be impacting upon the migratory and resident birds and dolphins have been the subject of increasing attention and measures have been taken to reduce any noise pollution caused by boats.

As mentioned above, the extensive monitoring programs in the lake have been continued and in places complemented in order to ensure that the biological, chemical and physical features of the Lake are maintained or improved in line with the objectives agreed through formal processes and with stakeholders. This

Fig. 2.7 The Chilika wetland training centre: (**a**) and (**b**) the main buildings and (**c**) and (**d**) training activities. (Photographs © CM Finlayson)

has been particularly important when considering the changes to the channel from the Lake to the Bay of Bengal. This has led to largescale improvements in the ecological character of the lake, as shown by the 10 indicators used in the Chilika ecosystem health report cards for the years 2012, 2014 and 2016: (http://www.chilika.com/documents/publication_1507881562.pdf), but it is not known conclusively whether the outlet to the sea will remain open, although the initial modelling indicated that it would, or if the fisheries yields will be maintained. The assessment and monitoring of these major issues is not only ongoing but are critical if the major benefits that have been obtained are to be sustained. The emerging pressure of global climate change has also been assessed and appropriate management responses, including monitoring, have been developed.

Given the complex nature of the lake ecosystem and its importance to local people the CDA has organised or participated in a number of workshops and other meetings to discuss progress with the various management activities and seek further expert advice. This included hosting the Asian Wetland Symposium in Bhubaneswar in 2005 as well as many further technical meetings. Such technical or expert reviews have been essential for guiding the management approaches that have been used in the Lake. The continued participation of the CDA and its partners in the meetings of the Ramsar Convention will provide an opportunity to enhance the management approaches, and importantly, to provide feedback and guidance to other governments and experts that need to similarly address the complex issues associated with the management of large Ramsar sites where biodiversity and livelihoods issues are juxtaposed.

Given the limited responsiveness of many Ramsar parties to the requirements of the Convention, as outlined by Finlayson (2012) in an analysis of the national reports on implementation of the Convention's strategic plan, the example of Lake Chilika and the Chilika Development Authority, supported by the Odisha and Indian Governments, stands out. In particular it epitomises both the challenges faced by wetland managers when livelihoods and biodiversity issues are confronted together, as well as an approach for bringing these together. Its success undoubtedly lies with those responsible for managing and supporting the Authority with their own success guided by the policy and technical guidance provided through the Convention, in particular that developed by the Scientific and Technical Review Panel and presented in the Ramsar Handbooks (Ramsar Convention Secretariat 2010a).

The success story of Lake Chilika reflects the success story of the policy-technical domain created by the Ramsar Convention based as it has been in a unique partnership between technical experts, non-governmental organisations, and governments. It is also very important not to forget that such success has been underpinned by commitment to, and respect for, the many people and communities that benefit from the wise use of wetlands. The latter is well illustrated by the following text taken from *The Chilika Statement* that was issued on the occasion of the Asian Wetland Symposium 2005 held in Bhubaneswar, Odisha India:

> That Chilika Lagoon in India is an outstanding example of wetland conservation and wise use following the principles of integrated management with strong emphasis on local people's participation and shared decision-making through networking of local, national and international experiences; restoration measures adopted have led to significant improvements in socio-economic conditions of communities dependent on Chilika Lagoon for livelihoods while maintaining ecological integrity. Ramsar Center Japan (2005)

References

Finlayson CM (1996) The Montreux Record: a mechanism for supporting the wise use of wetlands. Proceedings of the 6th meeting of the conference of the Contracting Parties of the Convention on Wetlands (Ramsar Convention Bureau, Gland, Switzerland). Technical sessions: reports and presentations, Brisbane, Australia. Vol. 10/12 B, pp 32–37

Finlayson CM (2012) Forty years of wetland conservation and wise use. Aquat Conserv Mar Freshwat Ecosyst 22:139–143

Finlayson CM, Kurshid N, Kaul S, Pattnaik AK, Trissal CL (2001) Ramsar Advisory Mission No. 50: India (2001). Removal of Chilika Lake Ramsar site, India, from the Montreux Record. Ramsar Wetlands Convention, Gland, Switzerland, 18 pp

Finlayson CM, Davidson N, Pritchard D, Milton GR, MacKay H (2011) The Ramsar Convention and ecosystem-based approaches to the wise use and sustainable development of wetlands. Journal of International Wildlife Law and Policy 14:176–198. https://doi.org/10.1080/13880292.2011.626704

Horwitz P, Finlayson CM (2011) Wetlands as settings: ecosystem services and health impact assessment for wetland and water resource management. Bioscience 61:678–688

Kumar R, Pattnaik AK (2012) Chilika – an integrated management planning framework for conservation and wise use. Wetlands International – South Asia/Chilika Development Authority, New Delhi/Bhubaneswar

Ramsar Center Japan (2005) The Chilika statement, Bhubaneswar, 9 February 2005, Tokyo

Ramsar Convention Secretariat (2010a) Wise use of wetlands: concepts and approaches for the wise use of wetlands. Ramsar handbooks for the wise use of wetlands, vol 1, 4th edn. Ramsar Convention Secretariat, Gland

Ramsar Convention Secretariat (2010b) Managing wetlands: frameworks for managing wetlands of international importance and other wetland sites. Ramsar handbooks for the wise use of wetlands, vol 18, 4th edn. Ramsar Convention Secretariat, Gland

Ramsar Convention Secretariat (2010c) Coastal management: wetland issues in integrated coastal zone management. Ramsar handbooks for the wise use of wetlands, vol 12, 4th edn. Ramsar Convention Secretariat, Gland

Ramsar Convention Secretariat (2010d) River basin management: integrating wetland conservation and wise use into river basin management. Ramsar handbooks for the wise use of wetlands, vol 9, 4th edn. Ramsar Convention Secretariat, Gland

Ramsar Convention Secretariat (2010e) Participatory skills: establishing and strengthening local communities' and indigenous people's participation in the management of wetlands. Ramsar handbooks for the wise use of wetlands, vol 7, 4th edn. Ramsar Convention Secretariat, Gland

Ramsar Convention Secretariat (2010f) Wetland CEPA: the Convention's Programme on communication, education, participation and awareness (CEPA) 2009–2015. Ramsar handbooks for the wise use of wetlands, vol 6, 4th edn. Ramsar Convention Secretariat, Gland

Chapter 3
Ecological Characterization of Chilika: Defining Strategies and Management Needs for Wise Use

Ritesh Kumar, C. Max Finlayson, and Ajit K. Pattnaik

Abstract Describing the ecological character of Chilika has enabled advances in the formal conceptualisations and definition adopted by the Ramsar Convention on Wetlands to be presented, in particular through the inclusion of ecosystem services and governance settings. The evolution of the concept of ecological character is described and related to the parallel concept of ensuring the wise use of wetlands. Framing the ecological character of Chilika as a social-ecological system allows for identification and prioritization of a number of wetland features from their ecological and social subsystems, as well as pathways for governance systems that can be guided by the needs of meeting the limits of acceptable change for these parameters. Uncertainty about the trajectory of changes in ecological character calls for continued efforts in knowledge building, experimental management design, bridging different knowledge systems, and ongoing monitoring and evaluation. The Chilika Development Authority can enable such by: (a) enhancing knowledge and understanding of critical ecosystem functions and their relationship with system dynamics (particularly species migration and exchange between riverine, lagoon and sea; sediment and nutrient dynamics; and factors influencing distribution of macrophytes); (b) feeding ecological knowledge into management decisions and actions (such as the regulation of interferences to migratory pathways or species habitats, regulation of tourism pressure on critical habitats); (c) supporting the development of multilevel governance systems, and polycentric institutional and organizational linkages (particularly linking the role of fisher cooperatives, tourism associations,

R. Kumar (✉)
Wetlands International South Asia, New Delhi, India
e-mail: ritesh.kumar@wi-sa.org

C. M. Finlayson
Institute for Land Water and Society, Charles Sturt University, Albury, NSW, Australia

IHE Delft, Institute for Water Education, Delft, The Netherlands

A. K. Pattnaik
Chilika Development Authority, Bhubaneswar, Odisha, India

C. M. Finlayson et al. (eds.), *Ecology, Conservation, and Restoration of Chilika Lagoon, India*, Wetlands: Ecology, Conservation and Management 6,
https://doi.org/10.1007/978-3-030-33424-6_3

village panchayats within the overarching governance regime); and, (d) developing capacity for dealing with perturbations and uncertainties, specifically in the context of climate change.

Keywords Ecological character · Wise use · Wetland management · Social-ecological system · Ramsar · Chilika

3.1 Introduction

The 'wise use' approach, adopted by the 170 Contracting Parties of the Ramsar Convention, is considered globally as a central tenet of wetland management (Gardner and Davidson 2011; Finlayson et al. 2011). As per a formal decision made by the Contracting Parties of the Convention, wise use of wetlands is the 'maintenance of their ecological character achieved through implementation of ecosystem approaches, within the context of sustainable development' (Ramsar Convention Secretariat 2010a). Assessment and reporting of ecological character and change in ecological character provides the basis for understanding the condition of wetlands, thereby informing policy development and priority setting within the Convention, as well as within the national jurisdictions of the Contracting Parties (Ramsar Convention on Wetlands 2005). An important function of site management planning is to outline an approach for the maintenance of the ecological character, and in doing so, retaining the essential ecological functions that underpin the delivery of ecosystem services and maintenance of biodiversity (Ramsar Convention Secretariat 2010a). Delivery of wise use commitments is therefore predicated on the extent to which wetland managers can define the site's ecological character, and use this information to design and implement effective management actions.

While the progenitors of the wise use approach encapsulated the need to incorporate the linkages between people and wetlands as a means to stem wetland loss, the construction of what comprised the 'ecological character' of a wetland, until recently, has remained primarily ecological. The definition of ecological character adopted in the 3rd Conference of the Parties to the Convention in 1987 included only ecosystem components and processes (Pritchard 2018). Ecosystem services were considered as being benefits that were derived from ecosystem functioning, as reinforced by the 1996 and 1999 amendments to the definition of ecological character, with the latter being expressed as 'the sum of the biological, physical and chemical components of the wetland ecosystem, their interactions, which maintain the wetlands and its products, functions and attributes' (Pritchard 2018). The Millennium Ecosystem Assessment (2005), however, brought to the fore a direct and formal articulation of the linkages between ecosystem functioning, ecosystem services and human well-being (Maltby and Acreman 2011), triggering a process of harmonization of existing wetland concepts with those used by the Assessment (Bridgewater 2008; Finlayson 2012).

A revised definition of ecological character, which included ecosystem services along with ecosystem components and processes, was adopted at the Ramsar Convention's Ninth Conference of Parties (Ramsar Convention on Wetlands 2005). In the intervening period, the links between ecological character description and management planning have been clarified by additional guidance on the constituents and purpose of describing the ecological character (Ramsar Convention on Wetlands 2008), use of conceptual models to hypothesize wetland functioning (Davis and Brock 2008), and development of national frameworks (e.g. Department of the Environment, Water, Heritage and the Arts 2008). More recent scholarship on the topic has focused on the characterization of change (such as Finlayson et al. 2016; Kopf et al. 2015) and use of palaeo-ecological information (Gell et al. 2016).

Despite the importance of systematically describing and assessing the ecological character of Ramsar Sites in delivering wise use commitments, practical examples of the application of this framework in deriving site management needs are limited and largely elusive. The existing Ramsar Convention guidance is used mostly as a checklist to collate information on wetland status and trends with reference to criteria for designating the site as a Wetland of International Importance, making limited connections of the analyses with site management. This is seen as a major gap in the guidance provided by the Convention especially given the expected adverse consequences of climate change on the ecological character of wetlands; further guidance within the context of climate change was provided by an analysis of Australian wetlands (Finlayson et al. 2011), but this has not been incorporated into that provided by the Convention.

This paper endeavours to extend the existing scholarship on wetland ecological character by demonstrating its use as a framework for defining management needs for Chilika, a Ramsar Site located on the east coast of India. In the process of developing an approach and method, the existing guidance on ecological character description and site management has also been reviewed and gaps identified. Such gaps and possible response options are also discussed through application in the case study. The paper concludes with recommendations for strengthening the application of ecological character as a framework for structuring site management.

3.2 Approach and Method

Ecological character, as per its existing definition by the Ramsar Convention, is 'the combination of the ecosystem components, processes and benefits/services that characterize the wetland at any given point in time' (Ramsar Convention Secretariat 2016). Ecosystem benefits/services, hereinafter ecosystem services, have been defined in accordance with the Millennium Ecosystem Assessment as 'the benefits that people receive from ecosystems'. Article 3.2 of the Convention calls for each Contracting Party to be 'informed at the earliest possible time if the ecological character of any wetland in its territory and included in the List of Wetlands of

International Importance has changed, is changing or is likely to change as the result of technological developments, pollution or other human interference' (Ramsar Convention Secretariat 2016).

The ecological character definition reflects the complexity and dynamic nature of wetlands by encompassing within its frame constituent biotic and abiotic components, the interlinking processes, as well as the diverse ways in which human societies benefit from these ecosystems (Maltby 2009). Wetland management planning entails defining strategies and actions for wetland wise use, and thereby maintenance of the ecological character (Ramsar Convention Secretariat 2010a). Guidance on management planning for wetlands involves the collation and synthesis of existing data on wetland features to enable the description of ecological character. Based on a set of evaluation criteria, wetland features are prioritized, keeping in view the overarching site management purpose (Ramsar Convention Secretariat 2010a). Management objectives are derived based on conditions required for maintenance of these priority features and setting operational limits for factors which influence these features. Operational limits are the range of values for each factor which can be considered acceptable and tolerable (Ramsar Convention on Wetlands 2012).

For Article 3.2 implementation, the obligation for 'maintenance' of ecological character pertains to human-induced adverse change (Ramsar Convention Secretariat 2010a), while recognizing that wetland restoration is one of the options to induce a positive change in ecological character. Natural variability in features, discerned from monitoring, is to be used as the basis for establishing a 'limit of acceptable change' (Ramsar Convention on Wetlands 2012), deviation from which may require management responses. The guidelines recommend assessing actual as well as likely change in ecological character, thereby affirming the significance of adaptive management, and use of precautionary approaches to defining site management objectives (Ramsar Convention Secretariat 2010a).

The wetland wise use approach is built around the premise that human use of these ecosystems on a sustainable basis is compatible with conservation (Finlayson et al. 2011), thereby aligning well with the fact that a certain level of natural variation and disturbance is necessary to maintain the integrity of wetland ecosystems (Ramsar Convention Secretariat 2010b). With ecosystem services included within the definition of ecological character, the goal of securing wise use and thereby 'maintenance of ecological character' can be examined regarding the wetland ecosystem's capacity to sustainably deliver the services on which humans depend (Bridgewater 2008).

It is increasingly appreciated that in a human-dominated world, ecosystem services are not generated by ecosystems alone, but by social-ecological systems of which humans form an integral part (Reyers et al. 2013; Levin et al. 2013). Assessment of such coupled systems is realistically framed by a 'humans-in-ecosystems' approach (Davidson-Hunt and Berkes 2003), and as an integrated complex adaptive system having feedbacks between the social and ecological subsystems (Olsson et al. 2006; Berkes 2010). Governance, including management and policy levels, provides a broad link between the two subsystems (Berkes 2017), as well as a mechanism to influence system trajectory (Chaffin et al. 2014). We refer to governance here as the social functions centred on steering collective behaviour towards desired outcomes and away from undesired outcomes (Young 2017), and a governance system as an ensemble of elements performing the function of governance in a given setting. Institutions, which are a collection of rights, rules, principles and decision-making procedures (North 1991) giving rise to social practices, assigning roles to participants and guiding interactions amongst participants, form a prominent feature of governance systems (Young 2017).

The social-ecological systems associated with wetlands can be separated by the following features, (a) connectivity or tight coupling amongst system components, (b) thresholds and non-linear patterns of change, and, (c) directional processes (Young 2017). The tight coupling between the subsystems requires focusing on the system as a whole, rather than treating social subsystems as subject to impacts of ecological systems, or vice-versa (Holling et al. 2002). The occurrence of thresholds and state changes often catapults the social-ecological system onto a dramatically different trajectory (Scheffer 2009). Unlike ecological systems which may shift to an earlier state (such as from being brackish-water dominated state to a freshwater-dominated state), social systems rarely shift to a prior state (such as reverting to a community managed state from a state-led management), making the change mostly directional (Scheffer 2009). Due to complex interactions taking place within such complex systems, surprise rather than predictability is the norm (Young 2017). The difficulty in separating development within a spatially delimited area from the impact of forces operating in the world at large begs consideration of interconnectedness from lenses wider than river basin or coastal zone scales, with which wetlands management has been mostly commonly linked to date. Thinking on the line of wetlands as settings, focusing on the interface rather than the spatially delimited area of impact, is one such example (Horwitz and Finlayson 2011).

For ecological character to provide a basis for wetland wise use, the ecological, as well as social subsystems, need to be considered along with the interlinking institutional and governance settings. The existing guidance, by limiting the human connection to the description of ecosystem services, underplays and obfuscates the significance of social processes and outcomes on wise use. From a hitherto ecological construct, the ecological character needs to be framed on the interactions of wetland ecosystem components and processes with social systems, building from rules and institutions that mediate human uses of wetlands as well as systems of knowledge and ethics that interpret wetland systems from a human perspective (Kumar et al. 2011).

Firstly, the use of the term 'ecological' character to refer to a hitherto social-ecological system is an apparent misfit, although there is no reason that such a definition should hinder progress, especially if ecological systems are taken as including people. We refrain from further discussing this aspect, and instead focus more on the application of the existing guidelines on using the ecological character as a framework to identify wetland management needs and identify at least three issues on which the current guidance fall short. The first and foremost pertains to how ecosystem services are to be described as part of an ecological character description. Ecosystem services are enlisted into the Millennium Ecosystem Assessment classes of provisioning, regulating, and cultural. The supporting services overlap with ecosystem process categories and thereby are not mentioned separately. The guidance requires a quantitative description of the service (for example, production levels for certain categories) and user population. There is a considerably developed body of research that links the relationships that people have with their natural resources, and the social forces, institutions and cultural values that sustain or undermine them (for example see Schlager and Ostrom 1992; Leach et al. 1999). Thus, merely describing production levels or user population linked with a set of ecosystem services hides underpinning elements of social differentiation, power relationship and polity which influence the wetland state (Ernstson 2013), and co-production of ecosystem services through nature, human labour, knowledge and technologies (Bruckmeier 2016). Preferences held for ecosystem services also implicitly impose a social decision process on underpinning wetland ecosystem components and processes (Menzel and Teng 2010). In several circumstances, the transformation of wetlands is linked with changing societal preferences for wetland ecosystem services within the larger developmental context.

The instrumental perspectives that underpin the ecosystem services concept limits consideration of relational values, or values beyond the inherent worth or satisfying preferences (Chan et al. 2012; Russell et al. 2013). Such relational values are rooted in identities, individual as well as societal interactions with nature, sense of place, physical and emotional health, provide meaning to the relationships, including the hitherto instrumental relationships captured within provisioning and regulating services (Pascual et al. 2017).

The second issue pertains to the choice of relevant 'baseline' or 'reference regime' against which change can be assessed (an issue also raised by O'Connell 2003; Gell et al. 2016). For a dynamic social-ecological system bearing directional changes, the state of a wetland at a single point in time is unlikely to provide a useful frame of reference. Further, given that in most circumstances there is insufficient data on past conditions, the definition of baseline condition is highly likely to be a socially constructed process subject to the interpretation of available knowledge within the perspectives of the interpreter. The choice of a baseline is thus not just a question of empirics, but of stakeholder perspectives, power relations and systems of knowledge and ethics in which natural systems are interpreted by humans (Berkes 2010). Description of 'natural variability' within intensely human-dominated setting may also be difficult given the complexities involved in isolating perturbations induced by human activity from ecosystem functioning.

The third issue relates to using of criteria for prioritization of ecological character features to be able to identify management focus. Prioritization criteria suggested in the existing guidance are mainly of an ecological nature (Ramsar Convention Secretariat 2010a), and do not address the social components.

This paper address the above mentioned issues through a step-wise description and evaluation of the ecological character of Chilika, suitably modifying and in certain cases adding new elements to existing Ramsar guidance. The analysis is presented in five steps. Firstly, a social-ecological systems approach is used to describe the ecological character, by mapping trajectories of the changes in ecological and social subsystems and their relating governance structures. Analysis of relational values and governance systems are included within the ecological character description for reasons discussed in the above paragraphs. Prioritization of ecological character elements follows next, based on user-defined criteria, to assist in identification of the boundaries of the area being considered for management action. A reference regime building on the prioritized ecological character elements is discussed. Risks of an adverse change in ecological character, concerning the subsystems and system as a whole, are then discussed. The analysis concludes with defining management needs for achieving wise use of Ramsar Sites.

This paper is based on the outcome of an ecological character description and evaluation workshop conducted by the authors with the team of scientists and experts of the Chilika Development Authority (CDA). The workshop was held during March 3–6, 2015 at the Wetland Research and Training Center of the CDA and was supported by the International Development Research Center under their Climate Change Adaptation Programme.

3.3 Applying the Ecological Character Framework in Chilika, India

3.3.1 Chilika, Odisha

Located on the east coast of Odisha State, Chilika is a pear-shaped coastal lagoon extending in its peak inundation to 1165 km^2 with a linear axis of 64.3 km and an average width of 20.1 km (Fig. 3.1). The lagoon opens into the Bay of Bengal through a 32 km long channel, separated from the sea by a narrow sandy spit. An extensive 400 km^2 marshy alluvial plains, cultivated during the dry season, flanks the lagoon's northern margins. There are 24 rocky islands dotting the eastern margins of Chilika (Fig. 3.1).

Chilika is an assemblage of shallow to very shallow marine, brackish and freshwater ecosystems (depth ranging between 50 cm to 3.7 m). The distributaries of the River Mahanadi (Daya and Bhargabi) and over 50 seasonal streams draining a direct basin area of 3929 km^2 bring in nearly 4900 million m^3 of freshwater into the lagoon (Kumar and Pattnaik 2012). The mixing of freshwater from catchment runoff and seawater brought in from the Bay of Bengal leads to the development of a salinity gradient, enabling the creation of diverse habitats suited for a range of species adapted to freshwater, brackish and marine environments. Based on assessments by a range of linked survey teams to date, Chilika is a habitat for 569 plankton, 22 algal, 726 plant, 136 mollusc, 29 crustacean, 317 fish (Mohanty et al. 2015), 225 bird, 7 amphibian and 19 mammal species, several of which are of high conservation significance globally and regionally (based on the Chilika Development Authority database). The wetland regularly hosts over a million wintering migratory birds and is one of the largest congregation of migratory waterbirds in the Central Asian Flyway (Balachandran et al. 2009). It is also one of the two lagoons in

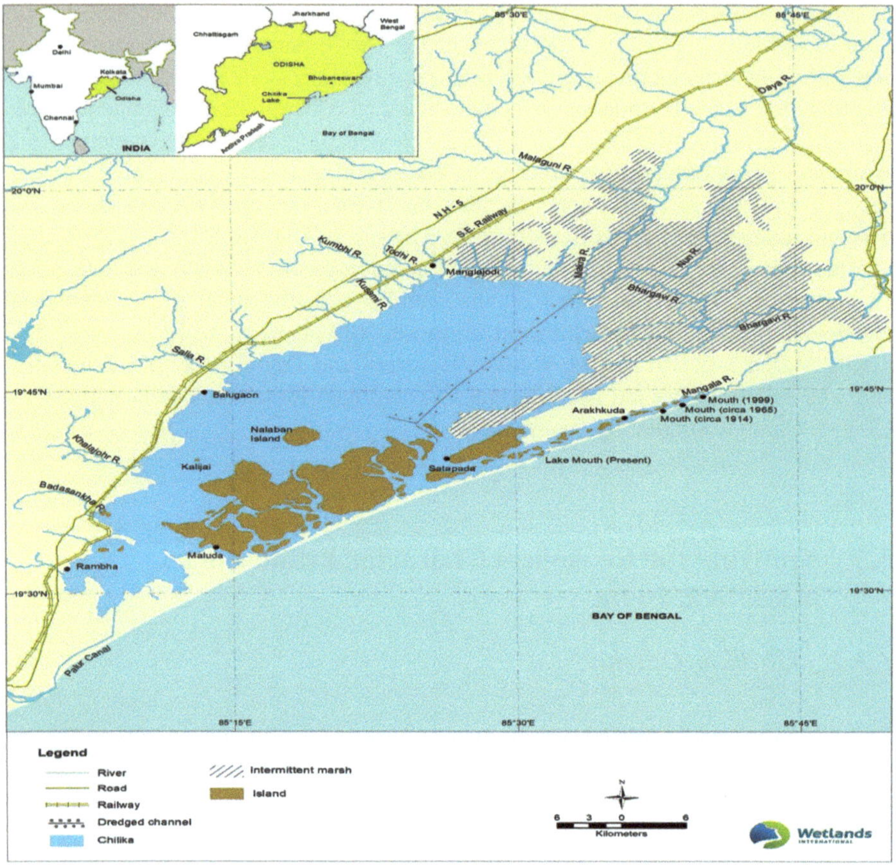

Fig. 3.1 Chilika and its environs. (Source: Redrawn with permission from Kumar and Pattnaik 2012)

the world inhabited by the Irrawaddy Dolphin (*Orcaella brevirostris*) (Reeves et al. 2008). *Barkudia insularis*, a limbless skink, is endemic to Chilika environs (Bauer et al. 2014).

Chilika is inextricably intertwined with the livelihoods of 0.14 million fishers inhabiting over 130 settlements along wetland fringes and its islands. The diverse and dynamic assemblage of fish, invertebrate and crustacean species provide the base of the rich fishery which includes 73 economically important fish, prawn and crab species with an average annual yield of 12,000 MT. The wetland is one of the most famous tourist destinations of the region. The temple of Kalijai and archaeological remains at Manikpatna, Palur and adjacent areas mark Chilika's rich maritime heritage (Tripati and Vora 2005). Chilika (along with Keoladeo National Park) was the first Indian wetland to be designated by the Government of India as a Wetland of International Importance under the Ramsar Convention in 1981.

3.3.2 Ecological Character Change

3.3.2.1 Changes in the Ecological System

About 13,500 years ago, Chilika formed a part of river delta with freshwater vegetation (Khandelwal et al. 2008). With increase in sea level some 9500 years ago, the area became an estuary with rich mangrove vegetation (Khandelwal et al. 2008). Chilika was part of the Bay of Bengal about 6000 years ago and during the Pleistocene constituted its gulf (Rao 1995). Its current form is attributed to a complex geologic process involving the deposition of beach ridges and spits enclosing a body of sea water within the Bay of Bengal (Tripati and Vora 2005). The separation of the lagoon from the sea is believed to have taken place during the last sea level rise, before 3750 $\pm$ 200 years BP followed by the emergence of land due to minor tectonic uplift (Venkatarathnam 1970). The inflowing tributaries of River Mahanadi, the Daya and the Bhargabi, bring in a considerable amount of silt draining a fertile alluvial tract and rendering the northern part of the lagoon shallow, as compared with the rest.

Records of the faunistic diversity of Chilika date back to the turn of the twentieth century (for example publications such as Annandale 1915). Assessments on species assemblages during the 1915–1940 period indicate significant exchange with riverine as well as marine environments, as confirmed by the presence of sizeable numbers of marine crustacean *Penaeus monodon,* sponges of marine genera, and freshwater fish within Chilika (for a historical resume refer Rama Rao 1995; Bandyopadhyay and Gopal 1991). The hydrological, material and species exchange between the lagoon and the Bay of Bengal is largely influenced by the dynamics of the inlet, which in turn are governed by the longshore currents in the form of a littoral drift causing a net transport of sediments towards the northeast and a northward movement of the inlet and subsequent choking (Chandramohan et al. 1993). This migration is impeded by cyclonic swells, and has required manual intervention to cut the mouth (Dujovny 2009; Bandyopadhyay and Gopal 1991). Owing to the northward drift, the inlet located 6 km northeast of Arkhakuda in 1914, shifted to 8 km northeast by 1965 (Tripati and Vora 2005).

Towards the 1970s the lagoon's connectivity with the Bay of Bengal was progressively impeded, causing a shift within the lagoon to a freshwater dominated state. The littoral drift led to further northward migration of the mouth, which by 1999 was another 2 km northeast of the position observed in 1965 (Tripati and Vora 2005). Channelization of deltaic floodplains, intensification of agriculture and decreasing forest cover in the direct catchments mobilized soil transport and increased the overall sedimentation in the lagoon (Das and Jena 2008; D'Souza 2002). Land use and land cover change for the Chilika basin for the period 1971–1990 indicated a decrease in area under dense forests, with a concomitant increase in areas under agriculture, plantations and settlements (Kumar and Pattnaik 2012).

Average salinity within the lagoon in 1999–2000 was recorded to be 8.5 ppt (reported in Barik et al. 2017). The lagoon's fish catch, predominantly constituted by species migrating from the Bay, dwindled from 8600 MT to 1702 MT (Mohapatra et al. 2007). Macrophytes, which covered only 20 km^2 of the lagoon areas in 1972 and about 100 km^2 in 1982, rapidly increased to over 440 km^2 by 1988 (Bandyopadhyay and Gopal 1991) and 523 km^2 in October 2000 (Ghosh et al. 2006). The lagoon became shallower, with the range of depths declining from 74–340 cm in 1992–93 to 42–142 cm in 1996–97 (Ghosh et al. 2006). Changes in faunal biodiversity were equally striking. Comparative studies of faunal diversity done during 1915–1924 and 1995 highlighted the predominance of freshwater species and corresponding losses in Porifera, Crustacea, Molluscs and Pisces, all correlated with a decline in salinity (Ghosh et al. 2006). Of particular concern was the spread of freshwater invasive species such as *Eichhornia crassipes* in the northern and central sector, which in the seventies was recorded as a casual visitor in the northern sector during monsoon, and disappearing subsequently with the rise in salinity (Biswas 1995). In 1993, the Government of India considered placing the Ramsar Site onto the Convention's Montreaux Record (a list of Wetlands of International Importance where changes in ecological character has occurred, is occurring, or is likely to occur).

Restoration measures put in place since 2000 included opening of a new mouth and dredging of a channel within the northern sector of the lagoon to ensure that riverine sediments are flushed out (Ghosh et al. 2006; Pattnaik and Kumar 2016). These interventions have been complemented by a basin scale participatory watershed management programme to contain silt loading from the catchments and enhance resources for community livelihoods (Pattnaik and Kumar 2016). The response of the hydrological intervention and basin management has been rapid and sustained.

After initial trophic bursts, the annual fish landing stabilized at nearly 13,000 MT (average landing for during 2003–2014). Annual censuses by CDA of Irrawaddy Dolphins within Chilika reported an increase from 89 to 158 individuals between 2003 and 2014, an increase in habitat use, and improved breeding, dispersal and decline in mortality rates. The sea grass meadows expanded from 20 km^2 in 2000 to 104 km^2 by 2014, along with a significant decline in freshwater invasive species. Based on the positive changes noticed in the ecological character, the Ministry of Environment, Forest and Climate Change (then the Ministry of Environment and Forests) of Government of India requested the Ramsar Convention for removal of the site from Montreux Record. Following an advisory mission in December 2001, the site was removed from the Record and the intervention recognized with the Ramsar Wetland Conservation Award and Evian Special Prize for "wetland conservation and management initiatives" (Kumar and Pattnaik 2012).

The Odisha coastline routinely experiences cyclonic storms due to seasonal depressions formed in the Bay of Bengal. One such very severe tropical cyclone Phailin (category 5 with an average wind speed of 259 km/h) made a landfall only 40 km from the wetland's southern shore on 12 October 2013. The ensuing cyclonic depression caused over 40% of the annual rainfall within 2 days of its landing.

The storm surge opened a new 125 m wide inlet south of an existing inlet, with cross section nearly double by November (Chilika Development Authority 2015). The lagoon experienced a prolonged lowering of salinity, increase in riverine species in the fish catch, and the presence of freshwater zooplankton species in the outer channel otherwise dominated by brackish and marine species. Sea grass beds were extensively damaged. Data from hydrobiological monitoring since October 2013 indicate that the lagoon maintained freshwater conditions until March 2014 (Barik et al. 2017). Subsequently, the salinity gradient was observed to re-establish. Recovery in sea grass beds was rapid, and the entire area is reverting to near pre-cyclone conditions. Monitoring records of the CDA indicate a recovery in the proportion of brackish and marine species in the fish catch.

3.3.2.2 Changes in the Social System

Chilika is fringed by over 130 fisher villages, with evidence of settlement dating back to few thousand years (Tripati and Vora 2005; Nayak 2014). The relationship these communities have with the lagoon range from being instrumental (such as fishing for livelihoods) to relational (sense of place, cultural identity and harmony) underpinning their well-being and having a bearing on the lagoon's ecology and linked governance systems.

Fishing as an economic activity was accorded a lowly status in the society undertaken by caste fishers, whereas the non-fishers undertook farming and other activities (Nayak and Berkes 2010). Clear and uncontested rights, customary practices, use of a range of crafts and gears and well-demarcated resource boundaries formed the basis of community-managed fisheries (Jones and Sujansingani 1954; Sekhar 2007; Nayak and Berkes 2010). Rapid transformation ensued in the seventies with an influx of Bangladeshi refugees who promoted nylon net fishing techniques, and the introduction of motorized boats (Biswas 1995). Towards the 1980s, an increase in fish harvesting capacity resulted due to mechanization of the fleet, and a gradual increase in active fishers (Samal and Meher 2003). During 1957–1980, the number of active fishers in Chilika increased from 8000 to 21,800 and the number of fishing boats from 2300 to 3100 (Biswas 1995).

A major shift in livelihood relationships occurred in the mid-1980s as prawn culture was introduced in Chilika as an income supplementation programme for low-income families (Samal 2002). Towards the 1990s several factors such as the increased demand for prawns in Japan and European markets, devaluation of Indian Rupee and trade liberalization increased profitability in prawn culture, thereby inducing traders and moneylender into Chilika fisheries (Iwasaki and Shaw 2008). Traditional fishers, not being able to muster the necessary capital to engage in prawn culture, were gradually coerced to give up their fishing rights to money lenders and traders. The shallow margins of the wetland, which were the net fishing grounds for traditional fishers, were near completely encroached for prawn enclosures (Samal 2002; Nayak and Berkes 2011). The traditional fishers, who in the past had significant influence over resources, were pushed into penury, with a declining resource base, and high level of indebtedness from sources financed by the non-fishers.

Post the hydrological intervention of 2000, the rapid increase in fish landing did not automatically translate into increased incomes and enhanced social conditions for the fishers. An assessment comparing selected statistics for the pre-hydrological intervention period (1999) to post-intervention (2007) indicated that despite a near sixfold increase in fish catch, the per capita income of the fishers registered an increase of only 34% over 1999, and nearly no impact on the amount of household debt (Kumar et al. 2011). A coercive nexus of middlemen and traders which had worked towards disruption of fisher cooperative societies thus cornered a disproportionate share of increased revenues. This promoted the CDA to respond through an organized programme of strengthening community-managed fisheries institutions, through the infusion of capital, enhanced access to infrastructure and capacity development support. Surveys conducted in 2014 indicated that 77% of the PFCSs had paid their outstanding loans and were commercially viable. Nearly 70% of the catch was being traded through the cooperatives, yielding their members at least 30% higher revenues as compared with those paid by middlemen (Kumar et al. 2016).

Ecological restoration of Chilika also brought into focus the value of Chilika as a recreational amenity. As per the statistics of the State Government, the number of annual arrivals to the wetland steadily increased from 0.24 million in 1999 to 0.54 million in 2013, creating a sizeable economy for the community. An increasing population and use of habitats by dolphins were added attractions. An economic assessment conducted in 2015, placed the value of tourism associated with Chilika at US$ 51.8 million, 2.3 times higher than that of fisheries (Kumar et al. 2016). Management of tourism in Chilika, however, was marked by phases of conflict and unsustainable behaviour owing to increasing competition to corner economic returns. By 2007–2008, functional tourist societies emerged along the major tourist entry points of the wetland. Considering the ecological sensitivities associated with tourism, CDA, in 2014, initiated the development of a tourism action plan to ensure that the ecological integrity of the wetland was maintained, while the community was able to benefit from the lake's resources.

More recently, evidence of local level recognition of the role of Chilika in buffering extreme events have also been coming to fore. The Indian Ocean Tsunami of 2004 brought to the fore the role of coastal wetlands, such as mangroves in buffering extreme events (Unnikrishnan et al. 2013; Marois and Mitsch 2015). Community-level efforts for restoring mangroves in abandoned prawn farms were taken up in 2006 in some island villages and have been successful. The 2013 cyclone made evident the role of Chilika in buffering extreme events. Surveys indicated that the villages located on the shore of the wetland received less damage from the cyclone, as much of the impact was taken by Chilika's immense waterspread. This has prompted several communities, especially those located on the coast, to take into account the role of healthy wetland ecosystems within community-level disaster risk reduction planning processes.

3.3.2.3 Changes in Governance Systems

The trajectory of change in governance systems in Chilika has been one of establishment and breakdown of the community managed fisheries, giving way to state-led architecture of collaborative governance. Governance systems in Chilika evolved based on its diverse and rich fisheries which were a source of sustenance to a large fishing community living on the lagoon's shorelines and islands. These traditional fishers evolved a system of fishing resource partitioning enabled by norms setting spatial (areas to be used for fishing), temporal (seasons in which fishing was permitted) and gear restrictions (Sekhar 2007). Diverse crafts and gears (Jones and Sujansingani 1954), use of which was linked with the caste of the fishers (Nayak 2014; Sekhar 2004), formed an integral part of such an arrangement. Dating back to the fourteenth century, the erstwhile rulers of Chilika marked and delimited fishing grounds within the lagoon, which were leased out to these fishers for customary fishing on payment of lease rentals (Nayak and Berkes 2011; Nayak 2014).

Post-independence, the rights to administer fishing grounds stood vested with the Revenue Department of the Government of Odisha. In furtherance of the preferential access rights to the fishers, the State Government constituted a Central Fishermen's Cooperative Marketing Society (CFCMS) in 1959 with some village-level Primary Fishermen Cooperative Societies (PFCS) as members. The PFCS were given fishing leases, thus allowing access to their fishing grounds and also enabling decision making about such access (Nayak and Berkes 2011). Up to the late 1970s, fisheries governance tended to secure the rights of traditional fishers, granting them preferential access to fishing grounds within the lagoon (Samal 2002). In 1991, the State Government, in the backdrop of aquaculture development, introduced a policy creating aquaculture areas and thereby legalizing entry of non-traditional fishers in the arena (Samal and Meher 2003). Protests against the Integrated Shrimp Farm Project, a joint venture of the Government of Odisha and Tata (a business house), became a rallying point, leading to a Chilika movement led by a coalition of fishers and students (Samal 2002). Fishing rights in Chilika became contested ground between the traditional and non-traditional fishers (Martinez-Alier 2002; Dujovny 2009), resulting in a number of violent protests, litigation in courts and ultimately aquaculture being declared illegal with a Supreme Court Ruling of 1996 banning aquaculture in any form within the wetland and its 1000 m periphery.

In 1972, the Indian Parliament enacted the Indian Wildlife Protection Act, providing the regulatory framework for the protection of wildlife, and most importantly establishing schedules of protected animal and plant species, hunting or harvesting of which was declared illegal. The Irrawaddy Dolphin was accorded the highest level of protection by mention in Schedule I of the Act. Similarly, some animal species were accorded protection through inclusion in the schedules of the Act. Nalabana, a flat, marshy island of 15.53 km^2 located in the centre of the lagoon and a site of the high congregation of waterbirds and nursery ground for several fish species, was designated as a wildlife sanctuary in 1987 under the provisions of the Act. The Odisha Marine Fisheries Regulation Act was introduced in 1988 setting limits on the use of fishing gear within the wetland ecosystem. Enforcement, however, remained a challenge, and rampant instances of wildlife poaching and decline in the population of Irrawaddy Dolphins were evident until the turn of the century.

The rapid decline in ecosystem condition and livelihoods of dependent communities prompted the State Government to constitute the CDA as a formal institution mandated by the Government of Odisha to undertake conservation and management of Chilika. Constituted in 1991, the Authority has a mission to "restore and sustainable management of lagoon and its drainage basin based on sound scientific principles through participatory processes" (as stated on its website www.chilika.com). Decision-making within the Authority is steered by its Governing Body, headed by the Chief Minister with elected representatives, secretaries of concerned government departments, administrators, and experts being members. Beginning from small-scale interventions for treating catchments and removing invasive species with funds provided by Government of India, the Authority laid down a strategy for ecological restoration of the lagoon. The most significant of the interventions was the opening of a new mouth of the lagoon in September 2000 and since then implementation a basin scale programme for reducing silt loads, conserving biodiversity habitats, enhancing fisheries, communication and outreach, and monitoring and evaluation (Pattnaik and Kumar 2016). Informally, the Authority works with a network of over 50 international, national and local organizations which support delivery of its various programmes. An integrated management plan (Kumar and Pattnaik 2012) serves as a basis for coordination of activities of the Authority. Despite being a state agency, the CDA works as an enabling institution, convening stakeholders around wetland management, building capacity and outreach and providing the science and knowledgebase for making informed decisions. In 2017, the Authority demonstrated application of its regulatory functions to demolish nearly 120 km^2 if illegal prawn enclosures within the lagoon.

Community institutions, which in the past played a central role in the governance of Chilika, have acquired a prominent economic role in influencing benefit-sharing mechanisms. The PFCSs, which were rendered moribund by the turn of the twentieth century, a deliberate outcome of power asymmetries created by non-fishers, are being strengthened to ensure that traditional fishers garner a fair share of their enterprise. The surge of tourists has created an income diversification opportunity, and as many as nine tourist boat associations have emerged to ensure that the boat-owners get a fair opportunity to benefit from tourist spending (Kumar et al. 2016). CDA is proactively working with these institutions as a means to achieve community participation in various conservation and management initiatives. The degree of resource-use conflict, as witnessed in the 1990s, has reduced with the recovery of the ecosystem. Enhancement of dolphin habitats and gradual increase in the population of this endemic species is also attributed to the mutualism established between the species and fishers (D'Lima et al. 2014). In Mangalajodi, the CDA has successfully transformed such values to develop a model of community managed eco-tourism, run and administered by a group of former waterbird poachers (Kumar et al. 2016).

3.3.3 Prioritizing Wetland Features

The Ramsar Conference of Contracting Parties in 2008 passed Resolution X.15 which lists a set of core inventory fields for ecological character description, including 23 ecosystem components, 9 ecosystem processes and 27 ecosystem services, from now on termed as wetland features. While all of these features have a bearing on the functioning of Chilika, there is a need to prioritize these to be able to identify reference conditions and management needs.

As indicated in the above text, we have expanded the list of wetland features to include additional elements derived from the ecosystem services literature, particularly the IPBES multiple values framework (Pascual et al. 2017). For prioritization, a set of five regulatory, ecological and social criteria have been used. The regulatory criterion addresses the needs of meeting commitments under international conventions (such as the Ramsar Convention, Convention on Biological Diversity Convention on Migratory Species), national regulations (such as The Wildlife Protection Act, 1972; The Environment (Protection) Act, 1986; Wetlands (Conservation and Management) Rules, 2010; and The Indian Fisheries Act, 1987) and state regulations (such as The Orissa Marine Fisheries Regulation Act, 1981 and Supreme Court's 1996 verdict prohibiting aquaculture) are considered. The ecological criteria build on three sub-criteria. The first criterion uses species conservation and endemism status. The second sub-criterion prioritizes wetland features regarding their significance in supporting an important ecosystem component, process or service. The third sub-criterion adjudges the wetland feature regarding its representativeness of the ecosystem type. Under social criteria, a particular wetland feature is considered a priority if contributing to the well-being of communities, either in instrumental terms (direct contribution to livelihoods) or relational terms (principles, virtues and preferences associated with the relationship of communities with Chilika, which underpin resource stewardship opportunities). In Table 3.1, we list the priority wetland features, and summarize the available information on status and trends.

3.3.4 Reference Regime and Limits of Acceptable Change

To derive the focus of site management, the conditions required for maintenance of priority ecological character elements need to be stated. Such conditions, cumulatively for all priority wetland features drawn from ecological and social sub-systems, can assist in defining a reference regime for the wetland ecosystem as a whole, against which change can be assessed, and in particular, human-induced adverse change can be inferred for meeting Article 3.2 commitments under the Ramsar Convention. However, the guidelines do not dwell on how these conditions are to be derived. Intuitively, such condition may refer to a past state (possibly, unmodified by human action), or a 'desired' state (subject to existing knowledge on ecosystem functioning, and the values and preferences of the stakeholders). We use the latter interpretation for defining a reference regime.

Table 3.1 Prioritization of wetland features

Wetland feature (prioritization criteria indicated in brackets)	Prioritization rationale	Status	Trends	Limit of acceptable change	Justification for the limit of acceptable change
1. Ecological sub-system					
1.1 Ecosystem components					
Size (1)	The importance of a wetland feature is likely to increase with size. Viability of small, and isolated features is usually questionable	Chilika spans an area of 1165 km^2.	Peak inundation area declined from 860 km^2 to 605 km^2 during 1929–1988 (Tripati and Vora 2005), but have been restored, with seasonal fluctuations recorded between 906–1165 km^2. Since 1990s, the natural shorelines in the southern sector and the outer channel have been encroached by aquaculture farms, which isolate about 100 km^2 of natural inundation area. The CDA and the government have recently taken measures to demolish these farms, however, challenges of restoring hydrological connectivity and habitat conditions remain	No human induced reduction in wetland area No human induced alteration of natural shoreline, and land use within the intermittently inundated area	The area of wetland as indicated at time of designation needs to be maintained as per commitment under Ramsar Convention. Additionally, as per provisions of Wetlands (Conservation and Management) Rules, 2010 the wetland area cannot be used for alternate non-wetland usages

Freshwater inflows (2(c))	Freshwater inflows are a key component of lagoon hydrology. Flood pulses during monsoon help flush sediments and nutrients from the lagoon, and ensure species exchange, especially of riverine fish	Chilika receives freshwater inflows from distributaries of Mahanadi River, and eight major streams of western catchments. During 2011–2015, the total annual freshwater inflow to the lagoon was 4200×10^6 m^3	Inflows from Mahanadi River constitute on average 75% of total freshwater flows. Modelling studies have indicated a decrease in summer rainfall, and intensification of monsoon rainfalls	The Mahanadi Delta Rivers bring in atleast 1280 $m^3 \times 10^6$ freshwater in Chilika during monsoon months	This flow level is the minimum required to enable flushing of salinity to below 5 ppt in Northern Sector prior to monsoon. The salinity level is desirous for providing migration cues to several economically important fish species (The World Bank 2005)
Water level variation (2(c))	The relative difference of water level between the lagoon and the sea governs the direction of water flow between lagoon and the Bay of Bengal. Near complete submergence of Nalabana island during monsoon and subsequent emergence during winter and summer months regulates vegetation growth thus creating conducive habitat for waterbirds. Similar transition takes place along the shorelines	The water level at Satapada increase upto 1.4 m amsl during September, reducing to 0.4 m amsl during February – March.	Absence of long term datasets to assess trends.	Inter-annual variation for a normal monsoon year should not fall below 1.2 m for more than two consecutive years	An inter-annual water level variation during a normal monsoon year of 1.2 m is required to maintain submergence – emergence cycle for Nalabana Island, thus enabling transition of vegetation required for resident and migratory waterbirds
Water quality – Salinity (2(c))	A gradient of salinity ranging from freshwater	During 2015, the lagoon salinity ranged from	The average salinity of the wetland, which	Average salinity does not dip below 9 ppt for	A dip in salinity levels below this threshold is

(continued)

Table 3.1 (continued)

Wetland feature (prioritization criteria indicated in brackets)	Prioritization rationale	Status	Trends	Limit of acceptable change	Justification for the limit of acceptable change
	conditions in northern zone, mixohaline conditions within the central and southern parts, and progressively euhaline conditions in the outer channel allows for maintenance of diverse species habitats. The gradient also provides migration cues for fish	freshwater conditions (0 ppt) during monsoon, to euhaline condition (33.2 ppt) in Outer Channel during summer. The average salinity was assessed to be 9.92 ppt	raged between 13.7–17.5 ppt during 1957–61, gradually reduced to 4–5.8 during 1991–92. Post hydrological intervention, the average salinity has increased, and observed to range between 9.3 to 14 ppt during 2002–2015	more than three consecutive years.	an indicator of lagoon tending towards freshwater characteristics
Water quality – Dissolved Oxygen (2(c))	A highly oxygenated condition is required for survival of several aquatic species, notably fish	As per monitoring data of 2011–2015, Chilika is well oxygenated (levels ranging between 4–13.8 ppm), essential for supporting aquatic life. Pockets of low dissolved oxygen have been usually observed in Northern Sector, a predominantly macrophyte dominated area	Monitoring records of 2011–15 are similar to those of 1988 (reported in Bandyopadhyay and Gopal 1991)	Dissolved oxygen concentration remains ≥5 mg/L	A high oxygenated condition is required for several aquatic species, notably fish
Flora – Phytoplankton (2(c))	Phytoplankton form an important component of	Till 2015, 259 species of phytoplankton recorded in the wetland,	Surveys conducted during 2000–2001 indicated presence of 84 species	Chlorophyl –a concentrations are higher than 5 μg/L	Elevated phytoplankton levels can suppress dissolved oxygen

	food web and primary production	providing a rich base of primary producers (Srichandan et al. 2015)	(Rath and Adhikary 2008)		levels. Chlorophyll-a is a measure of phytoplankton biomass and is a good indicator of the health of an ecosystem 123
Flora- Aquatic Macrophytes (Extent) (2(c))	The extent of macrophytes and dominant assemblages serve as an indicator of regime change	Predominant aquatic macrophytes in the northern include species of Phragmites, Hydrilla and Vallisneria. Within the southern and central sector, submerged forms of genus Najas, Stuckina and Halophila are dominant. The outer channel having largely marine influence has mostly submerged stands of Stuckina	The gradual reduction in salinity created a favourable condition for colonization by freshwater macrophytes. Towards nineties, species of Eichhornia, Pistia, Azolla and Scirpus greatly expanded, and covered nearly half of the total wetland area. The area under macrophytes increased from 20 km^2 of the lagoon areas in 1972 to 523 km^2 in October 2000. Post hydrological intervention of 2000, there has been a significant reduction in the area under freshwater macrophytes. Eicchornia is presently confined to river confluence areas. Stands of Phragmites have come	Continuous stands of Phragmites are limited to northern sector.	Phragmites has a propensity to colonize habitats, and aggravate siltation. Its distribution therefore needs to be maintained as in the present conditions

(continued)

Table 3.1 (continued)

Wetland feature (prioritization criteria indicated in brackets)	Prioritization rationale	Status	Trends	Limit of acceptable change	Justification for the limit of acceptable change
			to colonize the northern sector and can also be seen on the fringes of central sector		
Flora-Seagrass (Extent) (2(b))	Distribution of seagrass in India is sparse, with Chilika harbouring one of the largest meadows in the country. Seagrasses are good indicators of coastal ecosystem health	The extent of seagrass meadows was recorded to be 104 km^2 in September 2015. Chilika has altleast six of 14 seagrass species recorded in India.	The extent of seagrass meadows has increased from 24.8 km^2 in 1999 to 102 km^2 in 2012. In the immediate months after October 2013 cyclone, large parts of the bed were uprooted or went under debris. By September 2014, however, the meadows had not only re-established, but marginally expanded to 104 km^2, and has maintained ever since	Not more than 10% reduction in peak extent of seagrass meadows recorded during 2001–2015	A lower bound of 10% is placed to account for reduction in sea grass extent on account of local conditions
Fauna-Irrawaddy Dolphin (Population Size and Habitat Use) (1, 2(b))	Chilika harbours a resident population globally vulnerable cetacean *Orcaella brevirostris*. Chilika is one of the two lagoons globally having	The Irrawaddy Dolphin population in Chilika has been assessed as 144 individuals in 2015. Their presence has been recorded in Outer Channel and	The number of dolphins reported to be around 20 in 1992 has steadily increased post restoration to 144 individuals in 2015. Found to be confined to outer channel in	Upto 10% interannual reduction in population. No deaths caused due to collision with boats, netting or due to human activity	This estimated lower boundary is to account for decline in population due to migration and natural deaths

	a resident population of Irrawaddy Dolphins		the 1992 assessment, dolphins have been reported from the central and southern sectors as well		
Fauna – Waterbirds (Total counts, congregation areas) (1, 2(b))	Chilika harbours one of the largest assemblages of migratory waterbirds within the Central Asian Flyway	During 2000–14, Chilika regularly supported 0.7–0.95 million waterbirds. A total of 124 species of waterbirds and wetland dependent birds (species which feed elsewhere and use wetlands for resting or roosting) have been recorded in Chilika amongst which 66 species are migratory. Observed counts for 45 species exceeded their known1% of the bio-geographical population. Fourteen species of birds listed in IUCN Red List (2015) have been recorded in Chilika of which the Near-Threatened River Tern *Sterna aurantia* is recorded to breed in Chilika	There has been a drastic decline in the population of major diving ducks such as Tufted Duck and Common Pochard. Declining wetland area under mudflats is believed to be one of the key reasons of delining population of key inhabitants such as Greater Flamingo. The population of waders as Lesser Sand Plover *Charadrius mongolus*, Curlew sandpiper *Calidris ferruginea*, Little Stint *Calidris minuta* and Ruff Philomachus pugnax has also been declining in the recent times. River Terns seem to have completely abandoned Nalabana as a nesting ground and have shifted to Panchakudi.	Total count range within a 20% of the average count during 2002–2015	A variability within 20% will ensure that Chilika continues to meet the international site designation criteria

(continued)

Table 3.1 (continued)

Wetland feature (prioritization criteria indicated in brackets)	Prioritization rationale	Status	Trends	Limit of acceptable change	Justification for the limit of acceptable change
			Since 2000, the marshes of Mangalajodi have been attracting a sizeable number of waterbirds. The numbers recorded within these regions have increased from a dismal 5000 during 1998–99 to 0.15–0.29 million during 2011–2015		
1.2 Ecosystem processes					
Processes that maintain animal populations – fish breeding and migration (2(c))	Over 70% of fin and shell fish found in Chilika are migratory in character. Maintenance of migration is therefore important for sustaining species diversity and commercial landings	Till 2015, 314 fish, 29 prawn and 35 species of crab have been recorded from the wetland	The most noticeable shift is in the breeding ground of Hilsa from middle and upper reaches of Daya and deltaic branches of Mahanadi to the areas near mouth of Makara. In general, a shift in breeding and nursing grounds towards Magarmukh is observed for the species previ-ously using Northern	No limits derived	

			Sector. *M. cephalus* which used to breed only near wetland mouth is reported to use the entire outer channel habitat		
Sedimentation (2 (c))	Increased sediment loading leads to reduced water holding capacity and enhances terresterialization processes in shallow wetlands as Chilika	Chilika receives annually one million MT of sediments. Three-fourth of the total sediments are received from the Mahanadi distributaries	Analysis of sediment cores indicates that the rates of sedimentation in 2005 were 3–5 times higher than the rates observed 100 years ago (CWRDM 2005). During the last four decades (1972–2011), there have been significant changes in the land use and land cover within the basin. Dense forests have been mostly converted into open forests. Similarly, agricultural lands have been converted into settlements (147.79 km^2) and aquaculture (13.28 km^2). Parts of floodplains with seasonally flooded agricultural land have transformed into marshes	Sedimentation rates do not exceed the current rates.	The current rates, though elevated, can serve as the boundary condition for limiting land use change within the Chilika basin
Inlet dynamics (2 (c))	The inlet of Bay of Bengal into Chilika	By December 2015, the sea-inlet had migrated	The opening to the Bay of Bengal located 6 km	The cross section of the sea inlet does not fall	Reduction in cross-section of the mouth

(continued)

Table 3.1 (continued)

Wetland feature (prioritization criteria indicated in brackets)	Prioritization rationale	Status	Trends	Limit of acceptable change	Justification for the limit of acceptable change
	undergoes spatio-temporal variation due to erosion and accretion processes. A northwards littoral sediment drift leads to migration of sea inlet, thus rendering the inlet unstable	3.3 km northeast relative to the position wherein hydrological intervention was made in September 2000	northeast of Arkhakuda in 1914, shifted to 8 km northeast by 1965, and further 2 km by 1999. Increasing distance of the inlet from the central sector of the lagoon greatly reduced tidal flux and the overall salinity gradient. The distance of the inlet to the lagoon was therefore shortened by opening of an inlet opposite village Sipakuda, thus reducing the distance from the central sector to 8 km from over 30 km. Major monsoon outflows and cyclonic storms are known to cut open multiple inlets. In 2008, such an inlet was made near Gabakunda, 1.2 km north of the inlet. Another inlet was cut open during Phailin in	below 1000 m^2 Northwards migration of sea-inlet does not cross the known peak (360 m) for over two consecutive years. The intertidal variation at Satpada during normal monsoon years do not fall below 25 cm for two consecutive years	below the threshold renders instability and ultimately closure. Similarly, high rates of northward migration would lead to elongation of channel, creating a risk of reduction of salinity gradient

			October 2013. The two inlets have since merged		
2. Social subsystem					
2.1 Instrumental linkages					
Capture fisheries (3)	Chilika is one of the major fishery of the state of Odisha, and source of livelihoods of 0.14 million fishers	Commercial landing is constituted by 73 fin and shell fish and crab species. During 2011–2015, the total annual landing averaged 12,465 MT, with fish, prawn and crabs constituting 57%, 40% and 3% respectively. The gross economic value of annual fish landing at 2015 prices was Rs. 1463 million	Available data indicates a decline in landing during 1958–1968 from 5707 MT to 2601 MT, and a subsequent gradual increase to 8926 MT by 1987. The landing since then continuously decline to 1755 MT by 2000. The average annual landing during 2001–2015 at 11,958 MT, is 236% of the average catch reported during 1981–1999. The proportion of prawn landing to total landing was nearly 30% till 1980, but dipped to 10% by 1999, and has increased to 40% at present. There has been an increase in crab landings since 2000. The current landing is however close to the estimated maximum annual sustainable fish	Total landing is upto 80% of assessed maximum sustainable yield (CIFRI 2007) Over 90% of the species landed are above sustainable size limit (*M. cephalus*:219-461 mm; *P.monodon*: 116–197 mm; *S.serrata*: 87 mm)	Landing beyond maximum sustainable limits is an indicator of resource overharvesting

(continued)

Table 3.1 (continued)

Wetland feature (prioritization criteria indicated in brackets)	Prioritization rationale	Status	Trends	Limit of acceptable change	Justification for the limit of acceptable change
			yield of 12,636–13,896 MT indicating the need for careful management. There has been an increase in fishing effort. During 1958–2015, the number of active fishers fishing in the lagoon increased five times (from 8000 in 1958 to around 40,000 in 2015), and the number of boats by 2,53 times (from 2300 in 1958 to 5900 in 2015). Mechanized boats were introduced in Chilika since 1975, and presently form 67% of the fleet. The annual landing per boat has hovered around 2000 kg, with substantial dip during 1991–2000, and progressive increase subsequently. Catch per unit boat and catch per unit fisher have tended		

			to stagnate over the years (Kumar et al. 2016)		
Tourism (3)	Chilika is a major tourist destination of the State of Odisha	During 2011–15, Chilika was annually visited by 0.53 million tourists, generating an economy worth US$ 52 million per annum	Post hydrological intervention, there has been a surge in tourist arrivals at Chilika. The annual arrivals for the period 2001–15 was 85% higher than the 1994–99 period. Eight new tourist motorboat associations have been registered since 2000, operating nearly 1400 boats in 2015	Operation of tourist boats does not deteriorate habitat of dolphins and waterbirds (limits in terms of number of boats or tourists have not been attempted)	Adverse impact of tourism on habitat quality justify regulation
Buffer for extreme events (3)	With its high interannual variation of water levels (of about 0.8 meter) and a large waterspread area, Chilika is able to reduce impacts of storm surges, and hold considerable proportion of floodwaters	During cyclone warnings, fisher communities place their boats inside Chilika rather than the open sea as the likelihood of damage is lesser	With increasing frequency and intensity of tropical cyclones, the value of Chilika as an extreme event buffer is increasingly being realized by the neighbouring communities and government agencies. Mangroves have been restored in 100 ha of Outer Channel area within abandoned shrimp farms. Community level plans in most villages includes	No limits derived	

(continued)

Table 3.1 (continued)

Wetland feature (prioritization criteria indicated in brackets)	Prioritization rationale	Status	Trends	Limit of acceptable change	Justification for the limit of acceptable change
			maintenance of wetlands health as a strategy for reducing disaster risk, alongwith maintenance of physical infrastructure as cyclone shelters		
2.2 Relational linkages					
Sense of place (3)	Communities living in and around Chilika, especially fishers bear strong cultural linkages with the lagoon	Chilika is referred to by the fishers as 'mother' and has several mentions within folklores, customary and religious practices		Cultural symbols are protected, and identities preserved	
Social cohesion (3)	Fishers in Chilika comprise two major groups, i.e., traditional and non-traditional fishing. Traditional fishers historically had tenure fishing rights and have been fishing in the lagoon for atleast two generations. Non-traditional fishers are new entrants to fishing in the lagoon, had other occupations, and	Of the 154 villages that dot Chilika shorelines, traditional fishers inhabit 103 villages. The fishers access markets and credit through Primary Fisher Cooperative Societies	Since the introduction of aquaculture in Chilika, the influence of non-fishers on the fishing grounds and market chain has increased substantially. The over 100 Primary Fisher Cooperatives Societies became completely moribund towards mid-nineties. On account of efforts made	Atleast 70% of capture fisheries is traded directly by the Primary Fisher Cooperatives	A higher proportion of catch traded by the Primary Fisher Cooperative Society indicates higher accruals to the traditional fishers

	did not have traditional tenure rights over lagoon resources. Interactions between the two groups are key to social cohesion within management of fisheries resources, and overall lagoon in general		to review the PFCS since 2008, nearly 70% of the catch is traded through these bodies. However, middlemen remain a significant source of credit to the traditional fishers. The number of violent conflicts between the traditional fishers and non-fishers has come down, though stray incidents do take place		

Codes used to indicate prioritization criteria
1: Maintaining wetland feature in a particular state required to fulfil regulatory commitments
2 (a): Species or ecological community has a high conservation value globally or nationally
2 (b): Species or ecological community is a characteristic feature of Chilika, and is required to maintain site's uniqueness
2 (c): Wetland feature supports or has a significant influence on a prioritized ecosystem component, process or service
3: Wetland feature supports wellbeing of communities in instrumental or relational terms

Available information on the ecological sub-component of Chilika indicates that the wetland alternates between a brackish and freshwater state, depending on the balance of the interaction between coastal and riverine hydrological processes. The high annual littoral sediment drift causes the sea inlet to continually move northwards. As the channel is elongated, reduction in the tidal prism renders the inlet unstable, allowing freshwater processes to predominate. Extreme events such as high catchment rainfall or coastal storms cut the sand berm, thus enabling restoration of the salinity gradient. The periodicity of such change cannot be defined in the absence of robust hydrodynamic models. However, the gradual clogging of the opening of the lagoon to the sea is an indicator of such a phenomenon. Maintaining the salinity gradient, as it exists at present post hydrological restoration, is an apparent desired state.

Nested within the longer cycles of alternating changes in Chilika are at least three hydrological processes which govern species habitats and ecosystem services. These are the intra-annual variability of water levels, connectivity of surface waters and functioning of ephemeral floodplains bordering northern shorelines of the wetland. Available monitoring records indicate that during a year with normal monsoon rainfall, the water level in the lagoon varies by up to 0.8 m. Such water level variation influences the inundation regime along the wetland shorelines and the islands. The cyclical submergence-exposure pattern within Nalabana Island regulates the growth, survival and reproduction of invertebrates and submerged vegetation which serves as food for migratory waterbirds inhabiting the island during winters. The variability also underpins the flood buffering capacity of the wetland. Connectivity of the surface waters of Chilika with the inflowing rivers and the Bay of Bengal allow the development of the salinity gradient, with lower salinity in the zones where freshwater is received in Chilika, mixohaline conditions within the central and southern parts of the lagoon, and progressively euhaline conditions in the outer channel which leads to the Bay of Bengal. The salinity gradient provides migration cues to the fish species, as over 70% of the fin and shell fish recorded in the lagoon are migratory. The functioning of the ephemeral floodplains on the northern margin of the wetland, especially the macrophytes stands, have a significant influence on the distribution and transport of riverine sediments. A recent study on structure and carbon metabolic profiles of rhizosphere communities of *Phragmites karka* indicated the significance of their nutrient cycling function (Behera et al. 2018).

Within the social sub-component, an alignment of livelihood systems with the maintenance of key ecosystem components and processes is desirable. Suitable conditions for such alignment exist when the instrumental, as well as relational values of communities, especially the traditional fisher communities, is maintained, this incentivizing resource stewardship.

A reference regime for Chilika can thus be described regarding wetland condition which enables:

(a) Maintenance of the salinity gradient (freshwater conditions within the northern sector, brackish in the central and southern sector and marine towards the mouth region).
(b) Species, nutrient and hydrological exchange between the lagoon and the Bay of Bengal as well as inflowing Mahanadi Delta tributaries and catchment rivulets

(c) Creation of diverse habitats suited for a range of plant and animal species by maintaining pulsating water levels, which lead to inundation and re-emergence of the Nalabana island and the littoral margins of the lagoon
(d) Resource use systems which preserve the habitat conditions of dolphins, waterbirds, seagrasses and other components of biological diversity of high conservation significance locally, nationally as well as globally
(e) Resource value sharing system within fisheries and tourism which ensure that a fair proportion of economic benefits from fisheries and tourism accrue locally to the communities
(f) Recognition and enhancement of cultural identities that communities living in and around Chilika have with wetland ecosystem

Based on the aforementioned regime description, in Table 3.1 we list the 'limits of acceptable change' for the priority wetland features. The current understanding of the wetland ecosystem gives reasonable indication that by maintaining the priority features within such limits can help maintain the reference regime of the wetland ecosystem.

3.3.5 Risk of Human-Induced Adverse Change

Being located in intense human use settings, there are some activities creating and having the potential to limit achievement of the reference regime. Participants of the workshop listed such threats and associated qualitative risk ranking (high, medium and low) based on the degree of intensification of threat and the degree of impact on any one or more of the reference regime.

Key threats to the wetland ecosystem include pollution (medium), use of destructive fishing gears (high), aquaculture (high), and unmanaged tourism (high). Within the drainage basin, threats are in the form of intensification of nutrient use in floodplains (low), regulation of inflowing rivers for meeting upstream freshwater needs (medium), and siltation inducing land use intensification in the floodplain (medium). Threats related to governance systems include weak ability of fisher cooperatives to secure fair economic returns to the traditional fisheries (medium), limited capacities of tourist boat associations to self-organize (medium), and weakening capacity of fisher communities to enforce restrictions in biologically significant habitats, such as use of stake-nets in migratory routes (medium).

3.3.6 Identifying Management Needs

The current framework for management of Chilika broadly aligns with the principles of adaptive management. This is evidenced in the form of a focus on learning (setting up of a state of art wetland monitoring system and creating human capacities to apply the learnings to management), policy experimentation (by creating diverse programmes related to maintaining lagoon-sea connectivity, watershed management, strengthening fisher cooperatives, communication and outreach), and creating mechanisms for participation of a broad range of stakeholders in taking management decisions (primarily through decision making systems of the CDA, including its informal networks). The crisis in the late nineties in the form of the rapid decline in fish catches, violent conflicts between user groups, colonization of the aquatic environment by invasive macrophytes, and listing under the Montreux Record were some of the key trends that provided a setting for the management system to emerge and guide recovery of the lagoon.

Moving into the near future, there are several trends that indicate increasing uncertainty in key ecological components and processes in Chilika. In particular, maintaining the existing salinity gradients is likely to be difficult owing to factors such as high rates of sea level rise in the Bay of Bengal as compared to other parts of the Indian coastline (Chowdhury and Behera 2015; Unnikrishnan et al. 2015), increasing frequency of cyclones (Unnikrishnan et al. 2011; Mishra 2014), lagoon surface warming (Schneider and Hook 2010), and high likelihood of flow reduction from Mahanadi River (Rao 1995; Mondal and Mujumdar 2015; Raje and Mujumdar 2010).

The adaptive management approaches adopted thus far need to be backed by a social and institutional framework, within the architecture of 'adaptive governance' (Chaffin et al. 2014; Bodin 2017). Stakeholder dialogues (scientist and subject matter specialists included), mix of institutional types, complex and layered institutions, and designs that facilitate experimentation, learning and change are critical elements of such governance system (Dietz et al. 2008). Pitfalls, such as focusing only on science without transforming learning to modify management, lack of leadership and action, excessive focus on planning, not taking action, and risk-averse decision making, need to be avoided to ensure that the governance agenda suits the uncertain environment (Allen and Gunderson 2011; Rist et al. 2013). Lack of leadership and presence of powerful stakeholder groups benefitting from the status-quo are known to induce inertia in social-ecological systems (Scheffer et al. 2003).

The CDA, as the nodal agency mandated by the government for the managemnt of Chilika, can enable such as change by:

(a) Enhancing knowledge and understanding of critical ecosystem functions and their relationship with system dynamics (particularly species migration and exchange between riverine, lagoon and sea; sediment and nutrient dynamics; and factors infleuncing distribution of macrophytes). The analysis of such functions should be linked with ecological as well as social sub-systems.
(b) Feeding ecological knowledge into management decsions and actions (such as the regulation of interferences to migratory pathways or species habitats, regulation of tourism pressure on critical habitats). For all such management decisions, a learning and evaluation system should be put in place to assess the effectiveness of management choices. Implementing decisions may require 'rule-setting', based on the management threats (such as capping fishing efforts and tourist boats, and/or temporally and/or spatially zoning their use) or

requirements of regulatory regimes (such as specification of thresholds for activities that are permitted within wetlands as per the national regulatory framework of wetlands). Caution is drawn to an over-reliance on rules that can a create management paralysis, by uniform application to often different localities and conditions, without considering their applicability; or, blindly following the rule and deviating from the underlying governance premise (Frantzeskaki et al. 2010).

(c) Support the development of multilevel governance systems, and polycentric institutional and organizational linkages (particularly linking the role of PFCS, tourism associations, village panchayats within the overarching governance regime). Intermediaries, in the form of civil society organizations, expert networks, etc., should be considered as a part of such a governance system, given their role in bringing in new ideas, communication between entities and bridging in the flexibility necessary for management of the fluid social-ecological systems and relatively rigid institutional arrangements (Olsson et al. 2006; Folke et al. 2005). Investing in social capital development can significantly enhance cooperation, particularly amongst community groups and lower resource use conflicts (Pretty 2003).

(d) Develop capacity for dealing with perturbations and uncertanities, specifically in the context of climate change. Ways to achieve this would be through enhanced capacities for ecological and social vulnerability assessments, scenario development, identifying management choices with participation of stakeholders, and building the required capability and credibility to implement such options in collaboration with multiple agencies. The ability of existing management to deal with adverse change in ecological character is closely related to the degree of fit achieved between the attributes of resource use regime and those of ecosystem functioning (Folke et al. 2007). In the context of climate change, achieving such a fit would require the ability to integrate the full range of wetland ecosystem services and biodiversity values within climate change mitigation and adaptation planning at the state and national scales, and operational regimes to influence land use decisions within Mahanadi Delta and coastal zone.

3.4 Conclusion

Framing the ecological character of Chilika as a social-ecological system allows for identification and prioritization of a number of wetland features from the ecological and social subsystems, as well as pathways for governance systems that can be guided by the needs of meeting the limits of acceptable change for these parameters. Uncertainty about the trajectory of ecological character change calls for continued efforts in knowledge building, experimental management design, bridging different knowledge systems, and continuous monitoring and evaluation. The prioritization of wetland features and the limits as indicated in the analysis need to be seen as reflections of current knowledge of the wetland functioning, and interpreted within a set of values and inherent biases. Within the framework of adaptive governance, there is a need to periodically visit these priorities, and update the limits and justification as new knowledge emerges, including addition or deletion of new features. In the immediate future, the existing governance systems would need to have the capability of influencing regional development processes, in terms of ensuring that full range of ecosystem services and biodiversity values of Chilika are recognized in planning and decision making.

Application of available Ramsar Convention guidelines on description and evaluation of wetland ecological character indicates inherent biases towards ecological aspects, and underplaying of socio-economic aspects, as well as those related to governance systems which bridge the ecological and social sub-systems. It is not out of place to conclude that while the wise use approach encompasses the role of humans as a part of a wetland ecosystem, the underlying definitions and assessment protocols for ecological character take only a partial view. The future of wetlands cannot be seen as separate from that of communities which are related to the wetlands, in instrumental as well as relational terms, and it is time that management approaches adopted a broader and pluralistic outlook. A beginning can be made in this direction by an explicit acknowledgement of ecological character as a social construct. Inclusion of ecosystem services within the definition of ecological character is a good beginning, yet deeper efforts are required to encompass the diverse ways in which people relate to wetlands. The present endeavour to describe and evaluate the ecological character of Chilika using a social-ecological systems approach is intended to trigger such a process.

Acknowledgements Financial support received from the International Development Research Center, New Delhi under project on 'Strengthening Livelihood Security and Adapting to Climate Uncertainty in Chilika Lagoon, India' is gratefully acknowledged. The authors also thank all scientists from the Wetlands Research and Training Center of the Chilika Development Authority for their contributions to the Chilika Ecological Characterization Workshop held from March 3–6, 2015 at the Center. Kamal Dalakoti (Wetlands International South Asia) prepared the map. Dr. Gurdeep Rastogi, Dr. R.N. Samal, Dr. P.R. Muduli, Mr. Sujit K. Mishra provided specific data for setting limits of acceptable change.

References

Allen CR, Gunderson LH (2011) Pathology and failure in the design and implementation of adaptive management. J Environ Manag 92(5):1379–1384. Elsevier Ltd. https://doi.org/10.1016/j.jenvman.2010.10.063

Annandale N (1915) A naturalistic view of the Chilka Lake. Calcutta Rev:1–15

Balachandran S, Sathiyaselvam P, Panda S (2009) Bird atlas of Chilika. Chilika Development Authority and Bombay Natural History Society, Bhubaneswar

Bandyopadhyay S, Gopal B (1991) Ecosystem studies and management problems of a coastal lagoon: the Lake Chilika. In: Gopal B, Asthana V (eds) Aquatic sciences in India. Indian Association of Limnology and Oceanography, New Delhi, pp 117–172

Barik SK, Muduli PR, Mohanty B, Behera AT, Mallick S, Abhijit Das RNS, Rastogi G, Pattnaik AK (2017) Spatio-temporal variability and the impact of Phailin on water quality of Chilika Lagoon. Cont Shelf Res 136(February):39–56. Elsevier. https://doi.org/10.1016/j.csr.2017.01.019

Bauer A, Srinivasalu C, Roy AD, Murthy BHCK, Molur S, Pal S, Mohapatra P, Agarwal I, Sondhi S (2014) Barkudia insularis, Madras spotted skink. The IUCN Red List of Threatened Species. https://doi.org/10.2305/IUCN.UK.2014-1.RLTS.T2593A2788379.en

Behera P, Mohapatra M, Adhya TK, Suar M (2018) Structural and metabolic diversity of rhizosphere microbial communities of Phragmites Karka in a tropical coastal lagoon. Appl Soil Ecol:0–1. December 2017. Elsevier. https://doi.org/10.1016/j.apsoil.2017.12.023

Berkes F (2010) Devolution of environment and resources governance: trends and future. Environ Conserv Found Environ Conserv 37(4):489–500. https://doi.org/10.1017/S037689291000072X

Berkes F (2017) Environmental governance for the anthropocene? Social-ecological systems, resilience, and collaborative learning. Sustainability (Switzerland) 9(7). https://doi.org/10.3390/su9071232

Biswas KP (1995) Ecological and fisheries development in wetlands: a case study of Chilika Lagoon. Daya Publishing House, New Delhi

Bodin Ö (2017) Collaborative environmental governance: achieving collective action in social-ecological systems. Science 357(6352):eaan1114. https://doi.org/10.1126/science.aan1114

Bridgewater P (2008) A new context for the Ramsar convention: wetlands in a changing world. Rev Eur Community Int Environ Law 17(1):100–106. https://doi.org/10.1111/j.1467-9388.2008.00582.x

Bruckmeier K (2016) Social-ecological systems and ecosystem services. In: Social-ecological transformation: reconnecting society and nature. Palgrave Macmillan, London, London, pp 183–234

Chaffin BC, Gosnell H, Cosens BA (2014) A decade of adaptive governance scholarship : synthesis and future directions a decade of adaptive governance scholarship : synthesis and future directions. Ecol Soc 19(3):56. https://doi.org/10.5751/ES-06824-190356

Chan KMA, Satterfield T, Goldstein J (2012) Rethinking ecosystem services to better address and navigate cultural values. Ecol Econ 74:8–18. https://doi.org/10.1016/j.ecolecon.2011.11.011

Chandramohan P, Sanil Kumar V, Nayak BU (1993) Coastal processes along the shorefront of Chilika Lake, East Coast of India. Indian J Mar Sci 22(December):268–272

Chilika Development Authority (2015) A brief report on impact of severe cyclonic storm Phailin on ecosystem of Chilika Lake. Bhubaneswar, Odisha

Chowdhury P, Behera MR (2015) A study on regional sea level variation along the Indian coast. Proc Eng 116(1):1078–1084. Elsevier B.V. https://doi.org/10.1016/j.proeng.2015.08.348

CIFRI (2007) Assessment of fish yield potential of Chilika Lake. Kolkata

CWRDM (2005) Estimation of rate and pattern of sedimentation in Chilika Lake, Orissa using environmental isotope technique. Kozikode, Kerala

D'Lima C, Marsh H, Hamann M, Sinha A, Arthur R (2014) Positive interactions between Irrawaddy dolphins and artisanal fishers in the Chilika Lagoon of Eastern India are driven by ecology, socioeconomics, and culture. Ambio 43(5):614–624. https://doi.org/10.1007/s13280-013-0440-4

D'Souza R (2002) Colonialism, capitalism and debating the origins. Econ Polit Wkly 37(13):1261–1272

Das BP, Jena J (2008) Impact of Mahandi Basin development on eco-hydrology of Chilika. In: Sengupta M, Dalwani R (eds) Taal 2007: proceedings of 12th lake conference. Ministry of Environment and Forests, Government of India and International Lake Environment Committee, Japan, New Delhi, pp 697–702

Davidson-Hunt IJ, Berkes F (2003) Nature and society through the Lens of resilience. In: Berkes F, Colding J, Folke C (eds) Navigating social-ecological systems. Cambridge University Press, Cambridge, UK, pp 53–82

Davis J, Brock M (2008) Detecting unacceptable change in the ecological character of Ramsar wetlands. Ecol Manag Restor 9(1):26–32. https://doi.org/10.1111/j.1442-8903.2008.00384.x

Department of the Environment, Water, Heritage and the Arts, Commonwealth of Australia (2008) National framework and guidance for describing the ecological character of Australian Ramsar Wetlands. Department of the Environment, Water, Heritage and the Arts, Canberra. http://www.environment.gov.au/system/files/resources/6d7408dc-2519-4294-9820-f7b22848l6dd/files/module-2-framework.pdf

Dietz T, Ostrom E, Stern PC (2008) The struggle to govern the commons. Urban Ecology: An International Perspective on the Interaction Between Humans and Nature 1907(2003):611–622. https://doi.org/10.1007/978-0-387-73412-5_40

Dujovny E (2009) The deepest cut: political ecology in the dredging of a New Sea mouth in Chilika Lake, Orissa, India. Conserv Soc 7(3):192. https://doi.org/10.4103/0972-4923.64736

Ernstson H (2013) The social production of ecosystem services: a framework for studying environmental justice and ecological complexity in urbanized landscapes. Landsc Urban Plan 109 (1):7–17. https://doi.org/10.1016/j.landurbplan.2012.10.005

Finlayson CM, Davidson N, Pritchard D, Randy Milton G, MacKay H (2011) The Ramsar convention and ecosystem-based approaches to the wise use and sustainable development of wetlands. J Int Wildl Law Policy 14(3–4):176–198. https://doi.org/10.1080/13880292.2011.626704

Finlayson CM (2012) Forty years of wetland conservation and wise use. Aquat Conserv Mar Freshwat Ecosyst 22:139–143

Finlayson CM, Clarke SJ, Davidson NC, Gell PA (2016) Role of Palaeoecology in describing the ecological character of wetlands. Mar Freshw Res 67:687–694

Folke C, Hahn T, Olsson P, Norberg J (2005) Adaptive governance of social-ecological systems. Annu Rev Environ Resour 30(1):441–473. https://doi.org/10.1146/annurev.energy.30.050504.144511

Folke C, Pritchard L, Berkes F, Colding J, Svedin U (2007) The problem of fit between ecosystems and institutions: ten years later. Ecol Soc 12(1):30

Frantzeskaki N, Slinger J, Vreugdenhil H, van Daalen E (2010) Social-ecological systems governance: from paradigm to management approach. Nat Cult 5(1):84–98. https://doi.org/10.3167/nc.2010.050106

Gardner R, Davidson NC (2011) The Ramsar convention. In: LePage BA (ed) Wetlands: integrating multidisciplinary concepts. Springer, Dordrecht, pp 189–203. https://doi.org/10.1007/978-94-007-0551-7

Gell PA, Finlayson CM, Davidson NC (2016) Understanding change in the ecological character of Ramsar wetlands: perspectives from a deeper time - synthesis. Mar Freshw Res 67(6):869–879. https://doi.org/10.1071/MF16075

Ghosh AK, Pattnaik AK, Ballatore TJ (2006) Chilika lagoon: restoring ecological balance and livelihoods through re-salinization. Lakes Reserv Res Manag 11(4):239–255. https://doi.org/10.1111/j.1440-1770.2006.00306.x

Holling CS, Gunderson LH, Ludwig D (2002) Panarchy: understanding transformations in human and natural systems. In: Gunderson LH, Holling CS (eds) Panarchy: understanding transformations in human and natural systems. Island Press, Washington, DC, pp 3–22

Horwitz P, Finlayson CM (2011) Wetlands as settings for human health: incorporating ecosystem services and health impact assessment into water resource management. Bioscience 61 (9):678–688. https://doi.org/10.1525/bio.2011.61.9.6

Iwasaki S, Shaw R (2008) Fishery resource management in Chilika Lagoon: a study on coastal conservation in the eastern coast of India. J Coast Conserv 12(1):43–52. https://doi.org/10.1007/s11852-008-0022-y

Jones S, Sujansingani KH (1954) Fish and fisheries of the Chilka Lake with statistics of fish catches for the year 1948–1950. Indian J Fish 1(1–2):256–344

Khandelwal A, Mohanti M, Felipe G'a-R'g, Scharf BW (2008) Vegetation history and sea level variations during the last 13,500 years inferred from a pollen record at Chilika Lake, Orissa, India. Veg Hist Archaebotany 17:335–344. https://doi.org/10.1007/s00334-007-0127-5

Kopf RK, Finlayson CM, Humphries P, Sims NC, Hladyz S (2015) Anthropocene baselines: assessing change and managing biodiversity in human-dominated aquatic ecosystems. Bioscience 65:798–811

Kumar R, Pattnaik AK (2012) Chilika – an integrated management planning framework for conservation and wise use. New Delhi

Kumar R, Horwitz P, Milton GR, Sellamuttu SS, Buckton ST, Davidson NC, Pattnaik AK, Zavagli M, Baker C (2011) Assessing wetland ecosystem services and poverty interlinkages: a general framework and case study. Hydrol Sci J 56(8):1602–1621. https://doi.org/10.1080/02626667.2011.631496

Kumar R, Ambastha K, Kumar S, Chakraborty A, Dalakoti K, Meetie AY (2016) Economics of ecosystem services and biodiversity for conservation and sustainable management of Inland Wetlands. New Delhi

Leach M, Mearns R, Scoones I (1999) Environmental entitlements: dynamics and institutions in community-based natural resource management. World Dev 27(2):225–247. https://doi.org/10.1016/S0305-750X(98)00141-7

Levin S, Xepapadeas T, Crépin A-S, Norberg J, de Zeeuw A, Folke C, Hughes T et al (2013) Social-ecological systems as complex adaptive systems: modeling and policy implications. Environ Dev Econ 18(02):111–132. https://doi.org/10.1017/S1355770X12000460

Maltby E (2009) The changing wetland paradigm. In: Maltby E, Barker T (eds) The wetlands handbook. Wiley Blackwell, Oxford, pp 3–42

Maltby E, Acreman MC (2011) Ecosystem services of wetlands : pathfinder for a new paradigm. Hydrol Sci J 58(8):1341–1359. https://doi.org/10.1080/02626667.2011.631014

Marois DE, Mitsch WJ (2015) Coastal protection from tsunamis and cyclones provided by mangrove wetlands – a review. Int J Biodivers Sci Ecosyst Serv Manag 11(1):71–83. Taylor & Francis. https://doi.org/10.1080/21513732.2014.997292

Martinez-Alier J (2002) Mangroves versus shrimps. In: The environmentalism of the poor: a study of ecological conflicts and valuation. Edward Elgar Publishing, Cheltenham, pp 79–99

MEA (2005) Ecosystems and human Well-being: synthesis. Island Press, Washington, DC

Menzel S, Teng J (2010) Ecosystem services as a stakeholder-driven concept for conservation science. Conserv Biol 24(3):907–909. https://doi.org/10.1111/j.1523-1739.2009.01347.x

Mishra A (2014) Temperature rise and trend of cyclones over the eastern coastal region of India. J Earth Sci Clim Chang 5(9):1–5. https://doi.org/10.4172/2157-7617.1000227

Mohanty SK, Mishra SS, Khan M, Mohanty RK, Mohapatra A, Pattnaik AK (2015) Ichthyofaunal diversity of Chilika Lake, Odisha, India: an inventory, assessment of biodiversity status and comprehensive systematic checklist (1916–2014). Check List 11(6):1–19

Mohapatra A, Mohanty RK, Mohanty SK, Bhatta KS, Das NR (2007) Fisheries enhancement and biodiversity assessment of fish, prawn and mud crab in Chilika lagoon through hydrological intervention. Wetl Ecol Manag 15(3):229–251. https://doi.org/10.1007/s11273-006-9025-3

Mondal A, Mujumdar PP (2015) Regional hydrological impacts of climate change: implications for water Management in India. Proc Int Assoc Hydrol Sci 366(June 2014):34–43. https://doi.org/10.5194/piahs-366-34-2015

Nayak PK (2014) The Chilika lagoon social-ecological system: an historical analysis. Ecol Soc 19 (1). https://doi.org/10.5751/ES-05978-190101

Nayak PK, Berkes F (2010) Whose marginalisation? Politics around environmental injustices in India's Chilika lagoon. Local Environ 15(6):553–567. https://doi.org/10.1080/13549839.2010.487527

Nayak PK, Berkes F (2011) Commonisation and decommonisation: understanding the processes of change in the Chilika lagoon, India. Conserv Soc 9(2):132. https://doi.org/10.4103/0972-4923.83723

North DC (1991) Institutions. J Econ Perspect 5(1):97–112. https://doi.org/10.2307/2234910

O'Connell MJ (2003) Detecting, measuring and reversing changes to wetlands. Wetl Ecol Manag 11(6):397–401. https://doi.org/10.1023/B:WETL.0000007191.77103.53

Olsson P, Gunderson LH, Carpenter SR, Ryan P, Lebel L, Folke C, Holling CS (2006) Shooting the rapids: navigating transition to adaptive governance of social-ecological systems. Ecol Soc 11 (1):18. http://www.ecologuandsociety.org

Pascual U, Balvanera P, Díaz S, Pataki G, Roth E, Stenseke M, Watson RT et al (2017) Valuing nature's contributions to people: the IPBES approach. Curr Opin Environ Sustain 26:7–16. https://doi.org/10.1016/j.cosust.2016.12.006

Pattnaik AK, Kumar R (2016) Lake Chilika (India): ecological restoration and adaptive management for conservation and wise use. In: Finlayson CM, Milton GR, Prentice RC, Davidson NC (eds) The wetland book: distribution, description and conservation. Springer, Dordrecht, pp 1–14. https://doi.org/10.1007/978-94-007-6173-5

Pretty J (2003) Social capital and the collective management of resources. Science 302 (5652):1912–1914. https://doi.org/10.1126/science.1090847

Pritchard D (2018) Ecological character concept of the Ramsar convention. In: Finlayson CM, Everard M, Irvine K, McInnes RJ, Middleton BA, van Dam AA, NC Davidson NC (eds) The wetland book I: structure and function, management and methods. Springer Publishers, Dordrecht

Raje D, Mujumdar PP (2010) Reservoir performance under uncertainty in hydrologic impacts of climate change. Adv Water Resour 33(3):312–326. Elsevier Ltd. https://doi.org/10.1016/j.advwatres.2009.12.008

Rama Rao KV (1995) Appraisal. In: Ghosh AK (ed) Fauna of Chilika Lake (Wetland ecosystem series I). Zoological Survey of India, Calcutta, pp 1–10

Ramsar Convention Secretariat (2010a) Managing wetlands: frameworks for managing wetlands of international importance and other wetland sites. In: Pritchard D (ed) Ramsar handbooks for the wise use of wetlands, vol 18, 4th edn. Ramsar Convention Secretariat, Gland. https://doi.org/10.1126/science.212.4496.795

Ramsar Convention Secretariat (2010b) Wise use of wetlands: concepts and approaches for the wise use of wetlands. In: Pritchard D (ed) Ramsar handbooks for the wise use of wetlands, vol 1. Ramsar Convention Secretariat, Gland. https://doi.org/10.1007/s00267-008-9072-z

Ramsar Convention Secretariat (2016) An introduction to the convention on wetlands, 5th edn. Ramsar Convention Secretariat, Gland

Rao PG (1995) Effect of climate change on streamflows in the Mahanadi river basin, India. Water Int 20(4):205–212. https://doi.org/10.1080/02508069508686477

Rath J, Adhikary SP (2008) Biodiversity assessment of algae in Chilika Lake, East Coast of India. In: Mohanty PK (ed) Monitoring and modelling lakes and coastal environments. Springer, Dordrecht, pp 22–33

Reeves RR, Jefferson TA, Karczmarski L, Laidre K, O'Corry-Crowe G, Rojas-Bracho L, Secchi ER et al (2008) Orcaella brevirostris. The IUCN Red List of Threatened Species e.T15419A4. https://doi.org/10.2305/IUCN.UK.2008.RLTS.T15419A4579987.en

Reyers B, Biggs R, Cumming GS, Elmqvist T, Hejnowicz AP, Polasky S (2013) Getting the measure of ecosystem services: a social-ecological approach. Front Ecol Environ 11 (5):268–273. https://doi.org/10.1890/120144

Rist L, Felton RL, Samuelsson A, Sandström L, Rosvall O (2013) A new paradigm for adaptive management. Ecol Soc 18(4):63. https://doi.org/10.5751/ES-06183-180463

Russell R, Guerry AD, Balvanera P, Gould RK, Basurto X, Chan KMA, Klain S, Levine J, Tam J (2013) Humans and nature: how knowing and experiencing nature affect Well-being. Annu Rev Environ Resour 38(1):473–502. https://doi.org/10.1146/annurev-environ-012312-110838

Samal KC (2002) Shrimp culture in Chilika Lake: case of occupational displacement of fishermen. Econ Polit Wkly 37(18):1714–1718. http://www.jstor.org/stable/4412064

Samal KC, Meher S (2003) Fishing communities on Chilika Lake: comparative socio-economic study. Econ Polit Wkly 38(31):3319–3325. https://doi.org/10.2307/4413863

Scheffer M (2009) Critical transitions in nature and society. In: Critical transitions in nature and society. Princeton University Press, Princeton/Oxford, pp 240–257

Scheffer M, Westley F, Brock W (2003) Slow response of societies to new problems: causes and costs. Ecosystems 6(5):493–502. https://doi.org/10.1007/s10021-002-0146-0
Schlager E, Ostrom E (1992) Property-rights regimes and natural resources: a conceptual analysis. Source Land Econo 68(3):249–262. https://doi.org/10.2307/3146375
Schneider P, Hook SJ (2010) Space observations of inland water bodies show rapid surface warming since 1985. Geophys Res Lett 37(22):1–5. https://doi.org/10.1029/2010GL045059
Sekhar NU (2004) Fisheries in Chilika Lake: how community access and control impacts their management. J Environ Manag 73(3):257–266. https://doi.org/10.1016/j.jenvman.2004.07.006
Sekhar NU (2007) Social capital and fisheries management: the case of Chilika Lake in India. Environ Manag 39(4):497–505. https://doi.org/10.1007/s00267-006-0183-0
Srichandan S, Kim JY, Bhadury P, Barik SK, Muduli PR, Samal RN, Pattnaik AK, Rastogi G (2015) Spatiotemporal distribution and composition of phytoplankton assemblages in a coastal tropical lagoon: Chilika, India. Environ Monit Assess 187(2). https://doi.org/10.1007/s10661-014-4212-9
The Ramsar Convention on Wetlands (2005) Resolution IX. 1 annex a a conceptual framework for the wise use of wetlands and the. In: The ninth meeting of conference of parties. The Ramsar Convention on Wetlands, Kampala. http://www.ramsar.org/sites/default/files/documents/pdf/res/key_res_ix_01_annexa_e.pdf
The Ramsar Convention on Wetlands (2008) Describing the ecological character of wetlands, and data needs and formats for Core inventory: harmonized scientific and technical guidance. In: The Ramsar Convention on Wetlands (ed) 10th Meeting of the conference of the parties. http://www.ramsar.org/sites/default/files/documents/pdf/res/key_res_x_15_e.pdf
The Ramsar Convention on Wetlands (2012) The definition and operation of concepts and approaches for 'limits of acceptable change' which may be applicable to the Ramsar context of defining and detecting change in the ecological character of wetlands. In: 11th Meeting of the conference of parties. The Ramsar Convention on Wetlands. http://www.ramsar.org/sites/default/files/documents/pdf/cop11/doc/cop11-doc24-e-limits.pdf
The World Bank (2005) Scenario assessment of the provision of environmental flows to Lake Chilika from Naraj Barrage, Orissa, India
Tripati S, Vora KH (2005) Maritime heritage in and around Chilika Lake, Orissa: geological evidences for its decline. Curr Sci 88(7):1175–1181
Unnikrishnan AS, Ramesh Kumar MR, Sindhu B (2011) Tropical cyclones in the bay of Bengal and extreme sea-level projections along the East Coast of India in a future climate scenario. Curr Sci 101(3):327–331. https://doi.org/10.1002/grl.50636
Unnikrishnan S, Singh A, Kharat M (2013) The role of mangroves in disaster mitigation: a review. In: Filho WL (ed) Climate change and disaster risk management. Climate change management. Springer, Berlin/Heidelberg, pp 257–272
Unnikrishnan AS, Nidheesh AG, Lengaigne M (2015) Sea-level-rise trends off the Indian coasts during the last two decades. Curr Sci 108(5):966–970. http://www.currentscience.ac.in/Volumes/108/05/0966.pdf
Venkatarathnam K (1970) Formation of the barrier spit and other sand ridges near Chilka Lake on the East Coast of India. Mar Geol 9(2):101–116. https://doi.org/10.1016/0025-3227(70)90063-0
Young OR (2017) Governing complex systems. In: Governing complex systems: social capital for the Anthropocene. The MIT Press, Cambridge, MA/Cambridge, UK, pp 4–7

Chapter 4
Ecosystem Services: Implications for Managing Chilika

Ritesh Kumar, Ajit K. Pattnaik, and C. Max Finlayson

Abstract An ecosystem services-led management of Chilika encourages a progression from a siloed approach to conservation of species and habitats to explicit consideration of benefits humans derive from these ecosystems, enabling anticipation of a wide range of consequences that may result from different management regimes, and provide tools for identifying, negotiating, avoiding, and managing potential negative tradeoffs. Wetland management would stand to benefit by explicit recognition of intrinsic, instrumental and relational values of the Ramsar Site and contributions to human well-being at multiple scales and sectors. While the investments into the restoration of Chilika has high economic efficiency, the distributional aspects of benefit sharing need to be addressed through interventions such as reducing fishing effort, increasing value realization through strategies as product differentiation, and enhancing participation of fishers in the higher segments of the value chain. The financing arrangements for wetlands management in place are not linked with the costs of ecosystem services provision, especially the maintenance of critical ecosystem processes and functions. Institutional arrangements for the management of provisioning services and select cultural services (mainly tourism) have emerged over a period of time, however, there is a relative vacuum when it comes to the management of regulating services (such as water regime moderation, nutrient cycling, carbon sequestration and others). Much of management effectiveness is dependent on the extent to which the institutions responsible for managing various sectoral programmes (such as climate change, rural development, water and sanitation, disaster risk reduction) take into account the multiple ecosystem services of

R. Kumar (✉)
Wetlands International South Asia, New Delhi, India
e-mail: ritesh.kumar@wi-sa.org

A. K. Pattnaik
Chilika Development Authority, Bhubaneswar, Odisha, India

C. M. Finlayson
Institute for Land Water and Society, Charles Sturt University, Albury, NSW, Australia

IHE Delft, Institute for Water Education, Delft, The Netherlands

C. M. Finlayson et al. (eds.), *Ecology, Conservation, and Restoration of Chilika Lagoon, India*, Wetlands: Ecology, Conservation and Management 6,
https://doi.org/10.1007/978-3-030-33424-6_4

Chilika and the implication of development programmes for sustained provision of such services. A research and monitoring framework for measuring and managing ecosystem services of Chilika needs to be based on an understanding of how the multiple services are generated by coupled social-ecological systems, their interactions and interlinkages with human well-being, and how values for ecosystem services feed into stakeholder behaviour and attitudes towards wetlands conservation and wise-use.

Keywords Ecological character · Ecosystem services · Multiple values · Governance · Distributional equity · Economic efficiency

4.1 Introduction

The management of wetlands designated as Wetlands of International Importance (Ramsar Sites) strives to achieve 'wise use' by ensuring compatibility of human use of the ecosystem with the goal of maintaining ecological character (Pritchard 2016). Wetlands wise use remains to date one of the longest established examples amongst intergovernmental processes of ecosystem approaches for conservation and sustainable development of natural resources (Finlayson et al. 2011). The approach recognises the essential linkages that exist between people and sustainable development of wetlands and encourages community engagement and transparency in negotiating trade-offs and determining equitable outcomes for conservation (Finlayson 2012). In 2005, the Contracting Parties to the Ramsar Convention adopted a revised definition of wise use to include the goal of maintenance of ecosystem services alongwith maintenance of ecosystem components and processes. This revision conceptually conveys an increasing appreciation of the coupling between nature and society (Schoon and van der Leeuw 2015), and the fact that ecosystem services are not generated by ecosystems alone, but by social-ecological systems of which humans are endogeneous (Levin et al. 2013; Reyers et al. 2013).

Ecological restoration of Chilika has been noted for the use of community-led adaptive management approach towards the wise use of wetland (Finlayson et al. 2001; Ghosh et al. 2006). The Chilika Development Authority (CDA), instituted in 1991 by Government of Odisha as the nodal agency for the management of Chilika has an aim of 'conserving lagoon ecology and bringing an all-round development in and around the lagoon' (www.chilika.com). The management plans which have guided CDA's functioning over the years have sought to seek a balance between maintaining species habitats and human use of the lagoon for fisheries and tourism. Yet, setting of management priorities has not been consciously based on an explicit recognition of multiple ecosystem services, and their underpinning ecosystem components and processes, as well as drivers of change. The aim of this paper is to synthesize the available knowledge on Chilika ecosystem services, and map implications for managing the Ramsar site.

The paper is structured in eight sections. We begin by outlining our analytical approach for unpacking ecosystem services from the lens of wetlands management. The economic values of select ecosystem services benefits are discussed next, followed by an assessment of the economic efficiency of investment in wetland restoration, and issues related to distributional equity. Valuation of ecosystem functions underpinning ecosystem services through the case of fisheries is presented thereafter. The final section maps the relevance of ecosystem services information with management and governance, using the analytical approach as the framework of enquiry. Barring the second section, the paper largely delves on economic values of ecosystem services benefits, while recognizing that such values form only a part of multiple values of wetlands, instrumental as well as relational, and a discussion based on the full range of values can form a more meaningful basis of analysis (Kumar et al. 2017; Pascual et al. 2017).

4.2 Analytical Approach

The Ramsar Convention defines wetlands wise use as 'the maintenance of their ecological character, achieved through the implementation of ecosystem approaches, within the context of sustainable development'. Ecological character is 'the combination of the ecosystem components, processes and benefits/services that characterise the wetland at any given point in time'. With the inclusion of ecosystem services within the definition of ecological character, a formal bridging of natural and social science notions of wetlands has been made (Braat and de Groot 2012), thus further embellishing the case for their integrated management on the basis of the full range of ecosystem services and biodiversity values while framing management objectives (Parrott and Quinn 2016; Zsuffa et al. 2014).

Since the revision of the definition of ecological character (Conference of Parties (CoP) Resolution IX.1), several guidelines have been adopted at the Ramsar CoP meetings to support the incorporation of ecosystem services in core inventory and assessment fields (Resolutions IX.1 Annex E, X.15, XIII.13), reporting change in ecological character (Resolution X.16), management effectiveness evaluation (Resolution XII.15); and sectoral guidance such as that on water management (Resolution IX.1 Annex C, X.19), poverty reduction (Resolutions IX.14, X.28, XI.13), human health (Resolutions X.23, XI.12), environment impact assessments (Resolution X.17), climate change (Resolutions X.24, XI.14, XII.11, XIII.12, XIII.14), urbanization (Resolutions X.27, XI.11, XII.10, XIII.16), disaster risk reduction (Resolution XII.13), tourism (Resolution XI.7), and sustainable development (Resolution XI.21). No attempt is made to summarize these guidelines here. Yet, it suffices to say that none of these guidelines individually or collectively represent a consistent framing of a process, or lines of enquiry to enable systematic assessment of ecosystem services in wetlands management planning and decision-making processes.

While the generation of knowledge about ecosystem services is a useful starting point, the knowledge alone may be insufficient for incorporating them into decision-making (Primmer et al. 2015). Rather a grasp of decision-making processes of stakeholders, integration of research into institutional design and policy implementation; and policy interventions designed for performance evaluation and improvement over time may be required in an iterative and adaptive framework (Daily et al. 2009). Recognising diverse wetland ecosystem services, and the multiple values that stakeholders hold for these services forms a cornerstone of effective management (Kumar et al. 2017).

The provision of ecosystem services relies upon the complexity and functioning of ecosystems and landscapes (Raudsepp-Hearne et al. 2010). Seen from the perspective of biophysical sciences, ecosystem services are an outcome of ecological production functions (Daily et al. 2009), which in turn are underpinned by biophysical structures and processes, often included within the category of supporting services (de Groot et al. 2010). The distinction of ecosystem functions from ecosystem components and processes has been highlighted, as the former encapsulates not just the combinations of the latter, but also the potential that ecosystems have to deliver ecosystem services (Naeem et al. 1999). The scales at which ecosystem services are produced, used and accessed provide a context for interpreting societal values that are attributed to these services and tools applied for their management (Raudsepp-Hearne and Peterson 2016).

Wetlands are multifunctional, delivering a range of ecosystem services, several of which respond to a similar set of drivers or ecosystem processes, and are therefore best treated as clustered bundles (Gonzalez-ollauri and Mickovski 2017; Raudsepp-Hearne et al. 2010), rather than stand-alone services. Tradeoffs between various ecosystem services bundles are inherent as not all services co-vary in response to wetland use and management (MEA 2005; Raudsepp-Hearne et al. 2010). Management that attempts to maximise the production of one ecosystem service (often a provisioning service) often results in substantial declines in the provision of other ecosystem services (often regulating and cultural services) (Bennett et al. 2009; Russi et al. 2013). Realigning management systems which reward the production of marketed provisioning services, but not the provision of non-marketed ecosystem services, such as regulating and cultural services, remains a fundamental concern (Guerry et al. 2015). An important appraisal element for wetland management which follows herefrom is whether the diversity of ecosystem services (and bundles) are considered while framing management objectives; and whether the underpinning ecosystem functions that sustain these services are adequately addressed within management actions.

Ecosystem services represent a political framing of nature-society relationships, often creating new markets, property and power relationships for public goods, with such changes having distinct distributional consequences (Kull et al. 2015). The transformation of natural capital into ecosystem services is influenced by a suite of institutions that mediate these transformations at all levels (Duraiappah et al. 2014). These institutions mediate and influence social processes governing access, such as entitlements (Leach et al. 1999; Sen 1984), power asymmetries (Robards et al. 2011), social differentiation (Leach et al. 1999) and relative poverty. Power relations, embedded within institutional and governance systems, shape the ability of ecosystems to provide ecosystem services (Felipe-Lucia et al. 2015; Ribot and Peluso 2009). From a social dimension, it is thus important to understand how human actions lead to the generation of ecosystem services, who in society benefit from these services, and how are the values for these ecosystem services articulated for integration in decision-making (Ernstson 2013).

The interlinkages of ecosystem services with the biophysical and social system is often framed in terms of a cascade, with ecosystem properties, functions, services, benefits and values as building blocks (Nassl and Löffler 2015; Potschin-young et al. 2018). Use of this framework requires an understanding of the complexity of ecosystem components and processes, ecosystem functions, but also the pathways and scales of service flow, the diverse benefits and values, and importance of using appropriate evaluation procedures (Boulton et al. 2016). Value attribution to benefits derived from ecosystem services is subjective, based on criteria such as individual, stakeholder group, time, and location, or from normative criteria related to aspects such as culture, time and location set by institutions (Spangenberg et al. 2014). We use the cascade framework to reflect on the biophysical, social and governance elements related to the integration of ecosystem services within management of Chilika.

4.3 Ecosystem Services Within Management of Chilika

The ecosystem services cascade, representing the conceptual linkages between the Chilika social-ecological system, institutions and governance and contribution to human well-being, is presented in Fig. 4.1 We build on the description of Chilika social-ecological system discussed in Kumar et al. (this volume), and elaborate the remaining elements of the cascade here.

4.3.1 Institutions and Governance Settings

The CDA serves as the nodal government agency for the management of Chilika. The Authority's general superintendence is vested in its Governing Body, chaired by the Chief Minister of the Government of Odisha, and having elected representatives of the region around Chilika, heads of various government departments, and major scientific institutions as members. The Authority conducts its activities in line with an approved management plan, and secures funds for implementation of activities from the national government and partly from major donor agencies. The Governing Body also serves as a platform for coordinating sectoral development projects and taking decisions on various policy and regulatory matters. Based on this institutional structure, the CDA has been able to complement ecological monitoring and habitat management programmes with programmes on fisheries, rural development and tourism.

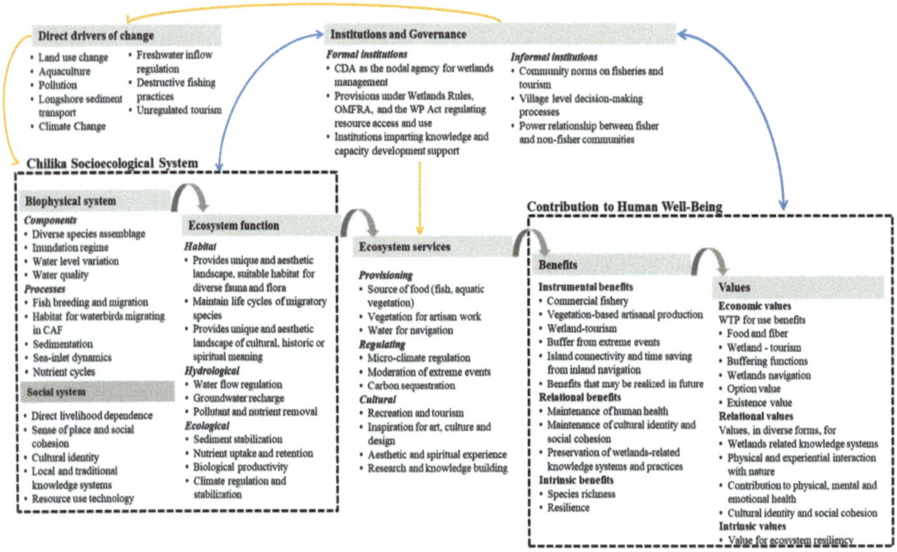

Fig. 4.1 The Chilika Ecosystem Services Cascade (based on Burkhard et al. 2010; Potschin-young et al. 2018)

Control, access and management of Chilika is based on several laws and regulation enacted by the national and the state governments. The Wetlands (Conservation and Management) Rules, 2017, notified under the Environment Protection Act (1986) sets several prohibitions, particularly on conversion of wetlands to non-wetland usages in Ramsar Sites, and requires management to be guided by an integrated wetland wise-use plan. In 2019, the Ministry of Environment, Forest and Climate Change (MoEFCC), Government of India issued the Coastal Regulation Zone notification, placing a range of development restrictions along the coastline. Seagrasses, biologically active mudflats and bird nesting grounds have been placed under the most stringent regulation under these rules. Management of Nalabana, a 15.5 km^2 island in the centre of Chilika, is guided by the provisions of the Wildlife Protection Act (1972) as the site was declared as a wildlife sanctuary in 1987. In 1984, the state government enacted the Orissa Marine Fisheries Regulation Act, under which fishing vessels, gears and fishing grounds are regulated by the State Department of Fisheries.

At the community level, village Panchayats are the formalised local self-government system entrusted with the responsibility of developmental planning and implementation in various spheres, including conservation of the local environment (Srivastava 2002). The formalisation of fishers access and use rights have evolved since the 1500s from a system of regulation by the king to vesting the rights in favour of the Odisha State post-independence (Nayak 2014). Since the 1960s, the Odisha State instituted a system of administering fishing area leases to the Primary Fishermen Cooperative Societies (PFCS). With the modernisation of fishing techniques, the introduction of aquaculture, and changes in lease policy in the nineties in favour of culture-based fisheries (Samal 2002), the lagoon witnessed an efflux of non-fishermen who gradually used their political and economic power to usurp the fishing grounds and convert shorelines into aquaculture areas (Dujovny 2009). While the shrimp culture was declared illegal in Chilika on the basis of an Odisha High Court Order of 1993 and Supreme Court order of 1996 (Nayak 2014), the power and economic asymmetries between the fisher and non-fisher communities have an important bearing on the benefit sharing from Chilika fisheries till date (Das 2018; Kumar et al. 2011).

Community institutions have historically been central to the management of Chilika fisheries. Based on a nuanced understanding of the resource, the fishers of Chilika evolved a system of resource partitioning by setting spatial limits (places to fish), temporal limits (seasonality), gear restrictions (what harvesting gear may be used), and physical limits (what sizes may be fished) (Sekhar 2004, 2007). Fishers also attach strong symbolic significance to the wetlands (revering it as mother nature), the dolphins (as a sign of good luck) and an abode of their goddess Kalijai, which has a temple on an island inside the wetland (D'Lima et al. 2014). With the resurgence of tourism since the opening of the new mouth in Chilika in 2000, fishers operating tourist boats have formed associations which allocate tourists to individual boats, and in the process reduce conflicts between boatowners.

4.3.2 Ecosystem Functions

Chilika provides diverse habitats suited for a range of species adapted to freshwater, brackish and marine environments. The lagoon is known to be inhabited by at least 259 phytoplankton (Srichandan et al. 2015), 77 zooplankton (CIFRI 2017), 44 macro-benthos (CIFRI 2017), 102 algae (Rath and Adhikary 2008), 726 plants, 126 molluscs (Mahapatro et al. 2016), 317 fish (Mohanty et al. 2015), 225 bird (Balachandran et al., this volume), 7 amphibian and 19 mammal species (Kumar and Pattnaik 2012), several of which are of high conservation significance globally and regionally. Chilika is also one of the two lagoons in the world inhabited by the Irrawaddy Dolphin (*Orcaella bervirostris*) (Reeves et al. 2008). *Barkudia insularis*, a limbless skink, is endemic to Chilika environs (Bauer et al. 2014). A population of about 30,000 water buffalo *Bubalus bubalis* has adapted to the saline condition of the lagoon and established as a separate breed (Singh et al. 2017) and even a source of geographical indication products such as Chilika curd (Nanda et al. 2013). Discovery of novel bacteria such as *Streptomyces chilikensis* (Ray et al. 2013) and *Halobacillus marinus* (Panda et al. 2018) indicate the bioprospecting potential of the lagoon.

Chilika also plays an important role in maintaining life-cycles of migratory species. For several waterbirds migrating along the Central Asian Flyway and East Asian Australasian Flyway, Chilika is an important stopover site (Balachandran et al. 2018; Palm et al. 2015). Of the 377 fish species documented in the lagoon thus far, 271 are migratory, and critically dependant on the wetland ecosystem to complete their lifecycle (Mohanty et al. 2015).

A sizeable waterholding capacity of nearly 1200 Mm^3 and an interannual tidal range of upto 0.6 m enables Chilika to absorb a significant proportion of monsoon flows. The mixing of freshwater from the Mahanadi River distributaries and seawater from the Bay of Bengal enables creation of a salinity gradient, with nearly freshwater conditions in the northern part, mixohaline conditions in the central and southern sector, and euhaline to hypersaline conditions in the outerchannel (Barik et al. 2017; Panda et al. 2015). The salinity gradient plays an important role in creating diverse habitats, regulating vegetation and providing migration cues to fishes (Kumar and Pattnaik 2012).

Chilika also serves as a sink for organic matter and nutrients, effectively recycling the inputs received through various transport processes resulting in regulation of nutrient and thus enhancing overall productivity (Amir et al. 2019; Ganguly et al. 2015). The dense *Phragmites karka* stands on the northern shoreline of the lagoon act as an ecological filter by trapping nutrients and pollutants and thus enabling maintenance of the water quality (Behera et al. 2018). Chilika harbours nearly one-fifth of the seagrass meadows of India (Geevarghese et al. 2018) which act as a carbon sink, sequestering annually 10.1–16.8 tCO_2 equivalent ha^{-1} in Chilika, and storing 22.4 tCO_2 equivalent ha^{-1} and 444 tCO_2 equivalent ha^{-1} in living biomass and soil organic carbon respectively (Ganguly et al. 2018).

The Chilika landscape has unique cultural, historical and spiritual significance. Archaeological investigations have indicated that the wetland was the site of important ports providing berthing facilities to ships travelling to Southeast Asian countries since 150 AD (Tripati and Patnaik 2008) and thus played an important role in the spread of Indian culture beyond India's shoreline (Tripati and Vora 2005). Chilika has also figured prominently in Oriya poetry (Mansinha 1960), and the works of noted Oriya poets such as Radhanath Ray and Gopabandhu Das. The lagoon has also been placed on the tentative list of sites under the World Heritage Convention (UNESCO 2014).

4.3.3 Ecosystem Services

Chilika, with 73 fish, prawn and crab species of commercial value (Mohapatra et al. 2007) is an important commercial fisheries for the state and the base of livelihoods of nearly 0.14 million fishers. The lagoon also contributes to off-shore fisheries, as many estuarine fish and prawn species use the wetland as spawning and breeding habitats. Several macrophytes are harvested for household and commercial use such as *Schoenplectus littoralis*, a cosmopolitan sedge (for making mats), *Phragmites karka* (for fuel and roofing material), and *Stuckenia pectinata* and *Naja sp* (for preserving fish catch). The extensive water expanse of Chilika allows for operation of inland navigation, providing connectivity to population living within the islands and leading to considerable time saving as compared with alternative road travel.

The immense water storage capacity within a densely populated area makes Chilika an important buffer for floods and cyclones which are known to frequently hit the east coast landscape. Evapotranspiration and heat storage enable large waterbodies as Chilika to regulate microclimates by taking away ambient heat and improving breeze circulation. Nutrient uptake and sediment retention within the lagoon prevents pollution within the coastal areas. Large swathes of sea-grasses and patches of mangroves enable storage and sequestration of carbon within biomass and sediments thus removing harmful greenhouse gases from the atmosphere.

With rich and biodiversity and scenic beauty, Chilika is a popular tourist destination on the Indian east coast, accounting for 8–10% of the total tourist arrivals in the state (Kumar and Pattnaik 2012). Balugaon, Satpada and Rambha receive the majority of tourists, who flock to watch waterbird congregations, Irawaddy Dolphins or just make a visit to the venerated Kalijai temple. The islands of the lagoon present a picturesque sight with the Khalikote hills as a backdrop. Chilika tourism forms the basis of a vibrant economy for the tourist boat owners, hoteliers and travel companies.

4.3.4 Benefits and Values

In line with the IPBES multiple values framework (Pascual et al. 2017), we classify benefits (and values) in three major categories, namely instrumental, relational and intrinsic. The instrumental benefits and values relate to Chilika being a source of food and fibre (through fisheries and aquatic plants), as a means of recreation which also provides livelihoods to a large population of dependent communities, and time-saving that result from the use of Chilika as a medium of inland transport. The category also includes benefits and values linked with the security of life and assets provided by buffering of floods and tropical cyclones, the avoided coastal pollution by filtering the runoff received from the direct catchments, and avoided impacts of climate change resulting from the carbon sequestered by the lagoon.

The relational benefits and values are related to the symbolic relationships that communities hold with Chilika, giving them a sense of identity and spirituality. Such values underlie the long-standing struggle of traditional fishers for fishing rights (Das 2018), the symbiotic relationship between the fishers and dolphins (D’Lima et al. 2014), and veneration of Chilika and Kalijai within various religious and cultural practices. The non-anthropocentric intrinsic benefit and value of Chilika are linked with the diversity of species and habitats within the lagoon, and the myriad ecosystem processes that connect the biotic and abiotic components of the ecosystem. Several elements of the intrinsic values have been explored through the ongoing monitoring programmes of the CDA and research carried out by different agencies, which also assist in managing wetlands placed in similar ecological and social contexts.

4.3.5 Drivers of Change

A range of natural and human induced direct drivers (directly influencing ecosystem processes) and indirect drivers (operating diffusely by altering one or more direct drivers) cause a change in ecosystem (Nelson et al. 2006). Direct physical and biological drivers of change operating in Chilika are changes in climate, coastal processes, land use in catchments, aquaculture and pollution loading. Climate change is manifesting in diverse ways including high rates of sea level rise in the Bay of Bengal as compared to other parts of the Indian coastline (Chowdhury and Behera 2015; Unnikrishnan et al. 2015), increasing frequency of cyclones (Unnikrishnan et al. 2011; Mishra 2014), lagoon surface warming (Schneider and Hook 2010), and high likelihood of flow reduction from Mahanadi River (Rao 1995; Mondal and Mujumdar 2015; Raje and Mujumdar 2010). The northwards littoral drift along the Bay of Bengal renders the coastal inlet prone to the impacts of shifting sand. During October 2000 to April 2018, the sea inlet at Sipakuda shifted northwards by 4.2 km. The mouth is also rendered dynamic, eroding and accreting at annual rates of 13.63 m and 13.9 m, respectively between 1988 and 2017 (Vivek et al. 2019).

Land use of Chilika catchment has a direct bearing on runoff and pollution received in the wetland. During 2011–2017, the built-up area in the basin increased from 6% to 17%, accounting for a decline in the area under forests (from 26% to 24.7%) and agriculture (36.3% to 29%) (CDA Unpublished Data).

The natural shorelines of Chilika, since the 1990s, have been encroached upon by shrimp aquaculture enclosures, despite being declared as illegal due to their adverse ecological and social impact (Galappaththi and Nayak 2017; Nayak and Berkes 2010). In 2018, 15,147 ha of shoreline were freed of illegal enclosures through an eviction action by the Chilika Development Authority (CDA, personal communication). Enhanced landscape aesthetics post-ecological restoration has led to a resurgence of tourism, particularly tourism linked with dolphin watching. However, there are indicators that this growth is fast reaching the carrying capacity of the wetland ecosystem, and if not well-managed, could turn into a driver of adverse change (Lima et al. 2018). Studies on petroleum hydrocarbon for the lagoon have exhibited higher concentrations in areas surrounding the jetties, attributed to the operation and maintenance of motorised boats, although the concentrations were found to be low and mostly benign to the aquatic environment (Mohanty et al. 2016).

4.3.6 Feedback Systems

The CDA maintains a network of hydrometric, tide gauging and water quality stations to assess the hydrological condition of Chilika on a real-time basis. The Wetlands Research and Training Center of the CDA researches ecological dimensions of the wetland, and publishes, on a bi-annual basis, an ecosystem health report card. The Annual General meetings of the authority are a means of sharing the information on the status of the lagoon to different development sectors. On an informal scale, the press and media regularly publish articles and clippings on issues related to Chilika. However, feedback mechanisms for social systems are relatively under-developed. Thus, information on the human well-being outcomes resulting from Chilika management is currently only peripherally included in the monitoring system.

4.4 Economic Values of Ecosystem Services Benefits

In this section, we present the economic values of select ecosystem services benefits of Chilika, namely commercial fisheries, aquatic vegetation for economic use, water transport, tourism and recreation, carbon sequestration and existence value.

4.4.1 Commercial Fisheries

CDA, since 2001, has been monitoring fish landings, marketing channels, prices at various trading locations and select biological paramters within the overall wetlands monitoring framework. An analysis of data for 2011–2015 indicates an average annual landing of 12,465 MT, of which fish, prawn and crabs constituted 57%, 40% and 3% of the quantity respectively. Prawns are the most valued component of Chilika fisheries. Of the total prawn landing, 43% is exported to international markets, with the trade almost restricted to three species, i.e. *Penaeus monodon, Fenneropenaeus indicus* and *Metapenaeus monoceros.* About a quarter (26%) is exported to other states, the rest traded around Chilika and adjoining districts. The fish landing is mostly traded as fresh fish (98.14%), and a minor proportion as live fish (1.03%) and dry fish (0.83%). Nearly half (47%) of total fish landing is exported to at least eight states, i.e. West Bengal, Jharkhand, Delhi, Madhya Pradesh, Tamil Nadu, Gujarat, Kerala and Andhra Pradesh. Local consumption, which occurs through markets around Chilika and consumption by the fishers forms the next major category (40%). 14% of the total fish catch is also traded within the western and southern districts of Odisha State. Of the total crab landing, 52% is reported to be exported to other states, with the rest being traded in markets within the state.

The gross economic value of Chilika fish, based on price and quantity data across various market segments and trading agents is presented in Table 4.1. Prices used for each market segment have been quantity weighted.

Table 4.1 Estimation of gross economic value of Chilika fisheries

	Fish	Prawn	Crab	Total
Amount traded (in MT)				
(a) within Chilika	2610.36	995.07	–	3605.43
(b) within Odisha state	859.40	326.27	185.18	1370.85
(c) Exported outside Odisha state	3677.76	1341.75	154.02	5173.53
(d) Exported to international markets	–	2315.75	–	2315.75
	7147.52	4978.84	339.21	12,465.57
Quantity sold to (in MT)				
(a) Retailers	765.08	536.08		1301.16
(b) PFCS	5079.73	3529.99		8609.72
(c) Intermediaries/commission agents	1302.70	912.78	339.21	2554.69
Quantity weighted prices (Rs. per kg)				
(a) Retailers	94.12	214.28	165.60	
(b) PFCS	78.43	178.57		
(c) Commission agents	62.13	131.58	138.00	
Gross value (in Rs. Million)	551.35	865.33	46.81	1463.48

4.4.2 Aquatic Vegetation for Economic Use

A household survey of 4074 households conducted during September to November 2012 indicated that 8400 MT of *Phragmites karka* was harvested annually for use as fuel and as thatch, 3836 MT of *Schoenplectus littoralis,* and 1900 MT of *Potamogeton pectinatus and Naja sp.* for use as packing material by fishers. Valuation of use of *Schoenplectus littoralis* for mat-making is based on the price of the final product. Valuation of use of *Phragmites* has been derived using the opportunity cost of time-based on the prevailing rural wage rate. Similarly, the opportunity cost of time spent in transporting the harvest of packing material to shoreline is used as a proxy price. Using these prices, the economic benefit from use of aquatic plants has been assessed to be Rs. 34.31 million.

4.4.3 Water Transport

Water transport in Chilika caters primarily to two segments, the first being the island villages having limited road connectivity, and the second being the tourists. Benefit to the tourists have been included within the consumer surplus estimates for tourism in the latter section. The CDA operates a passenger ferry between the islands on a no profit-no loss basis. During 2003–2014, water transport in Chilika was annually availed by 35,600 persons, with an average time cost saving of 4.5 h per person when compared with an alternate road route. Assuming that the proportion of working population within the passengers is similar to that of the regional average (42%), the opportunity cost of time saved based on the average rural wage rate is assessed to be Rs. 13.6 million.

4.4.4 Tourism and Recreation

Individual Travel Cost Method (ITCM) has been used to estimate tourism and recreational benefits from Chilika. Demand curves relating the annual site visitation rate (every 10 years) to the per capita visit costs, income, and other socioeconomic characteristics have been estimated separately for the domestic and foreign tourists. A questionnaire survey of tourists to elicit the overall economic value attributed to wetland based tourism was carried in and around Chilika during September – November 2012. Overall, 433 tourists responded to the survey, of which 36 respondents were of foreigners and the rest Indian nationals. Of the total responses received, the survey forms of 179 of the domestic tourists and 31 the international tourists were complete in all respects and used for estimating consumer surplus. Individual consumer surplus was aggregated to the total site arrival for estimation of the overall consumer surplus for the site. Following model was estimated:

$$\overline{CS} = \left(e^c * \overline{trip\_dur}^{\beta_1} * \overline{dist}^{\beta 2} * \overline{jour\_pur}^{\beta 3} * \overline{gsize}^{\beta 4} * \overline{income}^{\beta 5} * \overline{age}^{\beta 6} \int\limits_{tc\,\min}^{tc\,\max} tc^{\beta 7} d(tc) \right.$$

$$= \left(e^c * \overline{trip\_dur}^{\beta_1} * \overline{dist}^{\beta 2} * \overline{jour\_pur}^{\beta 3} * \overline{gsize}^{\beta 4} * \overline{income}^{\beta 5} * \overline{age}^{\beta 6} * \left[\frac{(tc)^{\beta 7+1}}{\beta 7 + 1} \right]_{tc\,\min}^{tc\,\max} \right.$$

(trip_dur: Trip duration in days; dist: Distance travelled to Chilika (km); jour_pur: dummy variable indicating purpose of journey; gsize: Number of persons accompanying group; income: Annual income of the household (in Rs. for domestic tourists and US$ for international); age: age of the respondent (years); tc: average trip cost per person (in Rs. for domestic tourists and US$ for international)) (Table 4.2)

The predictors explain 54% and 45.5% of the variability in the visitation rate for domestic and international tourists, respectively. For domestic tourists, the visitation rate was found to be negatively related to distance, group size and per person trip cost. For international tourists, trip cost per person was the only variable which was found to be significantly and negatively related to visitation rate. The annual average consumer surplus based on the demand curve was estimated to be Rs. 5806.82 for domestic tourists and US$ 2686.56 (equivalent to Rs. 170,597 at 2015 exchange rate). The aggregate consumer surplus, estimated based on average annual arrivals during 2010–2014, has been estimated to be Rs. 3027.34 million for domestic tourists and Rs. 351.77 million for international tourists. The two categories sum to Rs. 3379.11 million annually.

Table 4.2 Regression model for estimation of travel cost

Parameter		Coefficients modelled for Domestic Tourists	Coefficients modelled for International Tourists
Adjusted R^2		.540	.455
DW statistic		1.948	2.240
N		179	31
F statistic		28.674**	32.05*
Ln (trip_dur)	Natural logarithm of duration of trip (days)	−.108	−0.004
Ln (dist)	Natural logarithm of distance travelled to Chilika (km)	−.420**	@
Ln (jour_pur)	Natural logarithm of dummy variable indicating purpose of journey	.147	−0.008
Ln(gsize)	Natural logarithm of number of persons accompanying group	−.173**	−0.0064
Ln (income)	Natural logarithm of annual income of the household (in Rs. for domestic tourists and US$ for international)	0.016	−0.126
Ln (age)	Natural logarithm of age of the respondent (years)	.102	0.273
Ln (tc)	Natural logarithm of average trip cost per person (in Rs. for domestic tourists and US$ for international)	−.181**	−0.225*
Constant		4.303**	0.718

**significant at 99% confidence interval, *significant at 95% confidence interval, @not used as predictor in regression model

4.4.5 *Carbon Sequestration*

The economic value of blue carbon sequestered by seagrass in Chilika has been estimated using the following equation:

$$VC = SQ * A * SCC$$

Wherein VC: Economic value of carbon sequestered, SQ: Rate of carbon sequestration (in t CO2 equivalent ha^{-1} year^{-1}); A: Area under seagrass (in ha); SCC: Social cost of carbon (Rs per t CO2).

Ricke et al. (2018) based on climate model projections, climate-driven economic damage estimation and socio-economic projections have estimated India's Social Cost of Carbon to be between US$ 49–157 per t CO2, with an average of US$ 86, equivalent to Rs 5693 at 2015 exchange rate. With a seagrass extent of 8660 ha and a rate of carbon sequestration ranging between 10.1–16.8 t CO2 equivalent ha^{-1} year^{-1} (Ganguly et al. 2018), the economic value of blue carbon in Chilika has been estimated to range between Rs. 498–828 million year^{-1}.

4.4.6 Existence Value

Closed-ended Willingness to Pay (WTP) data was obtained from a survey of 984 residents around Chilika carried out during September – November 2012. The WTP was assessed using a logit model to identify the determinants of the responses to the question: "Yes, I am willing to pay Rs. X for conservation and wise-use of Chilika" or "No, I am not willing to pay Rs. X for conservation and wise use of Chilika", where X refers to the amount of closed bid in each case. The model relates the 1 (yes) and 0 (no) response variable to the bid levels faced by each respondent.

The general form of the model is expressed by the following equation (Cox 1958):

$$P_i = E(Y = 1|X_i) = \frac{1}{1 + e^{-(\beta_1 + \beta_2 X_i)}}$$

Wherein, P_i is the probability of an individual i willing to pay the stated bid amount X_i. Using a logit regression to relate individual responses to the bid values results in estimates of coefficients β_1 and β_2, which can be used to derive the mean WTP. The coefficients were estimated to be −0.005 and 3.829 respectively, both significant at 99% confidence interval. The coefficient β_2 is negative and significant, indicating that the probability of accepting a particular bid level decreased with an increase in the bid amount. The Hosmer and Lemeshow Test yielded a significance value of 0.005. The Cox and Snell R Square and Nagelkerke R Square values were estimated to be 0.382 and 0.509. The model estimated "no" and "yes" values 89.1% and 70.9% correctly, with an overall percentage correctness of 80.1%. The estimated mean WTP per respondent is Rs. 257.63. The aggregate existence value, by extrapolating the mean WTP to the total number of households living in and around Chilika has been estimated to be Rs. 17.32 million.

4.5 The Economic Efficiency of Investment in Wetland Restoration

The costs related to managing Chilika are currently met through the financing of specific projects by the Central and State Governments. The primary source of Central Government assistance is from the MoEFCC under its national scheme on wetlands, titled National Plan for Conservation of Aquatic Ecosystems. Funding to wetlands of national priority is at times also included as a Grant-in-Aid for special problems as per the recommendation of the Finance Commission of the Government of India, routed through the Ministry of Finance.

Based on the data provided in annual reports and account statements, the CDA during 1992–2014, entailed an expenditure of Rs. 1545.55 million (equivalent to US $ 22.78 million at 2016 exchange rate) for various restoration interventions. A major proportion of funding (76%) was received in the form of Grant-in-aid by the Finance Commission of Government of India (tenth, eleventh, twelfth and thirteenth). The balance of the funding was received from the MoEFCC and the State Government of Odisha (5% and 7% respectively). Nearly half (46%) of the expenditure has been on the maintenance of hydrological regimes (maintaining connectivity with the Bay of Bengal). Approximately, one fifth of the investment (19%) has been made on wetland monitoring and evaluation and another one fifth on fisheries development and livelihood improvement.

A benefit-cost ratio has been computed as an indicator of economic efficiency (Pearce 1998) of the investment made in the management of Chilika. Expenditure by CDA on different management components have been treated as public investments. Private investments for fisheries and tourism have been considered, as these form an integral component of total capital deloyed for accessing ecosystem services benefits. Data on capital costs incurred by the fishers were derived from a survey of fishers conducted in 2012, and the values extrapolated for the past years assuming an inflation rate of 6%. For tourism, it is assumed that 90% of the tourist expenditure spent locally for travel accommodation and food are invested (estimated from tourist expenditure data collected during ITCM survey). The per capita tourist expenditure estimated separately for domestic and international tourists for 2012 have been extrapolated for the previous years using an inflation rate of 6%, assuming that the proportion of local expenditure does not change over time.

Incremental benefits from fisheries and tourism were included in the benefit stream. In the case of fisheries, incremental landing for the period 2001–2014, over an average landing of period 1991–2000 has been used for analysis. In the case of tourism, incremental tourist arrivals since 2001, over average arrival for the period 1994–2000 have been assessed. The consumer surplus for domestic tourists estimated in 2014 was adjusted for various years using data on the consumer price index. Surplus for international tourists of 2014 was adjusted using a ratio of US$-Rupee exchange rate for a given year to that of 2014. The benefit-cost ratio on the basis of public investment is 16.2. When the private investment in also included, the ratio is 3.73 (Table 4.3).

4.6 Distributional Aspects of Benefits from Chilika Fisheries

One of the objectives pursued by Chilika management is to rejuvenate the PFCS to ensure better economic returns to the capture fishers as an incentive for responsible fishing. Since 2008, the CDA has been implementing a Fisheries Resource Management Plan (FRMP) (JICA and CDA 2009) which focuses on enhancing the capacity of the fishery cooperatives through measures as capital infusion, training in accounting, provision of ice boxes, creation of landing centres, and creating awareness on responsible fisheries.

Table 4.3 Composition of costs and benefits from Chilika restoration (Rs. Million)

Total Costs			9405.09
Public investment		2161.76	
Habitat management	320.08		
Wetland monitoring and research	73.59		
Wetland monitoring and evaluation	394.67		
Socioeconomic improvement and livelihoods	246.50		
Livelihoods	203.54		
Improvement of water exchange	923.38		
Private investment		7243.33	
Depreciated value of boats and machinery	1362.51		
Depreciated value of tourism infrastructure	5880.82		
Total benefit		35,039.74	35,039.74
Value of increased fish landing	14,261.39		
Value of increased tourism	20,778.34		

A comparative analysis of the distribution of economic benefits from Chilika fisheries has been conducted for the period of 2008 (prior to efforts placed for rejuvenation of PFCS) and 2015 (wherein major components of Fisheries Resource Management Plan had been implemented). The 2008 scenario has been constructed using data on fish landing, landing center prices and catch disposal accessed from the CDA. These data were complemented by sample survey of 4133 households on occupation pattern, asset ownership, pattern of catch disposition, point of sale, prices obtained, workforce participation, indebtedness, and ownership of fishing equipment. The situation of 2015 was assessed based on a survey of 8 PFCS (3877 fishers).

The gross revenue earned from fishing has been derived using data on quantity weighted prices at various points of sale (namely PFCS, commission agent, mahajan, retailer or direct to consumer) with quantities sold at various points. The net revenue has been estimated by reducing the capital expenses (depreciation of boats, nets and gear, costs of fuel for fishing fleet) from the gross revenue. To ensure comparability, the 2008 prices were adjusted to 2015 using the Consumer Price Index (Rural) data. In the case of catch handled by PFCS, the operational costs paid to the society (Rs. 5 per kilogram of fish and Rs. 7 per kilogram of prawn) have been deducted from gross revenue, in addition to costs of capital deployed. The gross revenue realized to the fisher has also been expressed in terms of percentage of the total value estimated from the highest landing centre price for the catch. This proportion is a proxy indicator of the share of fishers in the value of fish landed if sold at the local market. The gross and net revenues have been expressed in terms of per household income using the 2010 assessment of the number of fisher households (23,115) (Kumar and Pattnaik 2012). The daily wage rate earned for fishing activity has been derived by dividing net revenue by the number of fishing days.

Data from the surveys indicate a distinct change in prices and points of sale during the period 2008–2015. Since the FRMP was implemented, the PFCS offered

Table 4.4 Changes in gross and net revenue to fishers (2008 and 2015)

Particulars	Survey year 2008	2015
Total fish, prawn and crab landing (in '000 kg)	10,051.36	12,053.56
Gross value of the fish catch realized to fishers (in Rs. Million)		
(a) At current prices	645.00	1332.48*
(b) At 2015 adjusted prices	810.71	
Gross value at highest landing center price (in Rs. Million)	1009.51	1917.03
Value realized to fishers as a proportion of value estimated using maximum local prices	53.91%	69.51%
Gross annual income per fisher household (Rs.)		
(a) At current prices	23,502.33	57,645.72
(b) At 2015 adjusted prices	35,072.71	57,645.72
Net annual income per fisher household (Rs.)		
(a) At current prices	16,684.70	43,046.62
(b) At 2015 adjusted prices	24,898.71	
Daily wage rate earned per fishing day (Rs.)	109.84	195.69

*The gross value reflected here differs from the one reflected in Table 4.1 which is computed on the average catch for the period 2011–2015

higher prices to the fisher as compared with the middlemen. For prawns, the quantity weighted price offered by the cooperative to its member fishers was estimated to be Rs. 178.57, which was 35% higher than that paid by the middlemen. Similarly, the quantity weighted price of Rs. 78.43 per kilogramme of fish offered by the cooperative to its members was 26% higher than that paid by the middlemen. The surveyed fish cooperatives did not report trading in crabs. However, there is still a sizeable proportion of catch sold to commission agents, as the cooperatives handled only 71% of the fish and prawn landing by its members.

Apart from changes in prices and trading points, the differences in per household gross and net income and the wage rate is also due to the fact the catch in 2015 was 19.9% higher. If the fish landing in 2008 were to be considered equal to that of 2015, the difference in gross annual household income (at prices adjusted to 2015 for comparability) within the two periods is of 24.5% (Rs. 57,645 in 2015 as compared with Rs. 46,298 in 2008). Similarly, the estimated wage rate in 2015 is 19.2% higher (Rs. 189.89 in 2015 as compared with Rs. 159.36 in 2008) (Table 4.4).

4.7 The Value of Ecosystem Components and Processes

A production function approach (Barbier 2007; Mäler 1991) has been used to analyse the contribution of ecosystem components and processes towards generating ecosystem services benefits from commercial fisheries. The production function has been specified as $q = q(m, n)$, wherein, q is the output (fish landing), m denotes the vector of manufactured and the human capital input and n denotes the vector of ecosystem components and processes as inputs.

The vector of ecosystem components and processes included in the model are in the form of two proxies, namely salinity and distance of sea inlet from the central sector. Within Chilika, salinity is an integrative indicator of ecosystem health, and provides cues for fish migration (Kumar and Pattnaik 2012). It also indicates the extent to which freshwater received from the Mahanadi Delta Rivers and seawater from the Bay of Bengal can mix (Panda et al. 2015). The distance of the mouth from the central sector of Chilika impacts key ecosystem processes such as tidal prism, tidal flux, and exchange of species between sea and lagoon. The vector of human and manufactured capital input into fisheries is described by the number of active fishers, number of boats and extent of fleet mechanization (the ratio of number of unmechanized boats to the number of mechanized boats). The first two variables are indicators of increase in human effort, while the latter has been used as a proxy for technology. The production function is estimated by the following stages:

$$mech_p = f(value\_curr, exd, prawn\_r, policy) \tag{4.1}$$

$$landing = f\left(mech_p, salinity, dist, fisher, boat\right) \tag{4.2}$$

It is assumed that (2) can be specified as a Cobb-Douglas function in the following form:

$$landing = \propto \left(mech_p\right)^{\beta 1}(fisher)^{\beta 2}(boat)^{\beta 3}(salinity)^{\beta 4}(dist)^{\beta 5} \tag{4.3}$$

With the coefficients, ß1, ß2, ß3, ß4 and ß5 representing output elasticity, and their sum determines returns to scale.

In Eq. (4.1), the extent of fleet mechanization has been estimated from the per fisher catch value at current prices (value_curr), exchange rate differential from the previous year (exd), the ratio of prawn landing to total landing (prawns) and the a dummy fisheries policy variable (policy). The per fisher catch value is a proxy for income generation, which in turn determines the ability to invest. The exchange rate differential is an indicator of export profitability, as a significant component of Chilika high-value prawns is exported to markets in Europe and Japan. The ratio of prawn to total landing is a proxy for landing composition, especially towards higher economic value species. The policy dummy variable captures the transition from a community-driven fisheries to prioritization for aquaculture and reversal thereof since the Supreme Court ordered a ban on aquaculture, and implementation of FRMP by CDA (Nayak 2017).

The function has been developed using annual time series data for the period 1957–2010. Data on landing, proxy for wetland's finfish and shellfish productivity is based on the data contained in Biswas (1995), CDA (2005) and monitoring records of CDA. The current value of landing has been derived using a quantity-weighted

Table 4.5 Regression estimates

Equation 1: $R^2 = 0.897$, DW Statistic = 1.398, F Statistic = 104.561**				
Variable	Description	N	Mean ± SD	Coefficient
$mech_p$[a]	Ratio of non-mechanized boats to mechanized boats	53	0.80 ± 0.17	
value _ curr	Current value of fish catch per fisher	53	0.80 ± 0.17	−1.195E-5**
exd	Difference between US$ to INR exchange rate in the current year with that of previous year	53	7534 ± 9836	−0.014*
prawn _ r	Ratio of prawn landing to total fin and shellfish landing	53	0.23 ± 0.07	0.804**
policy	Dummy variable (1 = policy favouring community fisheries, 2 = policy favouring aquaculture, 3 = policy favouring integrated management)	53	1.45 ± 0.66	−0.129**
Constant				0.896**
Equation 2: Adjusted $R^2 = 0.404$, DW Statistic = 2.219 , F Statistic = 4.746**				
Ln (landing) [a]	Natural logarithm of Total finsfish and shell fish landing (MT)	43	10.02 ± 0.34	
Ln (salinity)	Natural logarithm of Average lake salinity (in parts per thousand)	43	2.20 ± 0.36	0.264
Ln (fisher)	Natural logarithm of Number of active fishers (individuals)	43	10.02 ± 0.35	−0.774
Ln (boat)	Natural logarithm of Number of boats (number)	43	8.23 ± 0.27	0.217
Ln ($mech_p$)	Natural logarithm of Ratio of non-mechanized to mechanized boats (projected from equation 1)	43	−0.29 ± 0.22	−1.050*
Ln (dist)	Natural logarithm of Distance of the wetland mouth to the sea from central sector (in km)	43	2.88 ± 0.47	−0.861**
Constant				16.180

[a]Dependant variable, **significant at 99%,*significant at 95%

price data series, constructed from the information contained in (Biswas 1995; CDA 2005; Jones and Sujansingani 1954), and surveys conducted by authors in 2014. Trend data on the number of fishers, total boats and mechanised boats is based on linear interpolation for 1957 (Mitra and Mahapatra 1957), 1986–1987 (Satyanarayana 1999), 1996–2004 (CIFRI 2007; Mohapatra et al. 2007) and for 2007 based on surveys by authors. The series on salinity is based on data contained in (Biswas 1995), (CIFRI 2007) and CDA wetland monitoring database. Series on the exchange rate has been developed using the database from Reserve Bank of India at www.rbi.org. A linear specification of Eq. 4.1 gave the best fit, whereas Eq. 4.2 was modelled using log-linear specification. Details of regression estimates are presented in Table 4.5.

Both the regression models are statistically significant. Being time series, the regression did suffer from autocorrelation effects. For Eq. 4.1, the Durbin-Watson

Table 4.6 Estimation of the contribution of ecosystem variables to fisheries

	Pre-restoration period (1991–2000)	Post restoration period (2001–2010)	Change
Salinity (in ppt)	6.80	11.57	3.68
Distance (in km)	23.32	7.89	(16.52)
Modelled landing (MT) (controlling for all other variables)	3986.45	11,920.05	7933.59

(DW) statistics fell in an indeterminate zone, whereas the residual plot indicated randomness. The initial log-linear solution for Eq. 4.2, however indicated significant positive autocorrelation (DW = 0.705), thereby requiring application of Cochrane-Orcutt estimation procedure. The resultant model is able to explain 40.4% of the variability within the independent variable. The signs of coefficients are as expected. Since the mechanization ratio is coded inversely (a higher value indicating non-mechanization), it is indicated to be negatively related to current value of fish catch per fisher and policy changes in favour of community management. Within Eq. 4.2, landing is indicated to be negatively related to distance and mechanization. This is in line with the known fact that a decrease in mechanization ratio (and thereby an increase in number of mechanized boats) increased fish landings. Similarly, an increase in the length of channel has been observed to reduce landings significantly, due to its known impacts on migration and lagoon-sea connectivity. While an increase in boats is indicated to be positively related with landing, an increase in fisher is negatively related. This might be due to excess number of fishers not contributing to a commensurate increase in landing, or even reduced incremental landing.

To arrive at the incremental contribution of change in vector of ecosystem functions, the values in a pre-restoration period (pertaining to the period 1991–2000) have been contrasted with a post restoration period (2001–2010), while controlling for the variables representing human and manufactured capital. As can be seen below, the change in ecological parameters leads to an incremental landing of 7933.59 MT. This forms 72% of the average landing for the 2001–2010 period, and if valued at 2014 quantity weighted prices, comes to Rs. 1149.06 million (Table 4.6).

4.8 Managing Chilika for Multiple Ecosystem Services

4.8.1 The Relevance of Multiple Ecosystem Services for the Management of Chilika

The perspective of conserving multiple ecosystem services is complementary to the traditional framing of conservation strategies around biodiversity, habitat complexity and ecosystem processes (Ormerod 2014). The approach entails a progression from a siloed approach to conservation of species and habitats to explicit consideration of benefits humans derive from these ecosystems, enabling anticipation of a wide range of consequences that may result from different management regimes, and provide tools for identifying, negotiating, avoiding, and managing potential negative tradeoffs (Ingram et al. 2012).

Management of Chilika has historically been centered around fisheries, and since the site's designation as a Wetland of International Importance, for biodiversity values of global significance. The integrated management plan, formulated post hydrological restoration, seeks to achieve wise use of Chilika by meeting twin objectives of ecological security as well as livelihood improvement of local communities. The ecosystem services framework widens the scope of wetland management to not just include the instrumental relationships (such as providing food and nutritional security, security to assets, income generation and recreation opportunities) but also relational linkages (such as role of wetlands related knowledge systems; physical and experiential interactions with nature; contributions to physical, mental and emotional health; and cultural identity and social cohesion). Similarly, in terms of spatial scales, the framework enables the setting of management objectives not just in consideration with the local environment, but also to the wider basin and coastal zone (primarily through regulatory services), and even global scale (such as a role in carbon sequestration). At the same time, management strategies need to be based on a consideration of spatiotemporal variance of these ecosystem services bundles, because this variance underpins the resilience of the ecosystem that, if weakened, may affect its capacity to deliver ecosystem services (Boulton et al. 2016).

Hydrological processes and functioning are key drivers of the many physical and biochemical interactions within ecosystems, which in turn control the performance of the services beneficial to humans. From a management perspective, the snapshot information presented by economic values need to be interpreted alongwith information on status and trends of underpinning ecosystem functions as well as drivers of change. An assessment concluded in 2016 indicated that catches of three commercially important fish species (*Mugil cephalus*, *Daysciaena albida*, and *Eleutheronema tetradactylum*) were seriously declining, and a major proportion (65–88%) of specimens of five commercially important species were immature, indicating overfishing (CIFRI 2017). The analysis also raises serious concerns on fishing along the two migratory pathways leading to wanton destruction of post larvae and juveniles of commercially important fish and shrimp species (CIFRI 2017). The overall catch also hovers close to the maximum sustainable yield (CIFRI 2007).

There are tradeoffs inherent in managing Chilika. It is apparent that management aimed at enhancing provisioning services (such as fisheries) or cultural services (such as tourism) may be at the cost of regulating services (such as ability to buffer hydrological regimes, recycle nutrients and sequester carbon) or even ecosystem functions (such as habitat diversity). At catchment scale, intensification of land and water use may alter the state of wetland towards higher salinity or nutrient enrichment, with a cascading effect on several ecosystem services. The primary approach of the CDA to manage such tradeoffs is to maintain the state of wetlands as achieved after the opening of the new mouth to the sea in September 2000. There is an emphasis on permitting only capture fisheries in the lagoon. Aquaculture, which involves physical transformation of the shoreline is not permitted as a part of the management strategy. However, there are no mechanisms in place to regulate impacts of anthropogenic activities such as fisheries and tourism on ecological sensitive areas of the lagoon, such as the portion inhabited by sea-grass beds, fish migratory channels or used as habitat of Irrwadday Dolphins.

4.8.2 Addressing the Issue of Distributional Equity of Ecosystem Services

Chilika management has a developmental objective of enhancing economic returns to the primary fishers as an incentive for responsible fishing (Kumar and Pattnaik 2012). The analysis of distributional aspects of benefits from Chilika fisheries presented in the paper indicate the dampening impact of the existing market structure on the economic returns to primary fishers. Measures taken for strengthening the PFCS have led to a 26–35% increase in prices at which the member fishers are able to trade their landings. At comparable landing and current prices, during 2008–2015, the gross annual household incomes have increased by 25%, and the estimated wage rate per fishing day by 19%. The daily wage rate earned per fishing day wage rates that have been derived from the assessments are comparable with the minimum wage rates for unskilled labour in rural sector, yet are considerably lower than the minimum wages rates for semi-skilled and skilled categories. The overall value realization to the fishers remains low, as despite forming 93% of the workforce, their share in value generation remains only 70%. Beyond economic benefits, strengthening community institutions also has distinct social and institutional impacts, in the form of increased cohesion, reduced conflicts, representation capability and ultimately increasing the possibility of implementing responsible fisheries (Agrawal and Benson 2011; Allison et al. 2012; Jentoft 2000). However, purely from an economic perspective, the scope of management would need to include measures for reducing fishing effort, increasing value realization through strategies as product differentiation, and enhancing participation of fishers in the higher segments of value chain.

4.8.3 Capturing Economic Values of Ecosystem Services

The assessment of economic values presented in the paper indicate the value of Chilika as a natural capital, and the fact that investment in wetland restoration makes a strong economic sense. This calls for putting in place financial mechanism to ensure that management is sustained over time, and enough budgets made available for implementing various management actions. Core functions such as wetlands monitoring cannot be delivered through projects based financing alone, as these are prone to funding gaps.

Through the case of Chilika fisheries, the economic analysis underlines the significant contribution of ecosystem components and processes in the delivery of ecosystem services, and thereby the need to consider the joint production character in management decisions. The current economic model factors in the costs of human capital (such as fleet, crafts and gears) in private pricing decisions, whereas the costs of maintaining the ecosystem components and processes have been shifted to the public budget. Such a financing system is untenable in the long run, as the public funding for financing wetland restoration has several competing interests. It is thereby pertinent that the financial flows emanating from the ecosystem services of the wetland are linked with the costs of maintaining such services. Within commercial fisheries, it is important that the prices also signal the resource base quality, and thereby are able to attract a premium, part of which could be reinvested into the wetland management. Within tourism, a levy charged on tourists vehicles and hotels can generate resources for ensuring that core functions of wetlands management are sustained.

4.8.4 Governance for Multiple Ecosystem Services

The mapping presented in Sect. 4.3 of the paper indicates a maze of formal and informal institutions which influence the management of the Chilika social-ecological system. The institutional fit (Folke et al. 2007) of this arrangement with the social-ecological system properties of Chilika is a critical ingredient of successful management. Over a period of time institutional arrangements for management of provisioning services and select cultural services (mainly tourism) have emerged, however, there is a relative vacuum when it comes to management of regulating services (such as water regime moderation, nutrient cycling, carbon sequestration and others). While CDA has been mandated to ensure Chilika is managed for multiple ecosystem services, the organization's ability to influence underpinning ecosystem processes, particularly those nested within the basin and coastal zone scale land and water use is limited.

Even within provisioning services such as fisheries, not all management actions are complementary and mutually reinforcing. Not all elements of community-scale fisheries resources management by PFCSs are supported by a production driven approach of the state fisheries department. The Governing Body of Chilika performs an important role as a bridging organization by enabling links between a diverse set of actors across management levels and institutions boundaries, however the mechanism has become top-heavy over a period of time, with reduced participation of primary user groups (Nayak and Berkes 2011).

There is also a mismatch between the geographic scale of Chilika ecosystem functioning (operating at the scale of basin and coastal zone) and institutional arrangements for managing the wetland (which is directed mostly towards and within the boundary of the wetland). Much is dependent on the extent to which the institutions responsible for managing various sectoral development programmes (such as rural development, water and sanitation, disaster risk reduction) take into account the multiple ecosystem services of Chilika and the implication of development programmes for sustained provision of such services. The capability of CDA to be able to accommodate private or communal ownership of common pool resources (relevant for provisioning services), while at the same time providing conditions whereby public goods (relevant for regulating and cultural services) will not decline in unsustainable rates, is central to this arrangement (Fig. 4.2).

4.8.5 Monitoring Ecosystem Services

Over the years, the CDA has developed a sophisticated wetlands monitoring system which is able to track status and trends in various wetland ecosystem components (such as species and species assemblages, water and sediment quantity and quality), processes (such as sedimentation, species migration, and interaction with Bay of Bengal) and through sporadic research, interlinkages between the two (such as impact of emergent macrophytes on hydrological processes). Novel attempts to synthesize the monitoring information into communicable ecosystem health metrics have also been made recently, in the form of ecosystem health report cards, published biannually (CDA 2017). However, the current monitoring system tracks only a few ecosystem services indicators, mainly those related to fisheries, tourism and habitat services. Sampling protocols designed to monitor biodiversity and physical environment may not always be suited to generate indicators of ecosystem services (Geijzendorffer and Roche 2013).

Much can be gained by aligning current ecologically oriented monitoring towards a trans-disciplinary system which effectively bridges the divide between research and management (Steffen 2009). For measuring and managing ecosystem services, a social-ecological systems research and monitoring framework which can account for how these services are generated by coupled social-ecological systems (how different ecosystem services interact, how changes in the bundles of ecosystem services influence human well-being (Reyers et al. 2013), and how values for ecosystem services feed into stakeholder behavior and attitudes towards wetlands conservation and wise-use. Aspects such as the impact of changes in land use, nutrient mobilization, connectivity with the sea and rivers, species composition, and climate change on ecosystem services of the lagoon need to be systemically investigate for informing wetland management. An ecological research agenda for ecosystem services may be structured around four key areas: (a) identifying species, assemblages, or ecosystem processes that are key ecosystem services providers, and characterizing their functional relationships; (b) determining aspects of community structure that influence ecosystem functions within the Chilika basin and coastal zone; (c) assessing key environmental factors that influence the provision of services; and (d) measuring the spatio-temporal scale over which providers of ecosystem services operate (Kremen 2005). An empirical base for understanding thresholds of massive persistent changes in social–ecological systems, the factors that control

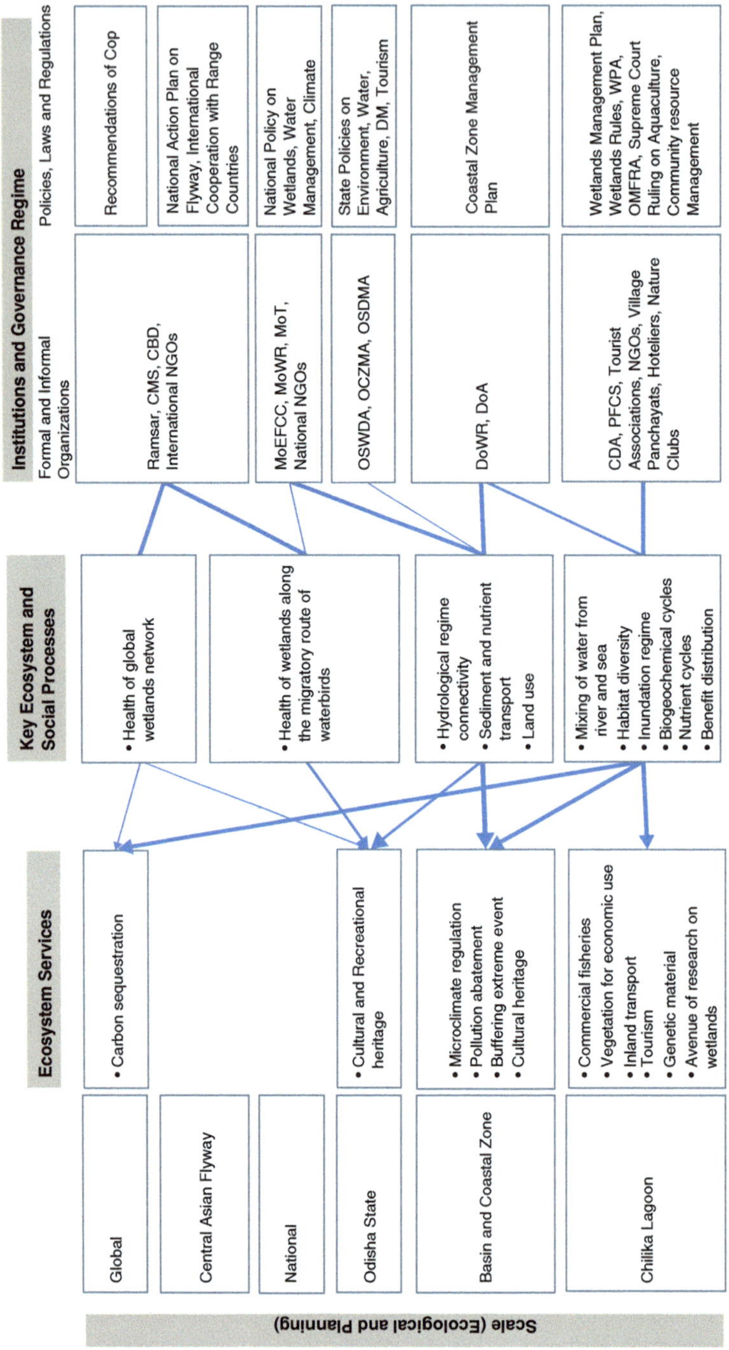

Fig 4.2 Chilika ecosystem services and institutions and governance regimes (Source: Authors)

probabilities of such changes, and leading indicators of thresholds (Carpenter et al. 2009; Steffen 2009) also needs to be developed as a priority.

Acknowledgements Financial support for conducting the research reported in this paper was received from International Development Research Center under 'Chilika Climate Change' project (Grant ID: 106703-001) and GIZ- India under Indo-German Bilateral Cooperation Project-'Incentives for Sustainable Management of Biodiversity and Ecosystem Services'. Satish Kumar and Kalpana Ambastha (Wetlands International South Asia) assisted in statistical modelling.

References

Agrawal A, Benson CS (2011) Common property theory and resource governance institutions: strengthening explanations of multiple outcomes. Environ Conserv 38(2):199–210

Allison EH, Ratner BD, Åsgård B, Willmann R, Pomeroy R, Kurien J (2012) Rights-based fisheries governance: from fishing rights to human rights. Fish Fish 13(1):14–29

Amir M, Paul D, Samal RN (2019) Sources of organic matter in Chilika lagoon, India inferred from stable C and N isotopic compositions of particulates and sediments. J Mar Syst 194:81–90

Balachandran S, Katti T, Manakadan R (2018) Indian bird migration atlas. Oxford University Press, Bombay Natural History Society, Mumbai

Barbier EB (2007) Valuing ecosystem services as productive inputs. Econ Policy 22(49):178–229

Barik SK, Muduli PR, Mohanty B, Behera AT, Mallick S, Abhijit D, Samal RN, Rastogi G, Pattnaik AK (2017) Spatio-temporal variability and the impact of Phailin on water quality of Chilika lagoon. Cont Shelf Res 136(February):39–56

Bauer A, Srinivasalu C, Roy AD, Murthy BHCK, Molur S, Pal S, Mohapatra P, Agarwal I, Sondhi S (2014) Barkudia insularis, Madras spotted skink. The IUCN Red List of Threatened Species

Behera P, Mohapatra M, Adhya TK, Suar M (2018) Structural and metabolic diversity of rhizosphere microbial communities of Phragmites karka in a tropical coastal lagoon. Appl Soil Ecol (December 2017):0–1

Bennett EM, Garry D, Gordon LJ (2009) Understanding relationships among multiple ecosystem services. Ecol Lett 12:1394–1404

Biswas KP (1995) Ecological and fisheries development in wetlands: a case study of Chilika lagoon. Daya Publishing House, Delhi

Boulton AJ, Ekebom J, Gíslason G m (2016) Integrating ecosystem services into conservation strategies for freshwater and marine habitats: a review. Aquat Conserv Mar Freshwat Ecosyst 26(5):963–985

Braat LC, de Groot R (2012) The ecosystem services agenda: bridging the worlds of natural science and economics, conservation and development, and public and private policy. Ecosyst Serv 1(1):4–15

Burkhard B, Petrosillo I, Costanza R (2010) Ecosystem services - bridging ecology, economy and social sciences. Ecol Complex 7(3):257–259. https://doi.org/10.1016/j.ecocom.2010.07.001

Carpenter SR, Mooney HA, John A, Capistrano D, Defries R, Diaz S, Dietz T, Duraiappah A, Oteng-Yeboah A, Pereira HM, Perrings C, Reid W, Sarukhan J, Scholes RJ, Whyte A (2009) Science for managing ecosystem services: beyond the millennium ecosystem assessment. Proc Natl Acad Sci 106(5):1305–1312

CDA (2005) Estimation of fish, prawn and crab landings in the Chilika lagoon. Bhubaneswar, Odisha

CDA (2017) Chilika Lake: 2016 ecosystem health report card. Bhubaneswar, Odisha

CDA (2019) Chilika Development Authority

Chowdhury P, Behera MR (2015) A study on Regional Sea level variation along the Indian coast. Procedia Engineering 116(1):1078–1084

CIFRI. 2007. Assessment of fish yield potential of Chilika Lake.. Kolkata

CIFRI (2017) Post restoration assessment of the ecology and fisheries diversity of Chilika Lake. Kolkata, West Bengal

Cox DR (1958) The regression analysis of binary sequences. J R Stat Soc 20(2):215–242

D'Lima C, Marsh H, Hamann M, Sinha A, Arthur R (2014) Positive interactions between Irrawaddy dolphins and artisanal fishers in the Chilika lagoon of Eastern India are driven by ecology, socioeconomics, and culture. Ambio 43(5):614–624

Daily GC, Polasky S, Goldstein J, Kareiva PM, Mooney HA, Pejchar L, Ricketts TH, Salzman J, Shallenberger R (2009) Ecosystem services in decision making: time to deliver. Front Ecol Environ 7(1):21–28

Das LK (2018) Social movements– judicial activism Nexus and neoliberal transformation in India: revisiting save Chilika movement. Sociol Bull 67(1):84–102

Dujovny E (2009) The deepest cut: political ecology in the dredging of a New Sea Mouth in Chilika Lake, Orissa, India. Conserv Soc 7(3):192

Duraiappah AK, Asah ST, Brondizio ES, Kosoy N, O'Farrell PJ, Prieur-Richard AH, Subramanian SM, Takeuchi K (2014) Managing the mismatches to provide ecosystem Services for Human well-being: a conceptual framework for understanding the new commons. Curr Opin Environ Sustain 7:94–100

Ernstson H (2013) The social production of ecosystem services: a framework for studying environmental justice and ecological complexity in urbanized landscapes. Landsc Urban Plan 109(1):7–17

Felipe-Lucia MR, Martín-López B, Lavorel S, Berraquero-Díaz L, Escalera-Reyes J, Comín FA (2015) Ecosystem services flows: why stakeholders' power relationships matter. PLoS One 10(7):1–21

Finlayson CM (2012) Forty years of wetland conservation and wise use. Aquat Conserv Mar Freshwat Ecosyst 22(2):139–143

Finlayson CM, Davidson N, Dave P, Randy Milton G, MacKay H (2011) The Ramsar convention and ecosystem-based approaches to the wise use and sustainable development of wetlands. J Int Wildl Law Policy 14(3–4):176–198

Finlayson C, Max N, Khursheed SK, Pattnaik AK, Trisal CL (2001) RAM no. In: 50 Ramsar advisory mission (RAM) removal of Chilika Lake Ramsar site, India, from the Montreux Record. Switzerland, Gland

Folke C, Pritchard L, Berkes F, Colding J, Svedin U (2007) The problem of fit between ecosystems and institutions: ten years later. Ecol Soc 12(1):30

Galappaththi EK, Nayak PK (2017) Two faces of shrimp aquaculture: commonising vs. decommonising effects of a wicked driver. Marit Stud 16(12):1–19

Ganguly D, Patra S, Muduli PR, Vishnu Vardhan K, Abhilash KR, Robin RS, Subramanian BR (2015) Influence of nutrient input on the trophic state of a tropical brackish water lagoon. J Earth Syst Sci 124(5):1005–1017

Ganguly D, Singh G, Purvaja R, Bhatta R, Paneer Selvam A, Banerjee K, Ramesh R (2018) Valuing the carbon sequestration regulation service by seagrass ecosystems of Palk Bay and Chilika, India. Ocean Coast Manag 159(May):26–33

Geevarghese GA, Akhil B, Magesh G, Krishnan P, Purvaja R, Ramesh R (2018) A comprehensive geospatial assessment of seagrass distribution in India. Ocean Coast Manag 159(January 2018):16–25

Geijzendorffer IR, Roche PK (2013) Can biodiversity monitoring schemes provide indicators for ecosystem services? Ecol Indic 33:148–157

Ghosh AK, Pattnaik AK, Ballatore TJ (2006) Chilika lagoon: restoring ecological balance and livelihoods through re-salinization. Lakes Reserv Res Manag 11(4):239–255

Gonzalez-ollauri A, Mickovski SB (2017) Providing ecosystem services in a challenging environment by dealing with bundles, trade-offs, and synergies. Ecosyst Serv 28:261–263 de Groot RS, Alkemade R, Braat L, Hein L, Willemen L (2010) Challenges in integrating the concept of ecosystem services and values in landscape planning, management and decision making. Ecol Complex 7(3):260–272

Guerry AD, Polasky S, Lubchenco J, Chaplin-Kramer R, Daily GC, Griffin R, Ruckelshaus M, Bateman IJ, Duraiappah A, Elmqvist T, Feldman MW, Folke C, Hoekstra J, Kareiva PM, Keeler BL, Li S, McKenzie E, Ouyang Z, Reyers B, Ricketts TH, Rockström J, Tallis H, Vira B (2015) Natural capital and ecosystem services informing decisions: from promise to practice. Proc Natl Acad Sci 112(24):7348–7355

Ingram JC, Redford KH, Watson JEM (2012) Applying ecosystem services approaches for biodiversity conservation: benefits and challenges. Sapiens 5(1):1–10

Jentoft S (2000) The community: a missing link of fisheries management. Mar Policy 24(1):53–60

JICA and CDA (2009) Technical cooperation project for conservation and wise use of natural resources of the Chilika lagoon through community participation in India

Jones S, Sujansingani KH (1954) Fish and fisheries of the Chilka Lake with statistics of fish catches for the year 1948–1950. Indian J Fish 1(1–2):256–344

Kremen C (2005) Managing ecosystem services: what do we need to know about their ecology? Ecol Lett 8(5):468–479

Kull CA, de Sartre XA, Castro-Larrañaga M (2015) The political ecology of ecosystem services. Geoforum 61:122–134

Kumar R, Pierre H, Randy Milton G, Sellamuttu SS, Buckton ST, Davidson NC, Pattnaik AK, Zavagli M, Baker C (2011) Assessing wetland ecosystem services and poverty interlinkages: a general framework and case study. Hydrol Sci J 56(8):1602–1621

Kumar R, McInnes R, Everard M, Gardner R, Kulindwa KAA, Infante Mata D (2017) Integrating multiple wetland values into decision-making. Ramsar Convention Secretariat, Gland

Kumar R, Pattnaik AK (2012) Chilika – an integrated management planning framework for conservation and wise use. Wetlands International South Asia and Chilika Development Authority, New Delhi

Leach M, Mearns R, Scoones I (1999) Environmental entitlements: dynamics and institutions in community-based natural resource management. World Dev 27(2):225–247

Levin S, Xepapadeas T, Crépin A-S, Norberg J, de Zeeuw A, Folke C, Hughes T, Arrow K, Barrett S, Daily G, Ehrlich P, Kautsky N, Mäler K-G, Polasky S, Troell M, Vincent JR, Walker B (2013) Social-ecological systems as complex adaptive systems: modeling and policy implications. Environ Dev Econ 18(02):111–132

Lima CD, Everingham Y, Diedrich A, Mustika PL, Marsh H, Lima CD, Everingham Y, Diedrich A, Mustika PL (2018) Using multiple indicators to evaluate the sustainability of dolphin-based wildlife tourism in rural India. J Sustain Tour 0(0):1–21

Mahapatro D, Panigrahy R, Panda S, Mishra RK (2016) Malcofaunal diversity of Chilika Lake, Odisha, India. Proc Zool Soc 71(3):272–280

Mäler K-G (1991) Production function approach in developing countries. In: Vincent JR, Crawford EW, Hoehn JP (eds) Valuing environmental benefits in developing countries. Michigan State University, East Lansing, pp 11–32

Mansinha M (1960) Radhanath, the poet of 'Chilika'. Indian Lit 3(1):20–22

MEA (2005) Ecosystems and human well-being: wetlands and water synthesis. Island Press, Washington, DC

Mishra A (2014) Temperature rise and trend of cyclones over the eastern coastal region of India. J Earth Sci Clim Change 5(9):1–5

Mitra GN, Mahapatra P (1957) Bulletin on the development of Chilika Lake: survey report on the fishing industry. Orissa Government Press, Cuttack

Mohanty B, Muduli PR, Behera AT, Barik SK, Nag SK, Samal RN, Pattnaik AK (2016) Assessment of petroleum hydrocarbon in a tropical brackish water lagoon: Chilika, India. Chem Ecol 32(7):653–658

Mohanty SK, Mishra SS, Khan M, Mohanty RK, Mohapatra A, Pattnaik AK (2015) Ichthyofaunal diversity of Chilika Lake, Odisha, India: an inventory, assessment of biodiversity status and comprehensive systematic checklist (1916–2014). Check List 11(6):1–19

Mohapatra A, Mohanty RK, Mohanty SK, Bhatta KS, Das NR (2007) Fisheries enhancement and biodiversity assessment of fish, prawn and mud crab in Chilika lagoon through hydrological intervention. Wetl Ecol Manag 15(3):229–251

Mondal A, Mujumdar PP (2015) Regional hydrological impacts of climate change: implications for water management in India. Proc Int Assoc Hydrol Sci 366(June 2014):34–43

Naeem S, Chapin FS III, Costanza R, Ehrlich PR, Golley FB, Hooper DU, Lawton JH, Neill RVO, Mooney HA, Sala OE, Symstad AJ, Tilman D (1999) Biodiversity and ecosystem functioning: maintaining natural life support processes, vol 4. Ecological Society of America, Washington, DC

Nanda DK, Rameshwar S, Tomar SK, Dash SK, Jayakumar S, Arora DK, Chaudhary R, Kumar D (2013) Indian Chilika curd – a potential dairy product for geographical indication registration. Indian J Tradit Knowl 12(4):707–713

Nassl M, Löffler J (2015) Ecosystem services in coupled social–ecological systems: closing the cycle of service provision and societal feedback. Ambio 44(8):737–749

Nayak PK (2014) The Chilika lagoon social-ecological system: an historical analysis. Ecol Soc 19(1):1

Nayak PK (2017) Fisher communities in transition: understanding change from a livelihood perspective in Chilika lagoon, India. Marit Stud 16(1):1–32

Nayak PK, Berkes F (2010) Whose marginalisation? Politics around environmental injustices in India's Chilika lagoon. Local Environ 15(6):553–567

Nayak PK, Berkes F (2011) Commonisation and decommonisation: understanding the processes of change in the Chilika lagoon, India. Conserv Soc 9(2):132

Nelson GC, Bennett E, Berhe AA, Cassman K, DeFries R, Dietz T, Dobermann A, Dobson A, Janetos A, Levy M, Marco D, Nakicenovic N, O'Neill B, Norgaard R, Petschel-Held G, Ojima D, Pingali P, Watson R, Zurek M (2006) Anthropogenic drivers of ecosystem change: an overview. Ecol Soc 11(2):29 [online]

Ormerod SJ (2014) Rebalancing the philosophy of river conservation. Aquat Conserv Mar Freshwat Ecosyst 24(2):147–152

Palm EC, Newman SH, Prosser DJ, Xiao X, Luo Z, Batbayar N, Balachandran S, Takekawa JY (2015) Mapping migratory flyways in Asia using dynamic Brownian bridge movement models. Mov Ecol 3:3

Panda AN, Mishra SR, Ray L, Das S, Rastogi G, Pattanaik AK, Adhya TK, Suar M, Raina V (2018) Taxonomic description and genome sequence of Halobacillus marinus sp. nov., a novel strain isolated from Chilika Lake, India. J Microbiol 56(4):223–230

Panda US, Mahanty MM, Ranga Rao V, Patra S, Mishra P (2015) Hydrodynamics and water quality in Chilika lagoon-A modelling approach. Procedia Engineering 116(1):639–646

Parrott L, Quinn N (2016) A complex systems approach for multiobjective water quality regulation on managed wetland landscapes. Ecosphere 7(6):1–17

Pascual U, Balvanera P, Díaz S, Pataki G, Roth E, Stenseke M, Watson RT, Dessane EB, Islar M, Kelemen E, Maris V, Quaas M, Subramanian SM, Wittmer H, Adlan A, Ahn SE, Al-Hafedh YS, Amankwah E, Asah ST, Berry P, Bilgin A, Breslow SJ, Bullock C, Cáceres D, Daly-Hassen H, Figueroa E, Golden CD, Gómez-Baggethun E, González-Jiménez D, Houdet J, Keune H, Kumar R, Ma K, May PH, Mead A, O'Farrell P, Pandit R, Pengue W, Pichis-Madruga R, Popa F, Preston S, Pacheco-Balanza D, Saarikoski H, Strassburg BB, van den Belt M, Verma M, Wickson F, Yagi N (2017) Valuing nature's contributions to people: the IPBES approach. Curr Opin Environ Sustain 26:7–16

Pearce D (1998) Cost – benefit analysis and environmental policy. Oxf Rev Econ Policy 14(4):84–100

Potschin-young M, Haines-young R, Görg C, Heink U, Jax K, Schleyer C (2018) Understanding the role of conceptual frameworks: reading the ecosystem service Cascade. Ecosyst Serv 29:428–440

Primmer E, Jokinen P, Blicharska M, Barton DN, Bugter R, Potschin M (2015) Governance of ecosystem services: a framework for empirical analysis. Ecosyst Serv 16:158–166

Pritchard D (2016) Wise Use Concept of the Ramsar Convention. In: Finlayson CM, Everard M, Irvine K, Mcinnes RJ, Middleton BA, van Dam AA, Davidson NC (eds) The wetland book I: structure and function, management and methods. Springer Netherlands, Dordrecht, pp 1–3

Raje D, Mujumdar PP (2010) Reservoir performance under uncertainty in hydrologic impacts of climate change. Adv Water Resour 33(3):312–326

Rao PG (1995) Effect of climate change on streamflows in the Mahanadi River Basin, India. Water Int 20(4):205–212

Rath J, Adhikary SP (2008) Biodiversity assessment of algae in Chilika Lake, East Coast of India. In: Mohanty PK (ed) Monitoring and modelling lakes and coastal environments. Springer, Dordrecht, pp 22–33

Raudsepp-Hearne C, Peterson GD, Bennett EM (2010) Ecosystem service bundles for analyzing tradeoffs in diverse landscapes. Proc Natl Acad Sci 107(11):5242–5247

Raudsepp-hearne C, Peterson GD (2016) Scale and ecosystem services: how do observation, management, and analysis shift with scale — lessons from Québec. Ecol Soc 21(3):16

Ray L, Suar M, Pattnaik AK, Raina V (2013) Streptomyces chilikensis sp. nov., a halophilic streptomycete isolated from brackish water sediment. Int J Syst Evol Microbiol 63:2757–2764

Reeves RR, Jefferson TA, Karczmarski L, Laidre K, O'Corry-Crowe G, Rojas-Bracho L, Secchi ER, Slooten E, Smith BD, Wang JY, Zhou K (2008) Orcaella brevirostris. The IUCN Red List of Threatened Species e.T15419A4

Reyers B, Biggs R, Cumming GS, Elmqvist T, Hejnowicz AP, Polasky S (2013) Getting the measure of ecosystem services: a social-ecological approach. Front Ecol Environ 11(5):268–273

Ribot JC, Peluso NL (2009) A theory of access. Rural Sociol 68(2):153–181

Ricke K, Drouet L, Caldeira K, Tavoni M (2018) Country-level social cost of carbon. Nat Clim Chang 8(10):895–900

Robards MD, Schoon ML, Meek CL, Engle NL (2011) The importance of social drivers in the resilient provision of ecosystem services. Glob Environ Chang 21(2):522–529

Russi D, Ten Brink P, Farmer A, Badura T (2013) The economics of ecosystems and biodiversity for water and wetlands. Chilika Development Authority, London/Brussels

Samal KC (2002) Shrimp culture in Chilika Lake: case of occupational displacement of fishermen. Econ Polit Wkly 37(18):1714–1718

Satyanarayana CH (1999) Hydrographic and phytoplankton characteristics of Chilka Lake: A brackishwater lagoon, on the east coast of India, Ph.D. thesis. Andhra University, Visakhapatnam, India

Schneider P, Hook SJ (2010) Space observations of inland water bodies show rapid surface warming since 1985. Geophys Res Lett 37(22):1–5

Schoon M, van der Leeuw S (2015) The shift toward social-ecological systems perspectives: insights into the human-nature relationship. Nat Sci Soc 23(2):166–174

Sekhar NU (2004) Fisheries in Chilika Lake: how community access and control impacts their management. J Environ Manag 73(3):257–266

Sekhar NU (2007) Social capital and fisheries management: the case of Chilika Lake in India. Environ Manag 39(4):497–505

Sen A (1984) Rights and capabilities. In: Sen A (ed) Resources, values and development. Basil Blackwell, Oxford, pp 307–324

Singh R, Mishra S, Rajesh C, Dash S, Niranjan S, Kataria R (2017) Chilika- a distinct registered buffalo breed of India. Int J Livest Res 7(9):259–266

Spangenberg JH, von Haaren C, Settele J (2014) The ecosystem service cascade: further developing the metaphor. Integrating societal processes to accommodate social processes and planning, and the case of bioenergy. Ecol Econ 104:22–32

Srichandan S, Kim JY, Bhadury P, Barik SK, Muduli PR, Samal RN, Pattnaik AK, Rastogi G (2015) Spatiotemporal distribution and composition of phytoplankton assemblages in a coastal tropical lagoon: Chilika, India. Environ Monit Assess 187(2):47

Srivastava TN (2002) Local 'self' government and the constitution. Econ Polit Wkly 37(30):3190–3198

Steffen W (2009) Interdisciplinary research for managing ecosystem services. Proc Natl Acad Sci 106(5):1301–1302

Tripati S, Patnaik AP (2008) Stone anchors along the coast of Chilika Lake: new light on the maritime activities of Orissa, India. Curr Sci 94(3):386–390

Tripati S, Vora KH (2005) Maritime heritage in and around Chilika Lake, Orissa: geological evidences for its decline. Curr Sci 88(7):1175–1181

UNESCO (2014) Chilika Lake. Tentative List. Retrieved April 16, 2019 https://whc.unesco.org/en/tentativelists/5896/

Unnikrishnan AS, Nidheesh AG, Lengaigne M (2015) Sea-level-rise trends off the Indian coasts during the last two decades. Curr Sci 108(5):966–970

Unnikrishnan AS, Ramesh Kumar MR, Sindhu B (2011) Tropical cyclones in the bay of Bengal and extreme sea-level projections along the East Coast of India in a future climate scenario. Curr Sci 101(3):327–331

Vivek G, Santonu G, Samal RN, Choudhury SB (2019) Monitoring of Chilika Lake mouth dynamics and quantifying rate of shoreline change using 30 m multi-temporal Landsat data. Data Brief 22(December):595–600

Zsuffa I, Van Dam AA, Kaggwa RC, Namaalwa S, Mahieu M, Cools J, Johnston R (2014) Towards decision support-based integrated management planning of papyrus wetlands: a case study from Uganda. Wetl Ecol Manag 22(2):199–213

Chapter 5
Sedimentologic, Chemical, and Isotopic Constraints on the Anthropogenic Influence on Chilika Lake, India

Torsten Vennemann, Laurent Decrouy, Géraldine Bourgeois, Kelly Delavy, Michèle Ecuyer, Pauline Lange, Samir Mishra, Gurdeep Rastogi, Ajit K. Pattnaik, and Mrutyunjay Suar

Abstract Given the population increase in the catchment to Chilika Lake and the related changes in land use policies, agricultural practices, and water resource management, this lake has been subjected to increasing anthropogenic influence. As a consequence, the unique biodiversity and primary production within the lagoon decreased, while eutrophication and siltation increased. As a counter-initiative it was decided to artificially open the lake to the sea by dredging. To help trace and quantify the anthropologic effects on Chilika Lake, a combined sedimentologic, chemical, and isotopic study of the lagoon and its sediments is in progress. The results from two campaigns during the monsoon and consecutive dry season suggest that the large gradients in salinity, sediment and nutrient inputs, as well as primary productivity within the lagoon are controlled by variable fluxes of water, sediment, and nutrients from the three separate catchments to the lagoon. Trends in changes of salinity, H- and O-isotope compositions of waters, but also of concentrations and C- and/or N-isotope compositions of the dissolved inorganic carbon (DIC), particulate organic matter (POM), and aquatic plants indicate that mixing in the lagoon occurs

T. Vennemann (✉) · L. Decrouy · G. Bourgeois · K. Delavy · M. Ecuyer · P. Lange
Institute of Earth Surface Dynamics, University of Lausanne, Lausanne, Switzerland
e-mail: Torsten.Vennemann@unil.ch

S. Mishra
Chilika Development Authority, Government of Odisha, Bhubaneswar, India

School of Biotechnology, KIIT University, Bhubaneswar, Odisha, India

G. Rastogi
Wetland Research and Training Centre, Chilika Development Authority, Balugaon, Odisha, India

A. K. Pattnaik
Chilika Development Authority, Bhubaneswar, Odisha, India

M. Suar
School of Biotechnology, KIIT University, Bhubaneswar, Odisha, India

C. M. Finlayson et al. (eds.), *Ecology, Conservation, and Restoration of Chilika Lagoon, India*, Wetlands: Ecology, Conservation and Management 6,
https://doi.org/10.1007/978-3-030-33424-6_5

between new freshwater inputs and evaporated water within the basin itself. Except for the outer channel, mixing with seawater is limited. In contrast, the C-isotope compositions of the organic matter in the sediments support a higher overall proportion of "marine" or estuarine POM during the past. The latter may be important during the dry season, coupling salinity increase to the changes in DIC and POM carbon isotope compositions. The salinity, DIC, H-, O-, and C-isotope compositions of water are compatible with evaporation as the main driver for a salinity increase, rather than admixtures with seawater.

Keywords Geochemistry · Stable isotopes · Sediments · Organic matter · Ecology · Palaecology

5.1 Introduction

Chilika Lake, the largest lagoon on the Asian continent and second-largest lagoon on Earth, is located on the east coast of India, just south of the Bay of Bengal. Between the wet (monsoon) and the dry (summer) season Chilika Lake has a surface area of 1160 km^2 or 900 km^2, respectively, with an average depth of only about 1.2 m (between 40 cm to 1.4 m maximum) (Jayaraman et al. 2007; Ghosh and Pattnaik 2005). It is separated from the Bay of Bengal by a sandbar of about 100 m to 1.5 km width, 30 km length, with a channel behind this sandbar that connects the lagoon naturally to the sea. This sandbar developed only during the Late Holocene (3000–4000 years B.P. and possibly as late as 350 years B.P.) The result was a change of Chilika Lake having been a bay open to the sea with abundant mangrove forests to a lagoon that became more isolated with time. While the lagoon thus became more influenced by abundant freshwater input during the monsoon season, it was finally closed off completely from the sea from 1992 to the beginning of 2000 (Khandelwal et al. 2008; Zachmann et al. 2009). Given the freshwater influence from three different drainage basins (northern, central, and southern sectors; Fig. 5.1) and the fact that the natural tidal mouth gets annually displaced towards the sea and hence is adversely affecting the tidal exchange with the sea, the lagoon is subjected to large temporal and spatial salinity gradients (Jayaraman et al. 2007). As such, the lagoon has a characteristic estuarine ecosystem that may still be influenced by seasonal variations of the seawater influx. It has a large and unique biodiversity and genetic ecological biodiversity, representing the largest wintering ground for migratory birds of the Asian subcontinent. It also has an important traditional fishing industry (with about 8000 to 12,000 tons of fish and prawns caught annually in the area) and is of touristic, cultural, religious and spiritual importance. However, by analogy with other lakes and estuarine systems, the anthropogenic influence on Chilika Lake has substantially changed over the past century (Ghosh and Pattnaik 2005): (A) the overall hydrology of the lake changed with the main supply of freshwater from the Mahanadi river basin (Northern Sector) being diverted for hydroelectric power production and because the river is used as a water resource

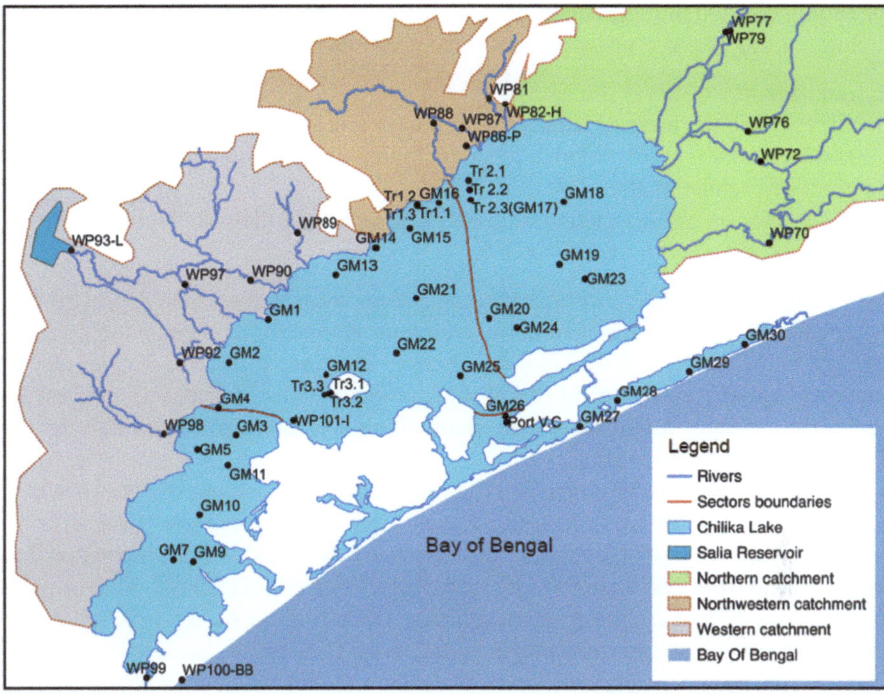

Fig. 5.1 Map of the locations sampled in Chilika Lake and its catchment. 56 sampling points are shown and have been georeferenced. The boundaries for the northern, central and southern sectors of the lagoon as well as the outer channel are also shown

for the cities of Bhubaneswar and Cuttack; (B) the partial compensation of this change in hydrology through the natural closure of the lagoon to the sea, with the adverse effect of a decrease in salinity during the monsoon period over the past few decades and increased silt sedimentation (Pal and Mohanty 2002), increased nutrient supply from the drainage basins, including untreated waste waters, and hence also increased growth of invasive freshwater macrophytes and local eutrophication; (C) changes in land-use policies and agricultural activity with increased rice production and use of pesticides in the main drainage basin (Northern Sector); (D) decreasing biodiversity within the lagoon including fish production with a decrease to about 2000 tons annually during the 1990s with the natural closure to the sea, but a dramatic increase in catch from the year 2000 onwards due to the dredged, man-made opening of the lagoon directly to the sea; (E) increased population pressure and resulting contamination with plastics, oil, waste waters, among others (Pal and Mohanty 2002; Kannan et al. 2005; Baliarsingh et al. 2014).

It is hence clear that the anthropogenic influence on Chilika Lake has substantially changed over the past century (Ghosh and Pattnaik 2005) and this opens questions on the changes in the ecology associated with this evolution. In view of the setting of Chilika Lake and in order to better evaluate the human impacts on the ecology of this shallow water lagoon, three priorities of research were identified:

A. To trace and quantify the hydrological cycle of Chilika Lake.
B. To evaluate the nutrient supply and extent of eutrophication along the shores of the lake, that is to distinguish the autochthonous from allochthonous contributions of organic matter.
C. To reflect on the sedimentologic and ecologic evolution of Chilika Lake by analysis of the sedimentary records.

A comparison of the geochemical compositions measured within the present-day water column to those measured in the sediments allows for a reflection on possible changes in the ecology of the lagoon with time. The three research priorities can be addressed via the geochemical and isotopic compositions of water as well as the particulate organic matter within the water column of the lagoon and the organic matter within the sediments.

The variations in the hydrogen and oxygen isotope compositions of water in a lake or a lagoon can be used as conservative tracers of the mixing processes of the different waters entering the lake or lagoon, as well as tracers for the effects of evaporation that may influence the lake itself (Clark and Fritz 1997; Hoefs 2009). The isotopic compositions of water in each of the drainage basins are, in turn, a function of the mean ambient air temperature of precipitation as well as the so-called "rain-out" effects on the moisture carried by the air mass that supplies the precipitation, including its source of moisture and the ultimate transport distance of the moisture in the air mass (distance from the sea or the point of origin of the evaporated moisture) and/or the atmospheric circulation patterns (Clark and Fritz 1997; Teranes and McKenzie 2001), that is the "continentality" of the precipitation, the altitude of condensation, as well as other factors such as relative local humidity, all of which may lead to condensation and rain formation and thus separation of liquid water from the residual water vapor in the air mass. Ultimately, it is the isotopic fractionation between liquid and vapor water molecules, which changes as a function of temperature, that is responsible for the changes in isotopic composition of rainwater and hence surface water. As a result, the isotopic composition of rainwater is largely positively correlated with mean air temperature and negatively correlated with the amount of rainout experienced by the moisture carried in the air mass (Clark and Fritz 1997).

The amount and the carbon isotope composition of dissolved inorganic carbon (DIC) are controlled by the source of the DIC (lithogenic – by dissolution of carbonate-bearing rocks – or biologic – by respiration of organic matter – or atmospheric uptake in surface and groundwater), the primary production within the water column that uses the DIC as a nutrient, and by equilibrium exchange with atmospheric CO_2 as a function of the partial pressures of CO_2 in the water column (Clark and Fritz 1997; Hoefs 2009).

The chemical and isotopic compositions of the organic matter within the water column and of organic matter within the sediment are commonly used for (paleo) ecological studies (Jeffrey et al. 1983; Altabet and McCarthy 1985; Kennicutt et al. 1987; Cifuentes et al. 1988; Bernasconi et al. 1997; Hodell and Schelske 1998; Schelske and Hodell 1995; Brenner et al. 1999; Meyers 2003; Lehmann et al. 2004; Michener and Lajtha 2007). The organic matter is a complex mixture of particulate organic remains and living organic tissue. The complexity derives from the multitude of sources and source organisms, the different biosynthetic pathways of the organisms, as well as transformations that occur during the decomposition of the dead organic matter. In aqueous systems in general, the carbon isotope composition of the particulate organic matter (POM) reflects largely that of the living plankton within the euphotic zone. However, depending on the depth of the water column (or residence time in the water column), the redox conditions, pH, and temperature, the POM may undergo changes in its chemical and isotopic composition due to biological reworking (Hodell and Schelske 1998; Bernasconi et al. 1997; Lehmann et al. 2004). This may also apply to the nitrogen isotope composition as well as the nitrogen concentrations suggesting that these effects may be related to the loss of, for example, the more labile amino acids that are relatively rich in ^{13}C and ^{15}N compared to the more refractory but isotopically light lipid fraction (Altabet and McCarthy 1985; Cifuentes et al. 1988; Hodell and Schelske 1998; Bernasconi et al. 1997). However, given that the magnitude of these effects is a direct function of the above-mentioned environmental conditions, a comparison of the chemical and isotopic composition of POM in the water column to that within the sediments allows for an evaluation of the trophic state of the aqueous system. As a corollary, any changes in the chemical and isotopic composition of the organic matter in the stratified sediments deposited over the history of the aqueous system allow for an interpretation on paleoecological changes.

In addition, despite these possible changes in the POM during its passage through the water column and early diagenesis within the sediment, in many aqueous systems a difference in the C/N ratios as well as the isotopic compositions of the POM produced in the euphotic zone of a lake and that of the POM introduced by terrestrial erosion is retained (Meyers 2003). This difference is related to the different carbon and nitrogen cycles in terrestrial and aqueous systems such that, for example, the marine primary production results in autochthonous organic matter with an isotopic composition that is about 7 permil higher in its $\delta^{13}C$ value compared to the terrestrially fixed carbon. Similar effects are also noted for the nitrogen system with generally higher $\delta^{15}N$ values for primary producers in aqueous compared to terrestrial systems (Michener and Lajtha 2007). Hence, these differences are often used to distinguish marine (or also freshwater) versus terrestrial organic matter sources in sediments, but also help to distinguish the autochthonous from the allochthonous contributions to the total organic matter in the water column and sediments, and hence allows for an interpretation of the local bio-productivity (Sackett and Thompson 1963; Hilton et al. 2008).

5.2 Methods

Samples for this study were taken from the same locations (within the precision of the GPS-coordinates for each location) during two consecutive seasons:

1. The monsoon season from the 7th to the 15th of September 2013, where a total of 188 samples from 56 locations have been collected; 94 of water, 32 of sediments, 53 of plant and 21 of fish (9 sampled during this season, another 12 sampled during the following dry season). 56 locations were sampled for water within the lake, as well as 15 locations for rivers.
2. The dry season during the first week of February 2014. About 46 locations were sampled for water only as some locations previously sampled fell dry during this season. Sediment sampling was also reduced to 32 stations only.

For the water samples all standard parameters (conductivity, temperature, pH, oxygen content) were measured directly in the field, while the chemical and isotopic composition of the water samples (major anions and cations, DIC, particulate organic matter in suspension (filtered to retain the fraction larger than 0.7 μm) was analyzed in the laboratories of the University of Lausanne using standard analytical procedures. In addition, the mineralogy, grain size, qualitative analyses of the ostracods and foraminifera, as well as the C- and N-concentrations and stable isotopic compositions of the organic (and inorganic) fractions within the sediments and sediment cores have been analyzed. In this chapter, the relevant isotopic compositions are discussed in preference, but all measurements are discussed in more detail as part of Masters project studies that focused on different aspects of the lagoon (Delavy and Ecuyer 2014; Lange 2014; Bourgeois 2015; Hostettler 2015).

5.3 Results

5.3.1 Chemical and Isotopic Composition of the Water Column

The hydrological cycle of Chilika Lake can be assessed through stable isotope measurements of water in conjunction with already existing routine measurements (coordinated by the CDA) of the major anions and cations, oxygen concentrations, salinity, and temperature. The latter are routinely analyzed for the 30 monitoring stations covering the different sectors of the lagoon (Northern, Central, Southern Sectors and the Outer Channel), but also for a number of surface and ground waters that enter the lagoon from three different catchments.

The different sectors of Chilika Lake have different average salinities and average stable H- and O-isotope compositions of their waters, both during the monsoon season (Fig. 5.2a) and during the consecutive dry season (Fig. 5.2b). During the monsoon season, this difference is also apparent for the catchments adjacent to the

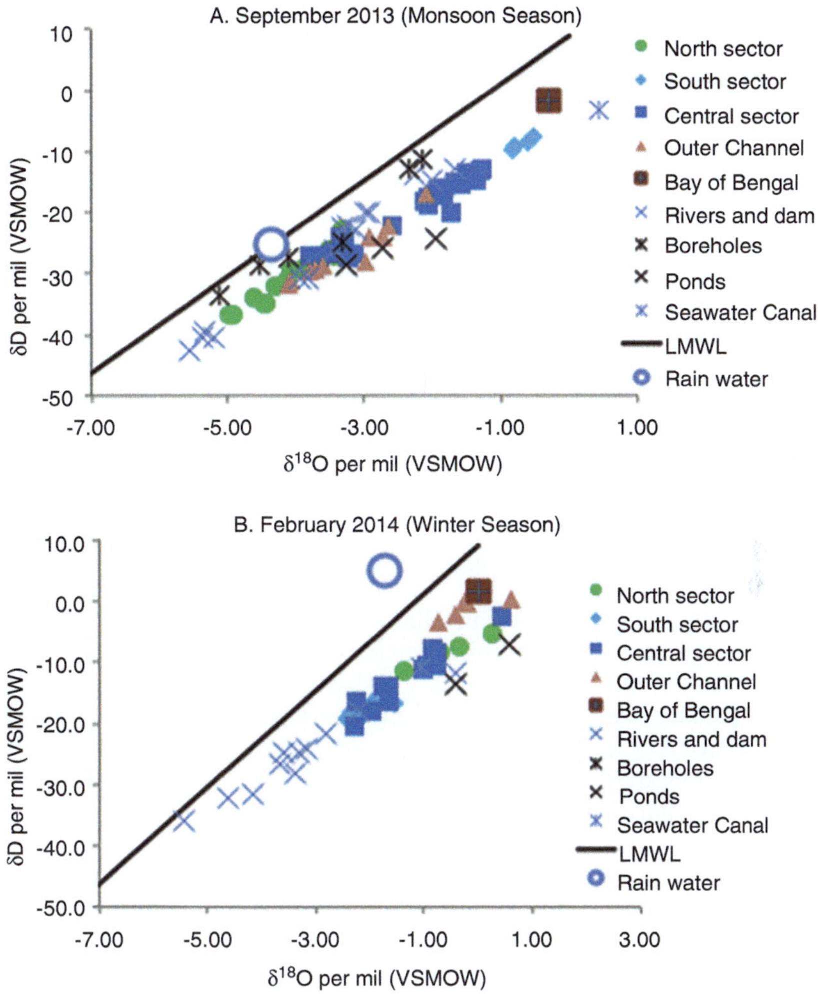

Fig. 5.2 Stable hydrogen and oxygen isotope compositions of lagoon waters and waters from the catchment of Chilika Lake as well as the local meteoric water line; (**a**) samples taken in September 2013, during the monsoon season and (**b**) samples for the winter season sampled during the first week of February

lagoon. The surface water (rivers and ponds) from the Mahanadi catchment that extends well into the interior of India towards the Himalayas, has lower values ($\delta^{18}O_{avg} = -4.3 \pm 1.2$‰, n = 8) compared to the local rainfall and hence surface waters draining into the lagoon in the western ($\delta^{18}O_{avg} = -2.9 \pm 0.8$‰, n = 8) and southern catchments (-1.2 ± 2.2‰, n = 2). Interestingly, this pattern changes somewhat during the dry season as the Northern Sector now has higher average values compared to the Central Sector, while the Southern Sector changes less and retains the highest average values. The order of relative enrichment in the heavy isotopes in the sectors of the lagoon is also paralleled by changes in the order of enrichment in the catchment surface waters ($\delta^{18}O_{avg} = -2.8 \pm 1.3‰$, n = 6; $-4.0 \pm 1.1‰$, n = 5; +0.6‰, n = 1, respectively for northern, central and southern catchments). It has been estimated that about 50% of the freshwater to the lagoon is derived from the northern catchment of the Mahanadi river, about 39% from the western, local catchment and only about 10% from direct precipitation runoff, all largely during the monsoon season, the same season that will also deliver the maximum sediment loads to the lagoon (Zachmann et al. 2009).

It is also clear from Fig. 5.2 that the waters do not plot along the local meteoric water line (LMWL; after Kumar et al. 2010) but rather along lines with a lower slope compared to the LMWL, suggesting that the waters were subjected to evaporation. The evaporation effect is similar for all of the waters in the lagoon as well as those in the catchments, and the regression line through all of these points does not pass through the measurements made for seawater sampled off the Bay of Bengal. An important implication of this is that the water within the lagoon does not represent a mixture of seawater and freshwater from the variable catchments. Instead, all waters have been driven towards higher values in δD and $\delta^{18}O$ through evaporative processes and that they are entirely of freshwater origin (Clark and Fritz 1997; Hoefs 2009).

This is also given by differences in salinity that parallel those of the stable isotope composition of water, as well as by differences in carbon isotope composition of DIC (Fig. 5.3). While the concentration and isotopic composition of DIC also change in parallel, the DIC also differs in its composition between the relative drainage basins: $\delta^{13}C_{DIC}$ values of –11.1‰, s.d. = 1.31, n = 6, for the northern, agriculturally intensively cultivated terrain, and –12.9‰, s.d. = 0.69, n = 7, for the western catchment rivers draining a largely forested terrain, and –4.8‰ for the Palur canal). Hence, the average C-isotope compositions of the DIC in the sectors are also different. In addition, as a nutrient in the carbon cycle, the DIC is consumed by photosynthetic activity of aquatic plants, further changing the isotopic composition of DIC towards higher ^{13}C content as the isotopically light fraction is preferentially taken up by the biologic material. This has further implications for the carbon cycle in each sector (see below).

The relationship between salinity and oxygen isotope composition and that between salinity and C-isotope composition of DIC, as well as the variations in hydrogen and oxygen isotope compositions of water all indicate trends that are not simple mixing lines with seawater (see the point for the Bay of Bengal), but rather evaporation and CO_2 degassing lines. For the C-isotope compositions the trends may also suggest increased primary productivity and biomineralization of the carbon within the lagoon, away from the freshwater sources that introduce the carbon and other nutrients. In the case of the Outer Channel, mixing with seawater is indicated by the trends in salinity, oxygen and hydrogen isotope composition and DIC (Fig.'s 5.2 and 5.3).

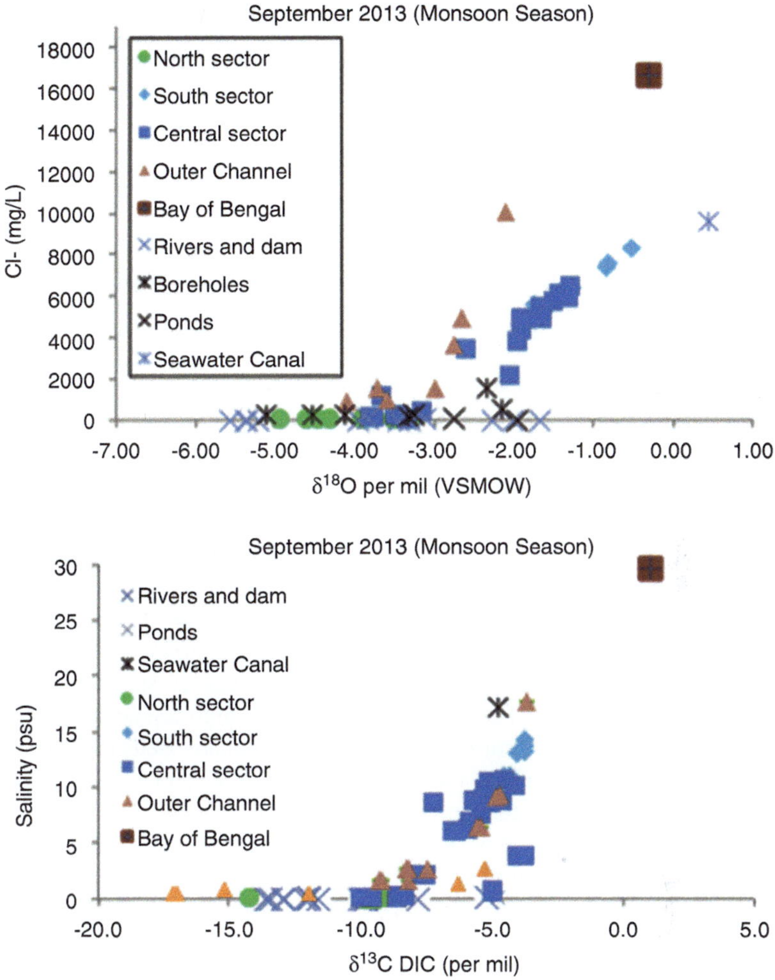

Fig. 5.3 Salinity relative to the oxygen isotope compositions of water (**a**), and relative to the carbon isotope compositions of DIC (**b**)

5.3.2 Chemical and Isotopic Composition of Particulate Organic Matter in the Water Column and Surface Sediments

To examine the sedimentological and ecological evolution of Chilika Lake the mineralogical, geochemical, and isotopic composition of the sediments as well as the geochemical and isotopic composition of the organic matter within sediments and also for reference that of the particulate organic matter within the water column of the lake were investigated in detail during two field seasons (Delavy and Ecuyer

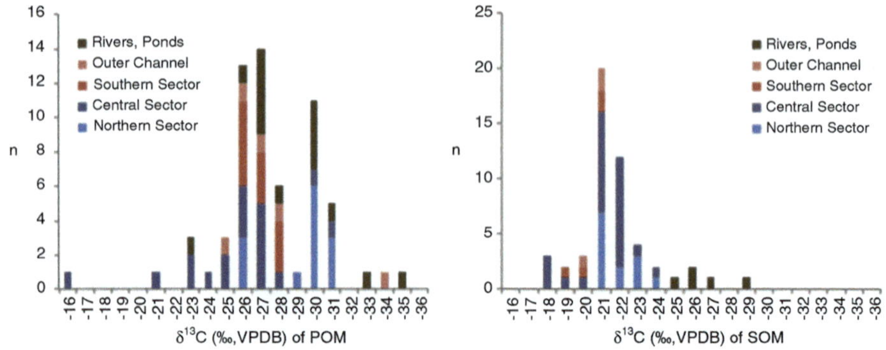

Fig. 5.4 (**a**) Carbon isotope composition of particulate organic matter and (**b**) carbon isotope composition of sediment organic matter (SOM) sampled during the campaigns of September 2013 and February 2014 of Chilika Lake

2014; Bourgeois 2015; Hostettler 2015). The mineralogy and grain size distribution of the sediments is directly related to the erosional evolution of the catchment and the dispersal of the sediment within the lake. As outlined above, the abundance, chemical and stable isotope composition of organic matter can also help to interpret the paleo-environmental and ecological conditions. In addition, calcareous fossils such as ostracods and/or foraminifera can be exploited as sensitive ecological indicators (Hostettler 2015). The ostracod species and their abundance, as well as the geochemical and isotopic composition of their carbonate shells (Sr, Mg, and Ca concentrations as well as C- and O-isotope compositions of calcite shells), have been shown to be indicators of oxygenation, salinity, and temperature of the water column in marine and brackish systems and have been studied for Chilika Lake too (Hostettler 2015), but because of limited space are not discussed further here.

The C-isotope compositions of the POM measured for the two seasons so far have a range that is typical for estuarine systems, notably because of the large range in compositions (Fig. 5.4) (e.g., Sackett and Thompson 1963; Altabet and McCarthy 1985; Cifuentes et al. 1988; Michener and Lajtha 2007). This is to be expected as terrestrially derived particulate carbon either as detritus or as living organic tissue formed within a freshwater-dominated system tends towards $\delta^{13}C$ values of about −25‰ for a typical C_3 type of vegetation (Sackett and Thompson 1963; Kennicutt et al. 1987; Finlay and Kendall 2007; Michener and Lajtha 2007). In contrast, autotrophic organic matter in freshwater and marine systems may have a wide range of values, from −16 down to −35‰ not being uncommon (Finlay and Kendall 2007). However, in marine systems values normally cluster closer to −23‰ for organic matter formed in surface waters, while lower values are more characteristic of deeper marine waters and or methanotrophic systems (Michener and Kaufman 2007). The principle reason for the large range in C-isotope composition of autotrophic organic matter in both fresh and marine systems is the range in nutrient type and its range in isotopic composition (CO_2 or DIC or even CH_4 as principle source of carbon and its respective origin – soil horizons, respiration, atmospheric, lithogenic – being the major control for the compositions).

The $\delta^{13}C$ values of the SOM (sediment organic matter) are, however, higher than those of the POM sampled during the monsoon and dry seasons in the water column (Fig. 5.4b). The reason for this difference could either be a dominant terrestrial organic carbon source for the sediment organic matter, or a higher proportion of "normal" marine-derived input of carbon that averages about −23‰. A more important marine influence on the organic matter would require that the bulk of the sediment-derived organic matter is relatively old, hence was introduced during periods where the potential marine influence could have been higher as the lagoon was still naturally open to the sea (Khandelwal et al. 2008). As was argued above, todays hydrologic system does not indicate an important entry of seawater to the lagoon, hence also excluding marine-derived nutrients to enter the system (with the exception of the outer channel). Alternatively, proportionally higher sedimentation of organic matter during the summer season (not sampled yet) and dry season could also be indicated by the higher $\delta^{13}C$ values of SOM. As indicated in Tables 5.1 and 5.2 and Fig. 5.5a, b, average $\delta^{13}C$ values of DIC, as an important nutrient during the dry season where limited terrestrially derived organic matter can enter the lagoon via the riverine input, are higher than those during the monsoon season. As such higher $\delta^{13}C$ values for POM using the DIC as major nutrient source during the dry season would be expected (Tables 5.1 and 5.2). Increased sedimentation of such organic matter in conjunction with a decreased freshwater input can also explain the trends in

Table 5.1 Average carbon isotope compositions of DIC and OM of Chilika Lake in September 2013

	Avg. $\delta^{13}C$ DIC s.d. n	Avg. $\delta^{13}C$ POM s.d. n	Avg. $\delta^{13}C$ SOM s.d. n	Avg. $\delta^{13}C$ Plants s.d. n	Δ DIC-POM
Northern Sector	−9.8	−30.3	−22.0	−27.4	20.5
	1.2	0.8	1.3	3.3	
	n = 15	n = 8	n = 8	n = 13	
Central Sector	−6.4	−25.4	−20.4	−23.5	19.0
	1.9	3.7	1.4	5.2	
	n = 30	n = 17	n = 16	n = 34	
Southern Sector	−4.4	−26.3	−20.4	−18.2	21.9
	0.4	0.5	0.7	3.0	
	n = 12	n = 4	n = 4	n = 4	
Outer Channel	−6.8	−28.3	−20.8		21.5
	2.0	3.7	0.3		
	n = 11	n = 5	n = 3		

Table 5.2 Average carbon isotope compositions of DIC and OM of Chilika Lake in February 2014

	Avg. $\delta^{13}C$ DIC s.d. n	Avg. $\delta^{13}C$ POM s.d. n	Avg. $\delta^{13}C$ SOM s.d. n	DIC-POM
Northern Sector	−7.6	−26.8	−21.8	19.1
	4.6	2.5	1.0	
	n = 6	n = 6	n = 6	
Central Sector	−8.9	−24.5	−22.2	15.6
	4.6	3.6	0.8	
	n = 12	n = 11	n = 11	
Southern Sector	−5.5	−27.0	−21.8	21.5
	0.3	0.7	0.8	
	n = 8	n = 8	n = 8	
Outer Channel	−1.9	−22.8	−22.6	20.9
	1.5	1.3	1.7	
	n = 5	n = 5	n = 3	

the salinity-$\delta^{13}C_{DIC}$ value relationship as well as that between $\delta^{18}O_{H2O}$ and salinity. Evaporation as the main driver for a salinity increase is thus accompanied by an increase in $\delta^{13}C_{DIC}$ due to continued autotrophic primary productivity and degassing of CO_2 during the warm summer season because of a decrease in the solubility of CO_2 with increasing water temperature.

The relationship between isotopic composition of DIC and that of the POM is illustrated for the two different seasons in Fig. 5.5a, b. For both seasons the isotopic compositions of the inorganic and organic pools of carbon more or less track each other, with an average difference between these carbon pools of 19–22‰, Δ(DIC-POM). However, there is a considerable spread of values across this line of constant offset. The reason for this spread in values across the line of 20‰ for Δ(DIC-POM) could be a matter of the actual time of measurement of the DIC relative to the period of biosynthesis of the POM. The DIC was sampled over the course of several days only for each season. In contrast, while the POM was sampled at the same time, the material likely represents several days or even weeks of bio-productivity in the water column in addition to an admixture of allochthonous POM from distinct terrestrial sources. Figure 5.5 also indicates that, except for parts of the Northern Sector for which the terrestrial inputs are also likely to be the highest, the 20‰ difference in average $\delta^{13}C$ values for the DIC and POM is more tightly correlated during the dry season, even though the large range in values is still preserved.

For the plants within the lagoon, a number of seagrasses were analysed (*Eichornia crassipes, Halophila ovalis, Hydrilla verticillate, Potamogeton Pectinatus, Salvinia molesta, Scirpus sp, Vallisneria spiralis*, (Delavy and Ecuyer 2014)), including transects in the coastal regions with abundant invasive macrophytes (*Phragmites karka*). The $\delta^{13}C$ values of plants have a range between −12 to −34‰; Fig. 5.6). A variation of several permil has also been measured for individual

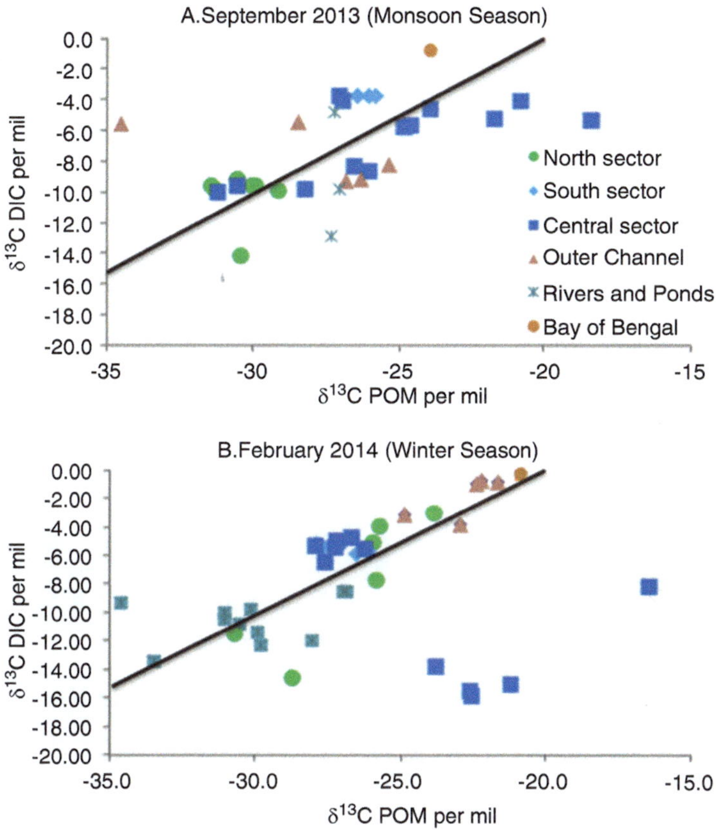

Fig. 5.5 Comparison of the isotopic composition of DIC to that of the POM sampled during the monsoon season (**a**) in September 2013 and the dry season (**b**) in February 2014 for Chilika Lake. The lines indicate a fractionation between DIC and POM of about 20‰

parts of the same plant (Delavy and Ecuyer 2014). This large range in values suggests large seasonal changes in nutrient sources, possibly also including localized sources of methanogenic carbon and complete reduction of nitrate to ammonia (see below) related to anoxic conditions measured in some shallow water coastal parts.

The N-isotope composition of organic-bound nitrogen in the sediment is similar to that of the plants within the lagoon, with a range in values of between +0.4 to +5.4‰ (+6.6‰; plants) for their $\delta^{15}N$ values (relative to Air; Figs. 5.7 and 5.8 (Delavy and Ecuyer 2014)). There is little or no difference between the different sectors though, which is unlike the variation for carbon. These values, together with the relatively low C/N ratios are typical for autochthonous, aquatic primary production of the organic matter (Cifuentes et al. 1988; Meyers 2003; Lehmann et al. 2004; Finlay and Kendall 2007, Michener and Kaufman 2007) rather than allochthonous organic matter of terrestrial origin. However, for the plants a number of transects were sampled in the coastal regions with abundant invasive macrophytes. In some

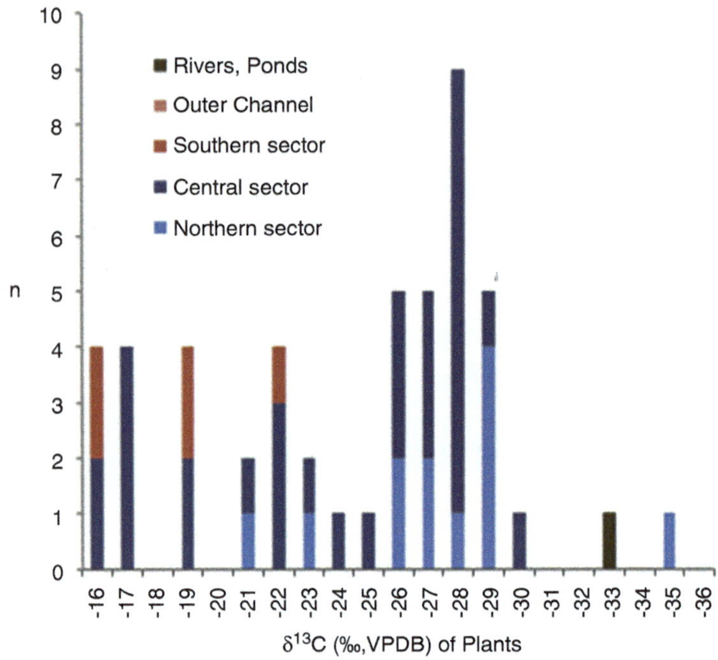

Fig. 5.6 Stable carbon isotope compositions of submerged and emerged plants as well as algaes sampled in Chilika Lake

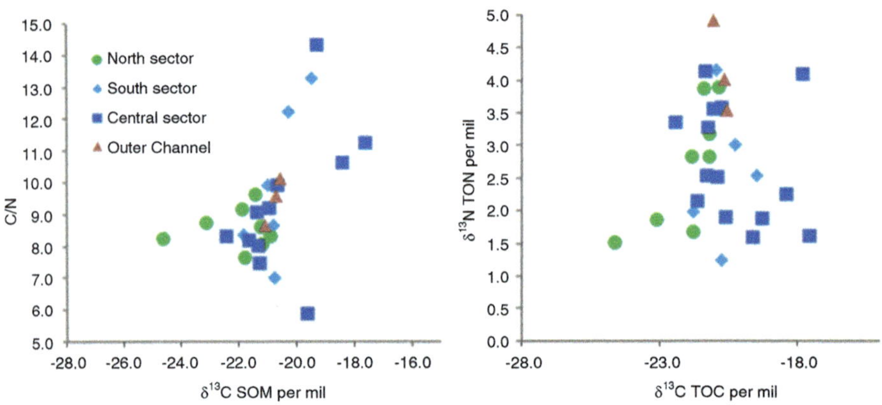

Fig. 5.7 C/N ratios and carbon and nitrogen isotope compositions of SOM sampled in September 2013 for Chilika Lake

transects, anoxic conditions lead to an abundance of ammonium in the shallow water coastal parts. The plants have values of between +4 up to +6.6‰ in their $\delta^{15}N$ values. This suggests large seasonal changes with possible local sources of methanogenic carbon and/or complete reduction of nitrate to ammonium during

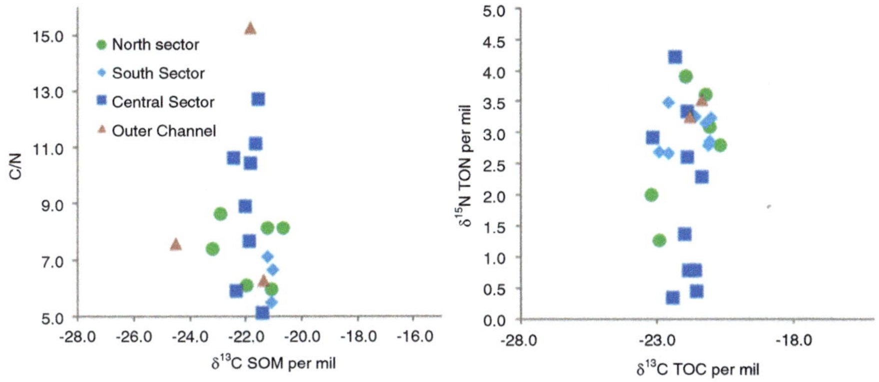

Fig. 5.8 C/N ratios and carbon and nitrogen isotope compositions of SOM sampled in February 2014 for Chilika Lake

the growth of the plants. Oxygen levels measured in these waters were below 2 mg/ltr. Given the density of the population along the western lake shores and the associated wastewaters as well as the agricultural activities, much of the nutrients causing the eutrophication along the lake shore may be of anthropogenic origin. This work on the nitrogen cycle is also of importance to an evaluation of the invasive macrophytes that proliferate along the shores and are seen as a potential environmental hazard but may actually be effective filters for excess nitrate and nutrient fluxes into the lagoon.

Plants and SOM from the Outer Channel have higher overall $\delta^{15}N$ values, indicative of different sources of N, likely also a larger influence of marine nitrogen sources from the Bay of Bengal. This is compatible with interpretations based on the H- and O-isotope compositions of water, the DIC and salinity relationships for the Outer Channel.

5.3.3 Chemical and Isotopic Composition of Organic Matter in the Sediment Core

A reconnaissance core was taken during the first sampling campaign in September 2013. The sediment core has been investigated for its sedimentological features, mineralogy, fossil contents of ostracods and/or foraminifera and attempts were made to date the top of the core with the radioactive tracer of ^{137}Cs. Given the information on previous cores by Zachmann et al. (2009) and Khandelwal et al. (2008), the present sediment core with about 1.5 to 2 m of sediment, corresponds ideally to the last 2000 years. The last several thousand of years are of particular interest to biogeochemical studies as previous work on pollen by Khandelwal et al. (2008)

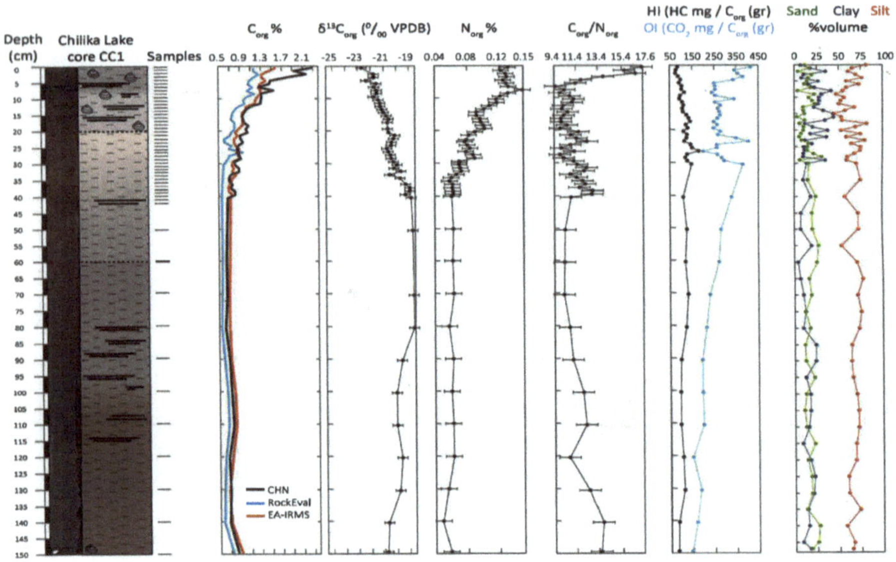

Fig. 5.9 Sediment core sampled from the Central sector of Chilika Lake in September 2013. Changes in geochemical parameters are given relative to the depth in the core that was sampled. (Figure taken from Bourgeois (2015))

has shown substantial changes in the biological communities, some of them likely anthropogenic in origin. In addition the agricultural practices have changed drastically and as a result processes of siltation and the nutrient cycles may have been impacted (Ghosh and Pattnaik 2005).

The geochemical results for this core are summarized in Fig. 5.9. The ^{137}Cs activities measured are compatible with a date back to about 1950 for the top 50 cm's of the core. Figure 5.9 illustrates that while most geochemical parameters remain relatively constant in values over the top 40 cm's, there is a gradual increase in the amount of organic carbon and nitrogen bound to organic matter. In parallel, the $\delta^{13}C$ values of the organic matter preserved in the sediments decrease towards the top while the C/N ratios increase. In view of the above discussion of similar values measured for the SOM in the surface sediments throughout the lagoon, these trends would be compatible with increasing terrestrial input of organic carbon and nitrogen as nutrients over the last 50 or more years (Meyers 2003; Finlay and Kendall 2007; Michener and Lajtha 2007). If the dates for the core can be confirmed, these changes in geochemical records could be related to changes in land use policies and increased agricultural importance within the Chilika Lake drainage basin. This may lead to an increased sedimentation rate, which is required in order to account for the relatively rapid accumulation of the top 40 cm's of the core in only 40 to 50 years (c.f. the work of Khandelwal et al. 2008).

5.4 Conclusions

Present results from two consecutive sampling campaigns representing the monsoon (wet) and the dry season, indicate that the chemical composition of the lagoon waters, including the nutrient supply, are largely controlled by terrestrial inputs from three different drainage basins to the Northern Sector, the Central Sector and the Southern Sector. Mixing of the different waters entering Chilika Lake and within the lagoon is very limited and a seawater influx is only important in the Outer Channel but very limited through the newly dredged seaward channel for the rest of the lagoon. All salinity changes as well as isotopic changes within the lagoon can be accounted for by evaporation and internal bio-productivity as well as degassing of the water column in CO_2. As a consequence, there are relatively large seasonal variations in both the isotopic composition of the waters as well as the dissolved organic and inorganic carbon content and the autochthonous produced organic carbon within the lagoon.

While average concentrations and isotopic compositions of both C and N of organic matter in the sediments are relatively constant at the present surface sediments compared to much larger variations in the particulate organic matter within the water column, larger changes are obvious over the last 50 years or so within sediment cores. Hence, in order to avoid unwanted changes in the ecological functioning of the lagoon, the three different drainage basins should be monitored/controlled if further environmental impact by the increased population and agricultural activities on the lagoon are to be limited.

Based on the limited exchange of water between the open marine system via both the Outer Channel and the newly dredged channel, any increased fish catch that may be related to the opening of the new mouth in 2000 is likely to be the result of marine fish that migrates into the lagoon and is readily caught within the shallow waters of the lagoon, rather than an increased nutrient supply and/or oxygenation or circulation of the lagoon waters themselves.

Acknowledgments This work would not have been possible without the dedicated investment of time and effort by the "team" of MSc students (Géraldine Bourgeois, Kelly Delavey, Michèle Ecuyer, Caroline Hostettler, Pauline Lange) and researchers (Prakash Kumar Sahoo), which profited from the financial support of the "Fonds d'Investissement" (FINV) of the Faculty of Geoscience and Environment of the University of Lausanne and a student/researcher academic exchange program established between KIIT University in Bhubaneswar and the University of Lausanne. The first FINV was granted to Dr. Jean-Luc Epard, whose help and support for this project is greatly appreciated. The study was also generously supported, financially and ***logistically***, by the Chilika Development Authority (CDA) of Orissa, India, as well as the staff and students of the School of Biotechnology of the KIIT University of Bhubaneswar, India. Finally, Dr. Jorge Spangenberg and Dr. Thierry Adatte are thanked for the assistance with many of the analyses made for these studies. The authors also appreciate the thorough and thoughtful review by an anonymous reviewer.

References

Altabet MA, McCarthy JJ (1985) Temporal and spatial variations in the natural abundance of 15N in POM from a warm-core ring. Deep Sea Res 32:755–772

Baliarsingh SK et al (2014) Oil pollution in Chilika lagoon: an anthropogenic threat to biodiversity. Curr Sci 104:516–517

Bernasconi SM, Barbieri A, Simona M (1997) Carbon and nitrogen isotope variations in sedimenting organic matter in Lake Lugano. Limnol Oceanogr 42:1755–1765

Bourgeois G (2015) Historical sediment record in Chilika Lake, India: geochemical, mineralogical and micropaleontological study. Master thesis, Faculty of Geoscience and Environment, University of Lausanne, 81 p

Brenner M, Whitmore T, Curtis J, Hodell D, Schelske C (1999) Stable isotope ($\delta^{13}C$ and $\delta^{15}N$) signatures of sedimented organic matter as indicators of historic lake trophic state. J Paleolimnol 22:205–221

Cifuentes LA, Sharp JH, Fogel ML (1988) Stable carbon and nitrogen isotope biogeochemistry in the Delaware estuary. Limnol Oceanogr 33:1102–1115

Clark ID, Fritz P (1997) Environmental isotopes in hydrogeology. CRC Press, New York

Delavy K, Ecuyer M (2014) Isotopic study of the spatial and temporal origin, source and dynamic of organic matter in Chilika Lake's estuarine ecosystem, Master thesis, Faculty of Geoscience and Environment, University of Lausanne, 161

Finlay JC, Kendall C (2007) Stable isotope tracing of temporal and spatial variability in organic matter sources to freshwater ecosystems. In: Michener R, Lajtha K (eds) Stable isotopes in ecology. Blackwell Publishing, Malden, pp 283–333

Ghosh KA, Pattnaik AK (2005) Chilika lagoon: experience and lessons learned brief. http://www.iwlearn.net/publications/ll/chilikalagoon_2005.pdf

Hilton RG, Galy A, Hovius N (2008) Riverine particulate organic carbon from an active mountain belt: importance of landslides. Glob Biogeochem Cycles 22:GB1017

Hodell DA, Schelske CL (1998) Production, sedimentation, and isotopic composition of organic matter in Lake Ontario. Limnol Oceanogr 43:200–214

Hoefs J (2009) Isotope geochemistry, 6th edn. Springer, Berlin

Hostettler C (2015) Ecology and geochemistry of living ostracods in Chilika Lake, Odisha, India. Master thesis, Faculty of Geoscience and Environment, University of Lausanne, 72 p

Jayaraman G, Rao AD, Dube A, Mohanty PK (2007) Numerical simulation of circulation and salinity structure in Chilika Lagoon. J Coast Res 23:861–877

Jeffrey AWA, Pflaum RC, Brooks JM, Sackett WM (1983) Vertical trends in particulate organic carbon 13C: 12C ratios in the upper water column. Deep Sea Res Part A Oceanogr Res Pap 30:971–983

Kannan K, Ramu K, Kajiwara N, Sinha RK, Tanabe S (2005) Organochlorine pesticides, polychlorinated biphenyls, and Polybrominated diphenyl ethers in Irrawaddy dolphins from India. Arch Environ Contam Toxicol 49:415–420

Kennicutt MC, Barker C, Brooks JM, DeFreitas DA, Zhu GH (1987) Selected organic matter source indicators in the Orinoco, Nile and Changjiang deltas. Org Geochem 11:41–51

Khandelwal A, Mohanti M, Garcia-Rodriguez F, Scharf BW (2008) Vegetation history and sea level variations during the last 13,500¬†years inferred from a pollen record at Chilika Lake, Orissa, India. Veg Hist Archaeobotany 17:335–344

Kumar B, Rai SP, Saravana Kumar U, Verma SK, Garg P, Vijaya Kumar SV, Jaiswal R, Purendra BK, Kumar SR, Pande NG (2010) Isotopic characteristics of Indian precipitation. Water Resour Res 46:W12548. http://sci-hub.tw/10.1029/2009WR008532

Lange P (2014) Water dynamics in Chilika Lake (India). Master thesis, Faculty of Geoscience and Environment, University of Lausanne, 66 p

Lehmann MF et al (2004) Seasonal variation of the $\delta^{13}C$ and $\delta^{15}N$ of particulate and dissolved carbon and nitrogen in Lake Lugano: constraints on biogeochemical cycling in a eutrophic lake. Limnol Oceanogr 49:415–429

Meyers PA (2003) Applications of organic geochemistry to paleolimnological reconstructions: a summary of examples from the Laurentian Great Lakes. Org Geochem 34:261–289

Michener RH, Kaufman L (2007) Stable isotope tracers in marine food webs: an update. In: Michener R, Lajtha K (eds) Stable isotopes in ecology. Blackwell Publishing, Malden, pp 238–282

Michener R, Lajtha K (2007) Stable isotopes in ecology. Blackwell Publishing, Malden

Pal SR, Mohanty PK (2002) Use of IRS-1B data for change detection in water quality and vegetation of Chilka lagoon, east coast of India. Int J Remote Sens 23:1027–1042

Sackett WM, Thompson RR (1963) Isotopic organic carbon composition of recent continental derived clastic sediments of eastern Gulf Coast, Gulf of Mexico. AAPG Bull 47:525–528

Schelske CL, Hodell DA (1995) Using carbon isotopes of bulk sedimentary organic matter to reconstruct the history of nutrient loading and eutrophication in Lake Erie. Limnol Oceanogr 40:918–929

Teranes JL, McKenzie JA (2001) Lacustrine oxygen isotope record of 20th-century climate change in central Europe: evaluation of climatic controls on oxygen isotopes in precipitation. J Paleolimnol 26:131–146

Zachmann DW, Mohanti M, Treutler HC, Scharf B (2009) Assessment of element distribution and heavy metal contamination in Chilika Lake sediments (India). Lakes Reserv Res Manag 14:105–125

Chapter 6
Modelling of Hydrodynamics and Salinity Characteristics in Chilika Lagoon

R. S. Kankara and U. S. Panda

Abstract Hydrodynamic circulation is a primary factor for most of the physical and ecological processes in lagoon environments. Chilika – Asia's largest brackish water lagoon is experiencing signficant transformations such as siltation, the growth of invasive macrophytes, northward migration of mouth and choking of the outer channel. These transformations are responsible for the reduced salinity, reduced water depth and weak lagoon-sea interaction, which in turn has led to decline in water spread area, increase in vegetated area and decrease in fish productivity. Chilika Development Authority (CDA) has taken up various initiatives to maintain the lagoon environment including improvement of Chilika mouth. Modelling is a useful tool to understand the influx of tides, wind stress and impact of freshwater influx into the lagoon and to analyse the hydrodynamic processes based on 'what if' scenario. A two-dimensional hydrodynamic model has been setup to investigate the changes in the hydrodynamics and salinity regime of the lagoon during the pre (1999) and post (2009) hydrological intervention period as well as the present scenario (2015). The study suggests that post-intervention period has significantly improved the lagoon-sea fluxes, seawater ingress and flushing of flood waters. With the advance of freshwater discharge during monsoon and post-monsoon, the mean salinity levels increased from 2.87 to 4.87 psu during 1999–2009 period, and subsequently reduced to the 3.4 psu in 2015.With all forcing factors, the annual increase in the mean salinity distribution during 2009 is 36% which has reduced to 18% in 2015. The increases in salinity underpin enhancement in several wetland ecosystem services and livelihood benefits to communities living in and around the lagoon. Reduction in salinity in later periods points to the need to maintain the inlets and dredged channels for sustainable management of the lagoon ecosystem.

Keywords Hydrodynamics · Salinity · Tidal exchange · Modelling · MIKE-21

R. S. Kankara (✉) · U. S. Panda
National Center for Coastal Research (NCCR), Ministry of Earth Sciences, Chennai, India
e-mail: kankara@nccr.gov.in

C. M. Finlayson et al. (eds.), *Ecology, Conservation, and Restoration of Chilika Lagoon, India*, Wetlands: Ecology, Conservation and Management 6,
https://doi.org/10.1007/978-3-030-33424-6_6

6.1 Introduction

Hydrodynamic modelling in lagoon environments is a primary requirement to understand the temporal variation in physical and environmental processes that lead to ecological changes. Chilika lagoon- the most significant brackish water tropical lagoon in Asia situated in the state of Odisha, along the east coast of India (19°28′ N – 19°54′ N and 85°06′ E – 85°35′ E) and oriented in the NE-SW direction (Fig. 6.1). The lagoon is about 65 km long and 3–32 km wide, with a water spread ranging from 906 km^2 during dry season (December–June) to 1165 km^2 during rainy season (July–October). The lagoon is separated from the sea by a 60 km long sandbar, spanning 323.62 km^2 area and acting as a barrier island between the lagoon and the sea. Chilika is a shallow coastal ecosystem with an average depth of 1.5 m and the maximum depth of 4.5 m in its southern part, closer to the Kalijai temple. The lagoon exchanges seawater with the Bay of Bengal through a 25 km long shore parallel to the outer channel. The average width of the channel is 900 m, and depth is of the order of 1 m. The gradual deterioration of the inlet channel had caused a substantial reduction in the tidal prism (Chandramohan and Nayak 1994; Chandramohan et al. 2001). On the south, the lagoon is connected to Rushikulya estuary by the Palur canal. The shallow depths and shoals along the channel offer

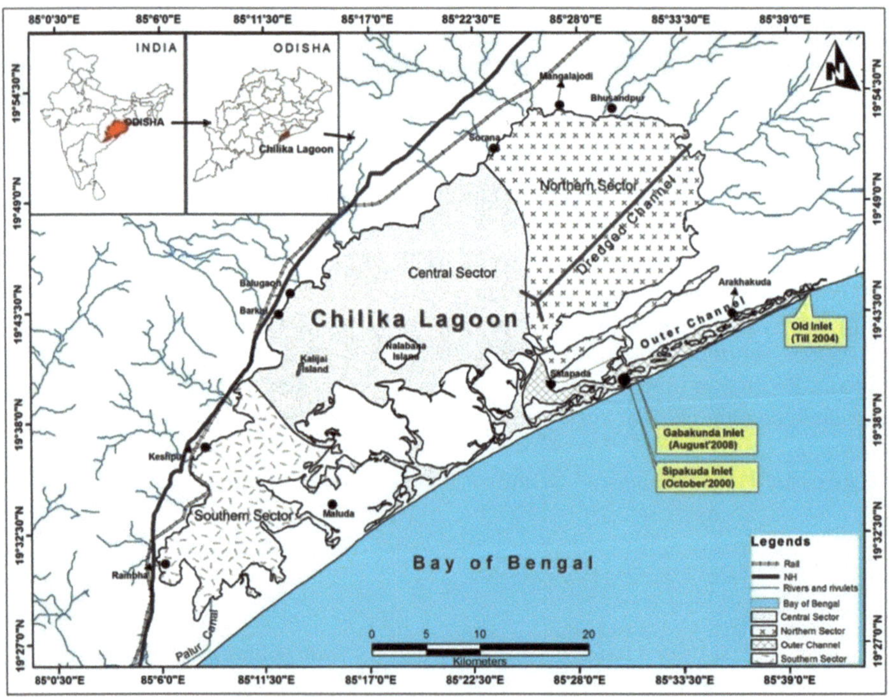

Fig. 6.1 Map of the Chilika Lagoon

considerable resistance to the flow resulting in the fall of tidal range and reduction in the tidal prism. Approximately 3.8×10^7 m^3/h of fresh water and 13 million tonnes of silt is drawn into the lagoon every year from more than 37 drains and tributaries of River Mahanadi during the monsoon season which lasts from June to September. The salinity in part of the lagoon on the northern side reduces upto 0 psu during the monsoon season due to high inflow of freshwater (Panda et al. 2015). However, the lagoon moderately regains the salinity levels during the fair weather season as a result of reduction in the discharge of fresh water into the lake and the ingress of seawater from the Bay of Bengal due to tidal circulation. About 65% of fresh water comes through Daya and Bhargavi, tributaries of river Mahanadi, which enter into the lagoon from the north (Panda 2008). The prolonged deposition of silt brought by rivers has caused the lagoon to become shallow over the years, in turn choking the Mugarmukh area, which is the the neck connecting the outer channel with the main body of the lagoon. Annual average salinity for the eastern part of the lagoon is 10–15 psu, for the central part 0.5–10 psu, and for the northern part 2–4 psu (Panda et al. 2013).

A new inlet was opened near Sipakuda in October 2000. Another channel was dredged for 200 m wide with 2.5 m depth in Magarmukh for free flow and mixing of seawater to maintain optimum salinity level inside the lagoon (CWPRS 1998). After opening of the new inlet mouth, the ecological conditions of the lagoon improved significantly (Pattnaik 2001; Mohanty et al. 2009). But, over the years, it has been observed that the inlet has migrated towards the north,and at present, it is situated about 4 km north of the mouth location in 2000.

Selected studies are available on the historical records and migration of Chilika inlets (Panda et al. 2013; Mishra and Jena 2014). Littoral drift and continuous deficient runoff of the Mahanadi basin for 2–3 years has helped forming large ebb deltas on both faces of the tidal inlet, inducing its closure. The tidal inlets of Chilika Lagoon are influenced by micro-tidal waves and the fine grained sandy coast with steeper slopes in north flank than the south. Sediment transport occurs in the surf zone, moving parallel to the coast. A part of the sediment gets deposited on the shore. Deposition is primarily due to the oblique waves breaking near the shore. Storm surges overtop and overwash a large quantum of water laden with sediment from Bay of the Bengal to the back barrier lagoon. After each storm, the evacuation of that volume through the channel section slows down. Consequently, the channel section either closes or shifts. The longshore transport is accountable for the closure of an inlet. The extreme climatic events lead to fluctuations in the Chilika shoreline, and the net annual shift during 1936–2005 at the rate of 1.09 m towards the Bay of Bengal (Mishra and Jena 2014).

Mathematical modelling is a useful tool to understand the existing hydrodynamic conditions of the Chilika lagoon, and to predict different scenarios for Chilika mouth. In the present study, a two-dimensional depth-averaged hydrodynamic model was applied to Chilika lagoon (Nayak et al. 1998; Chubarenko and Tchepikova 2001; Dias and Lopes 2006; Panda et al. 2013). This model was setup to describe the hydrodynamic changes in flow and salinity Chilika lagoon for pre (1999) and post (2009) intervention and current condition (2015) periods.

6.2 Modelling Approach

6.2.1 Hydrodynamic Modelling

A state-of-the-art fully integrated with the effects of tide, winds, source/sink, and time-dependent generalised hydrodynamic model, DHI- MIKE-21, was used for two-dimensional free surface flows. The model simulated unsteady flows in one layer (vertically homogeneous) fluids and based on the numerical solution of full non-linear equations of conservation of mass and momentum integrated over the vertical to describe flow and water level variations. The effective shear stresses in the momentum equations contain momentum fluxes due to turbulence, vertical integration and subgrid-scale fluctuations are included in the model to provide damping of short-wavelength oscillations and to represent subgrid scale effects (Madsen et al. 1989; Wang 1990).The Hydrodynamic (HD) model simulates important parameters such as water elevation, current speed and directions in different space and time scale. The advection-dispersion equation solves to compute flow and distribution of salt subjected to a variety of forcing, sources and boundary conditions. Being a shallow water system with significant influence of wind, which generates good mixing between surface and sub-surface waters, it is reasonable to assume that vertical water movements are negligible and a depth-averaged model like Mike 21 can be appropriate. Many researchers also justified the use of a two-dimensional vertically integrated model with respect to approaches that consider two or more vertical layers in the absence of stratification phenomena (Ramirez and Imberger 2002; Balas and Özhan 2002; Zacharias and Gianni 2008).The hydrodynamics of such a lagoon can be described by using a well known shallow water equations, which describe the evolution of an incompressible fluid in response to gravitational and rotational accelerations (Pedlosky 1987) and mass transport equation for salinity.

Shallow water equations for hydrodynamics: Integration of the continuity and horizontal momentum equations over depth requires the following two-dimensional shallow water equations (DHI 2007).

$$h = \eta + d$$

$$\frac{\partial h}{dt} + \frac{\partial h\bar{u}}{dx} + \frac{\partial h\bar{v}}{dy} = hS$$

$$\frac{\partial h\bar{u}}{dt} + \frac{\partial h\bar{u}^2}{dk} + \frac{\partial h\overline{uv}}{dy} = f\bar{v}h - gh\frac{\partial \eta}{dx} - \frac{h}{\rho_0}\frac{\partial p_a}{dx} - \frac{gh^2}{2\rho_0}\frac{\partial \rho}{\partial x}$$
$$+\frac{\tau_{sx}}{\rho_0} - \frac{\tau_{bx}}{\rho_0} - \frac{1}{\rho_0}\left(\frac{\partial s_{xx}}{\partial x} + \frac{\partial s_{xy}}{\partial y}\right) + \frac{\partial}{\partial x}(hT_{xx}) + \frac{\partial}{\partial y}(hT_{xy}) + hu_sS$$

$$\frac{\partial h\bar{v}}{dt} + \frac{\partial h\overline{uv}}{dx} + \frac{\partial h\bar{v}^2}{dy} = -f\bar{u}h - gh\frac{\partial \eta}{dy} - \frac{h}{\rho_0}\frac{\partial p_a}{dy} - \frac{gh^2}{2\rho_0}\frac{\partial \rho}{\partial y}$$
$$+\frac{\tau_{sy}}{\rho_0} - \frac{\tau_{by}}{\rho_0} - \frac{1}{\rho_0}\left(\frac{\partial s_{yx}}{\partial x} + \frac{\partial s_{yy}}{\partial y}\right) + \frac{\partial}{\partial x}(hT_{xy}) + \frac{\partial}{\partial y}(hT_{yy}) + hu_sS$$

Where, t is the time (s); x and y are the Cartesian co-ordinates (m); η is the surface elevation; d is the still water depth (m); $(h = \eta + d)$ is the total water depth (m); $\overline{u}$ and $\overline{v}$ are the depth averaged velocity components in the x and $y$directions; $f = 2\Omega \sin \varphi$ is the Coriolis parameter, Ω is the angular rate of revolution and φ the geographic latitude; g is the gravitational acceleration;ρ is the density of water; s_{xx}, s_{xy}, s_{yx} and s_{yy} are components of the radiation stress tensor; T_{xx}, T_{xy}, T_{yx} and T_{yy} are components of lateral stress; τ_{sx} and τ_{sy} are the components of the surface wind stress; τ_{bx} and τ_{by} are the components of bottom stress; p_a is the atmospheric pressure; ρ_0 is the reference density of water; ρ is the density of water; S is the magnitude of the discharge due to point sources; u_s and v_s are the velocity by which the water is discharged into the ambient water. The lateral T_{ij} includes viscous friction, turbulent friction and differential advection. They are estimated using an eddy viscosity formulation based on the depth averaged velocity gradients

$$T_{xx} = 2A\frac{\partial \overline{u}}{\partial x}, T_{xy} = A\left(\frac{\partial \overline{u}}{\partial y} + \frac{\partial \overline{v}}{\partial x}\right), T_{yy} = 2A\frac{\partial \overline{v}}{\partial y}$$

6.2.1.1 Mass Transport Equations for Salinity

The transports of temperature, T, and salinity, s, follow the general transport-diffusion equations as $\frac{\partial s}{dt} + \frac{\partial us}{dx} + \frac{\partial vs}{dy} + \frac{\partial ws}{dz} = F_s + \frac{\partial}{dz}\left(D_v \frac{\partial T}{dz}\right) + s_s S$.

Where, D_v is the vertical turbulent (eddy) diffusion coefficient, $\widehat{H}$ is a source term due to heat exchange with the atmosphere, and T_s and s_s are the temperature and the salinity of the source.

6.2.1.2 Bottom Stress

The bottom friction can be specified as the frictional velocity associated with the bottom stress which is given by $U_{rb} = \sqrt{c_f \left|\vec{u}_b\right|^2}$. For two-dimensional calculations $\vec{u}_b$ is the depth-average velocity and the drag coefficient can be determined from the Chezy number, $C\ (m^{1/2}/s)$. The specified values for the Chilika environment for Chezy number is 32 and a constant eddy viscosity of 0.5 m^2/s is used (Chapra 1997).

The spatial discretisation of equations is performed using a cell-centred finite volume method in the horizontal plane an unstructured grid is used comprising of triangle elements. An approximate Riemann solver is used for convective fluxes, which makes it possible to handle the discontinuous solutions. For the time integration, an explicit Euler method is used. Due to the stability restriction using an explicit scheme, the time step interval should be selected so that the Courant-Friendrich-Levy (CFL) number is less than 1. For the present study CFL number less than the critical CFL number (0.8) has been observed. The approach for treatment of the moving boundaries problem (flooding and drying fronts) is based on the work by

(Zhao et al. 1994). The depth of each cell is monitored, and the cells are classified as dry, partially dry or wet and flooded boundaries. As the Chilika lagoon is very shallow, the values considered for drying depth $h_{dry} = 0.005$ m, flooding depth $h_{flood} = 0.05$ m and wetting depth $h_{wet} = 0.1$ m which satisfy the rule $h_{dry} < h_{flood} < h_{wet}$.

6.2.2 Model Domain and Bathymetry

The numerical computation has been carried out on a spatial domain that represents the Chilika lagoon through an unstructured mesh. It allows high flexibility with its subdivision of the numericaldomain varying in form and size. It is especially suited to reproduce the geometry and the hydrodynamics of complex shallow water basins such as the Chilika lagoon with its narrow outer channel area, small islands, dredged channels and uneven complex boundary structure. Two model domains were generated from 85°04′ to 85°43′ East and 19°27′ to 19°55′ North coordinates, based on the information from British Admiralty Sea Maps (extracted from a DHI C-MAP in digital form), and toposheets of Chilika region prepared by the Survey of India and from various field observations. The bathymetry map has been validated with the GPS observations collected during field survey. The mesh files for the Chilika lagoon, pre-intervention period i.e., for the year 1999 (Fig. 6.2a, Domain 1: 7440 nodes and 7278 elements) and post-intervention period i.e., for the year 2009 (Fig. 6.2b, Domain 2: 3542 nodes and 5231 elements) and current condition i.e., year 2015 (Fig. 6.2c, Domain 3: 4084 nodes and 6112 elements) has been configured considering the computational unstructured mesh, water depths and boundary information. Subdivision of the continuum discretizes the spatial domain into

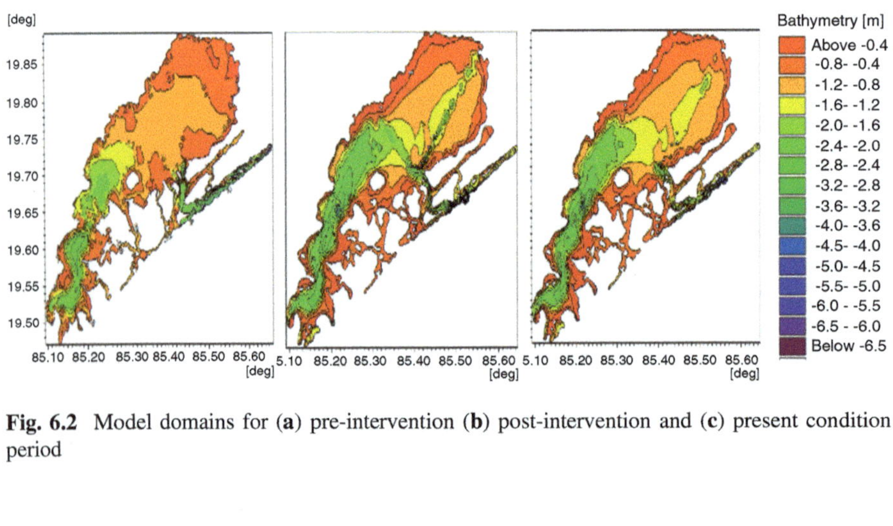

Fig. 6.2 Model domains for (**a**) pre-intervention (**b**) post-intervention and (**c**) present condition period

non-overlapping cells/elements. In the two-dimensional case the elements can be arbitrarily shaped polygons, however, in this study, only equilateral triangles were considered.

Large angles and high resolutions in a mesh contradict with the need for short simulation times. The resolution of the mesh, combined with the water depths and chosen time-step governs the Courant numbers in a model set-up. The maximum Courant number maintained to be less than 0.5. Such that simulation time dependency on the triangulation of the mesh relates not only to the number of nodes in the mesh, but also the resulting Courant numbers. The Courant number C_R expresses the number of computational points the information moves in one time step. $C_R = c\frac{\Delta t}{\Delta x}$. Where, c (celerity) $=\sqrt{gh}$, Δt is t time step and Δx is the grid spacing. As a result of this, the effect on simulation time of a fine resolution at deep water can be relatively high compared to a high resolution at shallow water.

6.2.3 Model Setup

Chilika is mainly influenced by tides, wind and fresh water inflow, therefore waves have been omitted in simulation. However, flooding and drying in shallow areas were considered to obtain accurate results, as model domain covers vast shallow patches. Total 4084, 6012 nodes were generated in the lagoon to represent the pre and post intervention bathymetry. The drying depth (minimum water depth allowed in a point before being taken out of calculation) was set at 0.01 m, and the flooding depth (water depth at which the point will be re-entered into the calculation) was set to 0.05 m. The model computational time step was set to 300 seconds and simulations were carried out for 28,800 time steps with CFL number less than 0.8. Considering the facts that Chilika receives considerable freshwater only in monsoon from 19 source points, the estimated quantity of fresh water was introduced during July–October, and negligible flux was imposed during fair season simulations. In the beginning, the seais set at rest by providing a constant water level uniformly in the model domain (i.e. z = 1.31 m, u = v = 0 att = 0) as an initial condition. Time series hourly data of water level was imposed at open sea boundary, i.e., at the mouth. Several short simulations were carried out toperformseveral sensitivity runs and tuning the calibration parameters (Chezy number (C) and Eddy viscosity) in order to develop confidence in 'what if 'scenario assessments. During the calibration, the hydrodynamic and water quality parameters were adjusted to get a satisfactory correspondence between model results and observed field data.To quantify the agreement between model and observations, method proposed by Willmott (1981) was used. Model was setup for 1 year for 3 different conditions year 1999, 2009 and 2015 corresponding to pre-intervention, post-intervention and current period, with three distinct forcing combinations i.e., (a) tide only, (b) tide + wind and (c) tide + wind + freshwater discharge. Finally, the model was run for 21 Dec 1998 to 31 Dec 1999, 21 Dec 2008 to 10 Nov 2009 and 21 Dec 2014 to 10 Nov 2015.

6.3 Results and Discussion

6.3.1 Water Level

The model was simulated for a 330–380 days period with initial and boundary conditions. Water level and flux at each nodal point were computed for each time step. Hourly time series data were stored to analyse the model results. Results of water level indicate the tendency of decreasing tidal amplitudes from the inlet mouth towards the Satpada. The model was able to capture the impact of the intervention on the water level. The simulated tidal range was 2.2 m at Sipakuda inlet, 0.7 at Satpada, 0.2–0.4 m in the main body of the lagoon during post-intervention period (2009) in comparison to significantly lower ranges of 0.4 m at Sipakuda, 0.2–0.3 m at Satpada and 0.1–0.2 in the main body of the lagoon during the pre-intervention period (1999). It may be noted that the major tidal variation was in the channel between Arakhakuda (Old inlet) upto Sipakuda during pre-intervention period. Water levels (near inlet) shows that the tidal amplitude is maximum at the inlet of the lagoon which decreases gradually as we proceed inward from the inlet, while the astronomical tidal range at inlet matched with observed data, at remote locations in the main body of the lagoon.

The model results were validated with field measurements (water level at Inlet, near Nalabana representing the central sector, on the lead channel representing the northern sector and near INS Chilika representing the southern sector) for the simulation period. Simulation of water levels (Fig. 6.3) shows that the tidal amplitude is maximum at the inlet which decreases gradually as we proceed inward from the inlet. The lunar principal constituent M2, observed to be dominant at Sipakuda inlet also the 'form number' depicts the semi-diurnal nature of the tide near the inlet. The M2 represents the tide due to a fictitious moon circling the equator at the mean lunar distance and with constant speed. Harmonic analyses made for four locations covering inlet and three sectors determined the tidal constituents (Amplitude and phase). Tide near inlets are purely semi-diurnal micro tide having tidal form number 0.18 and mixed semi-diurnal during monsoon with the tidal form number 0.30. The tides transform into mixed mainly semi-diurnal inside the lagoon, primarily when the complex bottom topography inside the shallow lagoon and dry period reduces the semi-diurnal component amplitudes through friction and non-linear effects (Dias 2001).

Figure 6.4 shows the hourly water level profile for a tidal cycle across the lagoon at six locations form Sipakuda>Satapada>Mugarmukha>central sector and two representative locations at centre of northern and southern sectors during post monsoon period. It is clearly seen that the water level of main lagoon are is much lower than outer channel region. A steep gradient from Sipakuda to Muggarmukha was also observed, which generates higher current velocity. During 1999–2009 period, the water level increased on an average by 20 cm, but the increase in 2015 was only marginal. (~5 cm).

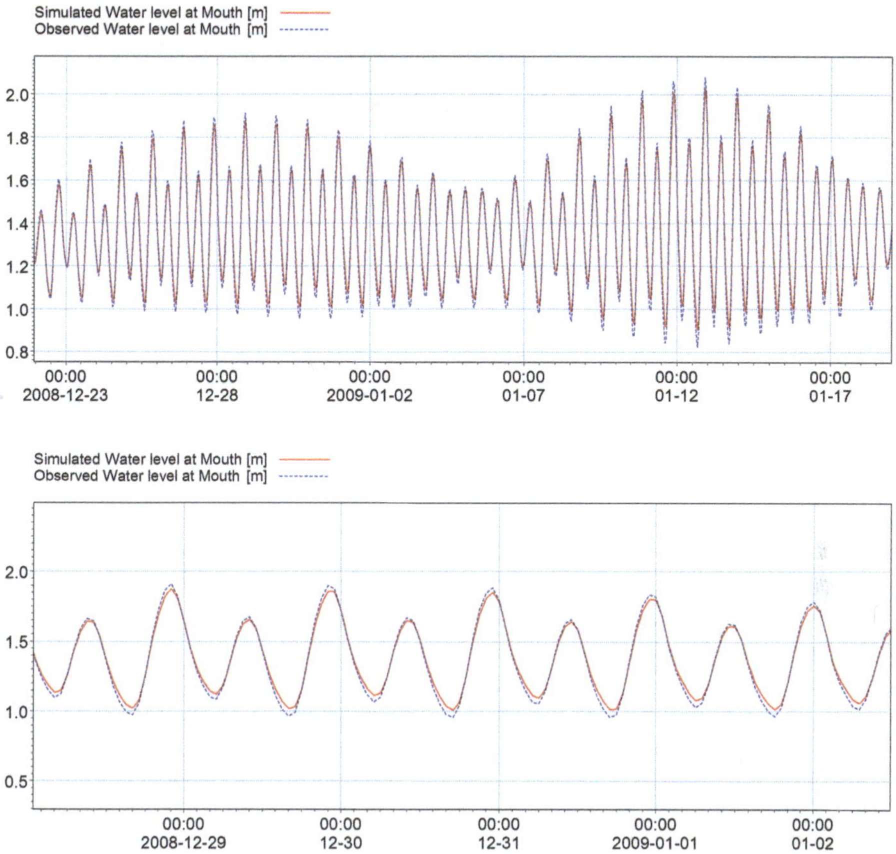

Fig. 6.3 Validation of water levels between Sipakuda inlet and Satapada during post intervention period (2009)

6.3.2 *Surface Water Circulation*

Hydrodynamics of the lagoon is governed by many forcing like bathymetry, wind stress, tides and freshwater influx from the rivers. In Chilika, wind-driven circulation dominates the density-driven circulation (Mohanty and Panda 2009). Due to the large open surface, the wind is a significant forcing factor in stimulating circulation. Seasonal changes in the wind magnitude and direction can cause large-scale changes in the circulation pattern of the lagoon. Apart from such seasonal changes, land and sea breeze, also affects the circulation in the lagoon. The role of wind in generating turbulence is more important in the lagoon zones distant from the sea, where tide-induced flows have no relevance (Cioffi et al. 1995). Apart from the surface wind stress, bottom topography and bathymetry is another crucial factor controlling the circulatory pattern in shallow lagoons. Tidal influx causes major changes in the

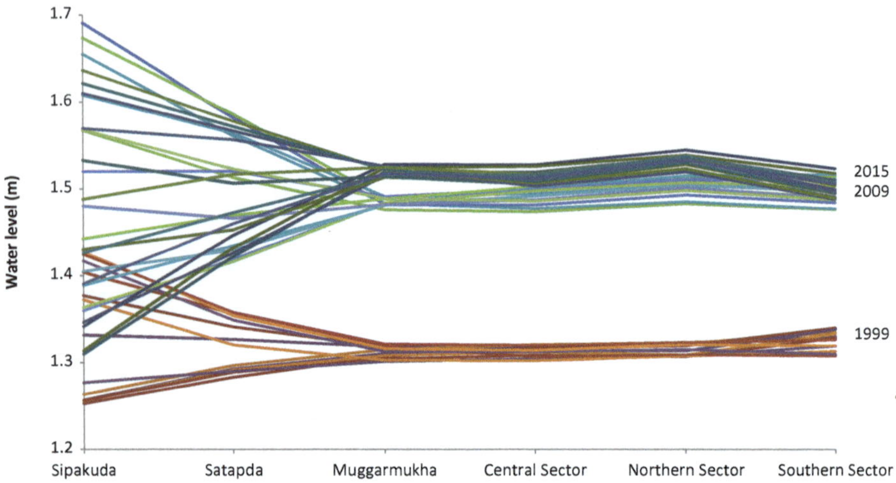

Fig. 6.4 Simulated hourly water level profiles from Sipakuda inlet towards the main body of the lagoon over period of 12 h

circulation in lagoonsbut its effect is limited to the region near the tidal inlet. Circulation in lagoon is also influenced by freshwater influx.

From the hydrodynamic simulations, it is seen that tide generated currentsbegin from the inlet of Chilika, crosses over the Muggarmukh channel and reach main body of the lagoon. The primary direction groups are easterly and westerly directions along the Outer Channel orientations. The current speed has significantly increased to the order of 30–40% at outer channel region with the intervention resulting in a better influx of seawater. Consequently, the saline water easily propagates in the main body of the lagoon. Simulated current velocity was observed to be maximum at inlet (0.5 m/s) and lower in the other parts of the lagoon (0.02–0.05 m/s). A significant portion of waters moves towards southern sector through the central sector with a speed of 0.04–0.06 m/s (Fig. 6.5).

Interestingly, eddies were observed in the northern sector and southern sector. During monsoon, water currents are higher in northern sector due to runoff than the southern sector. Flood currents could be seen between Muggarmukha to Satapada channel areas, which gradually decreases towards the main body of the lagoon. In the post-intervention period, the floodwaters were observed to enter into the lagoon through the Muggarmukha channel and diverges into two streams, one towards southern sector through central sector, and other towards northern sector, especially along the lead. Further, the northern sector stream diverges into to two parts and forms eddy like circulation pattern in its western and eastern parts (Fig. 6.5). Circulation patterns in the northern sector indicate the predominant role of the bottom topography.

The hydrodynamic simulations affirm the primal role of wind as the primary forcing factor on the water circulation in the main body of the lagoon. The tide determines the discharges through the connecting inlet and modulates the circulation

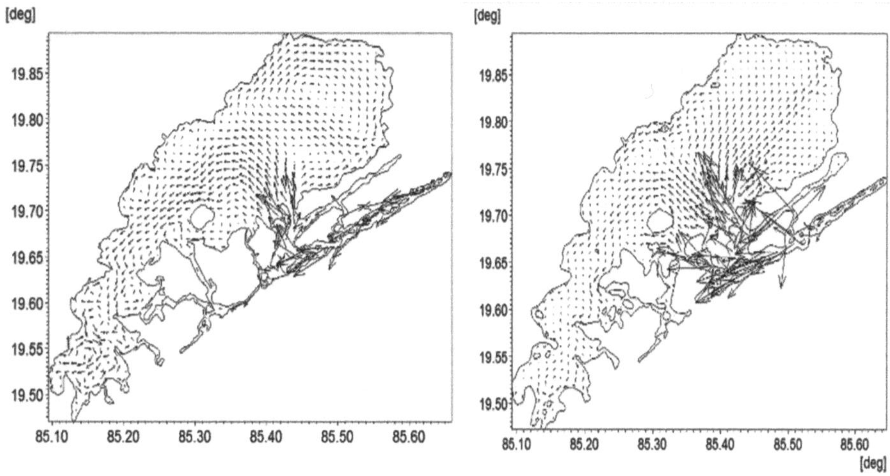

Fig. 6.5 Simulated flood current pattern during January (**a**) 1999 (pre-intervention) and (**b**) 2009 (post-intervention)

pattern set up by the wind. Topographies of the bottom along with width are responsible for the formation of gyre circulation subsystems in the north sector and some parts of the southernsector. However, with substantial freshwater influx after monsoon drives the water in unidirectional towards the sea through the inlet and changes the entire circulation pattern.

6.3.3 Salinity Simulation

Salinity was simulated by prescribing concentration at the inlet and zero concentration at 19 source points. Salinity pattern drives the lagoon ecosystem. Spatial-temporal variation of salinity in the lagoon is significant, ranging from fresh water regime (~5) to saline regime (~35) during the summer; whereas it reduced to entirely fresh in northern sector to brackish (~10) in the main body of the lagoon during wet period due to massive influx of freshwater discharge into the lagoon. The lagoon is well mixed in a vertical column and has negligible salinity stratification. Salinity gradient is lower in the southern sector and gradually increases in the direction of lagoon inlet (Panda and Mohanty 2008). For salinity simulation, a number of simulation scenarios were executed using different combinations of the four calibration factors in Dalton's and Angstrom's law (Zacharias and Gianni 2008) run,and the model sensitivity to each of the constants was checked out. Latent heat flux, wind coefficient and sun constants were calibrated for the regions. The shortwave penetration is dependent on the visibility, which has been specified with one as a light extinction coefficient. The heat exchange is included with air temperature, relative humidity and clearness coefficient. European Centre for Medium-Range Weather

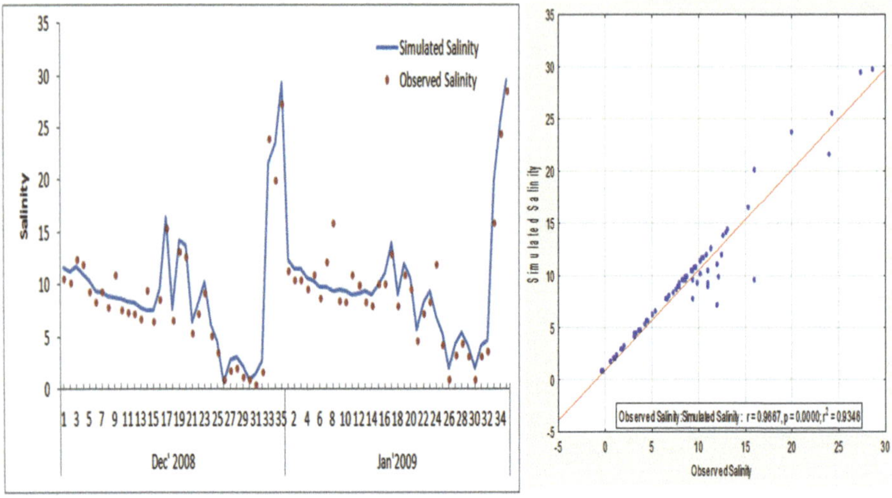

Fig. 6.6 (**a**) Validation of Salinity at 36 stations uniformly distributed over the lagoon (**b**) Regression analysis between observed and simulated salinity for the 36 locations

Forecasts (ECMRWF) data for 1999 and 2015 and AWS measurements (IMD) for 2019 were used. The initial conditions were defined with a surface distribution from various measurements.

Simulations are performed to understand and quantify different forcing factors – wind, tides and freshwater discharge – and their independent and combined roles in the hydrodynamics and salinity distribution in the lagoon. The results indicate that the signatures of salinity intrusion through inlet, low salinity in northern sector, and medium salinity in southernsector. Salinity validation made at 36 stations for the exact time of sampling (stations uniformly distributed over whole lagoon) during post-intervention period (2009) agrees well with observation with $r^2 = 0.93$ (Fig. 6.6). Results suggest that the temporal variability is mainly due to the disparity of the seasonal changes in salinity (Fig. 6.7).

The time series data indicates a significant increase in salinity during post-intervention periods. Experiments with tidal forcing alone show that tides have transported salinity in the channel upto the Muggurmukh (gut area) (Fig. 6.8). The wind forcing affects salinity redistribution mostly in the main body of the lagoon. The opening of the inlet at Satapara has helped in increasing the tidal influx and hence the salinity, the influence is visible in the main body of the lagoon (Figs. 6.9a and 6.9b). The month of June is the transition period for a change in salinity, being linked with the onset of monsoonal precipitations and consequently substantial freshwater influx into the lagoon. The simulations indicate that the the lagoon takes nearly 6–8 weeks to restore the salinity regime after the monsoonal flood.

The variation of mean salinity with the engineering interventions during October 2000 is shown in Table 6.1. It is noted that after the hydrological intervention, the influx of seawaterhas improved the salinity distribution in the lagoon. During

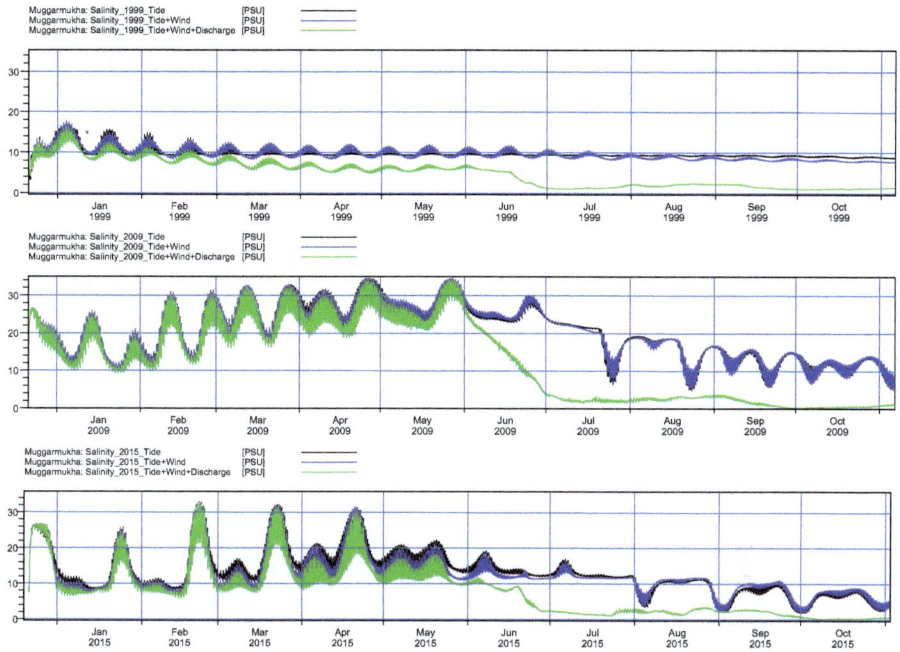

Fig. 6.7 Seasonal and annual mean salinity variation in pre- (1999) and post- (2009) interventions and current situation (2015) period with combinations of different forcing factors (Tide only, Tide + Wind and Tide + Wind + Freshwater discharge)

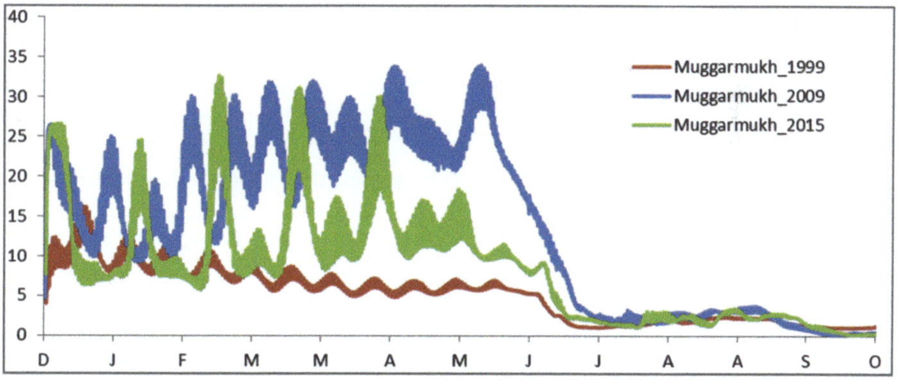

Fig. 6.8 Mean salinity at profile at Muggarmukh during pre- (1999) and post- (2009) interventions and current situation (2015)

summer (March–June), mean salinity has increased from ~6 psu in 1999 to ~17 psu in 2009 (163%), dipping to ~12 psu in 2015. In monsoon (July–Oct) with freshwater influx, salinity increased by 147% during 1999–2009, and 111% during 1999–2015. A similar pattern was observed during the post-monsoon period with tide and tide + wind forcing.

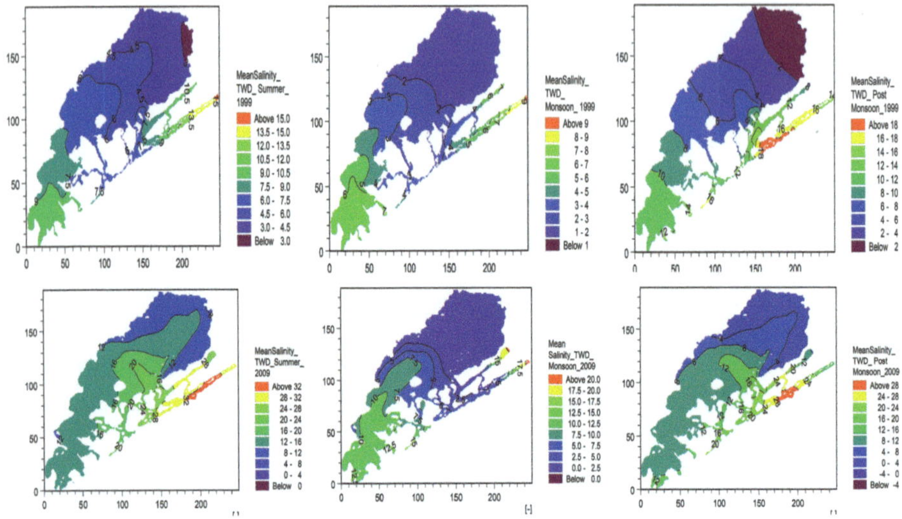

Fig. 6.9a Spatial variation of seasonal mean salinity during pre- (1999) and post- (2009) interventions with Tide + Wind + Freshwater discharge forcing

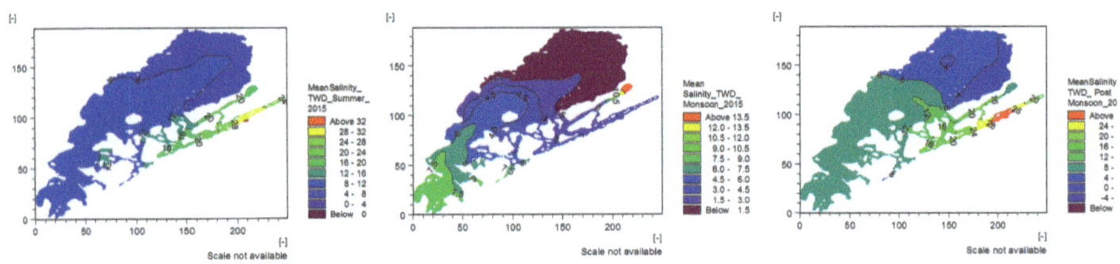

Fig. 6.9b Spatial variation of seasonal mean salinity during current situation (2015) period with Tide + Wind + Freshwater discharge forcing

The freshwater discharge has completely altered the lagoon's ecosystem. With the advance of freshwater discharge during monsoon and post-monsoon, the modelled mean salinity falls upto 2.87 psu, 4.87 psu and 3.4 psu in 1999, 2009 and 2015.Thisindicates reduced flushing of freshwater into the sea during 2009–2015 period.

However, during post-monsoon, the freshwater discharge continues and still decreases the salinity improvement at 41% even though the mean salinity increased nearly two-fold during the monsoon period. When modelled with all forcing factors, annual salinity was observed to increase by 36% during 1999–2009 period, however,reduced to 18% in 2015.

Comparing the salinity profiles in three periods (Fig. 6.8), it is clear that the hydrological intervention has led to improvement in salinity the regimes. The spatial mean salinity figures also reveal the same. The higher isohalines in the lagoon, especially in the northern sector, depicts the role of dredged mouth and lead channel which provides a clear passage for seawater into the lagoon (Fig. 6.9b). The funnel

Table 6.1 Seasonal and annual mean salinity variation in psu in pre- (1999) and post- (2009) interventions period with different forcing factors

Forcing factors	Season	Pre-intervention (1999)	Post-intervention (2009)	Current condition (2015)	Change in mean salinity 1999–2009 (%)	Change in mean salinity 2009–2015 (%)
Tide only	Summer (MAMJ)	6.64	17.43	12.02	163	132(−31)
	Monsoon (JASO)	6.8	16.81	10.78	147	111(−36)
	Post Monsoon (NDJF)	6.4	9.34	9.37	46	0
	Annual	3.79	8.53	6.64	125	123(−22)
Tide + Wind	Summer (MAMJ)	6.73	17.8	12.48	164	134(−30)
	Monsoon (JASO)	6.93	17.03	11.18	146	112(−34)
	Post Monsoon (NDJF)	6.44	9.35	9.54	45	47(+2)
	Annual	4.09	8.63	6.64	111	88(−23)
Tide + Wind + Freshwater Discharge	Summer (MAMJ)	6.1	15.09	10.37	147	116(−31)
	Monsoon (JASO)	2.87	4.87	3.4	70	40(−30)
	Post Monsoon (NDJF)	6.32	8.91	9.08	41	43(+2)
	Annual	4.14	5.65	4.64	36	18(−18)

type isohalines from Muggermukh towards the central sector and further higher gradient isohalines towards southern sectorshows the improvement in the brackish water characteristics of the lagoon during post-intervention (2009) period. In 2015, however, salinity has been observed to decline.

6.4 Conclusion

The two-dimensional hydrodynamic model was used to obtain nine synoptic views of hydrodynamic and salinity pattern for Chilika lagoon for three scenarios (1999, 2009 and 2015) and three significant forcings (tide, wind and freshwater influx). The model results are promising for simulating spatial variations of hydrodynamic propagation, and salinity conditions in response to multiple forcing mechanisms. It is concluded that the tide and wind are the primary forcing factor during the summer period, whereas freshwater influx during monsoon and post seasons influences the water mass and circulation. Wind is responsible for the formation of clockwise and counter-clockwise circulation subsystems in the main body of the lagoon. With monsoon onset, substantial freshwater inflow to the lagoon drives the water unidirectional towards the sea and alters the whole circulation pattern. Inlet and upland discharge control the salinity regime. Salinity gradients are diverse with different water masses of the lagoon, however, after monsoon lagoon tends to the freshwater regime. Hydrological intervention and concurrent restoration measures have facilitated better exchange with sea resulting in an improvement in salinity distribution and hence ecology of the lagoon. Shifting of the inlet(s) and siltation in the dredged channels are significant concerns, which has slowly trending lower saline influx in post-intervention to present conditions. The inlet is highly dynamics over a period of time and depends on freshwater discharge and northward longshore drift along the coastline. The present scenario of lagoon system suggests detailing modelling of the lagoon by incorporating the latest bathymetry, geomorphologic features around the inlet, flushing/dredged channels and Muggarmukh gut area.

References

Balas L, Özhan E (2002) Three-dimensional modelling of stratified coastal waters. Estuar Coast Shelf Sci 54:75–87. https://doi.org/10.1006/ecss.2001.0832

Chandramohan P, Nayak B (1994) A study for the improvement of the Chilka Lake tidal inlet, East Coast of India. J Coast Res 10:909–918

Chandramohan P, Jena B, Kumar V (2001) Littoral drift sources and sinks along the Indian coast. Curr Sci 81:292–297

Chapra S (1997) Surface water-quality modeling. Tufts University/Waveland Press, Inc, Illinois

Chubarenko I, Tchepikova I (2001) Modelling of man-made contribution to salinity increase into the Vistula Lagoon (Baltic Sea). Ecol Model 138:87–100. https://doi.org/10.1016/S0304-3800(00)00395-1

Cioffi F, Eugenio A, Gallerano F (1995) A new representation of anoxic crises in hypertrophic lagoons. Appl Math Model 19:685–695

CWPRS (1998) Mathematical model studies to assess the effect of the proposed channel flow circulation and salinity Chilika Lake, Odisha. Pune

DHI (2007) Mike21 & Mike 3 flow model FM. Hydrodynamic and transport module: scientific documentation

Dias J (2001) Contribution to the study of the Ria de Aveiro hydrodynamics. University of Aveiro, Aveiro

Dias J, Lopes J (2006) Implementation and assessment of hydrodynamic, salt and heat transport models: the case of Ria de Aveiro Lagoon (Portugal). Environ Model Softw 21:1–15

Madsen P, Rugbjerg M, Warren I (1989) Subgrid modelling in depth integrated flows. In: 21st international conference on coastal engineering. American Society of Civil Engineers, Spain, pp 505–511

Mishra S, Jena J (2014) Migration of tidal inlets of Chilika Lagoon, Odisha, India – a critical study. Int J Eng Technol 6:2453–2464

Mohanty P, Panda U (2009) Circulation and mixing processes in Chilika lagoon. Indian J Geo-Mar Sci 38:205–214

Mohanty R, Mohapatra A, Mohanty S (2009) Assessment of the impacts of a new artificial lake mouth on the hydrobiology and fisheries of Chilika Lake, India. Lakes Reserv Res Manag 14:231–245. https://doi.org/10.1111/j.1440-1770.2009.00406.x

Nayak B, Ghosh L, Roy S, Kankara R (1998) A study on hydrodynamics and salinity in the Chilika Lake. In: International workshop on sustainable development of Chilika Lagoon. Chilika Development Authority, Bhubaneshwar, pp 31–47

Panda U (2008) Modelling water quality and spatial information system on Chilika Lagoon. Berhampur University

Panda U, Mohanty P (2008) Monitoring and modelling of Chilika environment using remote sensing data. In: Sengupta M, Dalwani R (eds) Taal2007: the 12th world lake conference, Jaipur, pp 617–638

Panda U, Mohanty P, Samal R (2013) Impact of tidal inlet and its geomorphological changes on lagoon environment: a numerical model study. Estuar Coast Shelf Sci 116:29–40. https://doi.org/10.1016/j.ecss.2012.06.011

Panda U, Mahanty M, Rao V et al (2015) Hydrodynamics and water quality in Chilika Lagoon-a modelling approach. Procedia Eng 116:639–646

Pattnaik A (2001) Hydrological intervention for restoration of Chilika Lagoon. In: Chilika newsletter, vol 2. Chilika Development Authority and Wetlands International South Asia, pp 3–5

Pedlosky J (1987) Geophysical fluid dynamics, 2nd edn. Springer, New York

Ramirez I, Imberger J (2002) The numerical simulation of the hydrodynamics of Barbamarco Lagoon, Italy. Appl Numer Math 40:273–289

Wang J (1990) Numerical modelling of bay circulation. In: Méhauté B, Hanes D (eds) The sea: ocean engineering science. Wiley, New York, pp 1033–1066

Willmott C (1981) On the validation of models. Phys Geogr 2:184–194

Zacharias I, Gianni A (2008) Hydrodynamic and dispersion modeling as a tool for restoration of coastal ecosystems. Application to a re-flooded lagoon. Environ Model Softw 23:751–767

Zhao D, Shen H, Tabios G et al (1994) Finite-volume two-dimensional unsteady-flow model for river basins. J Hydraul Eng 120:863–883. https://doi.org/10.1061/(ASCE)0733-9429(1994)120:7(863)

Chapter 7
Assessment of Runoff and Sediment Yield from Selected Watersheds in the Western Catchment of the Chilika Lagoon

Bhabani S. Das, Priyabrata Santra, Rohit Ranjan, Hitesh B. Vasava, Rabindra Nath Samal, and Ajit K. Pattnaik

Abstract The knowledge of erosion-prone zones and its severity over a spatial scale is an important input into development sediment management strategies. In this chapter, hydrologic modelling of runoff and sediment load using Soil and Water Assessment Tool (SWAT) from two selected river basins (e.g., Badanai and Kansari) of western catchment has been used as examples to assess the capability of models for simulating runoff and sediment load into the lagoon and to implement suitable preventative measures. Calibration of SWAT model setup for both the river basins showed that effective hydraulic conductivity of the main channel (CH_K), base flow alpha factor (ALPHA_BF), curve number corresponding to antecedent moisture content II (CN2), and roughness coefficient of the main channel (CH_N) were most sensitive parameters. Nash-Sutcliffe coefficient of simulated flow during the calibration period was 0.76 in Badanai River Basin whereas it was 0.67 in Kansari River Basin. Estimation of runoff was even better during the validation period in both river basins (NSC = 0.88 and 0.69, respectively). The model results indicate that at least 50–60% of the total rainfall contributed to runoff, whereas only 10.4% contributes to groundwater. Furthermore, runoff water from these two basins carries a lot of eroded sediments with an average concentration of 0.12 kg m^{-3}. Such models provide vital information for estimation of the total runoff and sediment

B. S. Das (✉) · R. Ranjan · H. B. Vasava
Agricultural and Food Engineering Department, Indian Institute of Technology Kharagpur, Kharagpur, West Bengal, India

P. Santra
Division of Agricultural Engineering for Arid Production Systems, ICAR – Central Arid Zone Research Institute, Jodhpur, Rajasthan, India

R. N. Samal · A. K. Pattnaik
Chilika Development Authority, Bhubaneswar, Odisha, India

C. M. Finlayson et al. (eds.), *Ecology, Conservation, and Restoration of Chilika Lagoon, India*, Wetlands: Ecology, Conservation and Management 6,
https://doi.org/10.1007/978-3-030-33424-6_7

generation potential from different areas of the lagoon catchments and can provide significant inputs for formulation of soil and water conservation plan.

Keywords Chilika · Hydrological modelling · SWAT · Runoff and sediment load

7.1 Introduction

Located in the eastern coast of Odisha along the Bay of Bengal, Chilika is one of the important and highly significant Ramsar Sites of India. With more than 800 species of fauna, this rich and productive ecosystem provides a livelihood to more than 200,000 fishermen in the surrounding area besides being the largest wintering home for many migratory waterbirds of the Central Asian Flyway (Panda et al. 2008). The lagoon has an average water depth of 1.8 m and average water-spread area of about 765 km^2 spreading north-south over a stretch of 65 km (Kumar and Pattnaik 2012).

Chilika has a catchment area of about 4406 km^2. Sixty percent of this area is located on the western side of the lagoon (also, called Western Catchment), which consist of agricultural land and forest vegetation (Fig. 7.1). The remaining 32% is located in the north and constitutes a fraction of Mahanadi Delta (Kumar and Pattnaik 2012).

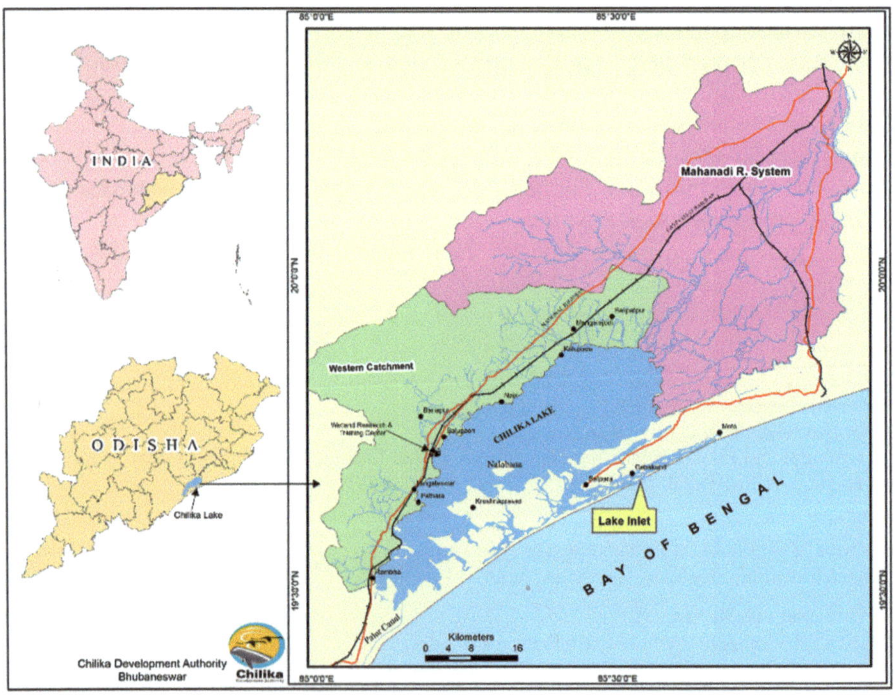

Fig. 7.1 Location of Chilika Llagoon and its catchment areas

Fifty-two rivers/rivulets (5 are from the Mahanadi basin and 47 from the Western Catchment) flow into Chilika. Estimation of freshwater inflow into the lagoon in 2010 showed that Western Catchment could contribute to the tune of 1.36×10^9 m^3 freshwater constituting about 42% of the total freshwater inflow into the lagoon (Kumar and Pattnaik 2012). The remaining 58% comes from the Mahanadi river system.

An estimated 571 km^2 of the catchment area has been identified to be highly erosion-prone (Chilika Development Authority 2012). Large sediment load in freshwater streams arising out of agricultural landscape continually enriches the lagoon water with nitrates, phosphates and Dissolved Organic Carbon (DOC) leading to extensive colonisation of macrophytes in Chilika (Panigrahi et al. 2007). Gupta et al. (2008) observed contrasting ratios for organic to inorganic carbon in the pre-monsoon and monsoon seasons, which was found to be controlled by the carbon loads in river waters and influx of seawater from the Bay of Bengal. High carbon and nutrient loads are linked with greater phytoplankton growth in the lake (Dube and Jayaraman 2008). Even the extent of light penetration through lagoon water is influenced by the sediment load, which is expected to affect ecosystem health (Panigrahi et al. 2007). Brackish water in the eastern side, carbon and nutrient-loaded runoff water from the western and northern side, high sediment loads, abundant macrophytes within the lagoon, and rapid development of peri-urban neighbourhood have significant influence on the functioning of Chilika ecosystem.

Historically, most lagoons have disappeared in geologic time (Barnes 1980) primarily as a result of siltation and plant colonisation eventually becoming swamps and marshy land. Chilika may not remain an exception unless adequate steps are taken. During last decade, shrinkage in the water-spread area, siltation and loss of salinity, extensive weed growth have emerged as some of the major threats to this lagoon (Ghosh et al. 2006; Samal 2011; Panda et al. 2013). Back in the early twentieth century, the total water-spread area for this lagoon was about 1165 km^2 in the wet season (monsoon) and about 906 km^2 in the dry seasons (post-monsoon), which has now reduced to an average of 760 km^2 (Ghosh et al. 2006). Similar to the water-spread area, the depth of water also has shown decreasing trend. The maximum water depth was observed to decrease from 340 cm in 1992 to 142 cm in 1997. Similarly, the minimum depth declined from 74 to 42 cm during the same period (Ghosh et al. 2006). The mouth of lagoon to the Bay of Bengal has witnessed an even more dynamism. Following the opening of an artificial mouth in 2000 at Sipakuda by the Chilika Development Authority (CDA), a new mouth has opened naturally at Gabakunda in 2008. The dynamic nature of the Chilika lagoon system necessitates a comprehensive assessment of the lagoon and the catchment system in a heuristic manner.

Fortunately, the Chilika Development Authority (CDA) at Bhubaneswar is implementing several interventions for managing Chilika and its catchment. In addition to opening an artificial mouth at Sipakuda in 2000, CDA has implemented several social forestry efforts, soil conservation measures, and improved agricultural practices among others (Chilika Development Authority 2012). It is currently monitoring runoff and sediment load in several streams flowing into the lagoon.

Physical interventions are bringing about significant changes in the Chilika lagoon system while the collection of data for both catchment and lagoon water is expanding our knowledge of how this ecosystem functions. The opening of the artificial mouth, for example, has resulted in (a) improved salinity flux resulting in favourable salinity gradient across the lake, (b) flushing-out of sediments from the lake, (c) reduction of water-logging in the paddy field of northern sector during monsoon, (d) improvement in the fishery resources, and (e) reduction in freshwater weed spread area in northern sector. An estimated 286 km^2 of the water-spread area continues to have an extensive infestation of macrophytes (Chilika Development Authority 2012). The CDA has also undertaken an extensive assessment of the lagoon water both in terms of water quality and water quantity.

If Chilika has to sustain its productivity and eco-diversity, the runoff and sediment load from its freshwater streams must be reduced at the source (i.e., catchment). Therefore, a comprehensive geohydrologic assessment of the Chilika lagoon system along with its catchments is an important step in this direction. The catchment for the Chilika lagoon is large, and linked with complex geohydrologic processes and human interventions. For such large systems, integrated system-level models hold the key to describe hydrologic states in a reasonably predictable manner. The objective of this chapter is to summarise a few modelling studies undertaken by our group to assess the contribution of the catchment to the runoff and sediment load to the lagoon. The premise is that the calibration and validation of a system-level model for the large catchment system may lead to the development of robust water and land management options such as construction of check dams or open wells to store excess water, judicious use of rainwater and soil moisture, conservation agriculture, development of a diversified cropping plan among others. In what follows, hydro-geological description of the Western Catchment is presented in conjunction with the distributed model of the Soil Water Assessment Tool (SWAT) (Arnold et al. 1998; Neitsch et al. 2002). In turn, the data required for calibrating a catchment-scale model, few preliminary results, and opportunities in using such calibrated models have also been discussed using two significant watersheds of the Western Catchment as example cases.

7.2 Chilika Lagoon and Its Hydrological Characteristics

7.2.1 Hydrology

The main source of freshwater for the Chilika lagoon is the rainwater on the lagoon surface and runoff water contributed from 52 rivers and rivulets interspersed on Western Catchment and Mahanadi delta fraction. The annual average freshwater influx for the period 1999–2007 was estimated to be 5.09×10^9 m^3 (Panda et al. 2008). Rivers on the north-east side (e.g., Bhargavi, Daya, Nuna, Makara) contribute 60–80% of total freshwater input, while the western rivers (e.g., Kansari, Kusumi, Janjira, Tarimi, etc.) and rivulets contribute the rest. No freshwater influx is possible from these rivers from north-east as well as a western side during the summer season (Sahu et al. 2014). Most streams from Western Catchment are torrential. Out of these, highest discharge is received from Kansari River (about 1838 $m^3 s^{-1}$ in 2002, which was 18.2% of the total discharge from Western Catchment) (Table 7.1). Discharge data of other gauged streams (e.g., Tarimi, Kusumi, Janjira, Badasankha, Manglajodi, Badanai, Kantabania) are given in Table 7.1. The contributions from the un-gauged streams were calculated using the rainfall-runoff coefficient of the gauged streams. All these streams are non-perennial and flow only during the monsoon season. The table shows that the Western Catchment contributed about 34% of the total discharge into the Chilika lagoon in the year 2002. The remaining is contributed by four tributaries (Makara, Daya, Nuna and Bhargavi) of the Mahanadi river system. Runoff water from Western Catchment carries a large volume of sediment load (about an average of 3×10^5 t of sediment per year) to the Chilika. Kansari carried the highest amount of sediment load (4×10^4 t). Monthly distribution of sediment load from several gauging stations in Western Catchment areas is presented in Table 7.2.

Apart from the sediment load in runoff water, sediment yield from the catchment area of the individual stream also indicates the magnitude of degradation of the watershed (Naik et al. 2008). Higher sediment yield is expected in case of absence of soil conservation measures in the catchment or higher erodibility of surface soil in the catchment. Table 7.3 shows sediment yields from various gauged streams of the Western Catchment collected during 2002. Maximum sediment yield may be observed from Kantabania (1.288×10^7 t/ha) followed by Manglajodi (6.81×10^6 t/ha). Large sediment yield may be attributed to highly degraded conditions of the catchment area of Kantabania and Manglajodi watershed. Rest of the streams' sediment yield varied from 0.0 to 2.85×10^6 t/ha, which shows a considerably better condition of their watersheds.

7.2.2 Weather

The catchments of the Chilika lagoon has a tropical climate with an average annual maximum temperature of 39.9 °C and a minimum temperature of 14 °C. During December and January, cold wave conditions prevail for a couple of weeks because of Western Disturbances. In the inland hilly tract, the climate is comparatively drier with higher temperature during the summer months and slightly cooler in winter. December to February is the winter season, which is followed by a hot season from March to May. Rainfall in the region is contributed by south-west and north-east monsoons from June to September and November to December, respectively. About 75% of the annual rainfall is received during the monsoon months from June to September. Rainfall generally decreases from north-east to south-west. The monsoon starts by about the second week of June and withdraws early in October. Wind speed is high from March to July and speed is low during the winter season. The direction of the wind is mostly from the north and northeasterly. However, during

Table 7.1 Monthly discharge from major streams from Western Catchment of Chilika lagoon during 2002

	[a]Discharge in 2002 (cumecs)							
Month	Mangalajodi	Tarimi	Kantabania	Kusumi	Badanai	Kansari	Janjira	Badasankha
January	0	0	0	0	0	0	0	0
February	0	0	0	0	0	0	0	0
March	0	0	0	0	0	0	0	0
April	0	0	0	0	0	0	0	0
May	0	0	0	0	0	0	0	0
June	83	61	80	55	0	0	36	0
July	505	61	95	88	2	537	145	6
August	906	288	214	376	20.34	510	517	364
September	303	223	169	215	–	562	380	179
October	24	113	97	187	–	229	293	18
November	–	–	–	–	–	–	–	–
December	–	–	–	–	–	–	–	–
Total	1821	746	655	921	22.34	1838	1371	567

[a]Total discharge from the western catchment in 2002 was about 10,098.9 $m^3 s^{-1}$

Table 7.2 Monthly sediment loads from major river basins in Western Catchment of Chilika lagoon during 2002

	Sediment load (metric ton)							
Month	Mangalajodi	Tarimi	Kantabania	Kusumi	Badanai	Kansari	Janjira	Badasankha
January	0	0	0	0	0	0	0	0
February	0	0	0	0	0	0	0	0
March	0	0	0	0	0	0	0	0
April	0	0	0	0	0	0	0	0
May	0	0	0	0	0	0	0	0
June	1082	811	1770	788	0	0	410	0
July	13,093	1167	1386	1214	92	17,150	221	191
August	22,711	3672	2373	12,109	330	16,744	7687	5646
September	4825	1871	1565	5256	0	9185	5660	1849
October	483	888	993	2404	0	2765	2078	130
November	–	–	–	–	–	–	–	–
December	–	–	–	–	–	–	–	–
Total	42,194	8409	8087	21,771	422	45,844	16,056	7816

Table 7.3 Sediment yield concentration of different streams from the Western Catchment of Chilika lagoon

Sl.no	Rivers	Sediment load (Mt)	Area (sq. km)	Yield (Mt/ ha)
1	Tarimi	8409	87	0.97
2	Kusumi	21,771	141	1.54
3	Kansari	45,844	161	2.85
4	Janjira	16,056	455	0.35
5	Kantabania	8087	6	12.88
6	Badasankha	7816	129	0.61
7	Manglajodi	42,194	62	6.81
8	Badanai	422	87	0.05

monsoon month it is primarily southerly and southwesterly direction because of the influence of south-west monsoon. Wind speed varies from 5.3 to 16.0 km h^{-1}.

7.2.3 Geology

The Chilika Lake is originated through a complex geologic process involving deposition of beach ridges and spits enclosing a body of seawater within the Bay. The lake was a part of the Bay of Bengal about 6000 years ago, and served to be its gulf during Pleistocene. The current form of Chilika is due to the successive recession of coastline aided by marine and fluvial dynamics over 6–7000 years (Kumar and Pattnaik 2012). It contains a wide range of sedimentary particles such as clay, silt, sand, gravel and shell banks but the significant part of the catchment area is silt (Chilika Development Authority 2012; Kumar and Pattnaik 2012). Main rock types seen around Chilika are khondalites, unclassified granites and gneisses, charnockites, anorthosites, granulites, laterites and alluvium. Khondalites are essentially garnet-sillimanite schist with varying quantities of garnet as well as some quartz and feldspar. The contact between the schist and gneisses appear to be gradual. A most common form of granite available around Chilika is usually coarse-grained consisting of quartz, feldspar, garnet and biotite. The charnockites show an intrusive relationship with khondalites. Some islands within Chilika such as Kalijai, Barakuda, Sanakuda, Somolo and Gopakuda also comprise of Eastern Ghat rocks and are rich in khondalites. The Bhasramundia represents charnockite, and Birds Island represents granite gneiss. Granite gneiss and charnockites occur in lesser abundance. The occurrence of anorthosite is seen in Banapur-Balugaon region. The overall trend of foliation of Eastern Ghat rocks is North-East South-West. However, local variations are also observed (Chilika Development Authority 2012; Kumar and Pattnaik 2012).

7.2.4 Soil

Table 7.4 shows a comparison of typical soil properties for Western Catchment and Mahanadi delta. These results were compiled using soil samples collected from the top 20 cm of soil layers from 241 geo-referenced locations of these two catchments. These data show that the soils in these two catchments are primarily coarse-textured, moderate to strongly acidic and have low soil organic carbon contents. With a horseshoe shape, the Western Catchment is surrounded by forests on all the three directions with the Chilika lagoon in the east. This may have been the reason for relatively higher organic carbon fraction in its soils than those of the Mahanadi basin. Assessment of soil series data (National Bureau of Soil Survey and Land Use planning, NBSS&LUP) for the Western Catchment, shows that these soils are generally thick, sandy loam in texture and falls under mainly five soil series (Sarkar et al. 2005). Detailed soil properties under these five soil series are presented in Table 7.4, which shows that soil under the Jamguda soil series was the thickest (~150 cm) whereas soils under Bandhadwar soil series were very shallow (~69 cm). Soils are slightly acidic except for Nuagarh soil series. The organic carbon content of surface soils is about 0.3–0.5%, however, for Jamguda soil series it is quite high (~1.82%).

7.3 Modelling Hydrological Behavior in Watershed Using SWAT

7.3.1 Hydrologic Modelling Approaches

Over the last few decades, a great stride has been made on developing physically-based and distributed-parameter hydrological models, e.g. Soil and Water Assessment Tool (SWAT), Agricultural Non-point Source Pollution (AGNPS), Système Hydrologique Européen (SHE) etc., which are capable of generating area-wise and hydrologic process-wise outputs over a watershed. Physical laws governing

Table 7.4 Measured soil properties of Western Catchment and Mahanadi delta region

Soil properties	Western Catchment (N = 209)		Mahanadi Basin (N = 32)	
	Average	Inter-quartile range	Average	Inter-quartile range
Sand content (%)	48.6	35.9–58.2	60.2	52.0–72.0
Clay content (%)	28.5	23.4–35.1	24.4	19.5–31.2
Bulk density[a] (mg m^{-3})	1.51	1.39–1.61	1.74	1.63–1.86
pH	6.35	5.65–7.07	5.61	5.08–5.71
EC (dS m^{-1})	0.73	0.21–0.55	0.10	0.04–0.07
SOC (%)	0.91	0.70–1.05	0.61	- 0.74

[a]For bulk density measurements, the number of samples for Western catchment and Mahanadi basin analysed were 115 and 20, respectively

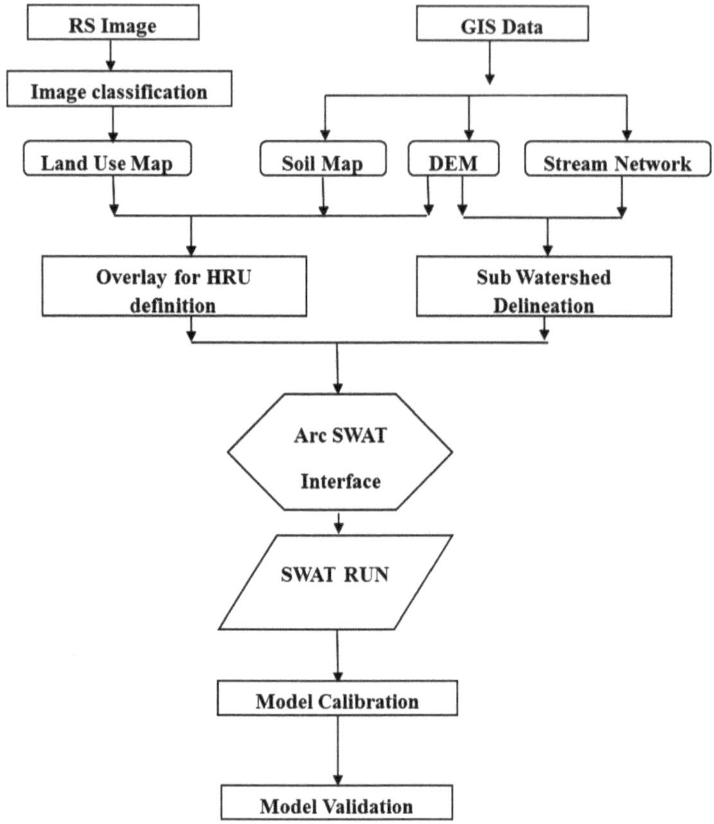

Fig. 7.2 Schematic diagram of working principle of SWAT model

hydrological processes are taken into account in these models. Once parameterised, such models may be directly applied to ungauged basins. The working principles and flowchart of SWAT modelling environment are presented in Fig. 7.2. Significant components of the model include hydrology, weather, erosion, soil temperature, crop growth, nutrients, pesticides, and agricultural management. These physically-based models are implemented by discretising the watershed into hydrological response units (HRUs), solving the physically-based governing non-linear hydrological equations for each zone, and by aggregating the outputs (Wood et al. 1988; Flügel 1995; Leavesley and Stannard 1995). The term 'distributed-parameter' stems from such segregation of watershed and parameterisation of each unit. Land use and soil characteristics for each HRU is assigned based on the most significant coverage of these two attributes within that HRU (Neitsch et al. 2002). For example, if loam covers the most significant area within an HRU, then this soil texture is assigned to the whole of that HRU irrespective of the presence of other soil textural classes within that HRU. A complete description of all components may be found in Arnold et al. (1998) and Neitsch et al. (2002).

The water balance equation is the driving force for the simulation of hydrological behaviour in a watershed using SWAT:

$$SW_t = SW_{t-1} + \sum_{i=1}^{t} \left(P_{day\_i} - Q_{surf\_i} - ET_{a\_i} - W_{seep\_i} - Q_{gw\_i}\right) \tag{7.1}$$

where SW_t is the final soil water content (mm), SW_{t-1} is the initial soil water content on day i (mm), P_{day_i} is the precipitation on day i (mm), Q_{surf_i} is the surface runoff on day i (mm), ET_{a_i} is the actual evapotranspiration on day i (mm), W_{seep_i} is the water entering into the vadose zone from the soil profile on day i (mm), and Q_{gw_i} is amount of return flow on day i (mm). In SWAT, the local hydrologic water balance in each HRU is provided by four storage volumes: snow (stored volume until it melts), soil profile (0–2 m), shallow aquifer (typically 2–20 m), and deep aquifer (>20 m). Soil water processes include infiltration, runoff, evaporation, plant uptake, lateral flow, and percolation to lower layers. Percolation from the bottom of the soil profile recharges the shallow aquifer (groundwater recharge). SWAT simulates the total groundwater recharge as: (a) water that passes past the bottom of the soil profile, (b) channel transmission losses and (c) seepage from ponds and reservoirs. Surface runoff from daily rainfall is estimated with a modification of Soil Conservation Service (SCS) curve number method (USDA, SCS 1972). In the curve number method, the daily rainfall is partitioned between surface runoff and infiltration as a function of antecedent soil moisture condition. Green and Ampt infiltration method is also available within SWAT to simulate surface runoff and infiltration (Green and Ampt 1911; Mein and Larson 1973). Erosion and sediment yield are estimated for each sub-basin with the Modified Universal Soil Loss Equation (MUSLE) (Williams 1975). The model either deposits excess sediment or re-entrains sediment through channel erosion depending on the sediment load entering the channel.

The SWAT model is calibrated with the observed data on runoff and sediment yield. Calibrated parameters are then validated in independent years to simulate a daily average runoff. Auto-calibration is generally performed with the daily observed data on runoff or sediment load or both. Before calibration, sensitive parameters are identified through sensitivity analysis. During sensitivity analysis, the sum of square on residual (SSR) is used as an objective function. Sensitive parameters are calibrated with parameter solution (ParaSol) optimisation method using auto-calibration option. Briefly, in the ParaSol algorithm as implemented with SWAT 2005, parameters affecting hydrology or pollution can be changed either in a lumped way (over the entire catchment), or in a distributed way (for selected sub-basins or HRUs). They can be modified by replacement, by addition of an absolute change or by a multiplication of a relative change. A relative change means that the parameter or several distributed parameters simultaneously are changed by a certain percentage. However, a parameter is never allowed to go beyond the predefined parameter ranges. The efficiency and performance of the

SWAT model calibration and validation is assessed according to Nash-Suthcliffe coefficient (Nash and Suthcliffe 1970) which is given by

$$NSC = 1 - \frac{\sum_{i=1}^{n} (M_i - O_i)^2}{\sum_{i=1}^{n} (O_i - \overline{O}_i)^2} \tag{7.2}$$

where, M_i is the modelled or simulated value of flow, O_i is the observed value of flow, $\overline{O}_i$ is the mean of observed values of flow, and n is the number of observations. The value ranges between $-\infty$ and 1 and the higher the value, the more efficient is the calibration. A negative value indicates that the mean of the observed value would have been a better predictor than the simulated values with the SWAT model.

7.4 Modelling Runoff from Western Catchment Using SWAT

7.4.1 Modelling Runoff in Badanai Stream

The catchment area of Badanai stream was delineated in ArcSWAT environment, which is known as Dengei Pahad Watershed (DPW). The watershed is located between 19°49′48″–19°52′8.4″ N and 85°13′55.2″–85°14′34.8″ E (Fig. 7.3) with an area of about 42 km^2. A detailed description of the DPW watershed is given in Santra and Das (2008). The area is a hilly terrain with the mean sea level varying from 5 to more than 451 m. The hills and isolated rocky knobs break the watershed into small but well-cultivated fields.

7.4.1.1 GIS Grids from Badanai River Basin

ArcSWAT version 1.0.7 was used to prepare the input database for SWAT run. The SWAT model requires three GIS data layers (digital elevation model (DEM), soils, and land use) and the weather data of the study area. The DEM acquired with the shuttle radar topographic mission (SRTM) with a spatial resolution of 90 m around the study area was downloaded from http://srtm.csi.cgiar.org/ website (Rabus et al. 2003). Before the DEM was used for modelling, it was projected to Universal Transverse Mercator (UTM) under appropriate zones. The study area falls in the UTM zone number 45. The DEM of the delineated watershed is shown in Fig. 7.4a. The southern portion of the watershed is low lying areas whereas the northern part of the watershed is dominated with high hill slopes. The difference in elevation from the watershed outlet to the highest hill point was found about 446 m within a horizontal distance of 7 km.

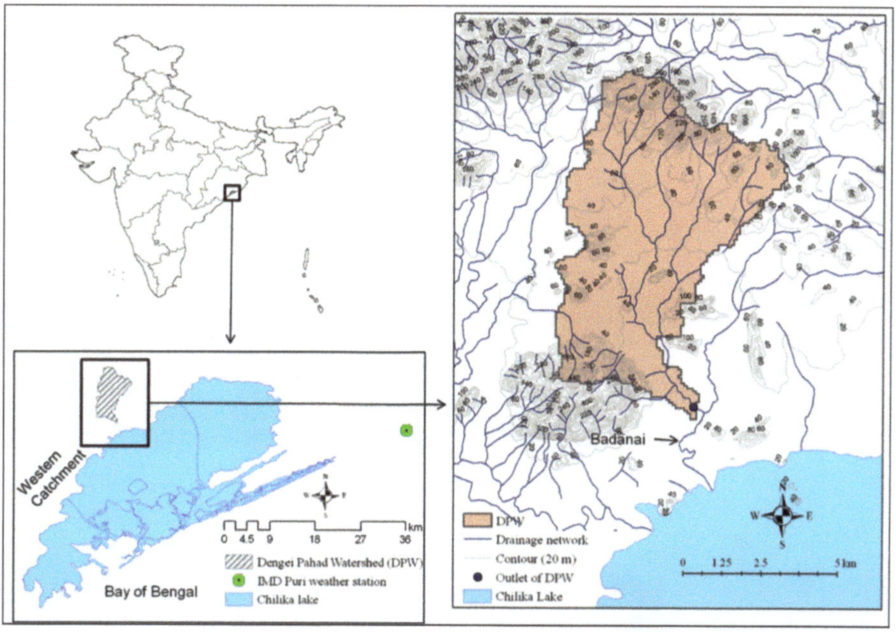

Fig. 7.3 Location of Badanai Stream catchment in Western Catchment of Chilika lagoon, which is also known as Dengei Pahad Watershed (DPW); bottom left frame indicates the location of Badanai Stream catchment in western catchment of Chilka Lake. Right frame shows the drainage network surrounding the watershed, 20 m contour lines and outlet of DPW watershed

The land use grid was prepared from the advanced wide field sensor (AWiFs) image acquired by Indian remote sensing satellite, IRS-P6. Raw image was classified into land use classes with the help of ground truth data using ERDAS IMAGINE version 8.0. The classified land use map of the watershed is given in Fig. 7.4b. The land use/land cover classes of the watershed are agricultural land (41.65%), forested area with evergreen and deciduous trees (21.72%), and wetlands with natural shrubs (36.63%). In agricultural land, rice is mainly grown during *kharif* season, which is totally rainfed and followed by fallow. Rabi crops are rarely grown in the area. The soil series map at 1:250,000 scale for Odisha state (Sarkar et al. 2005) was used as the source of soil grid. According to soil series map, 69% of the watershed is under a single soil series, Nuagarh.

7.4.1.2 Weather Data

SWAT requires daily values of weather data such as precipitation, maximum and minimum temperature, solar radiation, relative humidity and wind speed as inputs. To run SWAT one can either prepare a file that contains observed data or use daily values simulated by weather generator built within the SWAT model. Daily weather data which includes rainfall, maximum temperature, minimum temperature, relative

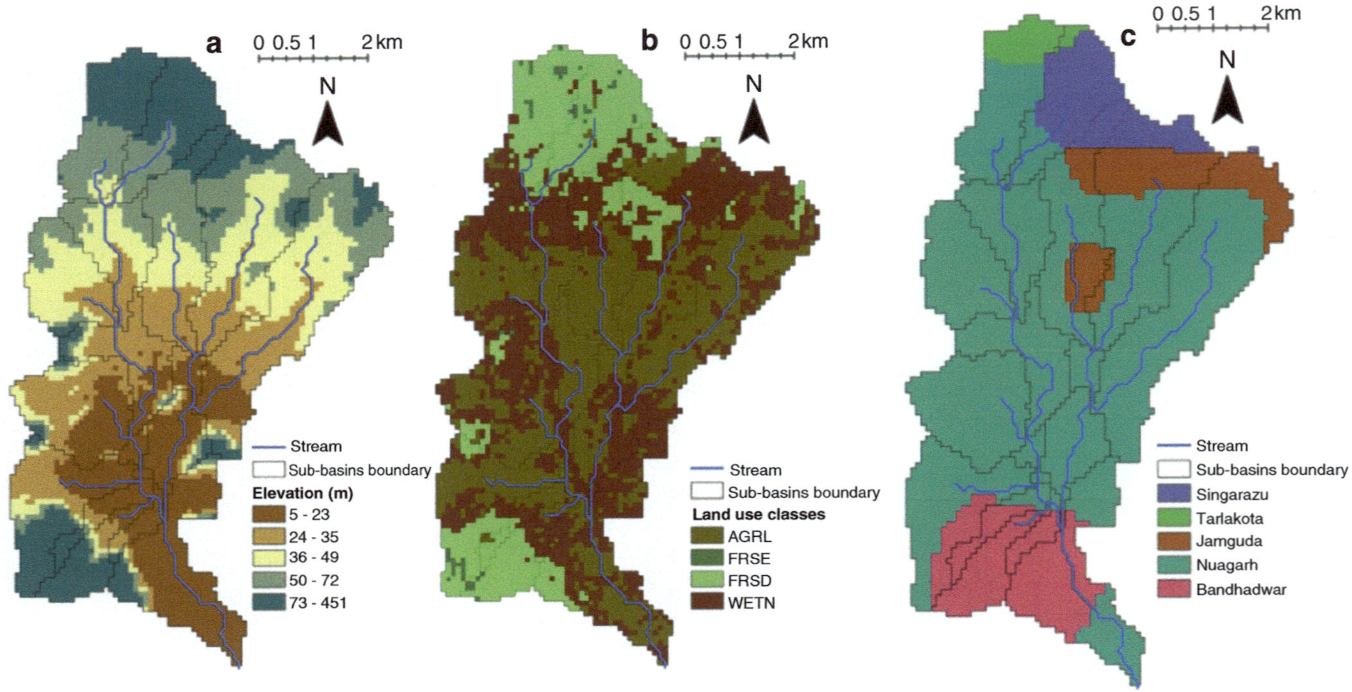

Fig. 7.4 GIS grids of Badanai Stream catchment (**a**) elevation grid prepared from Shuttle Radar Topography Mission (SRTM) digital elevation model data (90 m) (**b**) land use/land cover grid (AGRL-Agricultural land-Generic, FRSE-Forested-evergreen, FRSD-Forested deciduous, WETN-Wetland-non-forested) and (**c**) soil grid prepared from soil series map of Odisha published by NBSS&LUP, Nagpur, India

humidity, and wind speed were collected for 11 years (1996–2006) from the nearby IMD weather station located at Puri (Fig. 7.3). Although there is a difference in environmental conditions of the study area and the location of IMD weather station at Puri, no other weather station data are available close to the study site. Observed solar radiation data was not available for the Puri weather station and, therefore, was calculated from the maximum and minimum temperature (Hargreaves and Samani 1985). Monthly averages of these data were calculated for this weather station, and the weather parameter file was prepared, which is presented in Table 7.5. It is noted here that the annual potential evapotranspiration (PET) of the study area is about 1400–1500 mm. With no weather records available for the Western Catchment, efforts may be needed to establish weather stations for long-term monitoring of weather data to facilitate the assessment for catchment hydrology for Chilika.

7.4.1.3 Rainfall-Runoff Relationship in the Watershed

Rainfall-runoff relationship of the Badanai stream catchment is depicted in Fig. 7.5, which shows that rainfall and runoff are related to each other. During summer months and the initial period of monsoon, rainwater does not significantly contribute to runoff because a significant portion of rain is utilised to wet the soil profile. However, subsequent rainfall events significantly contribute to runoff because soil profile and vadose zone (unsaturated soil layer between the soil surface and groundwater table) almost reached to saturation stage at that time. Runoff from the watershed is generally observed from July to November in a year. During the rest of the periods, streams remain dry. Overall, it was found that runoff flow rate was higher during 2006 than during 2004 and 2005 (Figs. 7.6 and 7.7). This may be mainly because of higher rainfall (2115.6 mm) during 2006 than during 2004 and 2005 (1161.2 mm and 1669.2 mm, respectively).

7.4.1.4 Sensitive Parameters for Flow

Sensitive parameters for simulation of the mean daily flow averaged over a month in Badanai River Basin are given below. Out of 26 parameters included in SWAT for flow simulation, six parameters are related to snowmelt and, therefore, not crucial for our watersheds. Six most sensitive parameters for calibration purpose are listed below with their brief description. Two ground water-related parameters, threshold depth of water in the shallow aquifer required for return flow to occur (GWQMN) and Groundwater delay time (GW_DELAY) are also selected because of their comparatively high sensitivity.

(i) ALPHA_BF: Base flow alpha factor (days), (ii) CH_K2: Effective hydraulic conductivity in main channel alluvium (mm h^{-1}), (iii) CH_N: Manning's roughness

Table 7.5 Weather parameters for Puri weather station of Indian Meteorological Department

[a] Weather parameters	Months											
	Jan	Feb	Mar	Apr	May	Jun	Jul	Aug	Sept	Oct	Nov	Dec
TMPMX	27.36	29.25	31.01	31.84	33.10	32.22	31.62	31.42	32.09	31.88	30.37	28.18
TMPMN	17.84	21.63	25.15	26.71	27.81	27.55	27.02	26.91	27.02	25.55	21.68	17.77
TMPSTDMX	1.29	1.49	1.00	1.00	1.09	1.71	1.67	1.59	1.62	1.93	1.52	1.39
TMPSTDMN	2.54	2.71	1.69	1.63	2.00	1.72	1.49	1.25	1.22	1.56	2.25	1.75
PCPMM	13.10	37.97	26.72	26.72	55.71	410.95	394.91	347.55	259.50	206.29	45.75	17.79
PCPSTD	3.07	8.03	4.38	4.61	7.48	50.00	35.27	24.00	20.86	20.53	7.74	6.13
PCPSKW	7.00	7.00	6.59	6.69	6.54	6.97	5.13	4.31	4.47	4.94	6.38	7.00
PR_W1	0.04	0.06	0.08	0.1	0.15	0.29	0.36	0.51	0.36	0.2	0.06	0.02
PR_W2	0.32	0.18	0.28	0.26	0.2	0.63	0.71	0.73	0.69	0.58	0.44	0.4
PCPD	1.73	2.00	3.27	3.82	5.00	13.18	18.09	21.27	16.82	10.91	3.55	1.36
RAINHHMX	15.00	15.00	20.00	25.00	30.00	35.00	35.00	40.00	40.00	25.00	20.00	15.00
SOLARAV	15.72	15.76	15.82	16.10	16.86	15.96	15.70	15.22	15.05	14.97	15.35	15.71
DEWPT	18.42	22.00	24.94	25.00	25.00	25.00	25.00	25.00	25.00	25.00	21.10	17.55
WNDAV	2.21	2.38	3.14	3.64	3.58	3.45	2.96	2.90	2.40	1.76	1.43	1.76

[a]Details description of each weather parameter is available in SWAT manual (Neitsch et al. 2002). TMPMX = Average maximum temperature for a month (°C), TMPMN = Average minimum temperature for a month (°C), TMPSTDMX = Standard deviation for maximum temperature in month, TMPSTDMN = Standard deviation for minimum temperature in month, PCPMM = average amount of precipitation (mm H_2O), PCPSTD = Standard deviation of daily precipitation in month, PCPSKW = Skew coefficient of daily precipitation in month, PR_W1 = Probability of a wet day following a dry day, PR_W2 = Probability of a wet day following a wet day, PCPD = Average number of days of precipitation in month, RAINHHMX = Extreme half-hour rainfall for month, SOLARAV = Average daily solar radiation for month (MJ m^{-2}).DEWPT = average dew point temperature for month (°C), WNDAV = Average wind speed in month (m s^{-1})

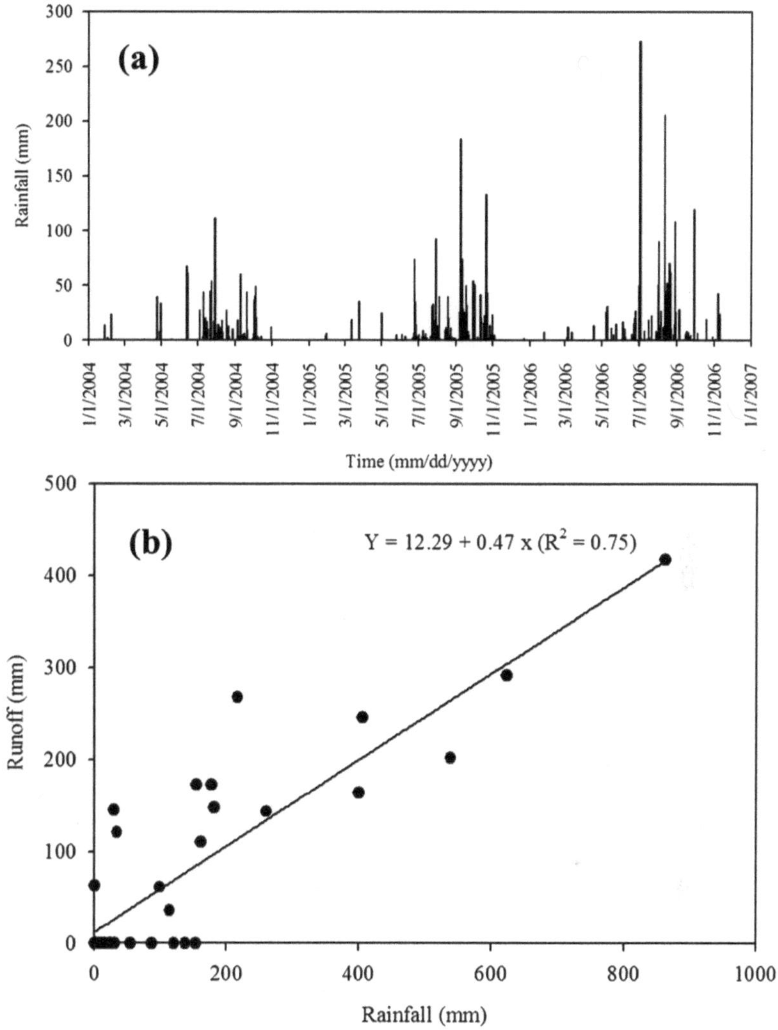

Fig. 7.5 Rainfall and runoff in Badanai Stream catchment during 2004–2006; (**a**) daily rainfall (mm) during the period 2004–2006 and (**b**) Relationship between monthly rainfall and runoff in the catchment

coefficient for the main channel, (iv) CN2: Initial SCS runoff curve number for moisture condition II, (v) ESCO: Soil evaporation compensation factor, (vi) SURLAG: Surface runoff lag coefficient, (vii) GW_DELAY: Groundwater delay time (days), (viii) GWQMN: Threshold depth of water in the shallow aquifer required for return flow to occur (mm H_2O).

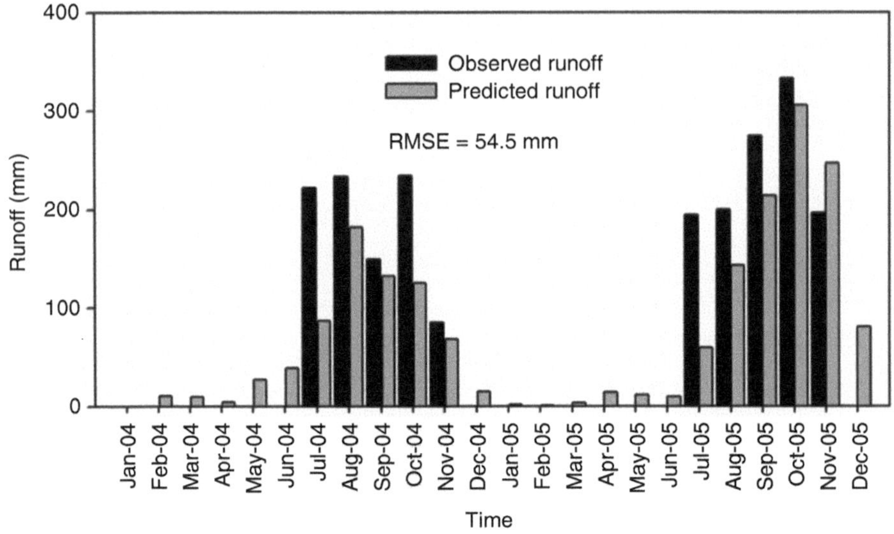

Fig. 7.6 Observed vs simulated monthly runoff in Badanai Stream in 2004–2005

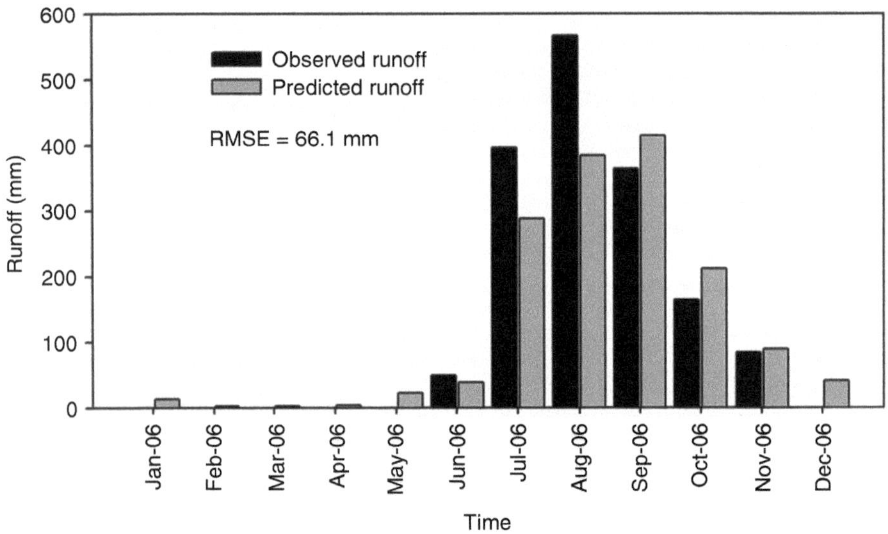

Fig. 7.7 Observed and simulated runoff from Badanai Stream during validation period in 2006

Table 7.6 Sensitive parameters for simulation of stream flow in Badanai River Basin using SWAT setup

Parameters	Rank of parameters in sensitivity analysis	Replacement method	Calibrated values
ALPHA_BF	2	1	0.1269
CH_K2	1	1	491.1000
CH_N	4	1	0.0300
CN2	3	3	+22.4800
ESCO	6	1	0.0226
GW_DELAY	13	1	480.6900
GWQMN	17	1	5000.0000
SURLAG	5	1	18.5630

+ sign in respect of CN2 value represents percent increase in CN value from its pre-calibrated set up, which is indicated by replacement method 3, whereas replacement method 1 changes the value from its old value to a new value during calibration iterations

7.4.1.5 Calibration of SWAT Model Parameters for Modelling Runoff in Badanai River Basin

Sensitive calibrated parameters obtained from auto-calibration of runoff from Badanai River Basin are presented in Table 7.6. The value of ALPHA_BF for the SWAT setup of Badanai River Basin was found to be 0.1269, which indicated a slow response of recharge to groundwater flow. High values of hydraulic conductivity (CH_K2) (491.1 mm h^{-1}) and roughness coefficient (CH_N) (0.03) for the main channel were observed. Presence of sand and gravels in the main channel of this hilly watershed leads to such high value. Specifically, the bed materials of channels at higher elevations with steep slopes mainly consist of sand and gravels. The main channels of sub-basins near to the outlet are high in silt and clay content and, therefore, the low value of CH_K2 and CH_N are expected for this zone. Since the main channel characteristics are almost similar throughout the watershed, we used only one value of these two parameters for all sub-basins and, hence, it represents the average for all sub-basins. The SCS curve number is a distributed parameter in SWAT and, therefore, each HRU was assigned with different CN2 numbers. Overall, after calibration CN2 was increased from its initial value by 22%. The final calibrated CN2 values ranged from 76 to 95. In case of agricultural land under Nuagarh and Bandhadwar soil series, CN2 was 95 whereas it was 88 under Jamguda soil series. CN2 value of forested land was 76 under Tarlakota, Jamguda and Singarazu soil series, whereas it was 88 under Nuagarh and Bandhadar soil series. The ESCO value was meager (0.0226), which indicated that most of the evaporative demand was extracted from deeper soil. The lag time to move water from the bottom of the soil profile to shallow aquifer, i.e. GW_DELAY was high (480.69 days). Therefore, deeper groundwater table and low contribution of groundwater to streamflow were expected. This was apparently visible from the high GWQMN value (5000 mm) (Table 7.6). The value of SURLAG was high (18.563), which indicated less amount of water held in storage, and a high portion of runoff was contributing to the stream.

7.4.1.6 Observed vs Simulated Runoff During Calibration and Validation Periods

Calibration of the SWAT setup in Badanai stream catchment improved the simulation performance of runoff from its pre-calibrated setup. Observed and simulated runoff for months during the calibration period is shown in Fig. 7.6. It was found that simulated values were higher than the observed value in most cases. High-intensity rainfall events generally occur in the study area during monsoon months and the SWAT setup under-predicted flow during these high-intensity rainfall events. Use of sub-daily rainfall data in place of daily rainfall data might have improved the simulation performance. NSC values of simulated runoff for months during calibration period are listed in Table 7.7. Simulation performance was better for the whole year than for monsoon months (June–September). It was expected because the stream flow in the watershed was torrential. During pre-monsoon and post monsoon months, stream flow was negligible. During the total calibration period of 2 years (2004–2005) NSC value was 0.76, whereas it was 0.45 when only monsoon months of each year were considered.

Observed and predicted values of runoff in Badanai stream for months during validation period are shown in Fig. 7.7. Overall, simulated runoff matched well with observed flow except during July and August. This may be because of high-intensity rainfall events during these 2 months and the probable reason of underprediction. The NSC values for simulation of runoff for months during validation period are listed in Table 7.7. In general, the NSC value is higher during validation period than calibration period. NSC value of simulated flows during total validation year was 0.88, whereas during monsoon months alone it was 0.76.

In addition to surface runoff, other hydrological components in Badanai catchment area such as evapotranspiration (ET) and groundwater recharge was also checked during SWAT simulation. On an average, simulated annual ET was 30–35% of total rainfall, simulated GW recharge was about 8% of total rainfall, and simulated runoff was about 60–65% of total rainfall. Because observed runoff is

Table 7.7 Nash-Sutcliffe coefficient (NSC) and root mean square error (RMSE) of the predicted monthly runoff (mm) using SWAT setup in Badanai River Basin

Time period	Season	NSC	RMSE (mm)
Calibration			
2004	Total year	0.69	54.8
	Monsoon	0.24	76.1
2005	Total year	0.81	54.3
	Monsoon	0.55	69.0
2004–05 (combined data)	Total year	0.76	54.5
	Monsoon	0.45	72.6
Validation			
2006	Total year	0.88	66.1
	Monsoon	0.76	91.4

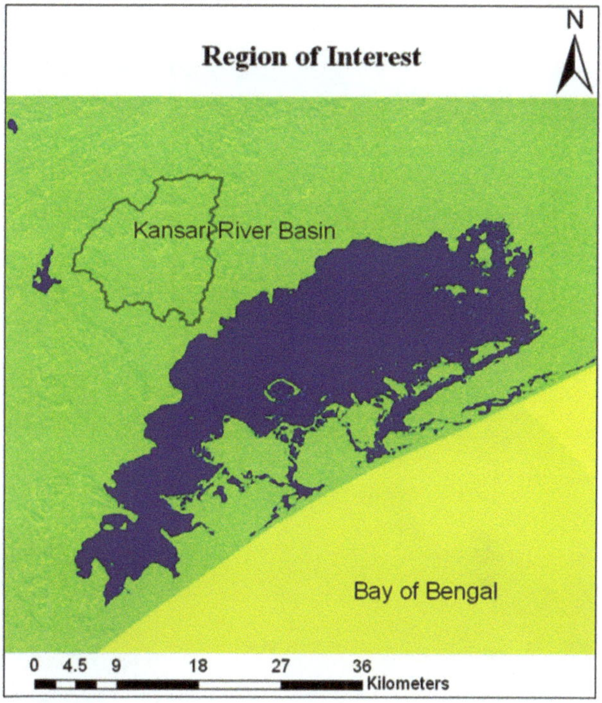

Fig. 7.8 Location map of Kansari River basin in Western Catchment of Chilika lagoon

about 53–59% of total rainfall, simulation result slightly overestimated such runoff. It has been observed that during 2004, with low annual rainfall (1161.2 mm), the contribution of annual ET was 42% of total rainfall, whereas, during excess rainfall year with occurrence of high-intensity rainfall events (Fig. 7.5a), contribution of ET to total rainfall was only 21%. The contribution of surface runoff was 71% of rainfall during excess rainfall year. Santra and Das (2013) have provided a more detailed discussion of runoff estimation through SWAT modelling approach. Because the sediment load data was not available for the Badanai, combined simulation of runoff and sediment yield was performed using the data from Kansari River Basin, which is discussed below.

7.4.2 Runoff and Sediment Load from Kansari River Basin

The Kansari River Basin is situated in the western catchment of the Chilika lagoon with an area of 174 km^2 and located between 85°5′11.94″ to 85°14′37.14″ E and 19°45′34.20″ to 19° 54′46.32″ (Figs. 7.8 and 7.9a) (Ranjan 2014). The topography of the watershed is very steep with elevations ranging from 0 to 791 m. The catchment area of Kansari stream largely contributes sediment in Chilika lagoon during monsoon periods (Tables 7.1 and 7.2).

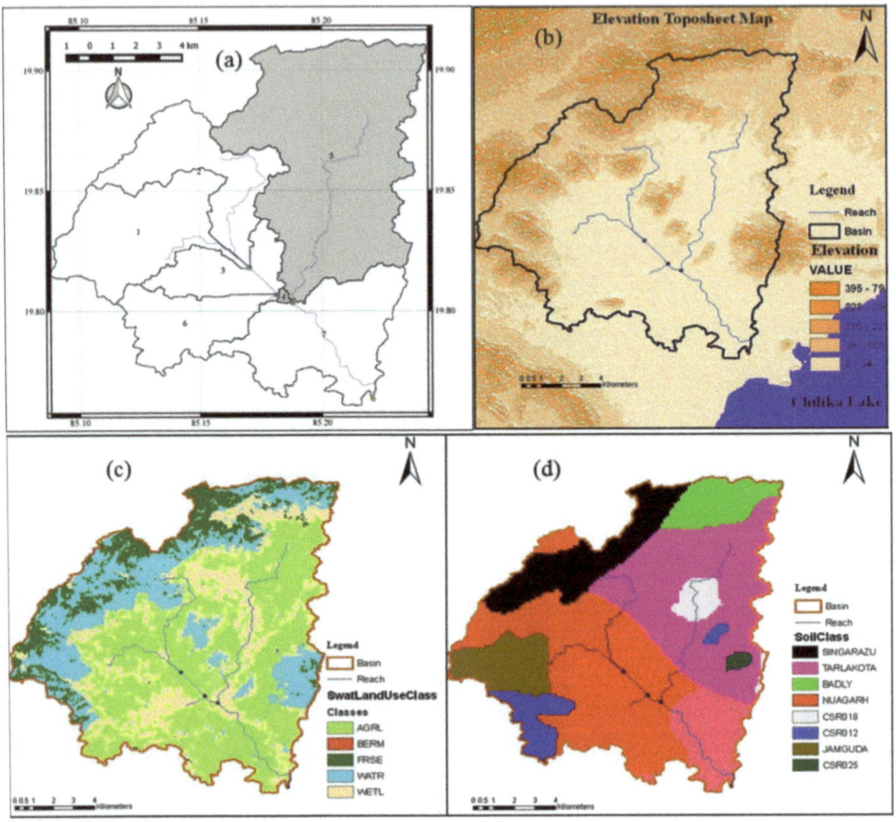

Fig. 7.9 GIS grids of Kansari River basin (**a**) elevation grid prepared from Advanced Space borne Thermal Emission and Reflection Radiometer (ASTER) global digital elevation model version 2 (GDEM V2) (30 m resolution) (**b**) land use/land cover (AGRL-Agricultural Land-Generic, FRSE-Forest-Evergreen, WATR-Water bodies and WETL-Wetlands-Mixed) and (**c**) soil grid prepared from soil series map of Odisha published by NBSS&LUP, Nagpur, India

7.4.2.1 GIS Grids from Kansari River Basin

The spatial data used in modelling runoff and sediment load of Kansari River Basin in SWAT environment includes the grid on digital elevation model (DEM), land use cover map and soil map. The Advanced Spaceborne Thermal Emission and Reflection Radiometer (ASTER) Global Digital Elevation Model Version 2 (GDEM V2) of 30 m resolution was used for hydrological modelling of Kansari River Basin using SWAT. The ASTER DEM for the study area was downloaded from the https://reverb.echo.nasa.gov. The elevation grid of the Kansari River Basin was prepared by extracting the DEM with Kansari River Basin boundary followed by re-projection in UTM projection system (Fig. 7.9b). North-western edge of the watershed is covered with steep hills with elevation ranging from 395 to 790 m above mean sea level whereas the southeastern part near the outlet are shallow lying areas (0–34 m). The

Table 7.8 Soil database of the watershed obtained from the soil series map of Odisha at 1:250,000 scale, published by NBSS&LUP

		Soil properties						
Soil series	Horizon depth (cm)	Organic carbon (%)	Sand (%)	Silt (%)	Clay (%)	Coarsefrag. Vol (%)	pH	EC (dS m^{-1})
Singarazu	0–13	0.36	66.9	14.4	18.7	–	5.3	0.08
	13–28	0.28	50	23.5	26.5	–	5.6	0.08
	28–60	0.17	50.5	19.1	30.4	–	5.8	0.04
	60–90	0.20	50.1	18.2	31.7	4	5.8	0.03
	90–127	0.12	55.6	17.3	27.1	7	5.8	0.03
Tarlakota	0–9	0.26	64.4	20.7	14.9	4	5.4	0.10
	9 to 41	0.20	60.3	21.1	18.6	12	5.8	0.08
	41–83	0.20	58.2	19.3	22.5	18	6	0.09
	83–102	0.16	57.1	18.1	24.8	25	6.3	0.05
Jamguda	0–18	1.82	33	37.7	29.3	–	6	0.95
	18–42	1.07	27	31.3	41.7	–	6	0.47
	42–68	0.67	26.8	30	45.2	–	6.1	0.38
	68–96	0.58	43.1	22.8	34.1	–	6.1	0.35
	96–124	0.47	45.3	23.1	31.6	–	6.2	0.27
	124–152	0.33	22.2	32	45.8	–	6	0.21
Nuagarh	0–21	0.6	14.4	48.5	37.1	–	7.3	0.42
	21–48	0.54	13.6	43.5	42.9	–	7.6	0.27
	48–82	0.35	35.6	15.4	49	–	7.9	0.13
	82–105	0.36	38.7	15.2	46.1	–	8	0.95
	105–155	0.2	40.2	14.8	45	–	8.2	0.76
Bandhadwar	0–14	0.80	47.6	15.6	36.8	50	5	0.05
	14–31	0.60	42.5	12.6	44.9	65	5	0.05
	31–69	0.40	44.5	10.0	45.5	75	4.9	0.05

land use grid of Kansari River Basin was prepared through supervised classification of AWiFs image acquired by IRS-P6 satellite and is presented in Fig. 7.9c. Five main land use/land cover classes were observed in Kansari River Basin, and they are Agricultural Land-Generic (AGRL), Forest-Evergreen (FRSE), Water (WATR) and Wetlands-Mixed (WETL). Land use statistics of the Kansari River Basin showed that about 42% area was covered by agricultural land, 25% by wetland, 21% by water bodies and 12% by forest area with evergreen and deciduous trees. The soil map of Kansari River Basin was prepared from soil series map of Odisha published by NBSS&LUP (http://www.nbsslup.in/downloads/NBSS_Catalogue_1601.pdf). About 63% of the watershed area of Kansari River Basin is occupied by Nuagarh and Tarlakota soil series, whereas Bandhwadwar soil series covers least area (8.46%) (Fig. 7.9d). Detail characteristics of these soil series profiles are discussed earlier (see Table 7.8).

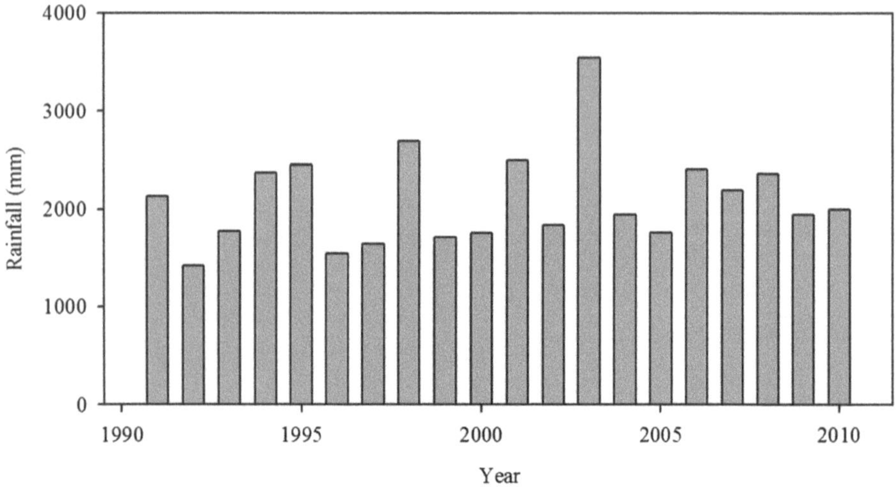

Fig. 7.10 Annual rainfall pattern in Western Catchment of Chilika Lagoon during the period from 1991 to 2010

7.4.2.2 Weather Data of Kansari River Basin

Weather data of the study area including daily rainfall, daily temperature and daily average wind speed during the period 1990–2010 were collected from Global weather data for SWAT and National Centers for Environmental Prediction (NCEP)- Climate Forecast System Reanalysis (CFSR) (http://cfs.ncep.noaa.gov/cfsr/). Average annual rainfall of the study area is shown in Fig. 7.10. The study area receives the rainfall during south-west monsoon from June to October. Other required meteorological data of the study area as given in Table 7.5 was used for modelling requirement in SWAT.

7.4.2.3 Sensitivity Analysis

Twenty-six (26) model parameters were initially used in the sensitivity analysis using global ranking methods proposed by (van Griensven 2006). The result of the sensitivity analysis is presented in Table 7.9. Parameters about global rank 1 are categorized as "very important", rank 2–10 "important", rank 11–16 as "slightly important" and rank 27 as "not important". Top 8 sensitive parameters are common for both Badanai and Kansari River Basin.

7.4.2.4 Calibration and Validation of Runoff Stream Flow in Kansari River Basin

Top 10 sensitive parameters were chosen for auto-calibration of the SWAT model (Table 7.10). As expected, the calibrated values for Kansari River Basin are different

Table 7.9 Sensitive parameters for runoff in Kansari River Basin

SN	Parameter		Rank	Sensitivity category
1	CN2	SCS runoff curve number for moisture condition II	1	Very important
2	ALPHA_BF	Baseflow alpha factor (days)	2	Important
3	ESCO	Soil evaporation compensation factor	3	
4	GWQMN	Threshold depth of water in the shallow aquifer for base flow (mm)	4	
5	SOL_AWC	Available water capacity of soil layer (mm/mm soil)	5	
6	SOL_Z	Soil depth	6	
7	REVAPMN	Threshold depth of water in the shallow aquifer for revap (mm)	7	
8	CH_N2	Manning coefficient for channel	8	
9	CH_K2	Effective hydraulic conductivity in main channel alluvium (mm/hr)	9	
10	GW_REVAP	Groundwater 'revap' coefficient	10	
11	GW_DELAY	Groundwater delay (days)	11	Slightly important
12	SURLAG	Groundwater delay (days)	12	
13	EPCO	Plant evaporation compensation factor	13	
14	CANMX	Maximum canopy index	14	
15	SOL_K	Soil conductivity (mm/h)	15	
16	SLSUBBSN	Average slope length (m)	16	
17	SLOPE	Average slope steepness (m/m)	17	
18	SOL_ALB	Moisture soil albedo	18	
19	BLAI	Leaf area index for crop	19	
20	BIOMIX	Biological mixing efficiency	20	Not important
21	SFTMP	Snowfall temperature (°C)	20	
22	SMFMN	Minimum melt rate for snow during the year (mm/°C/day)	20	
23	SMFMX	Maximum melt rate for snow during the year (occurs on winter solstice) (mm/°C/day)	20	
24	SMTMP	Snow melt base temperature (°C)	20	
25	TIMP	Snow pack temperature lag factor	20	
26	TLAPS	Temperature laps rate (°C /km)	20	

from those of Badanai River Basin (see Table 7.6). For example, hydraulic conductivity of the main channel was found to be 491.1 mm/h for Badanai River Basin whereas it was 54.58 mm/h in Kansari River Basin. It indicates that channel bed in Badanai River Basin contains more gravels and sand than Kansari River, which was also evidenced from field observations. In both the river basins, the calibration process increased the SCS CN number from its default initial value to the tune of 22.48% in case of Badanai River Basin and 24% in Kansari River Basin. ALPHA_BF was found very low in Kansari River Basin (0.0012) as compared to Badanai River Basin (0.1269), which means that the response of recharge water to

Table 7.10 Calibrated value of sensitive parameters for the Kansari River Basin

S. no	Calibrated parameters	Range Lower	Range Upper	Calibrated value
1	Baseflow ALPHA factor (days) (ALPHA_BF)	0	1	0.0012
2	Effective hydraulic conductivity in main channel alluvium (mm/h) (CH_K2)	0	150	54.58
3	Manning coefficient for channel (CH_N2)	0	1	0.93
4	SCS runoff curve number for moisture condition II (CN2)	−25%	25%	24
5	Soil evaporation compensation factor (ESCO)	0	1	0.60
6	Groundwater DELAY (days) (GW_DELAY)	0	500	239
7	Threshold depth of water in the shallow aquifer (mm) (GWQMN)	0	5000	2146
8	Available water capacity of soil layer (mm/mm soil) (SOL_AWC)	−25%	25%	24.36
9	Groundwater delay (days) (SURLAG)	0	10	5.46
10	Groundwater 'REVAP' coefficient (GW_REVAP)	0	0.036	0.018

Table 7.11 Statistical comparison between monthly observed and calibrated streamflow at the Kansari River Basin

	Stream flow (Monthly)		Stream flow (Monthly)	
	Calibration		Validation	
Statistical parameters	Observed	Simulated	Observed	Simulated
Mean ($m^3\ s^{-1}$)	13.02	8.67	10.41	6.50
Standard deviation ($m^3\ s^{-1}$)	21.14	11.62	14.61	6.96
Maximum ($m^3\ s^{-1}$)	62.10	32.22	35.84	19.98
Minimum ($m^3\ s^{-1}$)	0.00	0.00	0.00	0.00
Nash-Sutcliffe coefficient (E)	0.67		0.69	
Coefficient of determination (R^2)	0.86		0.87	

groundwater flow is prolonged. GW_DELAY was found very high for both the river basin; 239 days for Kansari River Basin and 480.69 days for Badanai River Basin. It indicates that rainwater after deep drainage takes more time to contribute to groundwater table from the bottom of the soil profile. GWQMN required for return flow in the channel as the base flow was found 2146 mm for Kansari River Basin whereas it was 5000 mm for Badanai River Basin. The surface runoff lag coefficient was lower in Kansari River Basin (5.46 days) than Badanai River Basin (18.563 days), which indicates a higher amount of runoff water storage in surface ponds in Kansari River Basin.

Comparison between observed and calibrated stream flow for Kansari River Basin is presented in Table 7.11. Mean monthly observed value of streamflow during calibration period was 13.02 $m^3\ s^{-1}$ (~194 mm) whereas the simulated streamflow was 8.67 $m^3\ s^{-1}$ (~129 mm). Such under-prediction was also observed during the validation period. Variation of observed streamflow was higher than the simulated one as seen from its higher standard deviation, which indicates the

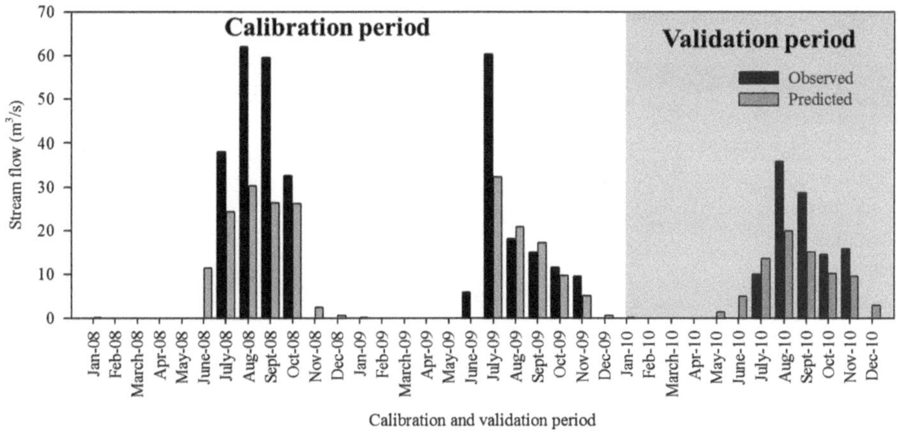

Fig. 7.11 Observed and simulated monthly average stream flow in Kansari River outlet during calibration (2008 and 2009) and validation period (2010)

smoothing effect while simulating the runoff from SWAT (Table 7.11). The gap between observed and predicted streamflow was quite high during peak streamflow, which has generally been observed during the heavy monsoon period. Average monthly calibrated streamflow at the outlet of Kansari river shows an NSC value of 0.67, whereas during validation period it was 0.69 (Table 7.11).

Observed and predicted values of monthly average streamflow during calibration (2008–2009) and validation period (2010) are presented in Fig. 7.11. It is notable here that whenever monthly average streamflow was >30 $m^3\ s^{-1}$, the gap between observed and predicted streamflow was wide (Fig. 7.11). It indicates that the calibrated model setup was unable to generate sufficient runoff during heavy rainstorms.

7.4.2.5 Calibration and Validation of Sediment Yield from Kansari River Basin

Calibration of sediment load in runoff water was also done simultaneously with the stream flow calibration because of interdependency between these two processes. Calibration was done for the period 2008–2009 at the outlet of sub-basin number 5 (see Fig. 7.9a). The statistical comparison between observed and calibrated sediment load are presented in Table 7.12. Mean monthly observed sediment load was 134.4 t whereas the simulated sediment load was 143 t. The highest monthly observed sediment load also closely matched the simulated sediment load. The calibrated sediment yield at gauging station shows an NSC of 0.69; it was 0.66 during the validation period. Observed and simulated monthly sediment load at sub-basin 5 of the Kansari River Basin during calibration and validation period are presented in Fig. 7.12. Simulated sediment load in the streamflow from Kansari

Table 7.12 Statistical comparisons between monthly observed and calibrated sediment load at the Kansari River Basin

	Sediment load		Sediment load	
	Calibration		Validation	
Statistical parameters	Observed	Simulated	Observed	Simulated
Mean (t)	134.34	143.02	110.17	142.95
Standard deviation (t)	251.04	224.08	162.51	173.87
Maximum (t)	978.42	818.34	440.41	533.91
Minimum (t)	0.00	0.00	0.00	0.00
Nash-Sutcliffe coefficient (E)	0.69		0.66	
Coefficient of determination (r^2)	0.69		0.74	

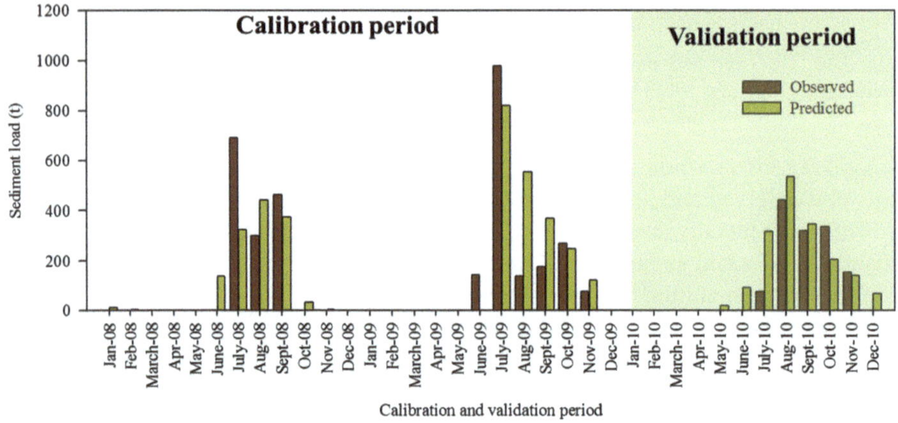

Fig. 7.12 Observed and simulated monthly sediment load (t) for the Kansari Catchment during calibration (2008–2009) and validation period (2010)

River Basin was found very close to the observed value both in calibration and validation period, however, was slightly overestimated. In contrast to sediment load, simulated runoff from the Kansari River Basin was lower than the observed runoff during most of the time. Sediment concentration in runoff water depends on both the amount and intensity of rainfall as well as on the initial soil moisture level. Therefore, sediment concentration in runoff water may vary as per the situation and hence the difference in runoff and sediment load pattern may be observed although these two are considered as interdependent. Maximum sediment load was observed in either July or August month during both calibration and validation period. Maximum sediment load was observed in July 2009 which was found to be 978.42 t whereas the corresponding predicted value was 813.34 t. During the calibration period, observed annual sediment load at sub-basin 5 was 1452 t and 1772 t in 2008 and 2009, respectively, whereas predicted sediment load was 1324 t and 2108 t, respectively. During the validation period in 2010, observed sediment load was 1322 t, whereas the predicted sediment load was found to be 1715 t.

7.5 Impact of Runoff and Sediment Load from Western Catchment on Chilika Lagoon

The water balance components from Badanai and Kansari River Basin as simulated through ArcSWAT model are presented below, followed by a discussion on their potential impacts on Chilika lagoon. As found from the geographical locations of Badanai and Kansari River Basin (see Figs. 7.3 and 7.8), runoff water generated from both these basins directly contributes to Chilika lagoon. From the modeling results, the estimated runoff values from Badanai River Basin were found to be 1509 mm during 2006 corresponding to an observed runoff of 1195 mm. Total annual rainfall during 2006 was 2115.6 mm. Thus, almost 56–71% of total rainfall was lost as runoff from Badanai River Basin. Similarly, the estimated runoff from Kansari River Basin in 2010 was 1117 mm corresponding to an observed value of 1563 mm. Total annual rainfall in 2010 was 1998 mm. Thus, from the Kansari River Basin also, almost 56–78% of total rainfall gets lost as runoff. Analysis of water balance component shows that Kansari River Basin had about 53.3% of rainfall contributed to runoff, 34.4% to ET, and 10.4% to the groundwater table. It has also been observed that 70–75% of annual runoff from both Badanai and Kansari River Basin is generated during the monsoon period from July to September. Apart from Badanai and Kansari stream, about 17–18 other streams from Western Catchment contribute runoff water to Chilika Lagoon, especially during the monsoon season. From the past record, it was found that the Badanai and Kansari streams contribute about 1% and 18.2% of total runoff from Western Catchment (Table 7.1). Thus, runoff from these two river basins may contribute as high as one-fifth of the total runoff volume from the entire Western Catchment Runoff water from Western Catchment carries a lot of sediment load within it. Among different river basins, Kansari contributes maximum sediment load to Chilika lagoon as observed from the measured sediment load data during 2002 (see Table 7.2) whereas Badanai contributes minimum. It has been observed that in 2002, about 30% of sediment load was contributed by Kansari River Basin whereas it was only 0.3% from Badanai River Basin (see Table 7.2). Sediment concentration in runoff water depends on the intensity and amount of rainfall and was observed higher during monsoon periods. For example, sediment concentration in runoff water from Badanai River Basin was 0.16 kg m^{-3} during 2006 because total annual rainfall and the number of high-intensity rainfall events were comparatively higher during 2006 than during 2004 or 2005. Moreover, sediment concentration in runoff water generated through first few rainfall events of the year (during the month of June) was higher (0.19–0.21 kg m^{-3}) and then gradually decreased to 0.07–0.09 kg m^{-3} during the month of November. Sediment load carried by runoff water is generally deposited in the lagoon and reduces the water depth. Moreover, along with sediments, runoff water also carries leached nutrients and other agricultural chemicals from agricultural fields in catchments areas to lagoon, which may affect the natural biota of the Chilika lagoon.

7.6 Conclusion

Hydrologic modelling of runoff and sediment load from two selected river basins (e.g., Badanai and Kansari) of Western Catchment of Chilika lagoon was used as examples to suggest the capability of hydrologic models for simulating catchment-scale processes for the Chilika lagoon in this study. Different water balance components were assessed using the SWAT modelling environment. Both the selected basins are located in the hilly topography of Western Catchments having an area of 42 km^2 and 174 km^2, respectively. The average discharge data showed that Badanai and Kansari River Basin contribute about 1% and 18.2% of total discharge from Western Catchment, whereas total catchment itself contributes 20–40% of the total water inflow into Chilika lagoon (Kumar and Pattnik 2012). All streams from Western Catchment areas were found to be torrential in nature and contribute runoff and sediment load during monsoon periods starting from July to October. Calibration of SWAT model setup for both Badanai and Kansari River Basin showed that effective hydraulic conductivity of the main channel (CH_K), base flow alpha factor (ALPHA_BF), curve number corresponding to antecedent moisture content II (CN2), and roughness coefficient of the main channel (CH_N) were most sensitive parameters. Sensitive parameters were calibrated using observed daily runoff data in Badanai stream from 2004 to 2005 and in Kansari River Basin during 2008–2009 followed by validation during 2006 and 2010, respectively. Nash-Sutcliffe coefficient of simulated flow during calibration period was 0.76 in Badanai River Basin whereas it was 0.67 in Kansari River Basin. Estimation of runoff was even better during validation period in both river basin (NSC = 0.88 and 0.69, respectively). Total estimated runoff water from the basins showed that at least 50–60% of the total rainfall contributes to runoff whereas only 10.4% contributes to groundwater. Furthermore, runoff water from these two basins carries a lot of eroded sediments with an average concentration of 0.12 kg m^{-3}. Such a large concentration of sediment load is harmful to the sustainability of the Chilika lagoon and must be prevented at the source, which is the catchment itself. Modeling results revealed that mean daily runoff flow for different months from Western Catchment may be estimated with sufficient accuracy. The calibrated SWAT model in the selected watershed from the Western Catchment may help to estimate the total runoff and sediment generation potential from different areas of the Western Catchment area and thus will help to formulate future soil water conservation plan to protect this important biodiversity hotspot of India.

References

Arnold J, Srinivasan R, Muttiah R, Williams J (1998) Large area hydrologic modeling and assessment. Part I: model development. J Am Water Resour Assoc 34:73–89

Barnes R (1980) Coastal lagoons – Cambridge studies in modern biology 1, 1st edn. Cambridge University Press, Cambridge

Chilika Development Authority (2012) Annual report 2010–2011. Bhubaneshwar

Dube A, Jayaraman G (2008) Mathematical modelling of the seasonal variability of plankton in a shallow lagoon. Nonlinear Anal Theory Methods Appl 69:850–865

Flügel W (1995) Delineating hydrological response units by geographical information system analyses for regional hydrological modelling using PRMS/MMS in the drainage basin of the River Bröl, Germany. Hydrol Process 9:423–436

Ghosh A, Pattnaik A, Ballatore T (2006) Chilika Lagoon: restoring ecological balance and livelihoods through re- salinization. Lakes Reserv Res Manag 11:239–255

Green W, Ampt G (1911) Studies on soil physics. Part I. The flow of air and water through soils. J Agric Sci 4:1–24

Gupta G, Sarma V, Robin R et al (2008) Influence of net ecosystem metabolism in transferring riverine organic carbon to atmospheric CO2 in a tropical coastal lagoon (Chilka Lake, India). Biogeochemistry 87:265–285

Hargreaves G, Samani Z (1985) Reference crop evapotranspiration from temperature. Appl Eng Agric 1:96–99

Kumar R, Pattnaik A (2012) Chilika: an integrated management planning framework for conservation and wise use. Wetlands International--South Asia; Chilika Development Authority, New Delhi

Leavesley G, Stannard L (1995) The precipitation-runoff modeling system – PRMS. In: Singh V (ed) Computer models of watershed hydrology. Water Resources Publications, Highlands Ranch, pp 281–310

Mein R, Larson C (1973) Modeling infiltration during steady rain. Water Resour Res 9:384–394

Naik P, Pati G, Choudhury A, Naik K (2008) Conservation of Chilika lake, Orissa, India. In: Sengupta M, Dalwani R (eds) Proceedings of Taal 2007: the 12th world lake conference, Jaipur, pp 1988–1992

Nash J, Suthcliffe J (1970) River flow forecasting through conceptual models part I – a discussion of principles. J Hydrol 10:282–290

Neitsch S, Arnold J, Kiniry J, et al (2002) Soil and water assessment tool user's manual. Version 2000. TWRI report TR-192. College Station

Panda S, Bhatta K, Rath K et al (2008) The atlas of Chilika. Chilika Development Authority, Bhubaneswar

Panda U, Mohanty P, Samal R (2013) Impact of tidal inlet and its geomorphological changes on lagoon environment: a numerical model study. Estuar Coast Shelf Sci 116:29–40. http://sci-hub.tw/10.1016/j.ecss.2012.06.011

Panigrahi S, Acharya B, Panigrahy R et al (2007) Anthropogenic impact on water quality of Chilika lagoon RAMSAR site: a statistical approach. Wetl Ecol Manag 15:113–126

Rabus B, Eineder M, Roth A, Bamler R (2003) The shuttle radar topography mission – a new class of digital elevation models acquired by spaceborne radar. ISPRS J Photogramm Remote Sens 57:241–262

Ranjan R (2014) A modelling approach to manage the Kansari River Basin watershed of western catchment of Chilika lagoon. Indian Institute of Technology Kharagpur, Kharagpur

Sahu B, Pati P, Panigrahy R (2014) Environmental conditions of Chilika Lake during pre and post hydrological intervention: an overview. J Coast Conserv 18:285–297

Samal R (2011) The transport of sediments in Chilika Lagoon, East Coast of India. PhD thesis, Berhampur University

Santra P, Das B (2008) Pedotransfer functions for soil hydraulic properties developed from a hilly watershed of Eastern India. Geoderma 146:439–448

Santra P, Das B (2013) Modeling runoff from an agricultural watershed of western catchment of Chilika lake through ArcSWAT. J Hydrol Res 7:261–269

Sarkar D, Sah K, Sahoo A, Gajbhiye K (2005) Soil series of Orissa. National Bureau of Soil Survey and Land Use Planning, Nagpur

USDA, Soil Conservation Service (1972) National engineering handbook, hydrology section 4, Chapters 4–10 van Griensven A (2006) Sensitivity, auto-calibration, uncertainty and model evaluation in SWAT2005

Williams J (1975) Sediment routing for agricultural watersheds. J Am Water Resour Assoc 11:965–974

Wood E, Sivapalan M, Beven K, Band L (1988) Effects of spatial variability and scale with implications to hydrologic modeling. J Hydrol 102:29–47

Chapter 8
Long-Term Analysis of Water Quality in Chilika Lagoon and Application of Bio-optical Models for Cyclone Impact Assessment

Abhishek Kumar, Sk. Md. Equeenuddin, and Deepak R. Mishra

Abstract A comprehensive analysis of sediment and phytoplankton dynamics in Chilika lagoon by synthesizing various remote sensing datasets is presented in this study. The goal of the study was to monitor and analyze the spatio-temporal variability of total suspended sediment (TSS) and chlorophyll-a (chl-a) concentration and associated environmental forcings in the coastal lagoon. NASA's Moderate Resolution Imaging Spectroradiometer (MODIS) surface reflectance cloud free data was used to develop a TSS and chl-a model. Finally, a case study showing implication of satellite based TSS and Chl-*a* models to assess the impacts of natural hazards such as cyclones on water quality of Chilika Lagoon is presented. This case study is based on comparing the effect of two anniversary very severe cyclonic storms (VSCSs): category-5 Phailin (12 October, 2013) and category-4 Hudhud (12 October, 2014) that impacted the lagoon. Analysis for 14 years (2001–2014) using MODIS 8-day composites (MOD09Q1) data indicated that the seasonal variability of TSS is dominant in all the three sectors of the lagoon compared to inter-annual variability. The main reason for large variations in the northern sector is the shallow depth and intrusion of large sediment discharge from Mahanadi River from the northern side, which is the largest fresh water distributary for Chilika Lagoon. Anniversary cyclone impact analysis revealed that *Phailin's* impact on Chilika Lagoon and its watershed resulted in unprecedented levels of precipitation and runoff before-during-after the landfall, which shattered the typical sectorial turbidity gradient. Exponential increase in turbidity because of a combination of run-off and wind driven re-suspension of fine sediments resulted in strong attenuation of light in water column post-*Phailin*. Limited light condition coupled with

A. Kumar · D. R. Mishra (✉)
Center for Geospatial Research, Department of Geography, University of Georgia, Athens, GA, USA
e-mail: dmishra@uga.edu

S. M. Equeenuddin
Department of Earth & Atmospheric Sciences, National Institute of Technology, Rourkela, Odisha, India

C. M. Finlayson et al. (eds.), *Ecology, Conservation, and Restoration of Chilika Lagoon, India*, Wetlands: Ecology, Conservation and Management 6,
https://doi.org/10.1007/978-3-030-33424-6_8

enhanced flushing rate due to flooded river and increased freshwater discharge reduced the Chl-*a* concentration after the passage of *Phailin*. In contrast, relatively farther landfall location, trajectory away from the lagoon, relatively lower wind intensity and short duration of stay of VSCS *Hudhud*, led to lesser precipitation and surface runoff compared to *Phailin*. Consequently, lagoon did not experience a drastic increase in turbidity and light attenuation. Sufficient light availability, stable wind, reduced flushing all favored the phytoplankton growth after passage of *Hudhud* and thus, Chl-*a* concentration increased almost threefold in all the sectors of the lagoon. The approach used in this study can be applied to other cyclone-prone coastal areas. Coupling of satellite based observation with modelling output from systems such as Giovanni can improve monitoring program implemented in numerous coastal estuaries and lagoons.

Keywords Phailin · Hudhud · MODIS · Giovanni · Chilika Lagoon · Suspended sediment · Chlorophyll-a · Algal bloom

8.1 Introduction

Coastal lagoons are among one of the most productive, complex, and dynamic ecosystems around the world as they are positioned at the interface of rivers and sea (Mishra et al. 2017; Srichandan et al. 2015). India has such a productive and complex lagoon named "Chilika" which is Asia's largest and world's second largest brackish water lagoon situated in Odisha state along the east coast at latitude 19°28′ – 19°54′ north, longitude 85°06′ – 85°35′ east. The lagoon has a spread of about 64 km in length from northeast to southwest along the Odisha coast and connected to the Bay of Bengal (BOB) (Fig. 8.1a). Based on salinity and depth, the lagoon is conventionally subdivided into four sub-regions (CDA 2008), namely northern sector (NS), central sector (CS), southern sector (SS) and outer channel (OC) (Fig. 8.1a). It receives fresh water mainly from three distributaries of Mahanadi River namely Nuna, Daya, and Bhargavi as shown in Fig. 8.1a.

Chilika is a shallow coastal lagoon with average depth of 1.8 m and exhibits variability in water coverage area from 1165 to 906 km^2 during monsoon and summer respectively (Siddiqi and Rama Rao 1995). Apart from being shallow and productive, Chilika is well known for its unique assemblages of fresh, brackish, and marine water ecosystem with estuarine character and supports more than 200,000 fishermen in the surrounding 132 villages (CDA 2008). Because of its rich biodiversity and socio-ecological values, the lagoon is designated as "Ramsar Site" a wetland of international importance (Srichndan et al. 2015). However, water quality of the lagoon has been degrading over the years due to factors such as siltation, change in salinity, increase in fresh water weeds, and eutrophication (Nayak et al. 2004; Jayaraman et al. 2005; Bramha et al. 2008; Panda and Mohanty 2008; Panigrahi et al. 2009). One of the main problems the lagoon is facing among all of the above is the decrease in overall salinity due to narrowing of the lagoon's mouth

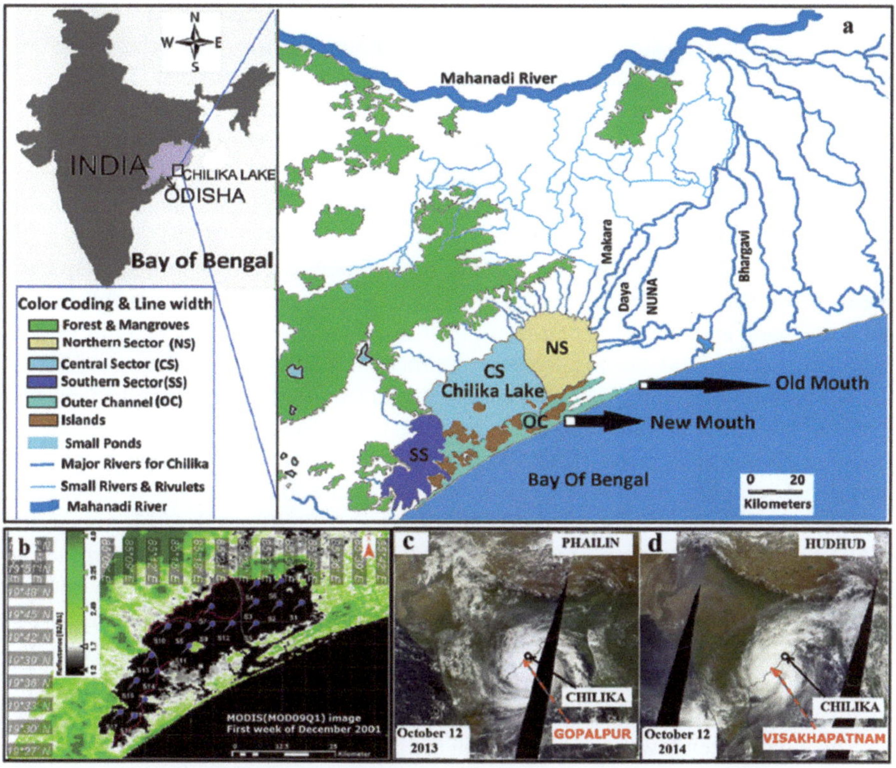

Fig. 8.1 Map of Chilika Lagoon showing the sub-regions of the lagoon and major tributaries (Daya, Nuna, Bhargavi, and Makara) of Mahanadi River (Largest fresh water source for lagoon) (**a**). The sub-regions are northern sector (NS), central sector (CS), southern sector (SS) and outer channel (OC). MODIS image (MOD09Q1) with sampling locations used for long-term quantitative analysis of TSS in different sectors of the lagoon (**b**). The forest cover and emergent vegetation in the west and upper part of northern side is shown in green, water area in black, and land area in white/gray (**b**). Areas which are typically covered by submerged vegetation and weeds are indicated by red line polygon (**b**). It should be noted that the sampling locations used in long-term TSS analysis were devoid of interference from submerged vegetation. The landfall locations of two VSCSs are indicated by red dashed arrow with respect to Chilika Lagoon (open circle) (**c** and **d**)

connected to the Bay of Bengal (Jayaraman et al. 2005; Panda and Mohanty 2008). The narrowing is mainly resulting from the accumulation of sediment entering through drainage basins surrounding the lagoon (Panigrahi et al. 2009). Approximately 1.5 million tons of sediment per year enters the lagoon from the northern part by the distributaries of the Mahanadi River and 0.3 million tons per year enters the lagoon from the western catchment (Patnaik 1998). Therefore, spatio-temporal monitoring of watershed surface runoff and its associated impact on total suspended sediment (TSS), a proxy of sediment load into the lagoon is of utmost importance for balance and proper management of the lagoon's ecosystem (Mishra and Gould 2016; Astuti et al. 2018).

Spatio-temporal information on distribution, dynamics, and trend of TSS in a large water body is difficult to obtain from routine *in-situ* monitoring programs because it is a spatially inhomogeneous and highly variable parameter (Dekker et al. 2001). Therefore, remote sensing techniques have been widely used to monitor TSS and other spatially variable optically active constituents (OACs) in the water column such as phytoplankton and cyanobacteria (Mishra et al. 2013, 2014a, b; Wang et al. 2016; Page et al. 2018), and colored dissolved organic matter (CDOM; Chaichitehrani et al. 2013; Ogashawara et al. 2017; Cao et al. 2018). However, the routine use of remote sensing for monitoring sediment dynamics in many environments (e.g., lakes, estuaries, coastal areas) has been limited due to several factors such as characteristics of remote sensing instruments (spatial and spectral resolution), data costs, and the availability of processing software (Miller and McKee 2004). Some studies have demonstrated that data from ocean color satellite sensors, such as National Aeronautics Space Administration (NASA)'s Moderate Resolution Imaging Spectroradiometer (MODIS), represent a cost-effective alternative to the traditional sampling methods (Hu et al. 2004; Miller and McKee 2004; Chen et al. 2007; Doxaran et al. 2009; Mishra and Mishra 2010). Atmospherically corrected MODIS surface reflectance products (e.g., MOD09GQ and MYD09GQ) from both Terra and Aqua satellite are well calibrated with high geolocational accuracy and can be used to monitor water quality parameters both effectively and frequently. For example, Cui et al. (2013) have used these products (MOD09GQ) from Terra and products (MYD09GQ) from Aqua satellite, for long-term monitoring (2001–2010) of suspended sediment concentration in Poyang Lake, China.

There have been few attempts made in the past to use *in-situ* remote sensing data (Rao et al. 1986) and satellite data (Sudhakar and Pal 1993; Pal and Mohanty 2002; Panda and Mohanty 2008) for water quality assessment of Chilika Lagoon. However, TSS concentration was not estimated in those studies. Mohanty and Pal (2001) tried to predict TSS concentration using LISS-I (IRS-IB) data by incorporating soil brightness index (SBI) for mapping TSS concentration but the coefficient used in their algorithm were obtained from other published literature. In addition, IRS data has reduced sensitivity and temporal frequency compared to MODIS data. Gupta (2013) used RESOURCESAT-1 AWiFS data for estimating TSS concentration by implementing chromaticity technique. This algorithm was based on a single date satellite image (26 November, 2003) and it can only predict TSS concentration up to 42 g/m^3. Further, MODIS data have several advantages over this product including high temporal resolution, high sensitivity (i.e., 12-bit radiometric resolution), cost effectiveness, and most importantly, free from complexities of atmospheric correction which is required for RESOURCESAT-1 AWiFS data.

In this study, both 1-day (MOD09GQ) and 8-days MODIS products (MOD09Q1) from Terra satellite were incorporated for short-term and long-term monitoring of TSS from surface of water column in Chilika Lagoon. The methodology adopted in this study to predict TSS concentration, though empirical in nature, the field data incorporated during TSS model development were from ten different dates, different months, and years. Therefore, the model was not affected by seasonal and temporal bias including changes in solar angles, a major source of uncertainty in remote sensing data, which makes it applicable in widely varying geographic regions regardless of variation in shape, size, color and type of sediment. To prove that, the TSS model was validated to similar lakes and estuaries in USA, China, and Argentina, by comparing result against published literature. This TSS model can be adopted as a standard procedure to frequently monitor the sediment dynamics in the lagoon because relationship between reflectance and TSS are generally transferable over time for the same geographic location as long as the source of sediment does not vary substantially (Ritchie et al. 2003; Dekker et al. 2002). Miller and McKee (2004) showed that the 645-nm band used in the algorithm provides considerably more detail about the horizontal distribution of suspended sediment than vertical volumetric distribution and that is why the effect of sediment grain size, shape, and texture is minimal at 645 nm band compared to shorter wavelengths. Therefore, the variability in the model performance because of variations in physical properties of the sediment types can be considered negligible.

The overall objective of this chapter is to analyze the trend and spatio-temporal variability of TSS in the lagoon for the last 14 years (2001–2014). The specific objectives are to (1) utilize a relationship established between *in-situ* measurements of TSS and atmospherically corrected reflectance in MODIS band 1 (R_{rs} at 645 nm) for routine application of MODIS 250 m surface reflectance products (MOD09GQ and MOD09Q1) for TSS monitoring, and (2) characterize the seasonal and inter-annual variability of TSS, and the impact of physical and meteorological parameters, and (3) finally a case study showing implication of satellite based TSS and chlorophyll-a (Chl-*a*) models to assess the impacts of natural hazards such as cyclone on water quality of Chilika Lagoon is presented. This case study is based on comparing the impact of two anniversary very severe cyclonic storms (VSCSs): category-5 *Phailin* (12 October, 2013) and category-4 *Hudhud* (12 October, 2014) on the water quality of Chilika Lagoon (Fig. 8.1c–d).

Estuaries, lagoons and surrounding watershed can experience extreme wind velocities, storm surges and rainfall during hurricanes or cyclones, resulting in strong water column mixing, variability in flow regime and modification in local geomorphology (Peierls et al. 2003). These episodic weather events also cause strong sediment re-suspension in the water column (Chen et al. 2009), which may temporarily alter the overall water quality of an aquatic system and associated biological, chemical and geomorphological processes (Miller et al. 2011). In addition, cyclones facilitate substantial nutrient loading from the surrounding areas that can trigger algal blooms, reduce water clarity and increase hypoxic zones in the lakes and estuaries (Peierls et al. 2003; Miller et al. 2006; Paerl et al. 2006). However, the question is "*Do all cyclones produce similar impact on coastal lakes/lagoons? If not, what factors play a role in determining the type of impact? An individual cyclone's characteristics is obviously one set of factors, what about watershed characteristics and lake optical property? Why some cyclones trigger algal blooms in a lagoon after their passage and some do not?*" These are some of the questions which shaped the framework for this comprehensive case study and we used satellite-based biophysical models and products to answer these questions.

8.2 Materials & Methods

The following workflow diagram summarizes overall process involved in this study to achieve all the objectives mentioned above (Fig. 8.2). The individual components of this workflow are explained in the following subsections.

8.2.1 MODIS Data and Processing

MODIS Terra sensor 250 m spatial resolution data were downloaded from the NASA Goddard Space Flight Centre's Level 1 and Atmosphere Archive and Distribution System (LAADS) (http://ladsweb.nascom.nasa.gov/data) FTP site. Terra satellite passes over Chilika Lagoon typically between 10:00 and 10:30

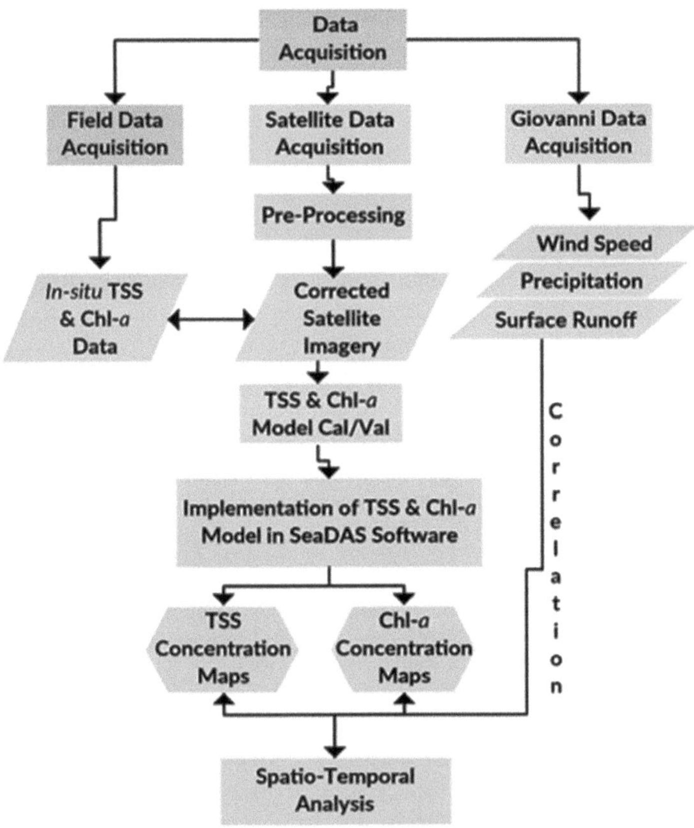

Fig. 8.2 Overall work flow showing individual process involved in this study to produce spatial maps of biophysical products (TSS and Chl-*a*) for spatio-temporal analysis and their variability with physical and meteorological parameters

A.M. local time. For long-term study, MOD09Q1 products covering an entire year were downloaded for 14 years (2001–2014) for weekly and monthly analysis. On the other hand, for assessing differential impacts of two anniversary severe cyclones, MOD09GQ products along with true color images were downloaded corresponding to cyclone months (October 2013 and October 2014). The true colour images were downloaded from NASA's GSFC World view website (https://earthdata.nasa.gov/labs/worldview/). High temporal resolution, precise geolocational accuracy and frequent revisit time, make MODIS an ideal satellite sensor to study high-frequency phenomenon such as cyclone impact. The two 250-m bands on MODIS sensor (band 1: 620 nm – 670 nm centered at 645 nm; band 2: 841 nm – 876 nm centered at 859 nm) have sufficient sensitivity to detect a wide range of changes in color of estuarine waters (Hu et al. 2004). Several water quality parameters such as TSS (Miller and McKee 2004; Binding et al. 2005; Chen et al. 2009; Zhang et al. 2010; Zhao et al. 2011; Doxaran et al. 2009; Madrinan et al. 2010; Mishra and Mishra 2010), turbidity (Chen et al. 2007; Petus et al. 2010; Dogliotti et al. 2015), and chlorophyll-a (Mishra and Mishra 2012; Prasad and Singh 2010; Katlane et al. 2012) have been estimated using these two bands. The standard MODIS products (MOD09GQ and MOD09Q1) used in this study contain atmospherically corrected surface reflectance value at bands 1 and 2. MODIS L1B data is used as primary input to obtain MOD09GQ product which is corrected for the effect of gaseous absorption, molecules and aerosol scattering, coupling between atmospheric and surface bi-directional reflectance function, and adjacency effect (Vermote et al. 1997). Doxaran et al. (2009) have described the atmospheric correction process in detail for this product.

To process the MODIS products, Sea viewing Wide Field of view Sensor (SeaWiFS) Data Analysis System (SeaDAS) software available at NASA's ocean colour website (http://seadas.gsfc.nasa.gov/) was used. Visual quality check was performed to flag out images with substantive missing data patches and cloud cover. All MODIS images were corrected for spatial distortion by re-projecting them from sinusoidal coordinate system to geographical coordinate system (WGS 84). A geometrical mask was used for extracting the Chilika Lagoon data from all MODIS images. An additional mask, which included a band ratio of remote sensing reflectance (R_{rs}) of band 2 and band 1 (R_{rs} (859 nm)/R_{rs} (645 nm) > 1.5), was used to mask out the islands present inside the lagoon which could not be eliminated using the first geometrical mask.

8.2.2 TSS Model and Long-Term Data Extraction

TSS model (Eq. 8.1) was developed for Chilika Lagoon by Kumar et al. (2016) by re- parameterizing Miller and McKee's linear TSS model (Miller and McKee 2004). The wide range of *in-situ* TSS samples (1.2–161.7 mg/L) from different months and years were used in re-parameterizing the Miller and McKee's TSS model. The calibration result revealed a significant relationship between TSS and the R_{rs}

($R^2 = 0.91$; $n = 54$; $p < 0.001$). In addition to an independent validation (RMSE = 2.64 mg/L; n = 16) at Chilika Lagoon, the model was also validated for different geographic locations (Taihu Lake, China; Poyang Lake, China; La-plata River Estuary, Argentina; Mobile Bay Estuary, USA) to test its broad geographic applicability (more details related to TSS model calibration and validation can be found in Kumar et al. 2016).

$$TSS_{Polynomial} = 13181 \times R_{rs}(MODIS\ B1)^2 - 1408.6 \times R_{rs}(MODIS\ B1) + 44.15 \quad (8.1)$$

MOD09Q1 data were used for creating TSS concentration maps using above polynomial model (Eq. 8.1) and extracting the weekly mean TSS data to analyze long-term spatio-temporal variability in Chilika Lagoon for 14 years (2001–2014). First, TSS dynamics in the lagoon was visually analyzed for 14 years (2001–2014) at weekly and monthly temporal frequency. MODIS MOD09Q1 product from the first week of months representing different seasons (January, March, May and October) were used to show the TSS spatial distribution patterns in different zones of the lagoon during past 14 years. To demarcate wind induced sediment re-suspension events, wind velocity and direction data were extracted over the Chilika Lagoon from the QuikScat satellite. Further, to quantify more comprehensively seasonal and inter-annual variability in TSS concentration, MODIS data from all months of each year were downloaded for 14 years (2001–2014). For each year, 46 satellite images (MOD09Q1) were downloaded and visualized for quality assessment. TSS model was implemented on cloud free images to derive spatial map products. Some of the images in which the lagoon was partly visible were also used because of a significant lack of cloud free images during monsoon season. Thus, a variation in number of satellite images per year was encountered between years in the long-term spatiotemporal analysis of the TSS. Six sampling points from each sector were used for extracting the TSS concentration from each map product (Fig. 8.1b). Extreme care was taken to avoid interference of emergent and submerged vegetation present in the upper part of northern sector, western part of central sector, and near the boundary of the southern sector. The emergent vegetation in upper part of northern sector were masked completely using a band ratio threshold of 1.5 ($B2/B1 > 1.5$). Similar to Miller and McKee (2004) and Lee et al. (2001), we also assumed that chlorophyll absorption and interference of submerged vegetation has a limited effect on the water leaving radiance at 645 nm (bandwidth: 620–670 nm) because of high attenuation by water itself and contribution from other OACs. In addition, presence of sediment induced turbidity minimizes the bottom reflectance contamination in reflectance data extracted from MODIS pixels and typically carries information from the first few inches of the water surface (Lee et al. 2001; Hu et al. 2004). For each month maximum 4 images were used and the TSS value was averaged to get the mean monthly TSS for each year starting from 2001 to 2014. Seasonal versus inter-annual variability was also compared after analyzing all 14 years TSS data. Finally, correlation of mean TSS (2001–2014) with different meteorological parameters was carried out for each sector of the lagoon.

8.2.3 Giovanni Data

The limited field-based observation of environmental factors invoked the need of satellite sensors and model-derived products. NASA's Giovanni system facilitates access to a range of remote sensing data and other earth science data sets, which helps researchers to implement selected data to a broad area of research field such as terrestrial, atmospheric and marine environment (Acker et al. 2014). It includes data from various NASA missions and projects. To analyze the effect of physical and meteorological forcing on the TSS variability in the lagoon, total surface precipitation and surface runoff data were downloaded from NASA's Giovanni database (http://giovanni.gsfc.nasa.gov/giovanni/). MATMNXLND 5.2.0 products from MERRA monthly history data collection were used for the specified catchment area of Chilika Lagoon. Result for each month was visualized from 2001 to 2014 and corresponding NetCDF file of each transect was downloaded for further analysis. The units of the variables were converted from $kg/m^2/s$ to $g/m^2/day$ using a multiplication factor of 86,400,000 to incorporate in the SeaDAS statistical tool. Further, three geometrical masks were used for analyzing the catchment area separately, northern (N) catchment (20° N – 20°48′ N, 85° E – 86° 24′ E), western (W) catchment (19°24′ N – 20° N, 84° E – 85° E), and overall catchment. Monthly average value was estimated for different catchment areas using corresponding geometrical masks. In addition, monthly wind stress magnitude data were also downloaded directly using Giovanni-4 seasonal time series plot for catchment area to complement precipitation and runoff data in the long-term analysis of TSS variability in the lagoon.

Further, to assess cyclone impact, high temporal frequency data was required. Therefore, 3-h Tropical rainfall measuring mission (TRMM) precipitation products (TRMM_3B42RT.007) were downloaded to capture the highest rainfall event around the lagoon catchment on the landfall day of the two VSCSs (*Phailin* and *Hudhud*). TRMM data from Giovanni was also used by Acker and Leptoukh (2007) to show the rainfall accumulation during Hurricane *Ivan* (16 September, 2004) around Gulf of Mexico, Florida and Alabama. In addition, Global Land Data Analysis System (GLDAS) was also used for visualising the GLDAS-derived rainfall rate, surface runoff and near-surface wind speed products to capture the surface level alteration, especially for analysing the variability in the catchment of the lagoon, caused by the anniversary cyclones. GLDAS is an interactive programme which incorporates both satellite and ground based observations at a spatial resolution of $0.25 \times 0.25°$ (Fox and Rowntree 2013). Rodell and Houser (2004) and Fang et al. (2008) have described about GLDAS products in detail in their studies. In this study, the 3-hourly product (GLDAS_NOAH025SUBP_3H.001) was used for analysing the short-term variations in physical (surface runoff) and meteorological (rainfall rate and wind speed) parameters. Further, data were time averaged to generate 1-day averaged data for all the physical and meteorological parameters. These products are based on NOAH model (http://disc.sci.gsfc.nasa.gov/datacollection/GLDAS_NOAH025_3H_V020.shtml). NOAH model incorporate near surface atmospheric forcing data as input such as soil moisture, soil temperature, canopy water content, energy flux and water flux terms of surface energy balance and water energy balance for simulation.

8.2.4 *Differential Impact Analysis of Anniversary Cyclones: A Case Study*

8.2.4.1 VSCSs Characteristics

Figure 8.3a shows the trajectories of VSCSs *Phailin* and *Hudhud*. *Phailin* struck the Odisha coast on October 12, 2013 around 22:30 h Indian Standard Time (IST) (17:00 h UTC) with the maximum wind velocity reaching up to 220 km/h (IMD 2013). This VSCS was classified as category 5 on the Saffir-Simpson scale as per the norms of National Oceanic and Atmospheric Administration (NOAA). It was the second strongest tropical cyclone in the recorded history to make landfall in India only after the super cyclone of Odisha which struck the same area in 1999 with wind speed up to 260 km/h (IFRC 1999). Initially, a low pressure was formed in South China Sea on October 06, 2013 which further intensified into cyclonic storm on October 09, 2013 moving west-northwestwardly. It further intensified into a VSCS by moving northwestwardly on October 10 and finally made landfall on October

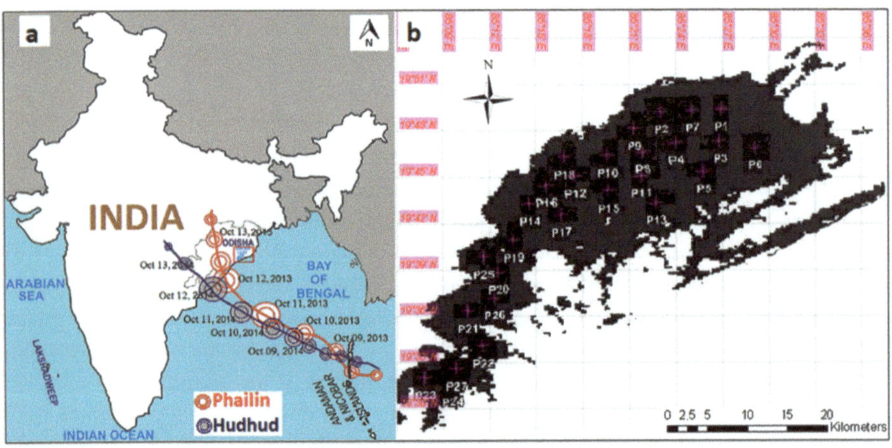

Fig. 8.3 Track of the two VSCSs (*Phailin*: red line and *Hudhud*: blue line) with respect to Chilika Lagoon (**a**). MODIS image (MOD09GQ) with random sampling locations (Total 27 random pixel locations (n = 27): 9 from each sector (NS, CS, and SS)) were used for quantitative analysis of TSS and Chl-*a* in different sectors of the lagoon for differential impact assessment of anniversary cyclones on the water quality of Chilika Lagoon (**b**)

Table 8.1 Comparative description of anniversary VSCSs (*Phailin* and *Hudhud*)

Parameters	*Phailin*	*Hudhud*
Origin location with date	Gulf of Thailand and adjoining North Andaman Sea (October 6, 2013) (9.5° N, 102.00° E)	Tenasserim coast and adjoining North Andaman Sea (October 6, 2014) (10° N, 96° E)
Landfall location	Gopalpur, Odisha (19.27° N, 84.92° E)	Visakhapatnam, Andhra Pradesh (17.69° N, 83.22° E)
Distance (direction) from Chilika	45 km (southwest of Chilika)	338 km (southwest of Chilika)
Duration (stay over land)	~5 days	~3 days
Landfall date with time	October 12, 2013 (22:30 IST)	October 12, 2014 (between 12:00 and 13:00 IST)
Category	Category-5	Category-4
Wind speed (maximum)	215 km/h (115 Knots)	185 km/h (100 Knots)
Rainfall near lagoon (landfall day)	160 mm	80 mm
Central pressure (minimum)	940 hPa	950 hPa
Pressure drop (maximum)	66 hPa	54 hPa
Storm surge (maximum)	2–2.5 m above astronomical tides	1.4 m above astronomical tides
Economic impact	USD $696 million	USD $3.4 billion
Total number of fatalities	45	124

The data were collected from reports of central and state government organisation (IMD 2013, 2014; GoO 2013)

12 near Gopalpur district in Odisha (IMD 2013). The landfall location of VSCS *Phailin* was only 45 km southwest from Chilika Lagoon. The landfall of *Phailin* brought torrential rain and storm surges up to 3.5 m to the eastern Indian states of Odisha and Andhra Pradesh (UNEP 2013). Important characteristics of both VSCS are presented in Table 8.1.

The second VSCS Hudhud, made landfall on October 12, 2014 near Visakhapatnam, Andhra Pradesh, exactly 1 year after VSCS Phailin. The landfall location was at a distance of 338 km southwest from Chilika Lagoon. This tropical cyclone was categorized as category 4 cyclone on the Saffir-Simpson scale. The maximum sustained wind speed during its landfall was about 185 km/h (IMD 2014). These VSCSs, Phailin and Hudhud, were very similar in their characteristics as both originated in the North Andaman Sea and moved almost in parallel path but there was a relatively small difference in strength, wind speed, landfall location, and passage within the catchment after the landfall (Table 8.1; Mishra and Panigrahi 2014). Both VSCSs produced high negative economic impact, and a large number of fatalities reported only during VSCS Hudhud compared to VSCS Phailin as it struck the areas with high population densities and larger cities (Table 8.1) (IMD 2013, 2014).

8.2.4.2 Differential Impact on TSS and Chl-*a*

MODIS true colour images corresponding to pre-VSCSs (7 October, 2013 and 2014: the nearest cloud-free image before the landfall of cyclones), landfall day (12 October, 2013 and 2014) and post-VSCSs (14 October, 2013 and 2014) were incorporated to show the differential change in water colour of the Chilika Lagoon and its surrounding region. However, MODIS 1-day revisit period limited its applicability for tracking the duration and path of cyclone on an hourly basis. Thus, we incorporated Giovanni-derived three-hourly rainfall and surface runoff transects on landfall day to show the track, duration and the maximum area of impact caused by both VSCSs. Data extracted from Giovanni transects corresponding to surface wind speed, rainfall rate and surface runoff for the catchment area of the lagoon was compared daily for the month of October, 2013 and 2014. Each of the 3-h products were time averaged for 24 h using Giovanni's time averaged map tool (http://giovanni.sci.gsfc.nasa.gov/giovanni/) and the averaged transect files were downloaded in NetCDF format. Further, a geometrical mask was used to extract the time averaged data corresponding to the catchment area of the lagoon.

The ultimate objective of this case study was to evaluate the differential impacts of VSCSs on the biophysical parameters that govern the water quality of the lagoon. To accomplish the objective, TSS and Chl-*a* concentration were extracted from 27 locations (P1-P27) marked in Fig. 8.3b using MOD09GQ data. Nine locations from each sector; NS (P1-P9), CS (P10-P18), and SS (P19-P27) were randomly selected to extract the TSS and Chl-*a* concentration in the lagoon pre- and post VSCSs. TSS and Chl-*a* spatial maps were created using MODIS based TSS model (Kumar et al. 2016) and Chl-*a* slope model respectively (Srichandan et al. 2015) (Eq. 8.2), both developed recently for Chilika Lagoon. The slope model was developed with the *in-situ* Chl-*a* samples corresponding to the cyclone month (October 2013). A significant relationship ($R^2 = 0.59$; $n = 29$; $p < 0.001$; RMSE $= 20.7\%$; bias $= 8.4\%$) was established between *in-situ* Chl-*a* concentration and slope of MODIS band 3 and band 4 as follows:

$$\text{Chl } a = 211.98 * \exp(6320.3 * \text{Slope Reflectance}) \tag{8.2}$$

where,

$$\text{Slope Reflectance} = \frac{R_{rs}\,(B3) - R_{rs}\,(B4)}{|\text{Band Center}(B3) - \text{Band Center}\,(B4)|} \tag{8.3}$$

$$\text{Reflectance(Band 3)} = 0.4796 * \text{Reflectance (Band 1)} + 0.0189 \quad (8.4)$$

$$\text{Reflectance (Band 4)} = 0.7186 * \text{Reflectance (Band 1)} + 0.0407 \quad (8.5)$$

A similar slope model was used to derive Chl-*a* concentration from MODIS data from Lake Pontchartrain, Louisiana, USA (Mishra and Mishra 2010). Accuracy was a major factor for selecting these published TSS and Chl-*a* models because both were calibrated and validated with *in-situ* samples from Chilika Lagoon itself. The accuracy of Chl-*a* model implemented in this study was compared with four other published models (Gitelson et al. 2003; Kahru et al. 2004; Zhang et al. 2011; El-Alem et al. 2012), and it was found to perform better compared with others (more details can be found in Srichandan et al. 2015). Finally, TSS and Chl-*a* maps were created and placed side by side to demarcate the spatio-temporal variation in these parameters pre- during post- VSCS periods.

8.3 Results and Discussion

8.3.1 Long-Term Trend of TSS- Visual Analysis

TSS composite from first week of different months representing different season (January, March, May, and October) were used to analyse the trend in inter-annual and seasonal variability (Fig. 8.4). From the qualitative analysis of these time-series TSS products, it was observed that the TSS concentration in northern sector have been consistently higher than that of the other two sectors except in summer (May) (Fig. 8.4). As the river discharge during this period remains low, high TSS throughout the lagoon in May could be due to the wind-induced sediment re-suspension events since wind speed was observed to be the highest during summer months (Bramha et al. 2008).

October is the month following the south-west monsoon (June-Sept) and April is the pre-monsoon period. Therefore, the influx of fresh water through the riverine system and land drainage to the lagoon is maximum during October and minimum during April (Pal and Mohanty 2002). This is the main reason for relatively low concentration of TSS observed in northern sector of the lagoon during pre-monsoon period (i.e., in March) (Fig. 8.4b). The central sector has shown intermediate variations in TSS concentration for different seasons as compared to other two sectors. This could be mainly due to less number of rivers connecting to this sector as compared to northern sector and comparatively more number of rivers than southern sector.

TSS concentration in the southern sector of the lagoon was generally lower as compared to other two sectors, except in the month of May (Fig. 8.4). Low depth and churning action of water by southerly wind is the primary cause of re-suspension of fine sediments from bottom during the month of May and increasing turbidity in the southern sector (Bramha et al. 2008). Chen et al. (2007) found that wind velocity and

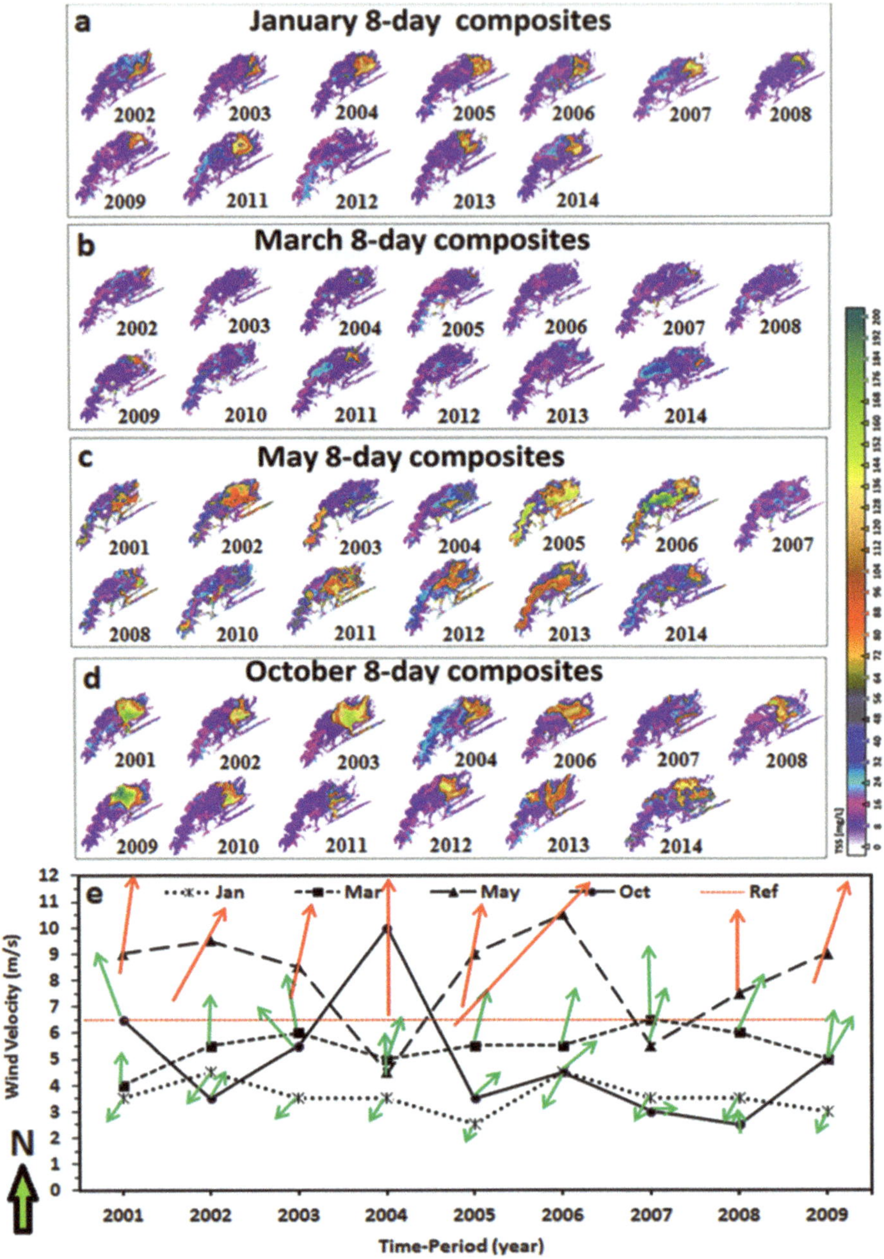

Fig. 8.4 Spatio-temporal distribution of TSS over Chilika Lagoon with MODIS 8-day composite images in different seasons across the years (**a**, **b**, **c**, and **d**). First week images were used for comparison as most of the images were found to be cloud free during this period. The gap in years among images indicates the lack of cloud free images for those years (**a**, **b**, **c**, and **d**). In addition, first week averaged wind velocity with direction derived from QuikScat for Chilika Lagoon

direction can play a major role in increasing the overall turbidity in water bodies particularly in the shallow areas through the re-suspension of bottom sediments. Zhao et al. (2011) described the event of re-suspension in shallow Mobile Bay estuary, Alabama using two MODIS images captured within the time gap of 2-days. To delineate the re-suspension event in this study, wind speed data from the QuikScat satellite were analyzed over the Lagoon for the first week of each month as shown in Fig. 8.4e.

It was observed that during summer (May), wind direction remains predominantly southerly and southwesterly over the lagoon, however, during winter (January), wind direction is either northerly or north-easterly (Fig. 8.4e). The wind direction results were found to be consistent with previous studies (Bramha et al. 2008; Mohanty and Panda 2009). It was observed that typically wind velocity during the month of May was high (8–10 ms^{-1}) as compared to other months and direction of wind was from south-west to north-east. In May 2004 and 2007, when average wind speed was below 6.5 ms^{-1}, inter-annual turbidity levels in the lagoon were the lowest because of the lack of sediment re-suspension events (Fig. 8.4c, e). However, when the average wind speed was more than 6.5 ms^{-1} in May during other years, re-suspension of sediment particularly in shallow areas increased the overall turbidity level. This threshold value (6.5 ms^{-1}) was chosen by observing TSS variability during October 2001 when slight resuspension was observed in southern sector at 6.5 ms^{-1}. Zhang et al. (2010) have also suggested that wind speed in the range of 5–6 ms^{-1} is always critical for resuspension events and they used >6 ms^{-1} as reference for Lake Taihu, China. The effect of re-suspension of bottom sediments in southern sector of the lagoon is clearly visible (Fig. 8.4a–d). Interestingly, during October (post-monsoon) 2004, the only October between 2001 and 2009, the wind exceeded the 6.5 ms^{-1} threshold reaching up to 10 ms^{-1}, sediment re-suspension event was observed throughout the lagoon in addition to the normal post-monsoonal river discharge induced high turbidity in the northern sector (Fig. 8.4e, d).

8.3.2 Long-Term Trend of TSS- Quantitative Analysis

To analyze the influence of precipitation and surface runoff on TSS in the lagoon, a comparative analysis was performed between the two major catchments, the northern (N) and the western (W) catchment (Fig. 8.5). Results for overall catchment area were also analyzed. It was observed that although the magnitude of precipitation in both catchment areas (N and W) was very similar, runoff from N catchment was

Fig. 8.4 (continued) corresponding to the months incorporated in long-term trend of TSS-visual analysis (**e**). The 6.5 ms^{-1} (dashed red line) indicates threshold limit for re-suspension induced turbidity and red arrow indicates wind speed above threshold limit while green arrows are below threshold limit (**e**)

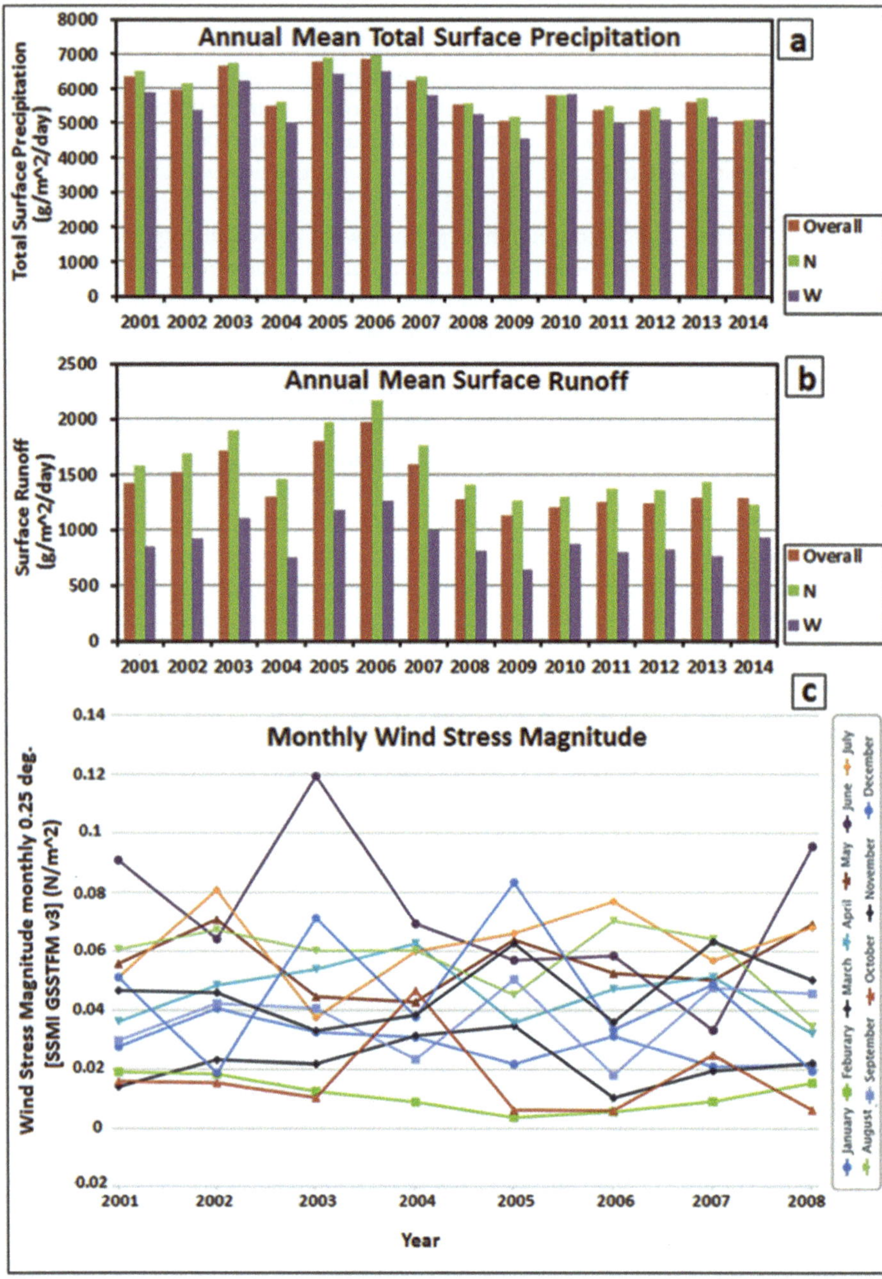

Fig. 8.5 Temporally and spatially averaged magnitude of meteorological and physical parameters: total surface precipitation (**a**); surface runoff (**b**) and wind stress magnitude (**c**) over the catchment area of Chilika Lagoon. N & W represents northern (20° N – 20°48′ N, 85° E – 86°24′ E) and western (19°24′ N – 20° N, 84° E – 85° E) catchment area. Overall catchment used for analysis covered area between 19°24′ N – 20°48′ N, 84° E – 86°24′ E

observed to be 24.69–49.59% (2014 and 2009 respectively) more than from the W catchment area (Fig. 8.5a, b).

The significant difference in runoff is mainly due to the large number of rivers and rivulets flowing into the lagoon from the north (Fig. 8.1a). Further, heavy forest cover in the western catchment may be playing a role in reducing the surface runoff. The above finding further justified the reason behind consistently higher TSS concentration in northern sector of the lagoon compared to central and southern sector. In addition to precipitation and runoff, monthly wind stress magnitude from Giovanni is shown in Fig. 8.5c. The monthly wind stress generated from Giovanni-4 seasonal time series plot (http://giovanni.gsfc.nasa.gov/) specified that during pre-monsoon (April, May) and monsoon (June, July and August) wind magnitude remain high (Fig. 8.5c) which plays major role in re-suspension of sediment mostly during pre-monsoon period as discussed in earlier section (visual analysis).

The mean TSS concentration data was obtained for each month and each sector. TSS data were extracted and averaged from pixel locations shown in Fig. 8.1b which covered the entire lagoon. Table 8.2 shows the complete statistics of the long-term analysis result on a monthly basis in different sectors of the lagoon from 2001 to 2014. There are variations in the number of satellite images used for different months mainly due to factors such as image quality and cloud cover. Maximum cloud cover was encountered during the monsoon (June, July, and August) which were also the months when precipitation and surface runoff were maximum. During the 14 years (7975 pixels), the minimum predicted TSS concentration was 6.54 mg/L which is the lower limit for the prediction model (Table 8.2). The highest TSS concentration was recorded during July 2006 (NS: 860.22 mg/L; CS: 435.94 mg/L; SS: 371.91 mg/L) due to the combination of maximum precipitation (17341.03 g/m^2/day), runoff (8019.735 g/m^2/day), and wind speed during monsoon of 2006, mainly in July. It clearly indicates the influence of these three factors in controlling the TSS dynamics in the lagoon. Overall concentration of TSS remained high during June, July, and August each year in all the sectors of the lagoon mainly due to heavy rainfall and high surface runoff during those months.

The monthly average of each parameter for 14 years, except wind stress (8 years), showed the dominance of monsoon induced precipitation, runoff, and wind stress and associated high TSS (Fig. 8.6a). The effect of precipitation and associated runoff on TSS was much more pronounced in NS followed by CS and SS (Fig. 8.6b). In addition, central and southern parts of the lagoon consistently experience the settling of sediment at a faster rate during post-monsoon period compared to the NS. This could be due to a combination of generally higher runoff and post-monsoon wind direction from north-east to south-west supporting the re-suspension event in NS of the lagoon.

Correlation analysis between monthly average TSS and precipitation and surface runoff for different sectors is shown in Fig. 8.6c–h. As observed in Fig. 8.6b, the TSS in the NS and CS of the lagoon is highly correlated with total precipitation (R^2: 0.852 and 0.927 for NS and CS respectively) and surface runoff (R^2: 0.910 and 0.772 for NS and CS respectively). However, TSS in the SS of the lagoon is weakly correlated with precipitation (R^2: 0.4295) and surface runoff (R^2: 0.1876) (Figs. 8.6c, d, e).

Table 8.2 Descriptive statistics showing the long-term (2001–2014) variability of TSS (mg/L) in different sectors of the lagoon

Sector	Statistics	Jan	Feb	Mar	Apr	May	Jun	Jul	Aug	Sep	Oct	Nov	Dec
NS	Min	6.55	6.54	6.55	6.55	6.63	7.61	15.87	13.39	6.55	6.55	6.54	6.55
	Max	172.16	113.05	230.64	217.77	270.79	226.54	860.22	530.05	431.67	276.81	200.77	185.72
	Mean	50.51	24.89	26.38	44.55	68.81	89.66	182.51	206.86	138.15	89.87	75.49	70.61
	Std. dev.	41.44	23.16	31.71	37.89	48.39	64.12	171.85	105.73	79.13	48.68	43.39	46.25
	Std. error	2.49	1.31	1.84	2.38	3.31	7.61	21.15	7.69	5.19	3.03	2.48	2.78
	N	276	312	298	253	213	71	66	189	232	257	306	276
CS	Min	6.55	6.55	6.54	6.55	6.55	6.58	6.56	6.55	6.54	6.55	6.54	6.55
	Max	52.26	64.99	251.38	149.63	355.42	220.79	435.94	464.47	462.99	206.9	122.49	104.24
	Mean	12.48	13.05	17.79	32.11	50.64	67.06	112.99	105.77	63.47	30.57	15.51	14.01
	Std. dev.	7.23	8.13	27.44	33.64	51.83	56.67	109.61	105.98	76.82	38.82	15.43	12.03
	Std. error	0.43	0.46	1.59	2.15	3.64	6.77	13.92	8.45	5.06	2.52	0.88	0.72
	N	272	312	295	243	202	70	62	157	230	237	304	273
SS	Min	6.54	6.55	6.55	6.55	6.55	6.56	7.49	6.59	6.55	6.55	6.54	6.54
	Max	34.91	41.77	204.96	185.72	167.46	236.18	371.91	186.64	211.48	52.57	30.37	44.15
	Mean	10.01	10.26	20.09	42.87	50.62	56.42	90.56	37.62	18.34	15.56	10.27	10.33
	Std. dev.	4.99	5.13	26.01	34.77	36.55	58.71	91.47	44.69	17.03	9.34	4.73	5.26
	Std. error	0.31	0.29	1.53	2.21	2.51	7.28	12.01	4.09	1.18	0.62	0.27	0.31
	N	270	303	286	248	212	65	58	119	206	223	305	274
Total no. of images used		47	52	50	43	37	12	11	33	41	43	51	46

Out of 46 MODIS 8-day composites (MOD09Q1) each year i.e. 644 images, total number of satellite images used for analysis was 466 and rest were discarded due to poor quality and cloud cover. The total number of MODIS pixels (N) used for TSS data extraction is 7975

This variation in the magnitude of correlation is mainly due to the variation in number of tributaries connecting to the different sectors of the lagoon as they act as the primary transporting medium controlling runoff into the lagoon. Wind stress magnitude on the other hand showed high correlation (R^2: 0.54) with mean TSS in SS compared to CS (R^2: 0.18) and NS (R^2: 0.42). The predominant direction of wind is from south and south-west to north and north-east during summer and that along with shallow depth in the SS could be the main reasons why that sector experiences more intense re-suspension events compared to other sectors.

Temporal variation of R^2 between TSS and above factors for different sectors was also assessed to cross validate the correlation analysis (Fig. 8.6f, g, h). The results indicated that NS of the lagoon was steadily well correlated with precipitation and runoff for all years (Fig. 8.6f). Also, CS was strongly correlated with runoff and precipitation though there was drastic drop in R^2 value for the year 2002 and 2010 (Fig. 8.6g). That was possibly caused due to the anomalously high wind speed (Figs. 8.3c and 8.4e) observed during May, 2002 which triggered a re-suspension event. A similar phenomenon was also observed in 2010. In the SS region, the TSS showed better correlation with wind speed compared to other two factors (runoff and precipitation) during the period (2001–2008), except for some years (2007 and 2008) (Fig. 8.6h). In general, it can be concluded that the spatio-temporal variability of TSS in different sectors of the lagoon is controlled by different environmental forcings and the pattern of variability in different sectors was observed to be similar (i.e. high in NS, moderate in CS, and low in SS) over the 14 years. However, the seasonal (monthly) variability is stronger compared to inter-annual variability for all parameters (TSS, precipitation, runoff, wind stress). Descriptive statistics of monthly and inter-annual variability in mean TSS concentration is provided in Table 8.3. It is apparent from the Table 8.3 that monthly (seasonal) variation in mean TSS for different sectors of the lagoon (Standard deviation: NS-58.36 mg/L, CS-35.92 mg/L, SS-25.24 mg/L) are very high compared to inter-annual variation (Standard deviation: NS-12.27 mg/L, CS-9.49 mg/L, SS-6.81 mg/L). Annual variability plot also highlighted the fact that TSS concentration is significantly more in NS compared to CS and SS (Fig. 8.6b).

8.3.3 Differential Impact Analysis of Anniversary Cyclones

MODIS true color images prior to both VSCSs indicated the usual turbidity regime of the lagoon as discussed earlier i.e., high turbidity in NS due to heavy freshwater influx and comparatively low turbidity in CS and SS (Fig. 8.7a–b). One important difference observed in true color images was that prior to *Phailin*, land pixels of the MODIS image appeared darker compared to pre-*Hudhud* (Fig. 8.7a–b). Also, the BOB appeared more turbid (light blue) in pre-*Phailin* MODIS image (Fig. 8.7a) compared to the pre-*Hudhud* image (dark blue) (Fig. 8.7b). The reason behind this difference is most likely the ~threefold high rainfall during the beginning of October 2013 (1st week total precipitation: 1.46 kg/m^2 for overall catchment 19.4° N – 20.4° N and 84.6° E – 85.8° E) compared to initial period of October 2014 (1st week total

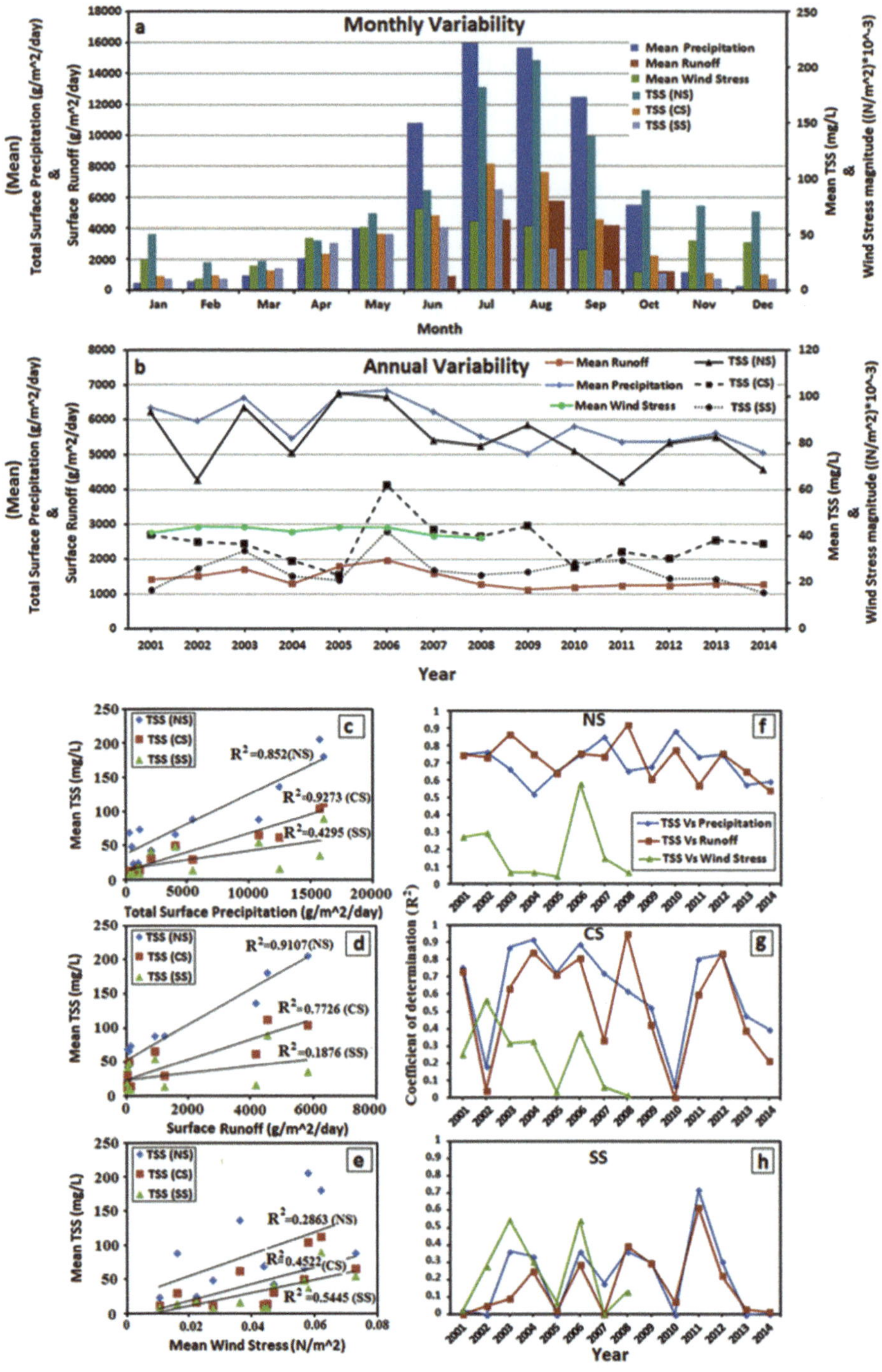

Fig. 8.6 Monthly (January–December) (**a**) and annual (2001–2014) (**b**) variability of various parameters (precipitation, runoff, wind Stress, and TSS) for Chilika Lagoon. Temporal variability

Table 8.3 Statistical comparison between seasonal and inter-annual variability in mean TSS concentration for different sectors of Chilika Lagoon from 2001 to 2014

	NS		CS		SS	
Statistics mean TSS (mg/L)	Monthly	Annual	Monthly	Annual	Monthly	Annual
Minimum	24.89	63.28	12.48	23.07	10.01	15.69
Maximum	206.86	101.29	112.99	62.21	90.56	42.03
Range	181.97	38.01	100.51	39.14	80.55	26.34
Std. deviation	58.36	12.27	35.92	9.49	25.24	6.81

precipitation: 0.55 kg/m^2 for the same catchment) that made the land surface saturated and appeared darker and also increased runoff to BOB making it more turbid. Angles et al. (2015) suggested that the environmental condition pre-and post-cyclone will primarily determine its impact on estuaries and lagoons. The landfall day image of VSCS *Phailin* revealed that its swath covered the entire Chilika Lagoon and its eye was clearly demarcated close to the lagoon (Fig. 8.7c). On the other hand, the outer band of VSCS *Hudhud* covered the Chilika Lagoon and its eye was comparatively at a larger distance from the lagoon (Fig. 8.7d). Chilika Lagoon appeared completely turbid (brown color) in post-*Phailin* MODIS image and a large sediment plume is also apparent that extended towards the BOB due to flooded river extracts (Fig. 8.7e). The flood intensity in Rushikuliya River below the Chilika Lagoon can be observed easily in MODIS images post-*Phailin* (Fig. 8.7e). In contrast, the post-*Hudhud* MODIS images revealed lesser impact on Chilika Lagoon in terms of turbidity levels, particularly in CS and SS, based on the visual analysis of the water color in MODIS true color data. Also, the spatial extent of the sediment plumes in BOB appeared a little thinner most likely due to comparatively lesser river discharge (Fig. 8.7f).

The path of both VSCSs and how long their impact lasted around the Chilika Lagoon can be observed in the 3-hourly Giovanni derived TRMM transects (Fig. 8.8). Previous studies discussed that characteristics of a cyclone such as landfall location, intensity, trajectory, speed of passage, are some of the important factors which can produce significantly different impact (Mallin et al. 2002; Mallin and Corbrett 2006; Srichandan et al. 2015). There was continuous rainfall observed over or near Chilika Lagoon (indicated by open circles in Figs. 8.8a–h) for 24 h on the landfall day of *Phailin*. In contrast, transects corresponding to landfall day of *Hudhud* revealed that the precipitation over or near the lagoon lasted only for 9–12 h (Fig. 8.8i–p). Swath of both the VSCSs and their progress can be observed in these transects. The landfall points of *Phailin* (Gopalpur) and *Hudhud* (Visakhapatnam)

Fig. 8.6 (continued) in mean TSS is shown by different symbols and color bar for different sectors (NS, CS and SS). Correlation between average TSS concentration for different sectors (NS, CS and SS) and meteorological and physical parameters (precipitation, surface runoff and wind stress) (**c**, **d**, and **e**). The mean value of each parameter for each month was averaged for 14 years (2001–2014) and correlated with mean TSS value for different sectors. Temporal variation is also shown in the form of R^2 for different sectors (**f**, **g**, and **h**)

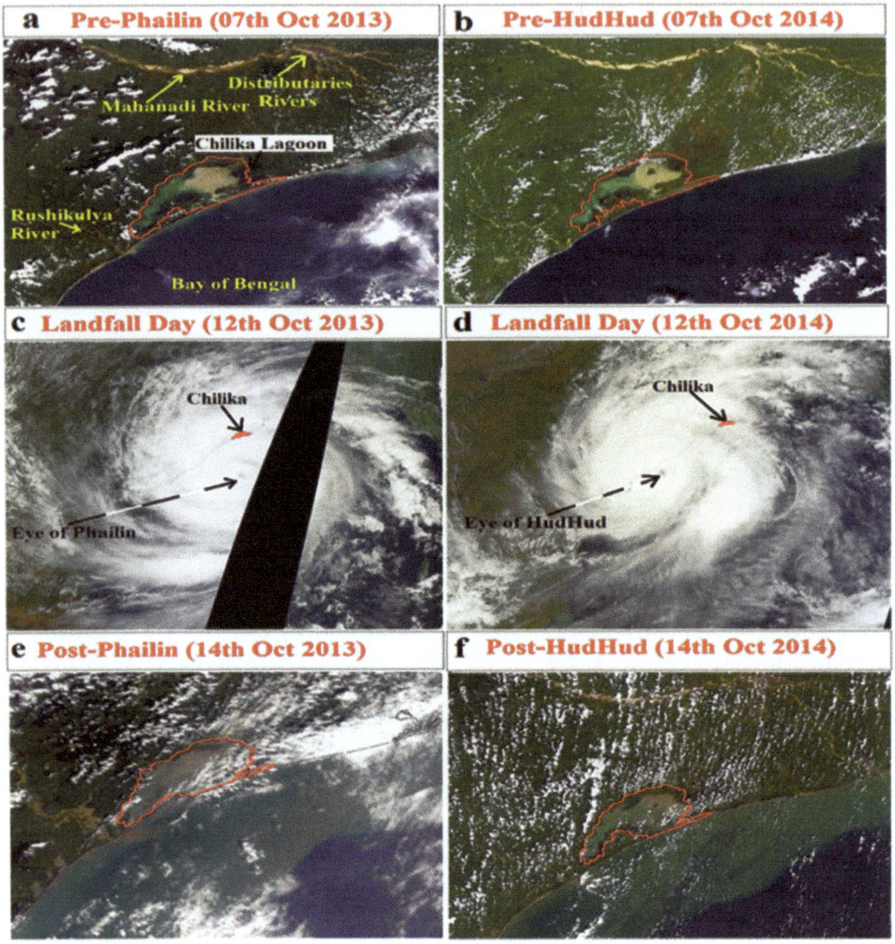

Fig. 8.7 Visual analysis of the impacts of VSCSs *Phailin* and *Hudhud* on Chilika Lagoon and surrounding region using MODIS true color images: pre-*Phailin* (**a**), *pre-Hudhud* (**b**), landfall day of *Phailin* (**c**), landfall day of *Hudhud* (**d**), post-*Phailin* (**e**), post-*Hudhud* (**f**). Chilika Lagoon is demarcated by the red polygon and the distributaries of Mahanadi River (major source of freshwater for the lagoon) connecting the lagoon are marked by solid arrow (**a**). The eyes of both cyclones are indicated by dashed arrow (**c–d**)

are shown by dashed arrows with respect to Chilika Lagoon (solid arrow) (Fig. 8.8g, l). The 24-h analysis revealed very high average rainfall (~100–110 mm) near the lagoon during *Phailin* compared to *Hudhud* (~55–60 mm).

The precise effect of heavy rainfall around Chilika Lagoon is further revealed by the 3-hourly surface precipitation and surface runoff transects (Fig. 8.9). 3-hourly surface runoff transects showed a close temporal matching with high rainfall locations (Fig. 8.9a–p, a'–p'). As suspected, the 24 h comparative result shown in Fig. 8.9q–r indicated high surface runoff during *Phailin* (NC: highest up to

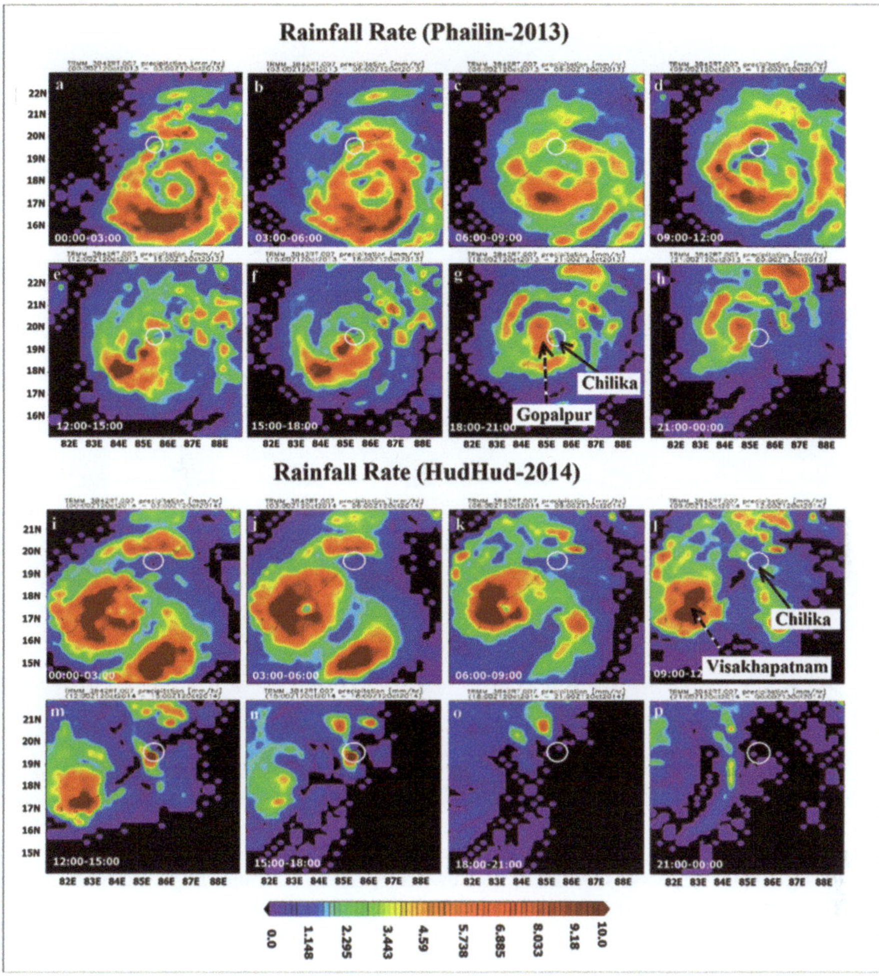

Fig. 8.8 Precipitation map produced by Giovanni web based application tool using TRMM 3-hourly product for 24 h on the landfall day of *Phailin* (October 12, 2013) and *Hudhud* (October 12, 2014). Chilika Lagoon is indicated by solid arrows with open circle and the landfall points are indicated by dashed arrows (**g** and **l**)

4379.45 g/m^2-h^{-1}; WC: highest up to 683.25 g/m^2-h^{-1}) due to high precipitation (NC: highest up to 17076.28 g/m^2-h^{-1}; WC: highest up to 9198.02 g/m^2-h^{-1}) in contrast to relatively lower surface runoff on the landfall day of *Hudhud* (NC: highest up to 532.48 g/m^2-h^{-1}; WC: highest up to 253.41 g/m^2-h^{-1}) due to comparatively less precipitation (NC: highest up to 5279.74 g/m^2-h^{-1}; WC: highest up to 4852.47 g/m^2–h^{-1}) and passage of *Hudhud* from sector with lesser tributaries and more vegetated areas.

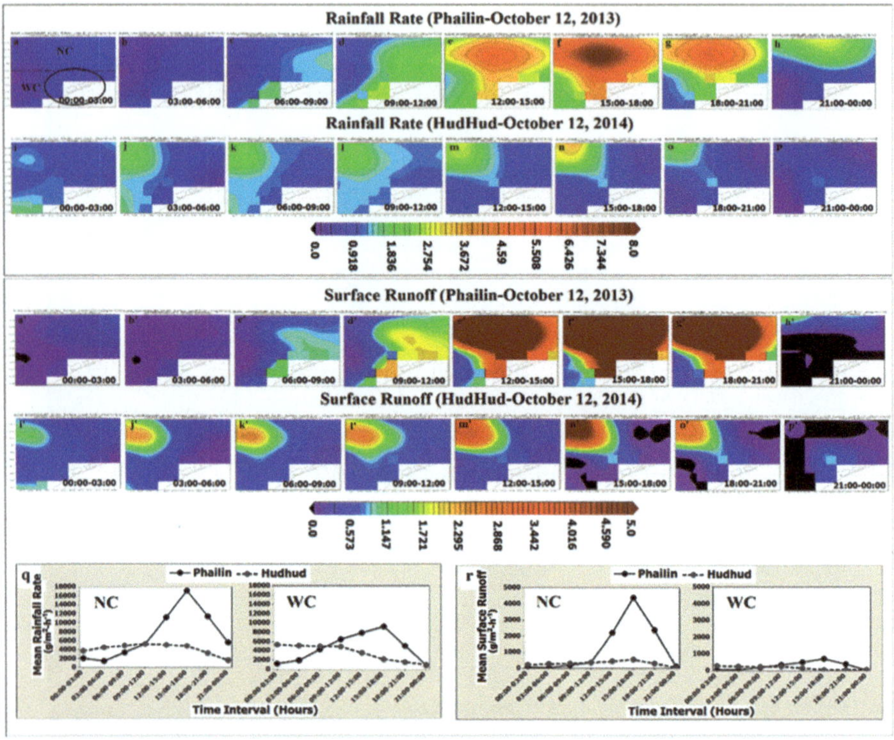

Fig. 8.9 Rainfall rate (**a–p**) and surface runoff (**a'–p'**) map produced by Giovanni web based application tool using GLDAS 3-hourly products for 24 h on the landfall day of *Phailin* (October 12, 2013) and *Hudhud* (October 12, 2014) near the catchment area. The two catchments northern (NC: 19.9° S – 20.4° N; 84.6° W – 85.8° E) and western (WC: 19.4° W – 19.9° N; 84.6° W – 85.2° E) used in comparative analysis are demarcated by dashed lines and Chilika Lagoon is marked with open circle (**a**). Catchment-wise comparison of rainfall rate (**q**) and surface runoff at every 3 h (**r**)

The surface runoff transects in both cases indicated that the maximum contribution was from northern and north-western zone (together considered as NC), where the lagoon receives highest surface runoff due to the heavy discharge from rivers and distributaries, compared to WC (Fig. 8.9; also, refer to Fig. 8.7a, b). The northern catchment includes agricultural land, bare ground, and urban areas, which primarily contributed to the surface runoff compared to the western watershed which is mainly comprised of forested land. Also, the downward sloping topography and major tributaries of Mahanadi River of NC supports higher surface runoff and the Khallikote forest present in the WC of the lagoon limits the surface runoff to a large extent. Previous studies also documented relatively higher runoff in agricultural and urbanized watershed which elevated nutrients and suspended sediments to nearby water bodies (Jordan et al. 2003; Mallin and Corbett 2006; Wetz and Yoskowitz 2013). The surface runoff transects revealed a significant increase in magnitude (~1 order of magnitude increase) after 12:00 p.m. on the landfall day of

Phailin and it persisted for continuous 9 h (Fig. 8.9e'–g', r). In contrast, there was no such order difference observed during VSCS *Hudhud* in the magnitude of surface runoff (Fig. 8.9i–p, r). The primary reason behind difference in runoff associated with the two cyclones was the speed of passage which is characterized as a determining factor associated with the magnitude of impact of such VSCSs in previous literatures (Mallin and Corbett 2006; Srichandan et al. 2015). For example, Mallin and Corbett (2006) observed that the fast moving hurricane *Andrew* resulted in low erosion compared to hurricane *Dennis* which lingered for several days over the watershed near North Carolina and resulted in enhanced runoff. The speed of passage for *Hudhud* was much faster (~9–12 h) compared to *Phailin* (>24 h) over the watershed of Chilika, which caused the difference in rainfall amount and duration. This difference led to the highly variable surface runoff from the watershed of the lagoon. The above results were primarily from the landfall day of both VSCSs. To quantify the variability of these parameters under normal environmental conditions, both pre-and post-cyclone data corresponding to entire month of October 2013 and 2014 were analysed (Fig. 8.10). Also, an additional variable, near surface wind speed was incorporated in further analysis.

Figure 8.10 shows the variability in mean surface wind speed, mean rainfall rate (surface precipitation), and mean surface runoff for overall catchment corresponding to the entire month before-during-after the cyclones (October 2013 and 2014). The gradual increase in surface wind speed started on October 9, 2013 (9.82 km/h) and reached to peak level on the landfall day of *Phailin* (October 12, 2013: 53.72 km/h) (Fig. 8.10a). Similar increase in surface wind speed started on October 7, 2014 (7.18 km/h) and continued to increase till the landfall day of *Hudhud* when it reached at its peak level (October 12, 2014: 35.55 km/h). The wind speed magnitude on the landfall day was 51% higher for *Phailin* compared to *Hudhud* (Table 8.4). Unlike wind speed, the rainfall rate and surface runoff increased at an exponential rate on the landfall day and decreased at the same rate after the passage of the cyclones. High sustained wind and runoff may have triggered the re-suspension of bottom sediments which increased the turbidity in the lagoon drastically on the landfall day of *Phailin*. An earlier study based on field measurement of turbidity in the lagoon has shown that before *Phailin* the turbidity of the lagoon was 32.8 NTU which increased to 60.4 NTU after *Phailin* (Srichandan et al. 2015). In contrast, comparatively lower mean surface runoff (243.97 $g/m^2\text{-}h^{-1}$) due to lower mean surface precipitation (3809.33 $g/m^2\text{-}h^{-1}$) and lower wind speed on the landfall day of *Hudhud* was not able to increase the turbidity of the lagoon to *Phailin* level (Table 8.4) (Fig. 8.7f). The magnitude and spatial distribution of turbidity in the lagoon on the aftermath of these cyclones are primarily determined by wind speed and runoff. While massive runoff resulting from heavy rainfall brings substantial amount of sediments to the lagoon and can be considered as the primary driver of turbidity, high wind speed triggers sediment re-suspension and is a secondary source of turbidity. Wind speed which plays the major role in sediment re-suspension during cyclonic events (Chen et al. 2009; Wetz and Yoskowitz 2013) was significantly higher during *Phailin* compared to *Hudhud* which may have triggered additional turbidity in the lagoon. It is important to break down and analyse the physical and meteorological factors

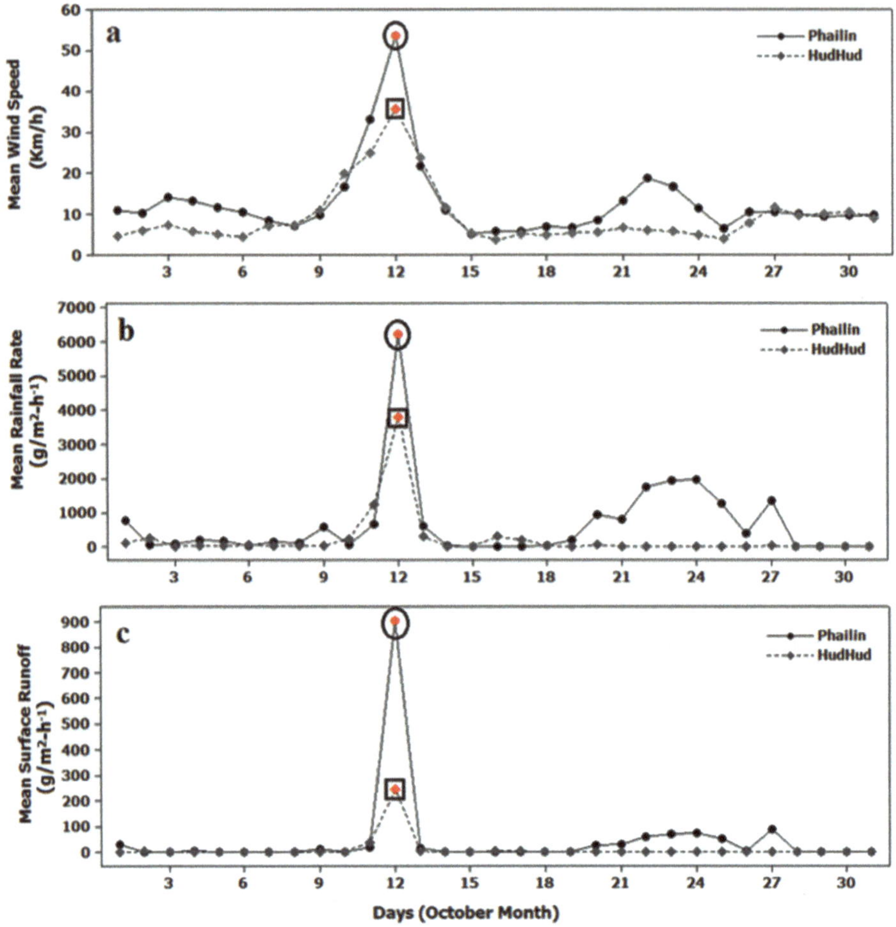

Fig. 8.10 Variation in meteorological parameters: mean surface wind speed and mean rainfall rate (**a–b**) and physical parameter: mean surface runoff (**c**) corresponding to overall catchment (19.4° N – 20.4° N and 84.6° E – 85.8° E) surrounding the Chilika Lagoon for October (X-axis: Days of October 2013 and 2014; Y-axis: mean surface runoff, mean rainfall rate and mean surface wind speed). The open circles and boxes indicated the unprecedented level of magnitude on the landfall day for both VSCSs, *Phailin* (October 12, 2013) and *Hudhud* (October 12, 2014) respectively

because they are strongly linked to the turbidity which is ultimately linked to the likelihood of an algal bloom after the passage of a cyclone. Wind induced bottom sediment re-suspension is also dependent on sediment types and size (Havens et al. 2011). The sediment size varies in different sectors of Chilika Lagoon such as fine sediments (silt and clay) in offshore (NS and CS) due to major river extracts, and coarser sediment (sand) in nearshore (SS and OC) because of exchange of sea water through the mouth (Raman et al. 2007; Ansari et al. 2015). Therefore, NS and CS which are dominated by fine sediment experienced higher degree of re-suspension

Table 8.4 Comparison between physical and meteorological parameters for the entire October and landfall day for both cyclones corresponding to overall catchment (19.4° N – 20.4° N and 84.6° E – 85.8° E) surrounding the Chilika Lagoon

Parameters	October – entire month			Landfall day		
	Phailin (October, 2013)	*Hudhud* (October, 2014)	%Higher (*Phailin*)	*Phailin* October 12, 2013	*Hudhud* October 12, 2014	%Higher (*Phailin*)
Mean surface wind speed (km/h)	12.79	9.33	+37.08%	53.72	35.55	+51.11%
Mean rainfall rate (g/m^2-h^{-1})	658.12	222.03	+196.41%	6235.21	3809.33	+63.68%
Mean surface runoff (g/m^2-h^{-1})	44.71	9.6	+365.72%	904.16	243.97	+270.61%

leading to high turbidity compared to SS. Havens et al. (2011) reported similar phenomena that fine sediments in central part of Lake Okeechobee were more susceptible to re-suspension after increases in wind velocity and caused more limitation of light compared to near shore coarse sediment.

Impact of anniversary VSCSs on the lagoon was further investigated with the MODIS derived TSS and Chl-*a* concentration. The mean TSS concentration was found to be significantly higher in NS (*Phailin*: 131.36 mg/L; *Hudhud*: 75.13 mg/L) and CS (*Phailin*: 35.32 mg/L; *Hudhud*: 13.53 mg/L) during October 2013 (*Phailin* month) compared to October 2014 (*Hudhud* month) (Fig. 8.11, Table 8.5). Difference in precipitation, surface runoff, and wind speed all combinedly created this differential impact. The mean TSS in SS was observed to be very similar in magnitude between October 2013 (16.24 mg/L) and 2014 (15.33 mg/L) and lowest among all sectors (Fig. 8.11, Table 8.5). This is because SS was the least affected section due to low surface runoff from western side of the lagoon. In contrast, the mean Chl-*a* concentration was found to be relatively higher for the month of *Hudhud* compared to *Phailin* in all the sectors (*Phailin*: NS: 4.21 μg/L, CS: 11.36 μg/L, SS: 21.94 μg/L; *Hudhud*; NS: 6.93 μg/L, CS: 18.44 μg/L, SS: 24.61 μg/L) of the lagoon (Fig. 8.11, Table 8.5). Massive increase in turbidity in the water column after *Phailin* reduced the transparency level and limited the availability of light (Srichandan et al. 2015). The limited light availability affected the Chl-*a* concentration, an indicator of primary productivity, which reduced significantly post-*Phailin* in all the sectors of the lagoon. Srichandan et al. (2015) also reported a significant decrease in Chl-*a* concentration immediately after *Phailin* in Chilika Lagoon and no sign of phytoplankton bloom after the passage. Previous studies also reported reduced phytoplankton biomass after the passage of cyclone due to significant light attenuation in water column (Paerl et al. 1998; Mallin et al. 2002; Mallin and Corbett 2006; Srichandan et al. 2015).

One important difference found from Fig. 8.11b, d is the lag effect of cyclones on phytoplankton growth in different sectors of the lagoon. It took more than a week

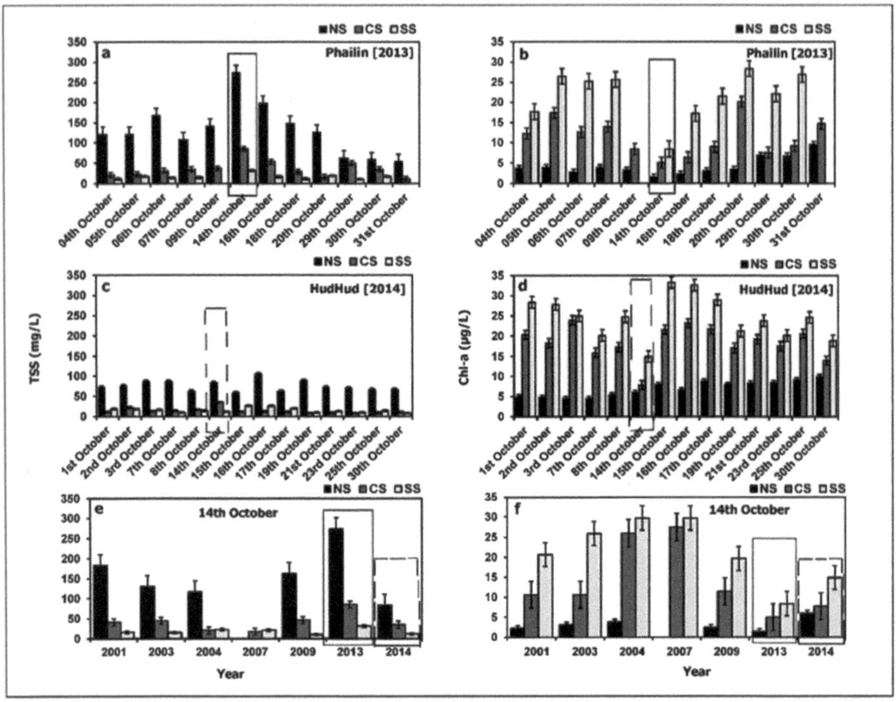

Fig. 8.11 Variations in mean TSS and mean Chl-*a* concentration across the three sectors of the lagoon (NS, CS, SS). The bars inside the solid box and dashed box represent the readings 2 days after the landfall of VSCSs *Phailin* and *Hudhud*. TSS and Chl-*a* data was not included from SS of the lagoon on October 09, 2013 and from NS of the lagoon on October 14, 2007 due to cloud cover

Table 8.5 Sector-wise statistics of TSS and Chl-*a* concentration measured during October (2013 and 2014) for both VSCSs

γ	Sector	Cyclone month	Minimum	Maximum	Range	Mean	Standard deviation
Mean TSS (mg/L)	NS	*Phailin:*	53.22	274.05	220.83	131.36	±62.84
		Hudhud:	58.05	104.48	46.43	75.13	±13.02
	CS	*Phailin:*	11.61	84.99	73.38	35.32	±19.94
		Hudhud:	7.67	35.04	27.37	13.53	±7.26
	SS	*Phailin:*	10.55	31.30	20.75	16.24	±5.97
		Hudhud:	7.82	26.99	19.17	15.33	±5.98
Mean Chl-a (µg/L)	NS	*Phailin:*	1.48	9.61	8.13	4.21	±2.35
		Hudhud:	4.46	9.91	5.45	6.93	±1.78
	CS	*Phailin:*	5.05	20.04	14.99	11.36	±4.56
		Hudhud:	7.81	23.91	16.10	18.44	±4.16
	SS	*Phailin:*	8.42	28.35	19.93	21.94	±6.10
		Hudhud:	14.93	33.31	18.38	24.61	±5.27

Data were extracted from 27 locations (9 from each sector (NS, CS, and SS)) and averaged out for each sector to obtain the mean TSS and mean Chl-*a* concentration as presented below after the passage of *Phailin* for the mean Chl-*a* concentration to recover to pre-*Phailin* level comparing the sector-wise values between October 7th and 20th, 2013 (October 7, 2013 – NS: 3.8 μg/L; CS: 13.89 μg/L; SS: 25.62 μg/L and October 20, 2013 – NS: 3.41 μg/L; CS: 20.04 μg/L; SS: 28.35 μg/L). However, only 2 days after the passage of *Hudhud*, i.e., on October 15, 2014 the mean Chl-*a* concentration (NS: 8.01 μg/L; CS: 21.59 μg/L; SS: 33.31 μg/L) crossed the pre-*Hudhud* level (October 7, 2014 – NS: 4.46 μg/L; CS: 15.9 μg/L; SS: 20.13 μg/L). This could be related to the combined effect of light availability and nutrient enrichment in the water column, just suitable enough to favor the phytoplankton growth. For example, field data collected from the lagoon suggested that there was about tenfold increase in nitrate concentration in Chilika lagoon after *Hudhud* compared to pre-*Hudhud* month. The turbidity of the lagoon in September 2014 was 138 NTU which decreased to 95.4 NTU in October 2014 after *Hudhud*. The reduced turbidity (increased transparency) along with other favourable meteorological, physical, and hydrological factors (calm wind, low precipitation, discharge, runoff, and low flushing rate) created suitable environment for phytoplankton growth as evidenced with the fact that average phytoplankton density before *Hudhud* (September 2014) was 2200 cells/mL which increased to 6216 cells/mL after *Hudhud* (October 2014).

Further, spatial maps corresponding to TSS and Chl-*a* concentration derived from MOD09GQ products followed similar pattern corresponding to true color MODIS images which verified the implication of biophysical models (Fig. 8.12). The maps were divided into two segments (red dash line); pre-VSCSs (pre- *Phailin* and pre-*Hudhud*) period (left) and post-VSCSs period (right) (Fig. 8.12). Visual analysis of the true color MODIS images clearly showed the highly turbid NS in all images (Fig. 8.12). The spatio-temporal pattern of TSS and Chl-*a* concentration pre-VSCSs period showed the usual gradients as discussed in the previous sections. There was consistent cloud cover over the lagoon starting from October 9 (2013 and 2014) which limited the availability of cloud free MODIS images close to the landfall date for both VSCSs. However, just 2-days after *Phailin*, on October 14, 2013, MODIS true color image (inside the circle: Fig. 8.12) showed a remarkable increase in turbidity throughout the lagoon and similar pattern was obtained when the TSS model was implemented on MODIS data. On the contrary, Chl-*a* map post-*Phailin* showed a substantial decrease throughout the lagoon on October 14, 2013 (inside the circle: Fig. 8.12) which could be due to the high TSS concentration that did not allow algal growth as discussed in earlier section. However, this result was in contradiction to many previous literatures that reported enhanced phytoplankton biomass after the passage of cyclones or a anthropogenic massive runoff (Angles et al. 2015; Sarangi et al. 2014; Huang et al. 2011; Mishra and Mishra 2010; Paerl et al. 2001, 2006; Mallin et al. 2006; Peierls et al. 2003). For example, similar increase was reported in recent studies by Sarangi et al. (2014) and Lotliker et al. (2014) in north-west region of BOB post-*Phailin*. Both studies correlated sea surface temperature (SST) with Chl-*a* concentration and concluded that enhanced nutrient supply due to mixing of water column and stratification of layer created a favourable condition for phytoplankton growth. However, Chilika is a shallow lagoon (depth range: 0.5–3.0 m) where stratification is not a major problem and there has been no evidence of

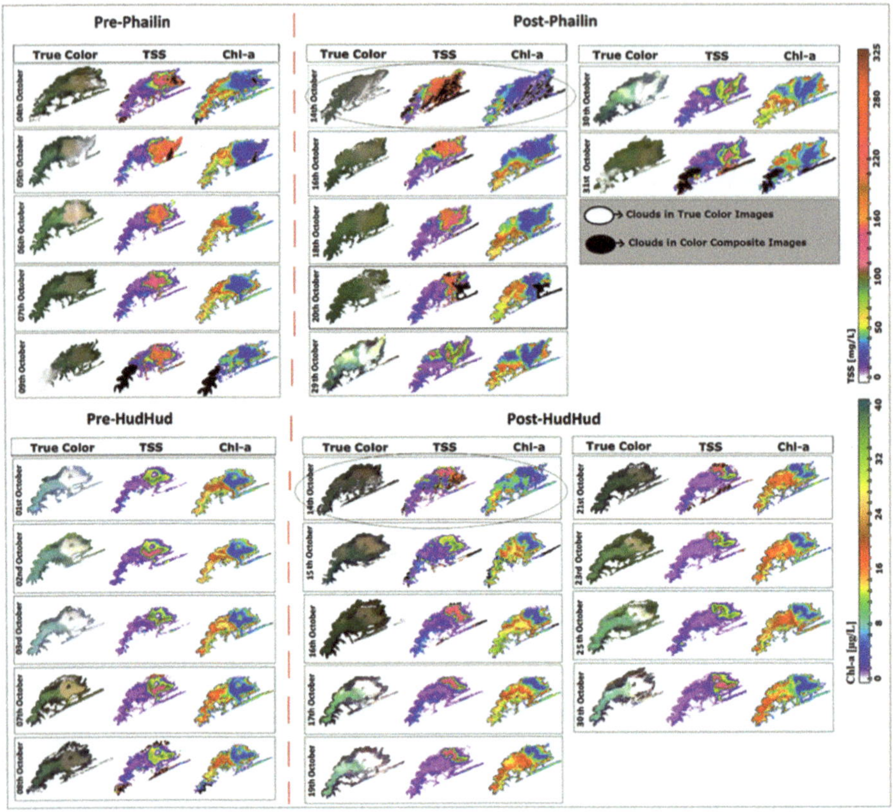

Fig. 8.12 Spatial and temporal variation of TSS and Chl-*a* concentration in Chilika Lagoon pre-during-post VSCSs (*Phailin*: October 2013 and *Hudhud*: October 2014) using MODIS true color images. The images inside the circles show the immediate aftermath of the VSCSs. Gaps observed in the dates are due to the unavailability of cloud free MODIS products

stratification in the lagoon based on 12 years of data (2000–2011) analyzed by CDA (Kumar and Pattnaik 2012). On the other hand, mixing of water column due to high wind action during cyclonic events may circulate large amount of nutrients from bottom of the lagoon to upper layers. These nutrients help proliferate phytoplankton growth if environmental conditions are supportive such as sufficient light, calm wind, and low flushing rate. These favourable environmental conditions were not available post-*Phailin* which restricted the phytoplankton growth. Additionally, Bacillariophyta and Dinophyta are the dominant phytoplankton species in the lagoon which require highly saline and transparent waters to grow but their growth was restricted post-*Phailin* because of very high freshwater discharge and storm water runoff which reduced salinity and increased turbidity to a great extent in the lagoon (Srichandan et al. 2015). Other studies have also concluded that re-suspension of the sediments increases the nutrient availability which triggers algal growth (Zhu et al. 2014; Sarangi et al. 2014). However, that was not the case after *Phailin*. Turbidity in the lagoon started reducing as the days progressed after the passage of *Phailin* because the sediment started settling down to the bottom and consequently, Chl-*a* concentration recovered back to pre-cyclone level after a week of the passage, but a phytoplankton bloom was not observed (Fig. 8.12).

On the other hand, post-*Hudhud* TSS maps did not show a significant change in TSS and a gradual increase in Chl-*a* concentration was observed throughout the lagoon (Fig. 8.12). There were several factors contributed for this differential impact post-*Hudhud*. For example, *Hudhud* made landfall relatively at farther distance from Chilika, stayed for shorter duration, and moved away from the lagoon after its landfall. The combination of landfall location, duration, and trajectory during *Hudhud* resulted in lower precipitation and surface runoff in the watershed of the lagoon. As a result, any sector of the lagoon did not experience an increase in TSS to the level of post-*Phailin*. However, through satellite imagery a phytoplankton bloom was observed with Chl-*a* concentration increasing three-times after the passage of *Hudhud*. Previous studies also suggested that when there is sufficient light available in the water column followed by tropical cyclone, stimulation of phytoplankton becomes more rapid (Miller et al. 2006; Wetz and Paerl 2008; Wetz and Yoskowitz 2013). Chilika Lagoon experienced a sudden increase in Chl-*a* just after 2-days of landfall of *Hudhud* and the upward trend continued for few more days. A gradual spread in the spatial coverage of increased Chl-*a* concentration from SS towards NS was clearly visible in post-*Hudhud* MODIS derived Chl-*a* maps (Fig. 8.12). This type of sudden increase in Chl-*a* is common in previous studies (Peierls et al. 2003; Paerl et al. 2010; Huang et al. 2011; Sarangi et al. 2014; Lotliker et al. 2014; Baliarsingh et al. 2015). For example, Huang et al. (2011) reported sudden increase in mean Chl-*a* from 5.3 to 14.7 μg/L in Pensacola Bay, Florida just after a day of passage of Hurricane *Ivan*. In another study, Baliarsingh et al. (2015) reported a significant increase in Chl-*a* just after 3-days of landfall of *Hudhud* in north-western BOB from 1.58–2.28 mg/m^3 to 2.57–6.62 mg/m^3. They attributed this increase to nutrient entrainment from river influx due to the VSCS.

In the above discussion, a combination of several factors was attributed to the differential impact of two cyclones on the water quality of the lagoon. However, isolating a single factor primarily responsible for the differential impact would not be useful because each individual factor has its own importance and often correlated with other factors. For example, distance from landfall location cannot be isolated from the trajectory and speed of the cyclone. Past studies have reported that even if a cyclone makes landfall close to study area but passes very quickly, then it would not bring as much amount of rainfall and runoff compared to a slow moving cyclone which stayed for a longer duration (Mallin and Corbett 2006). Similarly surface runoff cannot be isolated from precipitation, nutrient pulsing cannot be isolated from mixing of water column. The same principle is applicable to the phytoplankton growth in a lagoon which requires a combination of several factors favourable for primary production. Srichandan et al. (2015) suggested that several other factors such as nutrient, turbidity, water residence time, and flushing rate are equally important for phytoplankton biomass production along with geographic-geomorphologic-bathymetric setting of an estuary. For instance, high flushing rate in combination with high fresh water discharge limited the phytoplankton growth in Neuse River Estuary post-Hurricane *Fran* (Paerl et al. 1998). Similarly, rapid flushing for several weeks was reported post-*Phailin* in Chilika Lagoon that slowed the rate of phytoplankton growth (Srichandan et al. 2015). Flushing rate was comparatively less rapid post-*Hudhud* as the flood intensities in the distributaries were much lower compared to *Phailin*. Another factor suggested by Wetz and Yoskowitz (2013) was calm wind after the passage of cyclone supports the stratification and light condition for phytoplankton growth. The wind speed was stable after the passage of *Hudhud* but wind speed was dynamic post-*Phailin* that might have slowed down the phytoplankton growth in the lagoon. Therefore, lower rainfall, lower surface runoff, less turbidity, low flushing rate, and stable wind, all favored the phytoplankton bloom post-*Hudhud*. The same factors but at a different magnitude prevented a phytoplankton bloom post-*Phailin*. The above results and discussions validated the proposed hypothesis that the likelihood of a phytoplankton bloom or significant increase in phytoplankton biomass after a cyclone is dependent on the physical, meteorological, and geomorphological characteristics of the VSCS and the lagoon.

8.4 Summary and Conclusion

This study showed that MODIS daily surface reflectance products (MOD09GQ) and MODIS 8-day composite products (MOD09Q1) are well suited for monitoring the biophysical parameters (TSS and Chl-*a*) concentration in Chilika Lagoon. MOD09GQ product can be used for short-term monitoring purpose, while MOD09Q1 can be used for long-term assessment. The result of quantitative analysis for 14 years (2001–2014) using MODIS 8-day composites (MOD09Q1) data indicated that the seasonal variability of TSS is dominant in all the three sectors of the Chilika Lagoon compared to inter-annual variability. The main reason for large variations in the northern sector is the shallow depth and intrusion of large sediment discharge from Mahanadi River from the northern side, which is the largest fresh water distributary for Chilika Lagoon. Similar findings such as high turbidity in NS followed CS and SS (Mohanty and Pal 2001), significant seasonal variability in water surface area and water quality parameters (Pal and Mohanty 2002) have also been reported in the past using field based measurements. However, this study analyzed additional contribution of physical and meteorological factors from different catchment area of the lagoon in increasing the TSS level. It was also found that monthly mean TSS (2001–2014) for NS and CS is strongly correlated with total precipitation and surface runoff compared to SS. However, TSS in southern sector was highly correlated with wind stress compared to the other two sectors. Overall, results indicated that different factors have different level of impact on the TSS variability across the three sectors of the lagoon.

Further, a case study showed the day-to-day applicability of the biophysical models for tracking the effect of natural hazards and other physical and geomorphological changes on the water quality of the lagoon. This case study primarily dealt with questions such as why some cyclones trigger phytoplankton blooms in estuaries and lagoons and some don't? and what factors control the likelihood of a bloom after the passage of a cyclonic storm? A comprehensive comparative analysis of several factors was performed to isolate the causes of the differential impact of anniversary VSCSs, *Phailin* and *Hudhud*, on the water quality of Chilika Lagoon. The anniversary VSCSs allowed the verification of a theoretical concept widely discussed in previous literatures that characteristics of a cyclone such as close landfall location, high wind intensity, longer duration of stay, trajectories along the watershed of study area would support high precipitation and surface runoff which may lead to increased turbidity and a phytoplankton bloom in nearby water bodies such as estuaries and lagoons (Wetz and Yoskowitz 2013; Mallin and Corbrett 2006; Paerl et al. 2001). *Phailin's* impact on Chilika Lagoon and its watershed resulted in unprecedented levels of precipitation and runoff before-during-after the landfall, which shattered the typical sectorial turbidity gradient. Exponential increase in turbidity because of a combination of run-off and wind driven re-suspension of fine sediments resulted in strong attenuation of light in water column post-*Phailin*. Limited light condition coupled with enhanced flushing rate due to flooded river and increased fresh water discharge reduced the Chl-*a* concentration after the passage of *Phailin*. In contrast, relatively farther landfall location, trajectory away from the lagoon, relatively lower wind intensity and short duration of stay of VSCS *Hudhud*, led to lesser precipitation and surface runoff compared to *Phailin*. Consequently, lagoon did not experience a drastic increase in turbidity and light attenuation. Sufficient light availability, stable wind, reduced flushing all favored the phytoplankton growth after passage of *Hudhud* and as a result, Chl-*a* concentration increased almost threefold in all the sectors of the lagoon.

The frequency of tropical cyclones is expected to increase under the global climate change scenario which makes satellite based high spatial and temporal assessment very useful compared to field sampling program which are limited in spatial and temporal domain (Srichandan et al. 2015). Satellite data coupled with model derived products may become very useful in near future for the assessment of cyclone induced impact and predicting phytoplankton bloom prone areas. The approach used in this study can be applied to other cyclone-prone coastal areas. Coupling of satellite based observation with modelling output from systems such as Giovanni can improve monitoring program implemented in numerous coastal estuaries and lagoons. Susceptible watershed areas that contribute in high surface runoff can be isolated and management plan can be implemented like creating buffer zone or plantation to minimize the surface erosion rate which is a major factor in deteriorating water quality of any lake. Also, long-term monitoring ability of satellite based model will facilitate researchers and regulators to assess the changes in estuarine and lake system and associated watershed on a broader scale. The ability to predict these changes on estuarine and coastal environments might become an essential part of designing and implementing the management and restoration effort for a lake, estuary, or coastal region in future. Overall, high frequency monitoring of physical, biophysical, and meteorological parameters using satellite data can help in tracking several environmental phenomena in the lagoon such as siltation, effectiveness of dredging activities, areas with high probability of algal blooms, impact of high TSS on seagrass habitats, and the overall water clarity and productivity of the lagoon.

Acknowledgements AK, the lead author, thanks NIT Rourkela for financial assistance in the form of institute fellowship. This work is a part of Master's thesis submitted by AK to NIT Rourkela. QuikScat data used in this study are produced by Remote Sensing Systems and sponsored by the NASA Ocean Vector Winds Science Team. The authors wish to thank the Goddard Space Flight Center (GSFC), NASA and MODIS support team for providing region specific MODIS satellite data products and Giovanni datasets.

References

Acker J, Leptoukh G (2007) Online analysis enhances use of NASA earth science data. EOS Trans Am Geophys Union 88(2):14–17

Acker J, Soebiyanto R, Kiang R, Kempler S (2014) Use of the NASA Giovanni data system for geospatial public health research: example of weather-influenza connection. ISPRS Int J Geo Inf 3:1372–1386

Angles S, Jordi A, Campbell L (2015) Responses of the coastal phytoplankton community to tropical cyclones revealed by high-frequency imaging flow cytometry. Limnol Oceanogr 60:1562–1576

Ansari KGMT, Pattnaik AK, Rastogi G, Bhadury P (2015) An inventory of free-living marine nematodes from Asia's largest coastal lagoon, Chilika, India. Wetl Ecol Manag 23(2):881–890. http://dx.doi.org/10.1007/s11273-015-9426-2

Astuti I, Mishra DR, Mishra S, Schaeffer B (2018) A hybrid approach for deriving inherent optical properties in oligotrophic estuaries. Cont Shelf Res 166:92–107

Baliarsingh SK, Parida C, Lotliker AA, Srichandan S, Sahu KC, Kumar TS (2015) Biological implications of cyclone Hudhud in the coastal waters of northwestern Bay of Bengal. Curr Sci 109(7):1243–1245

Binding CE, Bowers DG, Mitchelson-Jacob EG (2005) Estimating suspended sediment concentration from ocean colour measurements in moderately turbid waters; the impact of variable particle scattering properties. Remote Sens Environ 94:373–383

Bramha S, Panda UC, Bhatta K, Sahu BK (2008) Spatial variation in hydrological characteristics of Chilika-A coastal lagoon of India. Indian J Sci Technol 1(4):1–7

Cao F, Mishra DR, Schalles J, Miller W (2018) Evaluating ultraviolet (UV) based photochemistry in optically complex coastal waters using the Hyperspectral Imager for the Coastal Ocean (HICO). Estuar Coast Shelf Sci 215:199–206

CDA (Chilika Development Authority) (2008) The atlas of Chilika. Chilika Development Authority, Bhubaneswar

Chaichitehrani N, D'Sa EJ, Osburn CL, Bianchi TS, Schaeffer BA (2013) Chromophoric dissolved organic matter and dissolved organic carbon from SeaWiFS, MODIS and MERIS sensors: case study for the northern Gulf of Mexico. Remote Sens 5:1439–1464

Chen Z, Hu C, Muller-Karger F (2007) Monitoring turbidity in Tampa Bay using MODIS/Aqua 250-m imagery. Remote Sens Environ 109:207–220

Chen S, Huang W, Wang H, Li D (2009) Remote sensing assessment of sediment re-suspension during Hurricane Frances in Apalachicola Bay, USA. Remote Sens Environ 113:2670–2681

Cui L, Qiu Y, Fei T, Liu Y, Wu G (2013) Using remotely sensed suspended sediment concentration variation to improve management of Poyang Lake, China. Lake Reservoir Manage 29:47–60

Dekker AG, Peters SWM, Vos RJ (2001) Comparison of remote sensing data, model results and in situ data for total suspended matter (TSM) in the Southern Frisian Lakes. Sci Total Environ 268:197–214
Dekker AG, Vos RJ, Peters SW (2002) Analytical algorithms for lake water TSM estimation for retrospective analyses of TM and SPOT sensor data. Int J Remote Sens 23(1):15–35
Dogliotti AI, Ruddick KG, Nechad B, Doxaran D, Knaeps E (2015) A single algorithm to retrieve turbidity from remotely-sensed data in all coastal and estuarine waters. Remote Sens Environ 156:157–168
Doxaran D, Froidefond JM, Castaing P, Babin M (2009) Dynamics of the turbidity maximum zone in a macrotidal estuary (the Gironde, France): observation from field and MODIS satellite data. Estuar Coast Shelf Sci 81:321–332
El-Alem A, Chokmani K, Laurion I, El-Adlouni SE (2012) Comparative analysis of four models to estimate chlorophyll-a concentration in case-2 waters using MODerate Resolution Imaging Spectroradiometer (MODIS) imagery. Remote Sens 4:2373–2400
Fang H, Hrubiak PL, Kato H, Rodell M, Teng WL, Vollmer BE (2008) Global Land Data Assimilation System (GLDAS) products from NASA Hydrology Data and Information Services Center (HDISC). ASPRS 2008 annual conference Portland, Oregon April 28 – May 2, 2008. http://www.asprs.org/a/publications/proceedings/portland08/0020.pdf
Fox RC, Rowntree KM (2013) Extreme weather events in the Sneeuberg, Karoo, South Africa: a case study of the floods of 9 and 12 February 2011. Hydrol Earth Syst Sci Discuss 10:10809–10844
Gitelson AA, Gritz Y, Merzlyak MN (2003) Relationships between leaf chlorophyll content and spectral reflectance and algorithms for non-destructive chlorophyll assessment in higher plant leaves. J Plant Physiol 160:271–282
GoO (2013) Memorandum on the very severe Cyclone Phailin and the subsequent flood 12-–15 October 2013. The Revenue and Disaster Management Department Government of Odisha. http://www.osdma.org/userfiles/file/MEMORANDUMPhailin.pdf. Accessed 10 Oct 2014
Gupta M (2013) Chromaticity analysis of the Chilika lagoon for total suspended sediment estimation using RESOURCESAT-1 AWiFS data- a case study. J Great Lakes Res 39:696–700
Havens KE, Beaver JR, Casamatta DA, East TL, James RT, Mccormick P, Phlips EJ, Rodusky AJ (2011) Hurricane effects on the planktonic food web of a large subtropical lake. J Plankton Res 33:1081–1094
Hu C, Chen Z, Clayton T, Swarnzenski P, Brock J, Muller-Karger F (2004) Assessment of estuarine water quality indicators using MODIS medium-resolution bands: initial results from Tampa Bay, Florida. Remote Sens Environ 93:423–441
Huang W, Mukherjee D, Chen S (2011) Assessment of Hurricane Ivan impact on chlorophyll-a in Pensacola Bay by MODIS 250 m remote sensing. Mar Pollut Bull 62(3):490–498
IFRC (1999) Cyclone: Orissa, India Information Bulletin No. 1. International Federation of Red Cross (IFRC) and Red Crescent Societies. http://www.ifrc.org/docs/appeals/rpts99/in005.pdf. Accessed 10 Oct 2014
IMD (2013) Very severe cyclonic storm, PHAILIN over the Bay of Bengal (08–14 October 2013): a report. http://www.imd.gov.in/section/nhac/dynamic/phailin.pdf
IMD (2014) Very severe cyclonic storm, HUDHUD over the Bay of Bengal (07–14 October 2014): a report. http://www.rsmcnewdelhi.imd.gov.in/images/pdf/publications/preliminary-report/hud.pdf
Jayaraman G, Rao AD, Dube A, Mohanty PK (2005) Numerical simulation of circulation and salinity structure in Chilika lagoon. J Coast Res 22:195–211
Jordan TE, Whigham DF, Hofmockel KH, Pittek MA (2003) Wetlands and aquatic processes: nutrient and sediment removal by a restored wetland receiving agricultural runoff. J Environ Qual 32:1534–1547
Kahru M, Mitchell BG, Diaz A, Miura M (2004) Modis detects a devastating algal bloom in Paracas Bay, Peru. Eos 85(45):465–472

Katlane R, Dupouy C, Zargouni F (2012) Chlorophyll and turbidity concentrations as an index of water quality of the Gulf of Gabes from MODIS in 2009. Teledetection 11(1):265–273
Kumar R, Pattnaik AK (2012) Chilika – an integrated management planning framework for conservation and wise use. Wetlands International – South Asia, New Delhi, India and Chilika Development Authority, Bhubaneswar, India
Kumar A, Equeenuddin S, Mishra DR, Acharya BC (2016) Remote monitoring of sediment dynamics in a coastal lagoon: long-term spatio-temporal variability of suspended sediment in Chilika. Estuar Coast Shelf Sci 170:155–172
Lee Z, Carder KL, Chen RF, Peacock TG (2001) Properties of the water column and bottom derived from Airborne Visible Infrared Imaging Spectrometer (AVIRIS) data. J Geophys Res Oceans 106(C6):11639–11651
Lotliker AA, Srinivasakumar T, Reddem VS, Nayak S (2014) Cyclone Phailin enhanced the productivity following its passage: evidence from satellite data. Curr Sci 106(3):360–361
Madrinan MJM, Hamdan ZAM, Rickman DL, Muller-Karger F (2010) Using the surface reflectance MODIS Terra product to estimate turbidity in Tampa Bay, Florida. Remote Sens 2:2713–2728
Mallin MA, Corbrett CA (2006) How hurricane attributes determine the extent of environmental effects: multiple hurricanes and different coastal systems. Estuar Coasts 29:1046–1061
Mallin MA, Posey MH, McIver MR, Parsons DC, Ensign SH, Alphin TD (2002) Impacts and recovery from multiple hurricanes in a Piedmont-coastal plain river system. Bioscience 52:999–1010
Mallin MA, Johnson VL, Ensign SH, MacPherson TA (2006) Factors contributing to hypoxia in rivers, lakes and streams. Limnol Oceanogr 51(2):690–701
Miller RL, McKee BA (2004) Using MODIS 250 m imagery to map concentration of total suspended matter in coastal waters. Remote Sens Environ 93:259–266
Miller WD, Harding LW, Adolf JE (2006) Hurricane Isabel generated an unusual fall bloom in Chesapeake Bay. Geophys Res Lett 33:L06612. http://dx.doi.org/10.1029/2005GL025658
Miller RL, Liu CC, Buonassissi CJ, Wu AM (2011) A multi-sensor approach to examining the distribution of Total Suspended Matter (TSM) in the Albemarle-Pamlico estuarine system, NC, USA. Remote Sens 3:962–974
Mishra DR, Gould R Jr (2016) Preface: remote sensing in coastal environments. Remote Sens 8 (8):665
Mishra DR, Mishra S (2010) Plume and bloom: effect of Mississippi river diversion on the water quality of Lake Pontchartrain. Geocarto Int 25(7):555–568
Mishra S, Mishra DR (2012) Normalized difference chlorophyll index: a novel model for remote estimation of chlorophyll-*a* concentration in turbid productive waters. Remote Sens Environ 117:394–406
Mishra SP, Panigrahi DR (2014) Storm impact on south Odisha coast, India. Int JAdv Res Sci Eng 3 (11). ISSN: 2319-8354 (E)
Mishra S, Mishra DR, Lee Z, Tucker C (2013) Quantifying cyanobacterial phycocyanin concentration in turbid productive waters: a quasi-analytical approach. Remote Sens Environ 133:141–151
Mishra S, Mishra DR, Lee Z (2014a) Bio-optical inversion in highly turbid and cyanobacteria dominated waters. IEEE Trans Geosci Remote Sens 52:375–388
Mishra DR, Schaeffer BA, Keith D (2014b) Performance evaluation of normalized difference chlorophyll index in northern Gulf of Mexico estuaries using the hyperspectral imager for the coastal ocean. GISci Remote Sens 51(2):175–198
Mishra DR, Ogashawara I, Gitelson AA (2017) Bio-optical modeling and remote sensing of inland waters, 1st edn. Elsevier, Waltham, p 332. ISBN: 9780128046449
Mohanty PK, Pal SR (2001) IRS data utilisation for estimation of turbidity level in Chilika lagoon, east coast of India. In: Proceedings of 22nd Asian conference on remote sensing, 5–9 November 2001, Singapore

Mohanty PK, Panda BUS (2009) Circulation and mixing processes in Chilika lagoon. Indian J Mar Sci 38(2):205–214
Nayak BK, Acharya BC, Panda UC, Nayak BB, Acharya SK (2004) Variation of water quality in Chilika Lake, Orissa. Indian J Mar Sci 33(2):164–169
Ogashawara I, Mishra DR, Gitelson AA (2017) Remote sensing of inland waters: background and current state-of-the-art. In: Mishra DR, Ogashawara I, Gitelson AA (eds) Bio-optical modeling and remote sensing of inland waters, 1st edn. Elsevier, Waltham, p 332. ISBN: 9780128046449
Paerl HW, Pinckney JL, Fear JM, Peierls BL (1998) Ecosystem responses to internal and watershed organic matter loading: consequences for hypoxia in the eutrophying Neuse River Estuary, North Carolina, USA. Mar Ecol Prog Ser 166:17–25
Paerl HW, Bales JD, Ausley LW, Buzzelli CP, Crowder LB, Eby LA, Fear JM, Go M, Peierls BL, Richardson TL, Ramus JS (2001) Ecosystem impacts of 3 sequential hurricanes (Dennis, Floyd and Irene) on the US's largest lagoonal estuary, Pamlico Sound, NC. Proc Natl Acad Sci U S A 98(10):5655–5660
Paerl HW, Valdes LM, Peierls BL, Adolf JE, Harding LW (2006) Anthropogenic and climatic influences on the eutrophication of large estuarine ecosystems. Limnol Oceanogr 51:448–462
Paerl HW, Rossignol KL, Hall SN, Peierls BL, Wetz MS (2010) Phytoplankton community indicators of short- and long term ecological change in the anthropogenically and climatically impacted Neuse River Estuary, NorthCarolina, USA. Estuar Coasts 33:485–497
Page B, Kumar A, Mishra DR (2018) A novel cross-satellite based assessment of the spatio-temporal development of a cyanobacterial harmful algal bloom. Int J Appl Earth Obs Geoinf 66:69–81
Pal SR, Mohanty PK (2002) Use of IRS-1B data for change detection in water quality and vegetation of Chilika Lagoon, east coast of India. Int J Remote Sens 23(6):1027–1042
Panda US, Mohanty PK (2008) Monitoring and modelling of Chilika environment using remote sensing data. In: Proceedings of Tail 2007: the 12th world lake conference, 28 October-2 November 2007, Jaipur, India, pp 617–638
Panigrahi S, Winker J, Panigrahy RC, Satpathy KK, Acharya BC (2009) Variability of nutrients and phytoplankton biomass in a shallow brackish water ecosystem (Chilika Lagoon, India). Limnology 10:73–85
Patnaik SS (1998) Estimation of sediment flow into the Chilika Lake. Abstract volume, workshop on sustainable development of Chilika Lake, Orissa (Bhubaneswar, India: Chilika Development Authority), pp 31–32
Peierls B, Christian R, Paerl H (2003) Water quality and phytoplankton as indicators of hurricane impacts on a large estuarine ecosystem. Estuaries 26:1329–1343
Petus C, Chust G, Gohin F, Doxaran D, Froidefond JM, Sagarminaga Y (2010) Estimating turbidity and total suspended matter in the Adour River Plume (South Bay of Biscay) using MODIS 250-m imagery. Cont Shelf Res 30:379–392
Prasad AK, Singh RP (2010) Chlorophyll, calcite and suspended sediment concentrations in the Bay of Bengal and the Arabian Sea at the river mouths. Adv Space Res 45:61–69
Raman AV, Kalavati C, Prabhakara Rao Y (2007) Modelling Chilka Lake: an ecosystem approach. Final report (2004–07) submitted to Ministry of Earth Sciences ICMAM Directorate Chennai. http://www.icmam.gov.in/CHIL.PDF
Rao DP, Subramanian SK, Sudarsana R (1986) Geomorphic analysis of Chilika Lake and adjoining area using remotely sensed data, Photonirvachak. J Indian Soc Remote Sens 14:53–60
Ritchie JC, Zimba PV, Everitt JH (2003) Remote sensing technique to assess water quality. Photogramm Eng Remote Sens 69(6):695–704
Rodell M, Houser PR (2004) Updating a land surface model with MODIS derived snow cover. J Hydrometeorol 5(6):1064–1075
Sarangi RK, Mishra MK, Chauhan P (2014) Remote sensing observations on impact of Phailin cyclone on phytoplankton distribution in northern Bay of Bengal. IEEE J Sel Top Appl Earth Observ Remote Sens 8(2):539–549

Siddiqi SZ, Rama Rao KV (1995) Fauna of Chilika Lake: limnology, zoological survey of India. Wetl Ecosyst Ser 1:11–136

Srichandan S, Kim JY, Kumar A, Mishra DR, Bhadury P, Muduli PR, Pattnaik AK, Rastogi G (2015) Interannual and cyclone-driven variability in phytoplankton communities of a tropical coastal lagoon. Mar Pollut Bull 101(1):39–52

Sudhakar S, Pal DK (1993) Water quality assessment of the Lake Chilka. Int J Remote Sens 14:2575–2579

UNEP (2013) Cyclone Phailin in India: early warning and timely actions saved lives. http://www.unep.org/pdf/UNEP_GEAS_NOV_2013.pdf

Vermote EF, El saleoql N, Justice CO, Kaufman YJ, Privette JL, Remer L, Roger JC, Tanre D (1997) Atmospheric correction of visible to middle-infrared EOS-MODIS data over land surfaces: background, operational algorithm and validation. J Geophys Res 102:17131–17141

Wang G, Lee Z, Mishra DR, Ma R (2016) Retrieving absorption coefficients of multiple phytoplankton pigments from hyperspectral remote sensing reflectance measured over cyanobacteria bloom waters. Limnol Oceanogr Methods 14(7):432–447

Wetz MS, Paerl HW (2008) Estuarine phytoplankton responses to hurricanes and tropical storms with different characteristics (trajectory, rainfall, winds). Estuar Coasts 31:419–429

Wetz MS, Yoskowitz DW (2013) An 'extreme' future for estuaries? Effects of extreme climatic events on estuarine water quality and ecology. Mar Pollut Bull 69(2013):7–18

Zhang Y, Lin S, Liu J, Qian X, Ge Y (2010) Time-series MODIS image-based retrieval and distribution analysis of total suspended matter concentrations in Lake Taihu (China). Int J Environ Res Public Health 7(9):3545–3560

Zhang Y, Lin S, Qian X, Wang Q, Qian Y, Liu J, Ge Y (2011) Temporal and spatial variability of chlorophyll a concentration in Lake Taihu using MODIS time-series data. Hydrobiologia 661:235–250

Zhao HH, Chen Q, Walker ND, Zheng QA, MacIntyre HL (2011) A study of sediment transport in a shallow estuary using MODIS imagery and particle tracking simulation. Int J Remote Sens 32 (21):6653–6671

Zhu M, Paerl HW, Zhu G, Wu T, Li W, Shi K, Zhao L, Zhang Y, Qin B, Caruso AM (2014) The role of tropical cyclones in stimulating cyanobacterial blooms in hypertrophic Lake Taihu, China. Harmful Algae 39:310–321

Chapter 9
Spatio-Temporal Variation in Physicochemical Parameters of Water in the Chilika Lagoon

Pradipta R. Muduli and Ajit K. Pattnaik

Abstract Assessment of physicochemical parameters of Chilika lagoon water is of vital importance as it guides to monitor the pollution status and formulate a management plan. Mixing of water from river and sea with different nutrient levels makes an ideal condition of salinity and nutrient stoichiometry supporting a unique benthic and pelagic biodiversity including fishery. The hydrology of the lagoon undergoes a significant change during the switchover between seasons as well as years. The rainfall pattern influences the discharge with different nutrient concentration and stoichiometry which ultimately alters the lagoon biogeochemistry. Apart from these, the spatiotemporal variability is also controlled by the exchange of seawater through the lagoon mouth which undergoes drastic change over time with respect to the cross-sectional area and its position. Overall brackish nature of the lagoon was sustained in most of the lagoon throughout the year due to the adequate inflow of saline water from the sea and fresh water from major Rivers.

The present study showed that the physicochemical parameters such as pH, dissolved oxygen (DO), biochemical oxygen demand (BOD), nitrite nitrogen (NO_2), nitrate nitrogen (NO_3), and phosphate phosphorous (PO_4) were within the threshold range suitable for the propagation of wildlife and fishery. The primary source of PO_4 and NO_2 were mostly from *in situ* mineralisation processes whereas, NO_3 and silicate (SiO_2) from the riverine influx. The overall nutrient stoichiometry indicated NO_3 and PO_4 were limiting with respect to SiO_2 throughout the year which favoured the growth of diatoms. During the monsoon period, PO_4 remained limiting due to dilution by fresh water from northeast rivers. Shallow regions of the lagoon get turbid during monsoon due to the inflow of river water with high SPM but during summer, the wind-induced bottom sediment churning becomes the key influencing variable. The variability of the flux of nutrients and suspended particulate matter (SPM) with respect to different rivers and season, re-suspension of sediment, and

P. R. Muduli (✉)
Wetland Research and Training Centre, Chilika Development Authority, Balugaon, Odisha, India

A. K. Pattnaik
Chilika Development Authority, Bhubaneswar, Odisha, India

C. M. Finlayson et al. (eds.), *Ecology, Conservation, and Restoration of Chilika Lagoon, India*, Wetlands: Ecology, Conservation and Management 6,
https://doi.org/10.1007/978-3-030-33424-6_9

autochthonous processes had a significant influence on the lagoon's biogeochemical cycle. Despite least photic depth (transparency), the appropriate condition of light and nutrient stoichiometry enabled highest primary productivity (PP) in the northern sector (NS) while southern sector (SS) had the lowest rate due to low nutrient content, even though it maintained the highest water clarity throughout the year.

Keywords Water quality · Biogeochemistry · Nitrite · Nitrate · Ammonia · Phosphate · Silicate

9.1 Introduction

Aquatic environments require appropriate maintenance through interventions which need systematic knowledge of the ecosystem functions controlling the biodiversity, fishery resources, habitats for several aquatic and benthic organisms and appearance of avifauna and mammals at higher aquatic food chain (indicators for good health of the ecosystem). In order to keep track of other biological variables, the water quality must be maintained in a healthy condition (Muduli et al. 2011). Since there are many factors responsible for the deterioration of water quality, all need to be monitored periodically for sustainable management. Hence, the study on the spatio-temporal variations of physicochemical parameters have become an essential aspect of the characterisation of aquatic biodiversity and the implementation of sustainable management strategies (Sarkar et al. 2007).

The water chemistry of any lagoonal ecosystem is mostly controlled by physical, geological and biological factors. There are several components of these factors comprising a change in climate, the morphology of ecosystem, biotic species diversity and abundance and the change in geology. Apart from the above factors, the physicochemical exchange reactions take place at surfaces and at boundaries, such as air-sea or, water-sediment interfaces also impact water chemistry. The dissolved constituents of such ecosystems are of two types, the conservative and the trace components (Paerl and Justic 2011). The conservative constituents are not influenced significantly by biological processes, and it takes a long time for changes in concentration due to chemical and geochemical processes. This is mostly controlled by the physical processes of advection and convection, turbulence diffusion, etc. whereas, the trace constituents, controlled by the physical processes as well as by the biological processes (uptake, excretion, and biodegradation etc.) (Paerl and Justic 2011). Aquatic ecosystems are influenced by two processes; the first one is anthropogenic and the second is a natural process. Contamination or pollution of river water coming from catchment areas, containing effluent from domestic sewage, imbalanced agricultural activities, and industrial settings come under the human activity (anthropogenic). The rainfall (precipitation), evaporation, weathering, and mixing of riverine freshwater from Rivers and saline water from Sea comes under the natural processes. Among the brackish water lagoons, Chilika is the largest one in

Asia and its water chemistry is influenced by both the anthropogenic and natural process.

Chilika lagoon being a hotspot for carbon-nitrogen biogeochemistry and livelihood support for more than 200,000 fishermen it warrants an extensive study on water quality. Hence the objectives of this chapter are to illustrate the variability of physicochemical parameters of Chilika lagoon under various environmental conditions and the functional relationships of the dynamics with saline and freshwater inputs. This study highlights the overall water quality of the Chilika lagoon and discerns the trend of water quality with respect to seasons from 1999 to 2015. These findings will be helpful to better understand the ecosystem's responses and finally contribute towards the sustainable management of the lagoon.

9.2 Environmental Characteristics of Chilika Lagoon

Chilika, a brackish water lagoon situated on the Indian east coast spans an area between 950 km^2 (summer) and 1165 km^2 (monsoon) (Gupta et al. 2008). The lagoon is connected to the Bay of Bengal through the dredged mouth near Satapada (Sipakuda) and Palur canal in the southern lagoon (through Rushikulya estuary) (Fig. 9.1). The inundation due to tidal impact is observed only near the outer channel

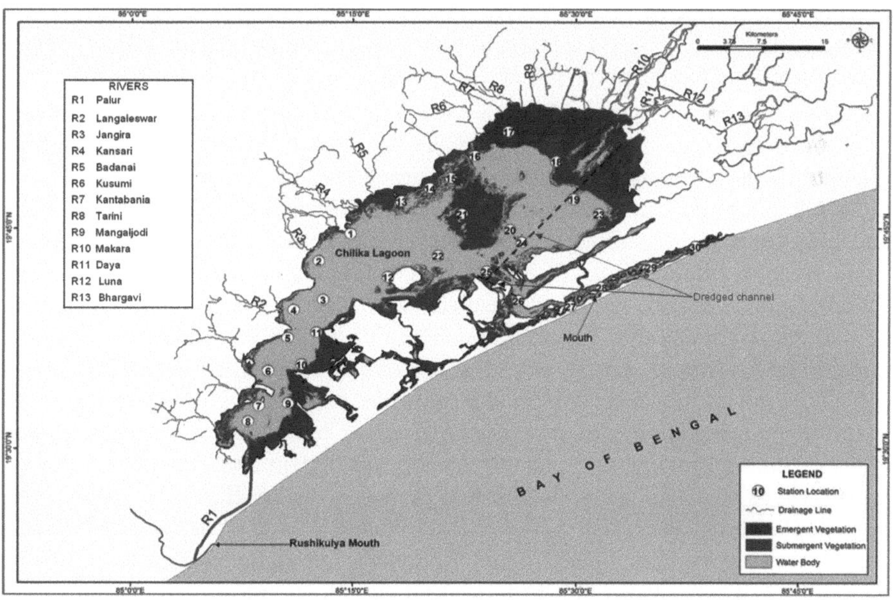

Fig. 9.1 Map of the Chilika lagoon showing the sampling locations (30 stations), inlets for seawater intrusion (mouth and palur canal through Rushikulya estuary) and regions covered with vegetation

region and is negligible towards the southern, northern and western end of the Chilika (Muduli et al. 2012). The tidal impact through the Rushikulya estuary observed to have an impact on the water chemistry of the southern region of the Chilika lagoon (Barik et al. 2017a).

The Chilika land use and land cover (LULC) during 2012 shows agricultural land, Chilika Lake, dense forest and rural settlement cover 26, 18, 14, 10% respectively. The LULC with cropping pattern, vegetation pattern and population can have bearing on the physicochemical parameters of the lagoon and has shown a considerable change from 1975 to 2012 (Ojha 2013). Apart from these, the presence of seagrass (southern sector) and macrophytes (submerged as well as emerged) also have major roles in the alteration of water quality on a seasonal basis (Gupta et al. 2008; Muduli et al. 2013). The area of Chilika Lagoon covered with such vegetation has been shown in the Chilika map prepared using GIS (Fig. 9.1). There are 52 rivers and rivulets drain into the lagoon, however, 13 of them are major contributors (Kumar and Pattnaik 2012) (Fig. 9.1). By the effort of Chilika Development Authority (CDA), channels have been dredged and are maintained periodically to ensure the proper inflow of freshwater and saline water into the lagoon and flushing of sediments. One of the dredged channels connects from the river Daya to the outer channel (lead channel) and another one from the outer channel towards Balugaon (called Balugaon channel) (Fig. 9.1). A recent study revealed the significant impact of such channel on the spatiotemporal variability of physicochemical parameters in the Chilika lagoon (Kim et al. 2016). There are 2342 nos. of registered mechanized boats are operated in the lagoon and the petroleum hydrocarbon (PHC) used for the maintenance and operational activities has also influenced on the water chemistry of the lagoon, However, the trace levels of PHC recorded in Chilika are within the safe limit (Mohanty et al. 2016, 2017).

The hydrological regime of Chilika is strongly influenced by riverine freshwater inputs and by local meteorological conditions such as winds and precipitation (Ganguly et al. 2015; Patra et al. 2016; Barik et al. 2017a). The variability in salinity is mostly determined by freshwater inputs, precipitation, evaporation, morphology and the exchange of seawater through the mouth located in the OC. The lagoon salinity is also influenced by El Niño/Southern Oscillation (Kim et al. 2015). The hydrology of river and seawater is the main driver (apart from biological processes) of nutrient dynamics in Chilika lagoon (Patra et al. 2016). Spatial and temporal variability in water chemistry in rivers and rivulets that drains into Chilika also influences the physicochemical parameters of the lagoon as rivers and streams are highly heterogeneous at different spatial scales. Chilika lagoon receives several types of inputs (through the catchment area) viz., urban and agricultural wastes, which results in a significant alteration in the water quality (Panigrahi 2006; Patra et al. 2016; Muduli et al. 2017). The change in water chemistry influences biodiversity as well as the composition of the ecosystem resources (Parida et al. 2017). Seasonal variations in precipitation, surface runoff, interflow, groundwater flow have a strong effect on riverine discharge and subsequently on the concentration of pollutants in coastal waters (Singh et al. 2005). A number of studies dealt with the spatial and temporal variations of surface water quality in other Indian ecosystems (Ganga: Chakrapani and Veizer 2005, Gomuti River: Singh et al. 2004; Godavari estuary: Sarma et al. 2009; Kochin estuary: Madhu et al. 2010).

The capability of rivers to export nutrients into Chilika is controlled by water discharge, residence time, water retention properties of the soils, geologic structure and extreme climatic change such as cyclone (Muduli et al. 2013; Srichandan et al. 2015a; Muduli et al. 2017). The chemical and physical composition of sediment can have a significant role in water chemistry (Mohanty et al. 2016, 2017) and biological components (Barik et al. 2017b). The nutrients that come through rivers, exchange with sea water or newly generated through mineralisation may or may not be enough for algal growth. The nutrient limitation (either P or N) in the ecosystems usually depend on its availability from different sources and it's uptake by phytoplankton (Ganguly et al. 2013; Mukherjee et al. 2018). Apart from these, some aquatic environment also experiences the limitation of trace elements responsible for photosynthesis (Zhang 2000). The variation in nutrient flux into the lagoon, dilution, wind-driven upwelling (shallow region) and seawater exchange may encompass the stoichiometry of nutrient in the Chilika lagoon (Zhou et al. 2012; Ganguly et al. 2015). Nutrient stoichiometry also has a significant role in plankton growth and productivity of the Chilika lagoon (Ganguly et al. 2013, 2015; Patra et al. 2017). Apart from nutrient variability, the change in the level of salinity also has a significant impact on the productivity of the coastal ecosystems. As per Kim et al. (2015), the level of seawater exchange has more influences on the productivity of the Chilika lagoon than local precipitation. The study also observed that the primary productivity of the lagoon increased gradually with the salinity after the restoration of the lagoon. Before the year 2000, the lagoon salinity was gradually decreasing due to chocking of the mouth through which sea water enters into the lagoon and the ecosystem was getting dominated by freshwater (Sahu et al. 2014). This change in water quality enabled the luxurious growth of freshwater macrophytes throughout the lagoon. In order to restore the brackish nature of the lagoon, an artificial mouth was dredged open during September 2000 by CDA, Govt. of Odisha, India. This effort was proved successful as the biodiversity and brackish nature of the lagoon restored within few months of the intervention (Sahu et al. 2014).

9.3 Water Quality Monitoring Methodology

Hydro-biological monitoring of Chilika Lagoon was initiated in 1999 as the Odisha state government assisted program. Thirty stations were identified providing comprehensive coverage of the four sectors of the lagoon (Muduli et al. 2013). Using the monthly salinity data of the last 17 years (1999 to 2015), the multi-dimensional scale (MDS) plot indicates there are 4 distinct sectors exist in the lagoon named as southern sector (SS), central sector (CS), northern sector (NS), and outer channel (OC) (Fig. 9.2). Most of the physicochemical parameters are covered in the monitoring program except NH_3, SiO_2, TN, TP, which was started being monitored from 2011 onwards. The physical parameters covered are AT, WT, depth, transparency, turbidity, and SPM. The chemical parameters include pH, salinity, DO, BOD, TA, and nutrients such as NO_2, NO_3, NH_3, PO_4, SiO_2.

Sub-surface samples (~0.3 m) were collected using a bucket taking utmost care to avoid contamination (Grasshoff et al. 1999). AT, WT, pH, salinity and DO were measured using water quality sonde (YSI, USA) after proper calibration. In a shallow ecosystem like Chilika, it is quite relevant to measure the transparency using a Secchi disk. Samples were placed carefully in the cleaned bottles for analysis of nutrient and other chemical parameters. NO_3 (precision: ± 0.02), NO_2 (precision: ± 0.01), PO_4 (precision: ± 0.01), and SiO_2 (precision: ± 0.02) were estimated using the nutrient autoanalyzer (SKALAR SANplus ANALYZER) following methodology by Grasshoff et al. (1999). BOD was determined by 5 days incubation at 20 °C and measuring DO (Grasshoff et al. 1999). Total alkalinity (TA) is measured using the titration method described by APHA (2005). Pearson correlation matrix was made by using PASW-18 statistics. The multivariate statistical techniques, such as

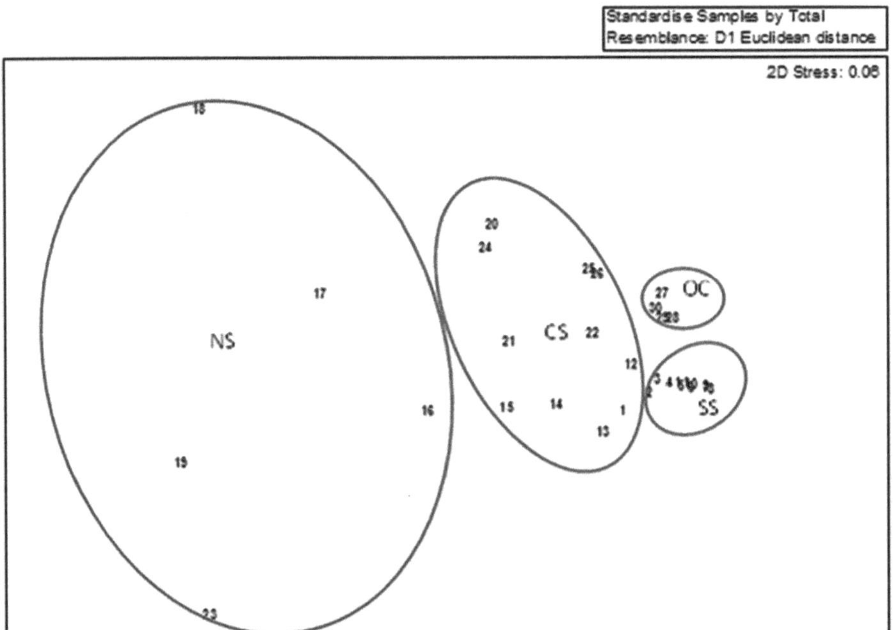

Fig. 9.2 Multidimensional scaling (MDS) showing grouping of 30 sampling stations into 4 sectors. (Salinity data of 30 stations right starting from March 1999 to December 2015 were considered as input data for MDS). *SS* Southern sector, *CS* central sector, *NS* Northern sector, *OC* Outer channel

multi-dimensional scale (MDS) was made using Primer-6. Column chart, scatter chart and trend lines were obtained using Microsoft Excel.

9.4 Variability in Physical Parameters

9.4.1 Air and Water Temperature

The temperature controls several physical and biogeochemical processes in the wetlands (Ganguly et al. 2013; Muduli et al. 2013; Barik et al. 2017a). The microbial activity and mineralisation process for decomposition of organic matter into nutrients can be significantly influenced with respect to change in temperature. Although the photosynthesis and growth of phytoplankton have been found alike for different species with respect to temperature, the growth rate observed to be lower over a certain level of light and temperature (Ganguly et al. 2013). Studies have been evidenced that the impact of temperature on carbon metabolism is much higher than that observed for nitrogen metabolism (Paerl and Justic 2011). During the period 1999 to 2015, the air temperature (AT) over Chilika was observed to vary between 17.4 °C and 37.8 °C (average 28.2 ± 3 °C), with lowest values during the peak winter season (December/January) and highest during peak summer (May/-June). Along with AT, the water temperature (WT) also varied proportionately. The surface WT ranged in between 18.9 °C and 35.9 °C with an avg. of 28.5 ± 2.9 °C. The WT of the lagoon recorded minimum during winter and maintained almost similar range during summer and monsoon (Figs. 9.3a and 9.3b). Irrespective of the rise and fall, the average WT remained almost the same during the study period (Fig. 9.4). Comparison with earlier estimates may indicate gradual warming, as records of 1960–61 showed the lowest range of WT between 24 to 26 °C (Ramanadham et al. 1964). A significant variation on the spatial scale also maintained in the lagoon as observed from Fig. 9.3a. According to the mean WT records, SS and OC maintained the highest as compared to other regions of the lagoon. The major seasonal variation could be attributed to the cooling of surface waters during winter (Satpathy et al. 2009) and intense solar radiation, which warms up the surface water during summer (Shenoi et al. 1999). WT was observed to be correlated with turbidity which could be due to the absorption of high solar radiation heat by the suspended particles (Abir 2014).

9.4.2 Water Level

The water level of the Chilika lagoon varied with respect to time and space. The water level reached the highest level during peak monsoon which subsequently went down, and the lowest level reached during the summer season (Fig. 9.3a). The inflow of highly suspended particulate matter (SPM) and settling in the lagoon can make a substantial change in-water level profile (especially in the NS) of the lagoon. The water level of the lagoon varied in the range of 0.27–7.48 m with an avg. of 1.79 ± 0.96 m was comparatively higher than earlier studies (e.g. Mohanty et al. 2009). However, the trend of the water level since 1999 to 2015 (considering the mean value of entire lagoon) showed there was no such significant change (Fig. 9.4) which might be due to effective outflux of suspended matter (sediment flow) from the river to sea (OC) through the dredged channel maintained by CDA. The difference could be due to the change in sampling locations (or/and distance from the shore) as well as time/ season of sampling when the water level showed a

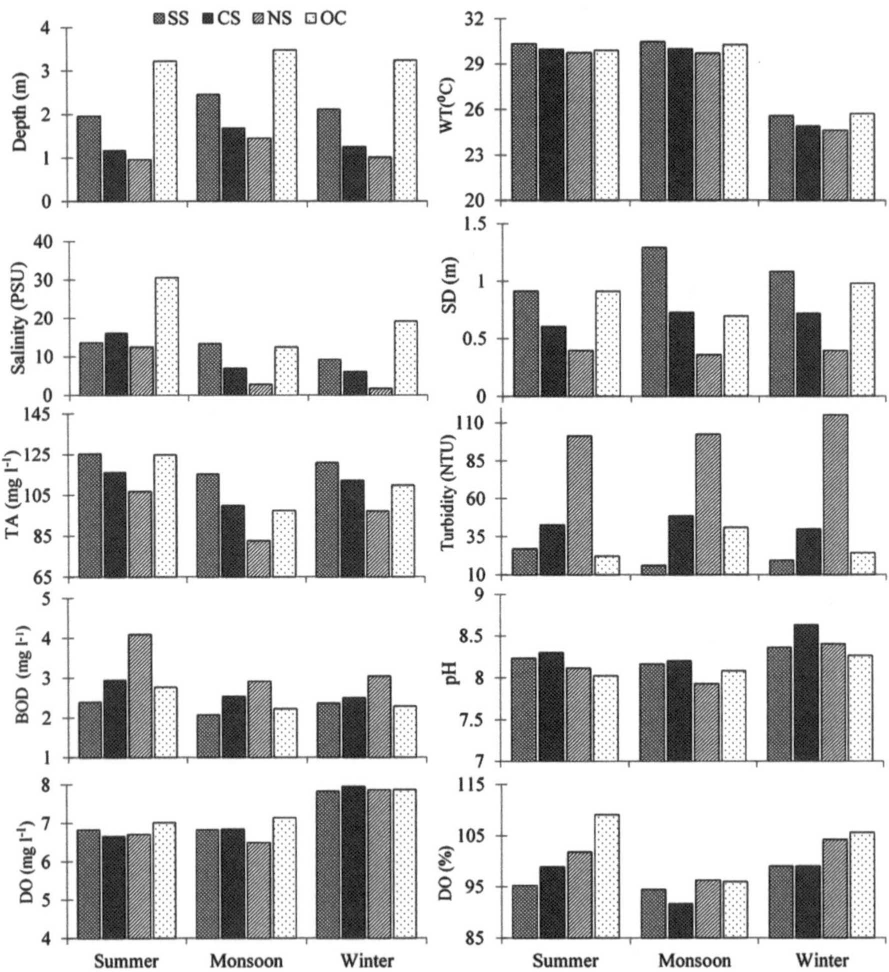

Fig. 9.3a Spatio-temporal variability of physico-chemical parameters in Chilika lagoon during March 1999 to December 2015. Summer (March–June), Monsoon (July–October), Winter (November to February). *SS* Southern sector, *CS* central sector, *NS* Northern sector, *OC* Outer channel

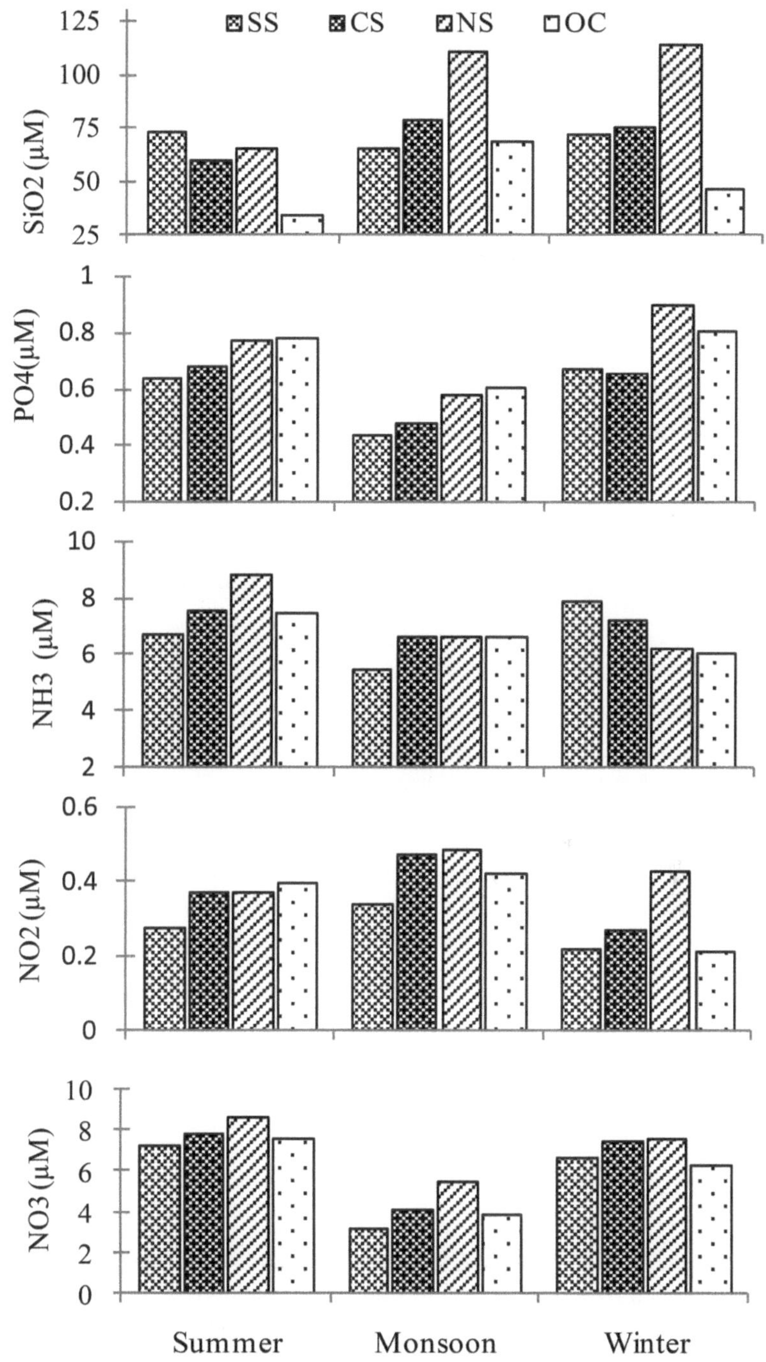

Fig. 9.3b Spatio-temporal variability of nutrients in Chilika lagoon during March 1999 to December 2015 (NH_3 and SiO_2 the data were considered from June 2011 to December 2015). Summer (March–June), Monsoon (July–October), Winter (November to February). *SS* Southern sector, *CS* central sector, *NS* Northern sector, *OC* Outer channel

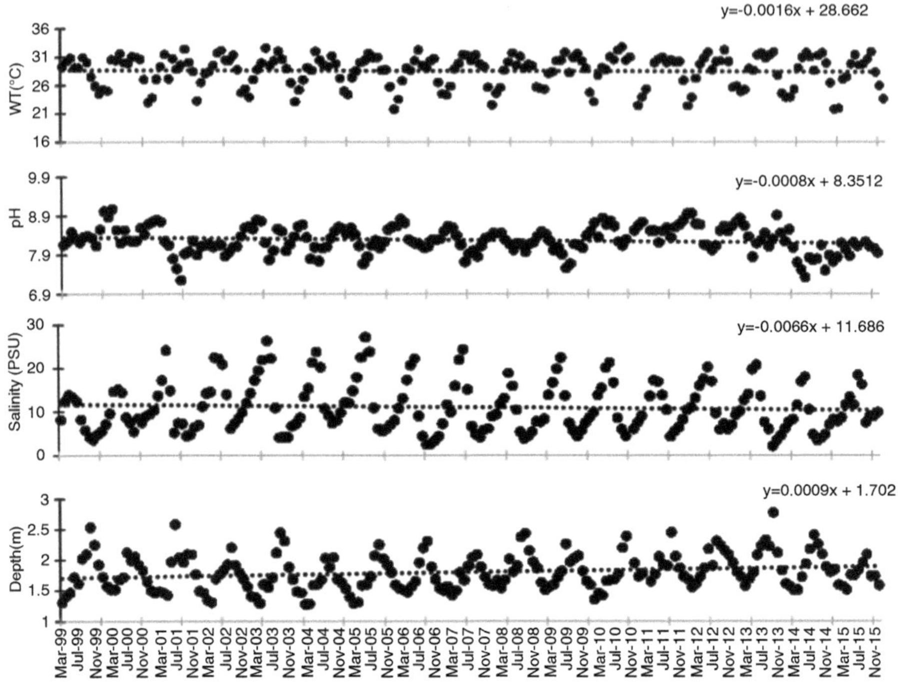

Fig. 9.4 Trend of variation of WT, pH, Salinity and the depth of the Chilika lagoon since 1999 to 2015. Though there is a significant variation of these parameters with respect to month/season, the overall values remain almost same over the time

significant variation. Irrespective of seasons, the NS with an average water level of 1.14 $\pm$ 0.43 m is the shallowest part of the lagoon. The OC is the deepest region, having a water level of 3.33 $\pm$ 1.37 m during monsoon due to increase in water level by the freshwater discharge from the river to sea through the dredged channel. The highest water level of 7.48 m also recorded in the station no. 28 which is very close to the sea. There has been no change in spatial patterns of water level variation during the study period and maintained consistently the order of OC > SS > CS > NS (Fig. 9.3a).

9.4.3 *Photic Depth*

The primary productivity of coastal ecosystem depends on the rate of photosynthesis which is controlled by available nutrients and PAR (photosynthetic active radiation) (Monteith 1972; Passarge et al. 2006; Ganguly et al. 2015; Pashiardis et al. 2017). The varying light intensity throughout the day, as well as weather condition, tend to influence the photosynthesis and thus the productivity of the lagoon. Since different phytoplankton species respond differently to differences in light intensity, the species diversity, abundance, and their spatial distribution in an ecosystem also vary significantly (Flöder et al. 2002). Light penetration can affect the other water quality parameters like WT, DO level, nutrient concentration and the primary productivity of coastal ecosystem (Srichandan et al. 2015a; Acharyya et al. 2012; Garg et al. 2017). Photic depth can either be represented in terms of Secchi disk depth or transparency, and depends on physical parameters such as SPM, turbidity, and biological parameters such as the concentration of pigments (e.g. chlorophyll) (Patra et al. 2016). The turbidity is mostly due to suspended sediments, chlorophyll (algal photopigments), and coloured dissolved organic matter content (Gallegos 1992) in the water column. Apart from these, the presence of submerged macrophytes or seagrasses also controls photic depth indirectly by avoiding resuspension of suspended matter from the benthic compartment of the lagoon. Since most of the shoreline along the lagoon is covered with vegetation (Fig. 9.1), it facilitates proper light penetration into the pelagic compartment as well as up to benthic sediments in shallow regions of the lagoon. For instance, the transparency of the lagoon was the controlling factor over the productivity of the Chilika lagoon after the very severe cyclonic storm (VSCS) Phailin (Srichandan et al. 2015b).

During 1999–2015, the transparency of the lagoon varied between 0.07 to 4.0 m with an avg. of 0.76 ± 0.48 m. Irrespective of seasons, the lowest transparency was observed in the NS (0.39 ± 0.26 m) and highest in SS (1.11 ± 0.47 m). SS was observed to be the least disturbed region and maintains the highest transparency as compared to other regions of the lagoon. During the summer period, the transparency level dropped to its minimum (0.7 ± 0.44 m) followed by monsoon and post-monsoon. Resuspension of bottom sediments by wind-driven forces during summer and the recipient of silt borne run-offs during monsoon (Fouilland et al. 2012) could be the possible factor for low transparency in the lagoon in the respective seasons.

9.5 Variability in Chemical Parameters

9.5.1 pH and Total Alkalinity

The pH of any aquatic ecosystem controls the biogeochemistry of carbon and nitrogen which ultimately lead to influence the biodiversity (Muduli et al. 2012; Srichandan et al. 2015a). The major controlling factor for pH in Chilika has been found to be the productivity, respiration, and mixing of fresh and saline water with different pH condition (Muduli et al. 2013; Ganguly et al. 2015; Robin et al. 2016). The variability in pH in different seasons could be due to CO_2 assimilation by phytoplankton and macrophytes (Srichandan et al. 2015a), and the release of CO_2 due to respiration and mixing of lagoon water with different external fresh water sources (Muduli et al. 2012).

During last 17 years, the pH of the lagoon varied from 6.1 to 10.35 (avg. of 8.27 ± 0.52) and are in the range noticed by earlier studies (Banerjee et al. 1998; Nayak et al. 2004) in Chilika lagoon and elsewhere (Ekeh and Sikoki 2003 in the

New Calabar River; Ansa 2005 in Andoni flats of the Niger Delta). The analysis of monthly data for 1999 to 2015 indicated that the lagoon is moderately alkaline with no significant temporal variation (Fig. 9.4). A fall of the pH level after the VSCS Phailin (October 2013 onwards) could be due to increasing of community respiration overproduction or, mixing of fresh water with lower pH because of cyclone induced precipitation and discharge (Muduli et al. 2012; Muduli et al. 2017). The average pH was within the threshold limit (6.5–8.5) assigned by the Central Pollution Control Board (CPCB), New Delhi, India (for the propagation of wildlife and fisheries). The sectoral variation of pH was significant and maintained in the order of CS > SS > NS > OC. The lowest pH observed during the summer could be attributed to higher CO_2 release by dominating respiration overproduction, the high residence time of water (Rajasegar 2003) and the highest in winter, possibly due to the higher productivity (Saravanakumar et al. 2008) supported by higher photic depth retained in the lagoon.

The TA is mostly controlled by the interplay of climatic and geological factors (Gorham et al. 1983), changing the type and amount of ions transported from the rivers and rivulets that drain into the lagoon. The recorded TA from 1999 to 2015 (20–304, avg. 109.7 ± 33.2 mg l^{-1}) was very low as compared to the previous study (Siddiqui and Rao 1995). The lowest concentration of TA was mostly observed in the NS and during monsoon which could be due to the low carbonate and bicarbonate content in the river discharged fresh water (Siddiqui and Rao 1995) and higher CO_2 saturation (Muduli et al. 2012; Gupta et al. 2008; Robin et al. 2016). The sector-wise variation of TA was in the order of SS > OC > CS > NS (121 ± 26, 111 ± 30, 109 ± 33, 95 ± 38 mg l^{-1} respectively). After the VSCS Phailin, there was a drop of TA which could be due to the above-mentioned reasons. This mirrors that the freshwater influxes from the river discharge and seawater from the mouth are the controlling factors for variation of TA in the Chilika lagoon.

9.5.2 Salinity

Salinity is a very crucial parameter which regulates the biogeochemical characteristics of the Chilika lagoon (Ganguly et al. 2015; Muduli et al. 2017). The biodiversity of the lagoon mostly distributed according to the salinity gradient and the change in species diversity occurs with the change in salinity in different seasons (Srichandan et al. 2015a). The salinity level of Chilika lagoon is mostly regulated by sea water through the mouth, river water discharge (Robin et al. 2016) and ElNiño/southern oscillation (Kim et al. 2015).

As per Ghosh et al. (2006), the lagoon salinity was in the range of 1.4 to 6.3 during 1995 which was the indication of the dominance of freshwater over saline water due to the chocking of mouth resulting in inadequate flow of saline water. To restore the ecosystem, CDA after vigorous modelling studies dredged a new mouth which ensured the brackish nature of the lagoon. From 1999 to 2015 the salinity varied in the range of 0 to 37 with an avg. of 11.09 and the trend showed that overall salinity maintained almost the same (Fig. 9.4). Till date, the lowest ever salinity has been recorded during October 2013 in all sectors of the lagoon due to massive freshwater discharge from rivers and heavy rainfall which accompanied Phailin (Srichandan et al. 2015a). The freshwater discharge measured during this period (964×10^6 m^3d^{-1}) was significantly different than the usual flow (691, 628, 732×10^6 m^3d^{-1} recorded during 2011, 2012 and 2014 respectively) which altered the salinity regime of the lagoon.

During the study period, the four sectors of the lagoon maintained different salinity according to the extent of mixing with saline sea water or riverine fresh water. On an overall, the lagoon maintained a mesohaline condition in all seasons, however, there was a difference with respect to sectors (NS: oligohaline (0.5 to 5), CS and SS: mesohaline (5–18), OC: polyhaline (18–30)). Though the SS has the least tidal influence, the connection with Palur canal (Fig. 9.1) followed by Rushikulya estuary and low intrusion of freshwater, enabled the SS to maintain a higher salinity than CS. The lowest salinity usually observed in the NS during monsoon season was due to major freshwater discharge, from north-east Rivers and the highest in the OC during the pre-monsoon (PRM) season, mainly due to low precipitation, high evaporation, and minimum freshwater dilution (Mohanty and Mohanty 2002).

Salinity in Chilika starts increasing after the monsoon and reaches its optimum level in peak summer. The rainfall by southwest monsoon results significant increase in freshwater flux from the rivers originated from western catchment and Mahanadi which drains into the lagoon and decreases the salinity level to the minimum (Srichandan et al. 2015a). Similar fall in salinity (0 to 30) also has been recorded by Martin et al. 2008 for Cochin backwaters, India. Other similar ecosystems overseas (Bach Dang Estuary, Vietnam) also experiences such change in salinity (Rochelle-Newall et al. 2011). The seasonal trend of avgerage salinity of the lagoon maintained in the order of summer> winter> monsoon (Fig. 9.3a). Periodic cyclone induced precipitation also influenced salinity dynamics of the lagoon for a long period (Barik et al. 2017a). The salinity exhibited variability on the temporal scale (over seasonal and inter-annual) as evidenced by earlier studies (Panigrahi et al. 2009; Muduli et al. 2012, 2017).

9.5.3 Dissolved Oxygen

Dissolved oxygen (DO) in an aquatic ecosystem is a vital indicator parameter of the health of the ecosystem. A DO level of 3.0 mg L^{-1} must be maintained for the protection of aquatic lives as per the environment protection rules, India (CPCB New Delhi 1986). During the study period, the average DO saturation (>98%) and DO in the lagoon (7.1 mg L^{-1}) falls in the threshold range as described in the guidelines for the propagation of wildlife and fisheries (CPCB New Delhi 1986) unlike other lakes in India which maintains low oxygen saturation (Vembanad Lake: Vincy et al. 2012; lagoons in Chennai coast: Jayakumar et al. 2013; Kolleru Lake: Kanuri et al. 2012).

Proper oxygen condition maintained in Chilika could be attributed to balanced productivity enabling near to neutral water quality and physical action such as wind flow which facilitates mixing of oxygen with water (Ganguly et al. 2015). As compared with phytoplankton, the macrophytes and seagrass could be more responsible for increasing the oxygen level in the water column (Murray and Wetzel 1987; Martin et al. 2015). However, the higher DO concentration due to photosynthesis by these (along with planktons) during the daytime also have the likelihood of alarming decrease during night hours, which again depends on the microbial community and their abundance in the particular region (Muduli et al. 2012).

During 1999 to 2015, the DO ranged between 0.3 to 14 mg L^{-1} which is similar to the latest study recorded 3.87 to 12.8 mg L^{-1} (Srichandan et al. 2015b) and also an earlier study by Nayak and Behera (2004) who recorded 3.9 to 13.9 mg L^{-1}. However, few reported values for DO (Patra et al. 2010; Mohanty et al. 2009; Jeong et al. 2008) found lower as compared to the present observations which could be due to the dissimilarity in sampling location and the respective environmental condition during the sampling period. DO concentration less than 3 mg L^{-1} is lethal to aquatic life in general and fish life in particular (CPCB New Delhi 1986) which was rarely observed in Chilika in few stations namely St. 13,14 and 16 (Fig. 9.1). The very low DO in these particular regions could be attributed to decomposed macrophytes and poor circulation of water (as per physical observation during the survey).

During the study period, no significant differences in DO on the spatial scale (with respect to sectors) were observed. This could be due to the fact that there are there is no specific dominant phenomena controls DO in Chilika rather several physical and biogeochemical processes. DO concentration in an aquatic system depends on the rate of photosynthesis by the phytoplankton and macrophytes, decomposition of organic matter by microbes (Granier et al. 2000) and chemical properties of water (Aston 1980). The DO level in the lagoon also gets controlled by several physical factors apart from biological phenomena (Barik et al. 2017a). Data showed the WT and wind maintains a linear relationship with DO and controls its concentration level in the water column. This could be the reason for the highest concentration DO in the winter season as compared to other seasons (Fig. 9.3a).

The Chilika lagoon remains well saturated with respect to DO, and hence the oxygen consumption for nitrification (signature of low oxygen) also observed to be negligible as compared to the consumption by bacterial respiration (Muduli et al. 2012). Earlier studies reported a negative correlation between DO and productivity with salinity which indicated a higher rate of production in the freshwater dominated region (Nayak et al. 2004). The negative relation was mostly due to the lower oxygen solubility in saline water (Sankaranarayanan and Panampunnayil 1979). During summer, the DO maintained a positive correlation with Chlorophyll-a which could be the indication of the major role of the primary productivity of the lagoon controlling oxygen distribution like in other tropical estuarine ecosystems (e.g. Sivasankar and Jayabalan 1994).

9.5.4 Biochemical Oxygen Demand

Biochemical oxygen demand (BOD) indicates the organic matter load in a particular ecosystem (Ndimele 2012). The biodegradation of organic materials exerts oxygen scarcity in the water body and gives an idea about the extent of pollution (Gupta 2009). During the study period, the BOD ranged between 0.04 and 14.52 mg l^{-1} with an avg. of 2.66 mg l^{-1} which is under the permissible limit of water quality criteria (3 mg L^{-1}) prescribed for tropical wetlands by CPCB environment protection rules, New Delhi (CPCB 1986). The sectoral variation indicated the NS with highest organic load and the SS sustained the lowest (NS > CS > OC > SS records 3.36 ± 2.22, 2.67 ± 1.61, 2.44 ± 1.44, 2.29 ± 1.31 mg L^{-1} respectively) (Fig. 9.3a). The highest BOD in the NS during summer could be due to the decomposition of weeds and macrophytes by elevated salinity and mixing of released decomposed organic matter from benthic compartment to water column. High organic load in the NS has also been observed by other studies (Gupta et al. 2008; Muduli et al. 2012). The lowest BOD was seen in the SS, as this region apparently receives a less organic discharge as compared with other sectors.

9.6 Nutrient Biogeochemistry and Influencing Factors

River waters generally deliver the most substantial fraction of the N and P loads to estuaries (Billen and Garnier 2007). For instance, the total nitrogen (TN), total phosphorus (TP), dissolved organic nitrogen (DON), and dissolved organic phosphorus (DOP) influx from major rivers (as named in Fig. 9.1) to the Chilika lagoon has been increasing since 2008–09 (Patra et al. 2016). As per the records, the influx of TN, TP, DON, DOP was 10,745, 1870, 4688, 904 mol d^{-1} respectively during 2008–09 which increased by ~2 times during 2014 (23,732, 3951, 7847 and 3219 mol d^{-1} respectively). In urban areas with high nutrient loads, point source inputs to catchment areas of the estuarine systems also contribute a considerable fraction of the nutrient load. Bottom sediments can also serve as a secondary source of nutrients to the water column especially in shallow ecosystems like Chilika lagoon. In estuarine sediments, nutrient concentrations maybe 10 to 100-fold higher than in the water column (Burkholder et al. 2006) which gets mixed with water column (upwelling) due to bottom churning action by the effect of wind flow. Inorganic N fertilizer use, NO_x emissions from fossil-fuel combustion, and nitrogen fixation in agricultural systems also may contribute as sources of the N through surface runoff and river discharge into the lagoon during the monsoon period. A recent study by Mukherjee et al. (2018) revealed the process of N_2 and atmospheric deposition has a significant role in the N dynamics of the Chilika lagoon.

The speciation of N in rivers and estuaries are generally poorly quantified, giving rise to major uncertainties regarding the factors affecting fluxes of N from estuaries to coastal seas. The relative contribution of the inorganic N ($DIN = NO_3 + NH_4 + NO_2$)

in TN from 2011 to 2015 showed 45% in average, and the seasonal variation showed 57.6, 48.5, and 28.9% during winter, monsoon, and summer respectively. These results indicated that the highest contribution of organic nitrogen during summer which could be attributed to the active mineralisation process that occurs in the lagoon than through the riverine discharge. There also could be a chance of absorption of organic nitrogen and phosphorous within the floodplains but there is no data to support the phenomena. The % of NH_3 in DIN during summer and monsoon (40.7 & 45.7% respectively) records higher than that observed during winter (27.9%). Organic matter mineralisation (both dissolved and particulate form) which is controlled by changes in nutrient species transformation (NO_3 to NH_3) (Dunn Ryan et al. 2012) could be the reason for the relative increase in an NH_3 fraction from winter to summer. The % of PO_4 from TP (inorganic+organic) during summer and monsoon (21.2 & 28% respectively) was lower compared to winter (76.6%) indicated the origin of phosphate was dominantly through *in situ* process during the winter season.

9.6.1 Spatio-Temporal Variation of Nitrite, Nitrate, and Ammonia

9.6.1.1 Variation of Nitrite

NO_2 is a species of inorganic nitrogen which is the intermediate oxidation state between NH_3 and NO_3. In the aquatic environment, the NO_2 is sourced from planktons through metabolic activity and gets released into the water (Chandran and Ramamoorthi 1984). The lagoon maintained NO_2 level ranging between 0.01–2.01 (average of 0.31 ± 0.27 μM) and on spatial scale, the variation was also significant: NS (0.38 ± 0.28 μM) > CS (0.33 ± 0.31 μM) > OC (0.29 ± 0.22 μM) > SS (0.23 ± 0.19 μM). As per Patra et al. (2016), the influxes of NO_2 from rivers were 200.2 mol d^{-1} during 2008–2009, which remained almost the same (192.7 mol d^{-1}) during the monsoon season of 2014. The lowest nitrite concentration observed in the SS could be due to the mixing with water from Rusikulya estuary (through Palur canal) with lower NO_2 concentration and least mixing of fresh water from River influx compared to other sectors. The highest concentration frequently observed in NS could be due to freshwater discharge or/and *in situ* production through extracellular release (Kanuri et al. 2013). The relationship between NO_2 and salinity (Fig. 9.5) indicated that the riverine discharge is not a significant source of NO_2 and the *in situ* active mineralisation processes could play a major role in it (Muduli et al. 2012; Ganguly et al. 2015).

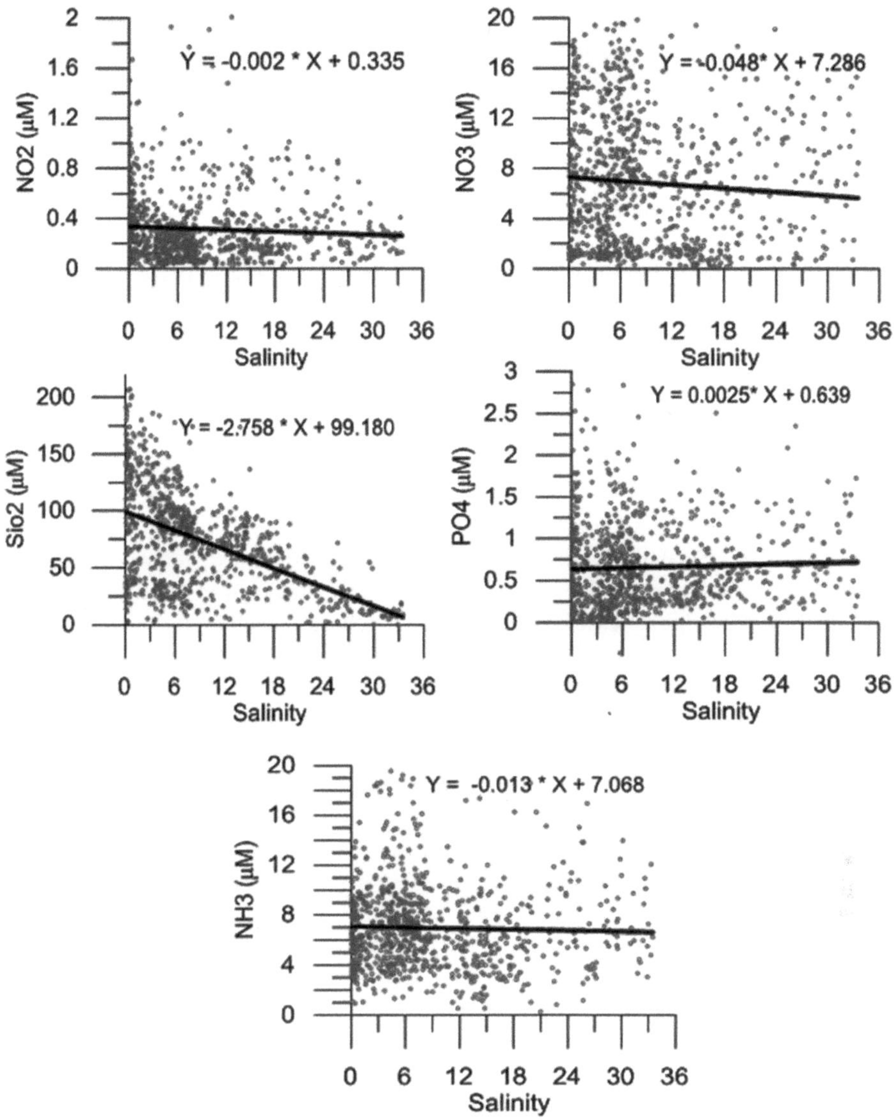

Fig. 9.5 Variation of nutrients: NO_2, NO_3, NH_3, PO_4, SiO_2 with respect to salinity. Monthly data used for the regression analysis: March 1999 to December 2015 for SiO_2 and September 2013 to December 2015 date used for NO_2, NO_3, NH_3, and PO_4

9.6.1.2 Variation of Nitrate

NO_3 is the final oxidation product of N compounds in fresh as well as marine water. It is an important nutrient apart from NH3 and urea for the photosynthesis and growth of plankton community in the presence of light (Ganguly et al. 2015). The concentration of NO_3 varies in the ecosystem with respect to space and time and is mostly controlled by the rate of uptake by primary producers and also an influx of the same through riverine discharge (Zepp 1997). NO_3 concentration in the Chilika lagoon varied in the range of 0.12–19.88, with an average of 6.86 ± 5.14 μM. Like NO_2 it also showed a major variation among sectors: NS; 7.77 ± 5.13 μM > CS; 6.99 ± 5.22 μM > OC; 6.44 ± 4.6 μM > SS; 6.17 ± 5.19 μM. However, the trend indicated the region with sea water influence (SS, OC) maintained lower concentration which could be due to dilution with seawater containing lower nutrient concentration. During the monsoon season (2008–09) the rate of NO_3 flux into the lagoon recorded 4465 mol d^{-1} (Patra et al. 2016) which is closer to the recent observation during 2014 (3112 mol d^{-1}). The regions with the influence of riverine discharge (NS) contain higher nutrient concentration (Fig. 9.3b). On a seasonal basis, the variation is also highly significant (Ganguly et al. 2015; Patra et al. 2016). As compared to other seasons, the summer (high residence time, Muduli et al. 2013) records the highest nutrient concentration which could be due to the mineralisation process (Panigrahi 2006). This observation is also in corroboration with findings of Muduli et al. (2012) and Gupta et al. (2008) who revealed the active mineralisation process in the NS of the lagoon. Some tropical estuaries also record high NO_3 concentration during monsoon (Patil and Anil 2008; Pednekar et al. 2014) and the difference may be due to lesser residence time in Chilika (Muduli et al. 2013) which helps the nutrient to flush out. Many studies report a significant negative relationship of NO_3 with salinity, indicates the major source of NO_3 was from freshwater discharge. However, the present observation from the regression equation indicated there were mixed responses i.e. the source was from *in situ* process as well as from riverine discharge (negative relationship with salinity) (Fig. 9.5). As per Barik et al. (2017a), the relationship of NO_3 with salinity not only varies in spatial scale but also in the temporal scale. The mixed response (Fig. 9.5) could be attributed to different seasons during which several factors influenced the NO_3 concentration (as discussed above).

9.6.1.3 Variation of Ammonia

NH_3 is one of the most important nitrogen sources for phytoplankton growth, though researchers also revealed that the urea could be preferred over NH_3 for uptake by phytoplankton (Ganguly et al. 2015). However, the preference for NH_3 is group-specific and generally observed in green algae and cyanobacteria. The increased concentration may result in a shift in phytoplankton community composition towards the dominance of cyanobacteria and green algae (Domingues 2011). Ammonia is the main excretory product of aquatic invertebrates and preferred over NO_3 by phytoplankton in certain environmental conditions (Gilbert et al. 1982). The potential sources of NH_3 into the aquatic systems could also be from terrestrial run-off, excretion by zooplankton or demineralisation of organic matter (Balls 1992).

During the study period, the lagoon recorded an average ammonia concentration of 6.95 ± 3.5 μM which is quite lower than the safe limit of 38 to 1629 μM beyond which, it act as toxic to freshwater organisms depending on the pH and temperature maintained in the ecosystem. In Chilika lagoon, the higher concentration of NH_3 in the NS and CS may be due to demineralisation of submerged macrophytes. The overall NH_3 concentration in the lagoon showed an insignificant relationship with salinity (Fig. 9.5) indicating that NH_3 is not sourced dominantly from the catchments but from the *in situ* biogeochemical processes inside the lagoon (Ganguly et al. 2015; Barik et al. 2017a). Stations having lower salinity levels (0 to 10) had higher NH3 concentration as compared with those with higher salinity 10 to 36 (Fig. 9.5). During monsoon of 2014, the NH_3 flux into the lagoon recorded 1892 mol d^{-1} which is 1.3 times higher than the record during 2008–09 (1455 mol d^{-1}; Patra et al. 2016). Due to lower NH_3 concentration in the river water, the NH_3 concentration remains lower during monsoon period (Fig. 9.3b) due to dilution. During monsoon, all the sectors maintained almost similar concentration which might be due to the mixing of water along with NH_3 throughout the lagoon except for SS which maintained the lowest.

9.6.2 *Spatio-Temporal Variation of Phosphate*

Phosphate is a vital inorganic nutrient for the growth of autotrophic phytoplankton, algae, and macrophytes. The limitation of the same can hinder the growth of plankton community and excess of the same also may lead to eutrophic condition (high productivity) of the water body in the availability of required DIN (Ganguly et al. 2015). Lagoon water with low P concentrations generally has not only very low PP, but also the low secondary production of invertebrates and fish. Conversely, extremely high P concentrations often lead to algal blooms, low water clarity, a decline of rooted plants, anoxic bottom waters, fish kills. Hence long-term monitoring of this parameter is a must to understand the health status of the lagoon. Such eutrophic status has not been recorded for Chilika lagoon due to the influx of turbid water through major rivers and making an ideal situation to control over excess productivity (Srichandan et al. 2015a).

As per the record since 1999 the average PO_4 in Chilika lagoon maintained in the range of 0.01 and 2.85 μM with an average of 0.66 ± 0.5 μM. The maximum recorded concentration of phosphate is almost 50% less as compared to earlier studies (Siddiqui and Rao 1995) during 1985–87. In comparison with 2008–09 values of the PO_4 influx into the lagoon (966 mol d^{-1}, Patra et al. 2016) the current levels have tended to have decreased ~3.7 times (259 mol d^{-1}). This may be due to decrease or control over the application of PO_4 based fertilizers in the agriculture fields in the catchment areas. The PO_4 concentration records lower during monsoon period as compared to rest of the period, which may be due to the dilution with fresh water containing low PO_4 content or absorption to suspended particulate matter in low saline condition (Fouilland et al. 2012).

The relationship of overall PO_4 content in the entire lagoon (throughout the year) with salinity (Fig. 9.5) specified the source of PO_4 could be from seawater or, biogeochemical process inside the lagoon or from riverine discharge (Barik et al. 2017a). Similar to NH_3 the PO_4 concentration also showed a mixed response with respect to salinity as all range of PO_4 concentration found scattered irrespective of salinity ranges (Fig. 9.5) (Ganguly et al. 2015; Muduli et al. 2017). Apart from the riverine input, the release of phosphate from sediments due to the churning of water by winds (Chandran and Ramamoorthi 1984) and release of PO_4 from SPM in higher saline condition may also significantly alter the concentration level in Chilika. Having less impact of riverine discharge and saline water on SS, it maintained the lowest concentration. Apart from this, the uptake of PO_4 by phytoplankton (Satpathy et al. 2009) or seagrasses observed in the SS could also be one of the reasons.

In most of the aquatic ecosystems, the predominant inorganic species mono- or diprotonated orthophosphate is the most bioavailable form of P. In such systems, ~60% of the PO_4 input to water may occur from the sediment-water exchange (Paerl and Justic 2011). Hence, it is also important to understand the different fraction of PO_4 available in the sediment as recorded by Barik et al. (2016) in the Chilika lagoon. Except for a few studies, the PO_4 dynamics in shallow waters are not well deciphered in view of its rapid biological (Admiraal and Werner 1983) and geochemical (Pomeroy et al. 1965) processes. The composition of rocks and the intensity of weathering of the same influences the influx of P content into the coastal lagoon. However, the availability of other elements controls the P available for the algae in the aquatic ecosystem.

9.6.3 Spatio-Temporal Variation of Silicate

Silicate is often considered as an essential element for the growth of diatoms (Wassmann 1999), and a requisite for the production of silica frustules. Silicate concentration (as per the records from 2011 to 2015) in Chilika varied between 0.1 and 363 μM (avg. 73.56 ± 45.62 μM). The maximum concentration of SiO_2 was observed in NS, followed by CS and SS and minimum in the OC. The high SiO_2 in the NS could be due to the influx of fresh water with silicate from northeast rivers which has increased over time [27,145 mol d^{-1} (Patra et al. 2016) to 40,051 mol d^{-1} during 2008–09 to 2014 respectively]. The lowest SiO_2 in the OC area might be due to the uptake of silicates by phytoplankton (especially diatoms and silicoflagellates) for their biological activity (Srichandan et al. 2015b). The lower concentration may also be due to adsorption into suspended sedimentary particles, chemical interaction with clay minerals and co-precipitation of soluble silicon with humic compounds and iron (Satpathy et al. 2009).

As compared to monsoon and winter, the SiO_2 concentration recorded highest in the summer season (Fig. 9.3b). It indicated that the SiO_2 in natural waters are strongly dependent on the WT and uptake by diatom which results in seasonal concentration patterns. Diatoms occur in the upper surface layer where light penetration enables photosynthesis. In summer, diatoms take up SiO_2 forming an outer shell which could be the reason for low SiO_2 levels during summer. After the diatom dies, Si is released again, along with other nutrients, through remineralisation. The higher residence time of Chilika water during summer (Muduli et al. 2013) enhances the sedimentation and burial of diatom frustules (SiO_2) in sediments which result further drop in SiO_2 concentration in the lagoon water. The regression of SiO_2 with salinity (Fig. 9.5) showed the source of SiO_2 was from the discharged freshwater rich in SiO_2. Unlike the pattern observed for N and P, SiO_2 was observed to decline with an increase in salinity. The highest range of SiO_2 recorded only in the range of salinity 0 to 6 (Fig. 9.5) which indicated that the origin could be from weathering of silicate in the catchments and discharge of the same into the lagoon during monsoon period (Patra et al. 2016).

9.6.4 Variation of Nutrient Stoichiometry

In order to take any decisions for the management of an ecosystem with respect to eutrophication, the knowledge on nutrient dynamics especially the variability of nutrient ratio with respect to changing environmental condition is very crucial. Seasonal changes and the biogeochemical process can have a strong potential to regulate the nutrient stoichiometry in the coastal and near-shore waters. Such a change in nutrient stoichiometry may result in shifts in the plant population and diversity (Loureiro et al. 2006). Alternatively, N and P have been reported as limiting nutrients for primary production in the coastal waters (Oviatt et al. 1995). The nutrient P becomes limiting when there is a surplus of N, and relatively less P gets added into the system during the monsoon period through surface runoff (Harrison 1990). Apart from N and P, the variability in SiO_2 concentration is also responsible for the balance of the Redfield ratio (Brzezinski 1985). Si becomes limiting when there is a surplus of N or P influx into the system either through autochthonous or allochthones sources which ultimately results in a change of biodiversity (Srichandan et al. 2015a).

In Chilika lagoon, the concentration of N and P depend on many ecosystem processes. During the periods of elevated freshwater runoff with a high level of N input, the P limitation are most evident (Sylvan et al. 2006; Patra et al. 2016) as the N/P ratio goes below the Redfield ratio (16). Apart from this, the possibility of P limitation also occurs due to adsorption of P to the sediment particles in low saline zones. P adsorption depends on the salinity of water as high saline water contents opposite ions and enables less adsorption as compared to low saline waters or fresh waters. Hence, the P which binds in a freshwater environment gets desorbed in saline or brackish environment (Ganguly et al. 2015; Barik et al. 2017a). Therefore, P limitation is more likely expected in the freshwater zone of Chilika (NS) and also during the monsoon period [dissolved inorganic nitrogen (DIN)/ dissolved inorganic phosphate (DIP) =59]. In other seasons, the Redfield ratio was observed as being maintained (Summer: 18.5 and Winter: 14.6).

In aquatic coastal ecosystems, the variability of nitrogen and nitrogen limitation has been found to be significantly controlled by fixation and release to the atmosphere through denitrification. During summer and winter, the Chilika lagoon exhibits nitrogen limitation but during monsoon, it is P limiting (Patra et al. 2016). The P limitation may also be due to nitrogen fixation by planktonic cyanobacteria. The records of nutrient stoichiometry N/Si < 1 and Si/P > 16 indicate N and P are limiting with respect to silicate (N: P: Si = 16:1:16) which favours the growth of diatoms. However, during summer as the Si concentration decreases, the species diversity shifts from siliceous based to non-siliceous based phytoplankton community (Srichandan et al. 2015b).

9.7 Summary

The change in trophic states (such as meso, oligo and eutrophic) of the Chilika lagoon was mostly controlled by the transparency and nutrient ratio maintained within the ecosystem in different seasons. The trophic state was dominantly influenced by non-algal light attenuation during monsoon season. The spatio-temporal variation or switch over of trophic states observed due to change in water quality by several physical factors such as rainfall followed by the freshwater influx and wind flow (mostly influences the lower depth regions). During the peak discharge period, the primary source of nutrients to the Chilika lagoon was through discharge from rivers and rivulets. Along with the nutrients, the SPM was also brought into the lagoon that resulted in decreased light penetration. During monsoon and winter, there was a decrease in productivity due to factors such as a decrease in photic depth and increase in flushing time. During summer the higher productivity (than monsoon and winter) was attributed to the high residence time of the water. During summer the nutrient concentrations were predominantly controlled by the mineralisation process except in the case of SiO_2 which was consumed by diatoms leading to a decreased concentration in the pelagic compartment. The nutrient stoichiometry indicated NO_3 and PO_4 used to be limiting with respect to silicate. The primary source of PO_4 in Chilika was from *in situ* processes whereas, nitrate and silicate were from the riverine influx. The nutrient dynamics in the lagoon considerably influenced by the mineralisation process in the system, variability of suspended matter and nutrient species influx through rivers, sediment churning by wind action especially during summer.

Though there are several studies on nutrient dynamics in Chilika lagoon, the information on the relationships between nutrients (NO_2, NO_3, NH_3, Urea), their retention, transport, and exchange with the sea are inadequate. The restoration of lagoon connectivity with the Bay appears to have a more drastic influence on productivity as compared with the riverine freshwater influx. The relative importance of these influences on the lagoon can be varied with the morphological trait of the outer channel. The changes in the riverine suspended matter influx can have a weighty influence on the nutrient dynamics as well as the topography of the ecosystem. The nutrient biogeochemical processes such as nitrification, denitrification, mineralisation, anaerobic ammonium oxidation (ANAMMOX), and dissimilatory nitrate reduction to ammonia (DNRA) need to be studied thoroughly to have a better understanding of the biogeochemistry of the ecosystem: Chilika. Apart from these, the budgeting of the nutrient flux in terms of river influx, *in situ* process and resultant flux from the sea would be helpful for the prediction of the productivity as well as changeover trophic condition of the lagoon. Accordingly, a management action plan can be formulated to regulate the material flux into the lagoon and maintenance of the ecosystem health.

Acknowledgments The authors are thankful to the World Bank for the funding support to the "Integrated Coastal Zone Management Project" through the State Project Management Unit, Odisha, India. to the authors also thank the scientific and supporting staff for the help during the sample collection as well as analysis in the laboratory. We also wish to thank Mr. Abhijit Das, CDA for data compilation and Mrs. Jagnyaseni Rout for preparation of the GIS map.

References

Abir S (2014) Seasonal variations in physico-chemical characteristics of Rudrasagar wetland – a Ramsar site, Tripura, North East, India. Res J of Chem Sci 4(1):31–40

Acharyya T et al (2012) Reduced river discharge intensifies phytoplankton bloom in Godavari estuary. India Mar Chem 132(133):15–22

Admiraal, Werner (1983) Utilization of limiting concentrations of orthophosphate and production of organic phosphates in cultures of marine diatoms. J Plankton Res 5:495–513

Ansa EJ (2005) Studies of the benthic macrofauna of the Andoni flats in the Niger Delta Area of Nigeria. Ph.D. Thesis, University of Port Harcourt, Port Harcourt, Nigeria p 242

APHA-AWWA-WEF (2005) Standard methods for the examination of water and wastewater, 21st ed

Aston SR (1980) Nutrients dissolved gasses and general biochemistry in estuaries. In: Olausson E, Cato I (eds) Chemistry and biogeochemistry of estuaries, Wiley, New York, pp 233–262

Balls PW (1992) Nutrient behavior in two contrasting Scottish estuaries, the forth and the Tay. Oceanol Acta 15(3):261–277

Banerjee RK et al (1998) Chilika Lake – present and past. (Bull. No. 80). Central Inland Capture Fishery research Institute, Barrackpore

Barik SK et al (2016) Sequential extraction of different forms of phosphorus in the surface sediments of Chilika Lake. Arabian J Geosciences 9(2):135

Barik SK et al (2017a) Spatio-temporal variability and the impact of Phailin on water quality of Chilika lagoon. Cont Shelf Res 136:39–56

Barik SK et al (2017b) Spatial distribution and potential biological risk of some metals in relation to granulometric content in core sediments from Chilika Lake, India. Environ Sci Pollut Res 25 (1):572–587

Billen, Garnier (2007) River basin nutrient delivery to the coastal sea: assessing its potential to sustain new production of non-siliceous algae. Mar Chem 106:148–160

Brzezinski MA (1985) The Si : C : N ratio of marine diatoms: interspecific variability and the effect of 5 some environmental variables. J Phycol 21:347–357

Burkholder JM et al (2006) Comprehensive trend analysis of nutrients and related variables in a large eutrophic estuary: a decadal study of anthropogenic and climatic influences. Limnol Oceanogr 51:463–487

Chakrapani, Veizer (2005) Dissolved inorganic carbon isotopic compositions in the upstream ganga river in the Himalayas. Curr Sci:553–556
Chandran, Ramamoorthi (1984) Hydrobiological studies in the gradient zone of the Vellar estuary, part I. Physicochemical parameters. Mahasagar-bulletin of the National Institute of Oceanography 17:69–78
CPCB New Delhi (1986) Environment protection rules. https://www.cpcb.nic.in/WaterqualityCriteria.php. Accessed 5 Jan 2016
Domingues RB (2011) Ammonium, nitrate and phytoplankton interactions in a freshwater tidal estuarine zone: potential effects of cultural eutrophication. Aquat Sci 73(3):331–343
Dunn Ryan JK et al (2012) Benthic metabolism and nitrogen dynamics in a sub-tropical coastal lagoon: Microphytobenthos stimulate nitrification and nitrate reduction through photosynthetic oxygen evolution. Est Coast Shelf Sci 113:272–282
Ekeh, Sikoki (2003) The state and seasonal variability of some physicochemical parameters in the New Calabar River, Nigeria. Suppad Acta Hydrobiologia 5:45–60
Flöder SJ et al (2002) The influence of fluctuating light intensities on species composition and diversity of natural phytoplankton communities. Oecologia 133:395–401
Fouilland E et al (2012) Vidussi impact of a river flash flood on microbial carbon and nitrogen production in a Mediterranean lagoon (Thau lagoon, France). Estuar Coast Shelf Sci 113:192–204
Gallegos CL (1992) Photosynthesis, productivity, and species composition in eutrophic sub-estuary: comparison of bloom and bon-bloom assemblages. Mar Ecol Prog Ser 81:257–267
Ganguly D et al (2013) Variable response of two tropical phytoplankton species at different salinity and nutrient condition. J Exp Mar Biol Ecol 440:244–249
Ganguly D et al (2015) Influence of nutrient input on the trophic state of a tropical brackish water lagoon. J Earth Syst Sci 124(5):1005–1017
Garg V et al (2017) Spectral similarity approach for mapping turbidity of an inland waterbody. J Hydrol 550:527–537
Ghosh AK et al (2006) Chilika lagoon: restoring ecological balance and livelihoods through re-salinization. Lakes Reserv Res Manage 11:239–255
Gilbert PM et al (1982) Utilization of ammonium and nitrate during austral summer in the Scotia Sea. Deep Sea Res 29:837–850
Gorham E et al (1983) The chemical composition of lakes in the northcentral United States. Limnol Oceanogr 28:287–301
Granier J et al (2000) Understanding the oxygen budget and related ecological processes in the river Mosel: the Riverstrahler approach. Hydrobiol 410:151–166
Grasshoff K et al (1999) Methods of sea water analysis, 3rd edn. Verlag Chemie, Weinheim, pp 89–224
Gupta GVM et al (2008) Influence of net ecosystem metabolism in transferring riverine organic carbon to atmospheric CO_2 in a tropical coastal lagoon, Chilika Lake, India. Biogeochem 87:265–285
Gupta et al (2009) CO2 supersaturation and net heterotrophy in a tropical estuary (Cochin, India): influence of anthropogenic effect. Ecosystems 12:1145–1157
Harrison NM (1990) Gelatinous zooplankton in the diet of the Parakeet Auklet: comparisons with other auklets. J Avian Biol 14:114–124
Jayakumar R et al (2013) An assessment of temporal variations in physicochemical and microbiological properties of barmouth.s and lagoons in Chennai (Southeast coast of India). Mar Pol Bull 70:44–53
Jeong KS et al (2008) Patterning limnological characteristics of the Chilika lagoon (India) using a self-organizing map. Limnol 9:231–242
Kanuri VV et al (2012) Spatio-temporal distribution of dissolved organic matter in a tropical fresh water lake: India. Int J Recent Sci Res 3(10):812–819
Kanuri VV et al (2013) Plankton metabolic processes and its significance on dissolved organic carbon pool in tropical brackish water lagoon. Cont Shelf Res 61–62:52–61

Kim JY et al (2015) Trends in a satellite-derived vegetation index and environmental variables in a restored brackish lagoon. Global Ecol Conservation 4:614–624

Kim JY et al (2016) Application of multivariate analysis to determine spatial and temporal changes in water quality after new channel construction in the Chilika Lagoon. Ecol Engg 90:314–319

Kumar and Pattnaik (2012) Chilika – an integrated management planning framework for conservation and wise use. Wetlands International – South Asia, New Delhi, India/Chilika Development Authority, Bhubaneswar. ISBN: 81-87408-05-7

Loureiro SC et al (2006) Toxicity assessment of two soils from Jales Mine (Portugal) using plants: growth and biochemical parameters. Arch Environ Contam Toxicol 50(2):182–190

Madhu NV et al (2010) Short term variability of water quality and its implications on phytoplankton production in a tropical estuary (Cochin Estuary – India). Environ Monit Asses 170:287–300

Martin GD et al (2008) Freshwater influence on nutrient stoichiometry in a tropical estuary, southwest coast of India. AEER 6:57–64

Martin et al (2015) Phytoplankton versus macrophyte contribution to primary production and biogeochemical cycles of a coastal mesotidal system. A modelling approach. Est Coas Shelf Sci 165:52–60

Mohanty, Mohanty (2002) Factors affecting fishes and shellfish biodiversity in Chilika lagoon. Fish Chimes 22(5):42–43

Mohanty RK et al (2009) Assessment of the impacts of a new artificial Lake mouth on the hydrobiology and fisheries of Chilika Lake, India. Lake Reservoir Manag 14(3):231–245

Mohanty B et al (2016) Assessment of petroleum hydrocarbon in a tropical brackish water lagoon: Chilika, India. Chem Ecol 32(7):653–668

Mohanty B et al (2017) Sources and variability of petroleum hydrocarbon residues in sediments of Chilika Lagoon, East Coast of India. Bull Environ Contam and Toxicol 99(1):100–107

Monteith JL (1972) Solar radiation and productivity in tropical ecosystems. J Appl Ecol 9:747–766

Muduli PR et al (2011) Spatial variation of hydrochemical characteristics in and around port Blair Bay Andaman and Nicobar Islands, India. World Appl Sci J 13(3):564–571

Muduli PR et al (2012) Spatio-temporal variation of CO_2 emission from Chilika Lake, atropical coastal lagoon, on the east coast of India. Est Coast Shelf Sci 113:305–313

Muduli PR et al (2013) Distribution of dissolved inorganic carbon and net ecosystem production in a tropical brackish water lagoon, India. Cont Shelf Res 64:75–87

Muduli PR et al (2017) The impact of tropical cyclone 'Phailin' on the hydrology of Chilika Lagoon, India. Int J Environ Sci Nat Res 4(2):555632. http://sci-hub.tw/10.19080/IJESNR.2017.04.555632

Mukherjee R et al (2018) Spatial variation of nitrogen uptake rates in the largest brackish water lagoon of Asia (Chilika, India). Est Coast Shelf Sci. http://sci-hub.tw/10.1016/j.ecss.2018.01.012. In press

Murray, Wetzel (1987) Oxygen production and consumption associated with the major autotrophic components in two temperate seagrass communities. Mar Ecol Prog Ser 38:231–239

Nayak BK et al (2004) Variation of water quality in Chilika Lake. Ind J Mar Sci 33:164–169

Nayak, Behera (2004) Seasonal variation of some physicochemical parameters of the Chilika lagoon (east coast of India) after opening the new mouth, near Sipakuda. Indian J Mar Sci 33 (2):206–208

Ndimele PE (2012) The effect of water hyacinth Eichhornia crassipes [Mart.] Solm. Infestation on the physicochemistry, nutrient and heavy metal content of Badagry Creek and Ologe lagoon, Lagos, Nigeria. J Environ Sci Techno l5:128–136

Ojha A (2013) Evaluation of land use/land cover dynamics of chilika catchment. Int j geom geosci 4 (2):388–396

Oviatt C et al (1995) An ecosystem-level experiment on nutrient limitation in temperate coastal marine environments. Mar Ecol Prog Ser 116:171–179

Paerl, Justic (2011) Primary producers: phytoplankton ecology and trophic dynamics in coastal waters. In: Wolanski E (ed) Treatise on estuarine and coastal science. Elsevier, Amsterdam, pp 28–32

Parida S et al (2017) Trace metal concentrations in euryhaline fish species from Chilika lagoon: human health risk assessment. Int J Environ Sc Tech 14(12):2649–2660

Passarge et al (2006) Competition for nutrients and light: stable coexistence, alternative stable states, or competitive exclusion? Ecol Monogr 76(1):57–72

Patra et al (2017) Isotopic composition (C & N) of the suspended particles and N uptake by phytoplankton in a shallow tropical coastal lagoon. Chem Ecol 33(8):708–724

Panigrahi S (2006) Seasonal variability of phytoplankton productivity and related physicochemical parameters in the Chilika Lake and its adjoining sea. Ph.D. thesis, Berhampur University, India

Panigrahi S et al (2009) Variability of nutrients and phytoplankton biomass in a shallow brackish water ecosystem Chilika Lagoon, India. Limnol 10:73–85

Pashiardis S et al (2017) Characteristics of photosynthetic active radiation (PAR) through statistical analysis at Larnaca, Cyprus. SM J Biometrics Biostat 2(2):1009

Patil, Anil (2008) Temporal variation of diatom benthic propagules in a monsoon influenced tropical estuary. Cont Shelf Res 28(17):2404–2416

Patra AP et al (2010) Seasonal variation in physicochemical parameters of Chilika Lake after opening of new mouth near Gabakunda, Orissa, India. World Journal of Fish and Marine Sciences 2(2):109–117

Patra S et al (2016) Influence of suspended particulate matter on the nutrient biogeochemistry of a tropical shallow lagoon, Chilika, India. Limnol 17(3):223–238

Pednekar SM et al (2014) Spatiotemporal distribution in phytoplankton community with distinct salinity regimes along the Mandovi estuary, Goa, India. Turk J Bot 38:800–818

Pomeroy LR et al (1965) The exchange of phosphate between estuarine water and sediments. Limnol Oceanogr 10:167

Rajasegar M (2003) Physico-chemical characteristics of the Vellar estuary in relation to shrimp farming. J Environ Biol 24:95–101

Ramanadham RM et al (1964) Limnology of the Chilka lake. J Mar Biol Assoc India 6(2):183–201

Robin RS et al (2016) CO_2 saturation and trophic shift induced by microbial metabolic processes in a river-dominated ocean margin (tropical shallow lagoon, Chilika, India). Geomicrobiol J 33 (6):513–529

Rochelle-Newall EJ et al (2011) Phytoplankton distribution and productivity in a highly turbid, tropical coastal system (Bach Dang Estuary, Vietnam). Mar Pollut Bull 62:2317–2329

Sahu BK et al (2014) Environmental conditions of Chilika Lake during pre and post hydrological intervention: an overview. J Coast Conserv 18(3):285–297

Sankaranarayanan, Panampunnayil (1979) Studies on organic carbon, nitrogen and phosphrous in sediments of cochin backwaters. Indian J Mar Sci 8:27–30

Saravanakumar A et al (2008) Seasonal variations in physicochemical characteristics of water, sediment and soil texture in arid zone mangroves of Kachchh-Gujarat. J Environ Biol 29:725–732

Sarkar SK et al (2007) Water quality management in the lower stretch of the river Ganges, east coast of India: an approach through environmental education. J Clean Prod 15:1559–1567

Sarma VVSS et al (2009) Influence of river discharge on plankton metabolic rates in the tropical monsoon-driven Godavari estuary, India. Estuar Coast Shelf Sci 85:515–524

Satpathy et al (2009) Phytoplankton community structure and its variability during southwest to northeast monsoon transition in the coastal waters of Kalpakkam, east coast of India. Int J Oceans Oceanogr 3:43–74

Shenoi SSC et al (1999) On the sea surface temperature high in the Lakshadweep Sea before the onset of the southwest monsoon. J Geophys Res 104:15703–15712

Siddiqui, Rao (1995) Limnology of Chilika Lake. In: Director of zoological survey of India (Calcutta), Fauna of Chilika Lake (Wetland Ecosystem Series I). Zoological Survey of India, Calcutta

Singh KP et al (2004) Multivariate statistical techniques for the evaluation of spatial and temporal variations in water quality of Gomti River (India) –a case study. Water Res 38:3980–3992

Singh KP et al (2005) Water quality assessment and apportionment of pollution sources of Gomti river (India) using multivariate statistical techniques –a case study. Anal Chim Acta 538:355–374

Sivasankar, Jayabalan (1994) Distribution of luminescent bacterium Vibrio harvesting Netravathi estuary, Mangalore. J Mar Biol Assoc India 36:251–259

Srichandan S et al (2015a) Interannual and cyclone drove variability in phytoplankton communities of a tropical coastal lagoon. Mar pollut Bulletin 101(1):39–52

Srichandan S et al (2015b) Spatio temporal distribution and composition of phytoplankton assemblages in a coastal tropical lagoon: Chilika, India. Environ Monit Assess 187(47):1–17

Sylvan JB et al (2006) Phosphorus limits phytoplankton growth on the Louisiana shelf during the period of hypoxia formation. Environ Sci Technol 40:7548–7553

Vincy MV et al (2012) Water quality assessment of a tropical wetland ecosystem with special reference to backwater tourism, Kerala, South India. Int Res J Env Sci 1(5):62–68

Wassmann P (1999) Spring bloom development in the marginal ice zone and the Central Barents Sea. PSZN I: Mar Ecol 20:321–346

Zepp RG (1997) Interactions of marine biogeochemical cycles and the photo degradation of dissolved organic carbon and dissolved organic nitrogen. In: Gianguzza A, Pelizzetti E, Sammarkano S (eds) Marine chemistry. Kluwer Acad Plub, London, pp 329–352

Zhang J (2000) Evidence of trace metal limited photosynthesis in eutrophic estuarine and coastal waters. Limnol Oceanogr 45(8):1871–1878

Zhou T et al (2012) An observational study on the latitudes where wave forcing drives brewer-Dobson upwelling. J Atmos Sci 69:1916–1935

Chapter 10
Geomorphology, Land Use/Land Cover and Sedimentary Environments of the Chilika Basin

Rajiv Sinha, R. Chandrasekaran, and Neeraj Awasthi

Abstract The Chilika is one of the largest brackish water lagoons in Asia and is well known for its biologic diversity. Designated as one of the Ramsar sites in 1981, Chilika was under serious threat in the late 1990s due to severe physical and ecological degradation. Some of the severe problems included large-scale siltation and reduction in the water level as well as salinity, threatening some of the rare species of fauna in this region. Following a significant intervention in the form of opening a new mouth in 2000, Chilika recovered significantly and is undoubtedly a model example of wetlands restoration. This chapter aims at documenting this success story based on geomorphic studies in and around Chilika using remote sensing images for the period 1980–2015. We have also mapped land use/and cover changes for two different periods 1980–2000 and 2000–2015 to understand the causal factors of the degradation. We conclude that despite the dynamic geomorphic environment around the Chilika, the intervention in 2000 has had a positive effect this far, but needs close monitoring. Anthropogenic impacts in the Chilika Basin are expressed in terms of significant increase in urban settlements at the cost of natural land cover such as scrub forest, agriculture plantation and wastelands. Some improvement in terms of increase in evergreen forest is recorded in the post-intervention period. However, a close vigil on this very sensitive ecosystem is desirable.

Keywords Lake geomorphology · Remote sensing · Wetland mapping · Ecosystem · Human impacts

---

R. Sinha (✉) · N. Awasthi
Department of Earth Sciences, Indian Institute of Technology Kanpur, Kanpur, India
e-mail: rsinha@iitk.ac.in; aneeraj@iitk.ac.in

R. Chandrasekaran
Department of Earth Sciences, Indian Institute of Technology Kanpur, Kanpur, India

Present Address: Achariya College of Engineering Technology, Villanur, Puducherry, India

C. M. Finlayson et al. (eds.), *Ecology, Conservation, and Restoration of Chilika Lagoon, India*, Wetlands: Ecology, Conservation and Management 6,
https://doi.org/10.1007/978-3-030-33424-6_10

10.1 Introduction

India is gifted with innumerable natural resources comprising mineral deposits, forests, vast expanses of fertile land for agricultural activities, rivers, wetlands, and water bodies such as ponds and lakes. Wetlands, either human-made or natural, freshwater or brackish play an imperative role in the terrestrial landscape and in maintaining environmental sustainability. Irrespective of their size, wetlands play a significant role in securing social well-being of people by providing drinking water, livelihood, improving groundwater conditions and controlling floods. These water bodies also play a substantial role in maintaining rich biodiversity. In recent years, along with natural forces, human activities have been identified as the dominant force responsible for adverse impacts on the wetlands ecosystems. Because of human encroachment, sewage disposal, eutrophication, and heavy metal pollution, many wetlands throughout the world (e.g. Okeechobee in Florida, Arre in Denmark, Balaton in Hungary, Biwa in Japan, Baikal in Russia, Victoria in Africa, Great Lakes of North America etc.) are showing varying degree of environmental degradation (Williams 2002). Land use/land cover (LULC) changes brought by a variety of social causes, and human activities have significantly affected the wetland environment and its biodiversity in different parts of the world (Zorrilla-Miras et al. 2014; Valdez et al. 2016).

Chilika (19°30′ – 20°30’N and 85°00′ – 86°00′E), situated on the East Coast of India, is one of the largest brackish water tropical lagoons in Asia (Fig. 10.1). The wetland, covering an area of 1165 km^2 during the peak monsoon season supports the livelihood of more than 1.5 lakhs fishermen living in 132 villages and is also well-known for its unique assemblage of marine, brackish, and freshwater ecosystem with estuarine characters (Kumar and Pattnaik 2012). In the Indian sub-continent, it is also one of the largest wintering grounds for the migratory birds within the Central and East Asian Australasian Flyway. During the peak migratory season, the Chilika hosts around 225 bird species alongwith a highly productive fishery (Kumar and Pattnaik 2012).

Given its rich biodiversity, the wetland was designated as a "Ramsar Site" in 1981 under the Ramsar Convention on Wetlands (Pattnaik 2002; Kumar and Pattnaik 2012). However, the Chilika started facing severe problems since 1990s arising from natural as well as anthropogenic activities (Sarkar et al. 2012). The major problems included (a) decline in the salinity of the lake (b) increased input of nutrients resulting in extensive macrophyte growth, eutrophication and their decomposition leading to anoxia and hypoxia-like conditions (Sahu et al. 2014), (c) changes in the strength and nature of hydrological regimes in the upstream, and (d) pollution from agriculture, aquaculture, and domestic waste.

The present study aims to map different landforms and LULC in the Chilika basin to evaluate spatial variability and temporal dynamics of geomorphic and anthropogenic processes during the last 35 years. Remote sensing and GIS are important tools to understand the global, regional, and local scale processes that affect the earth, and therefore, have a wide range of applications in the field of geology, agriculture,

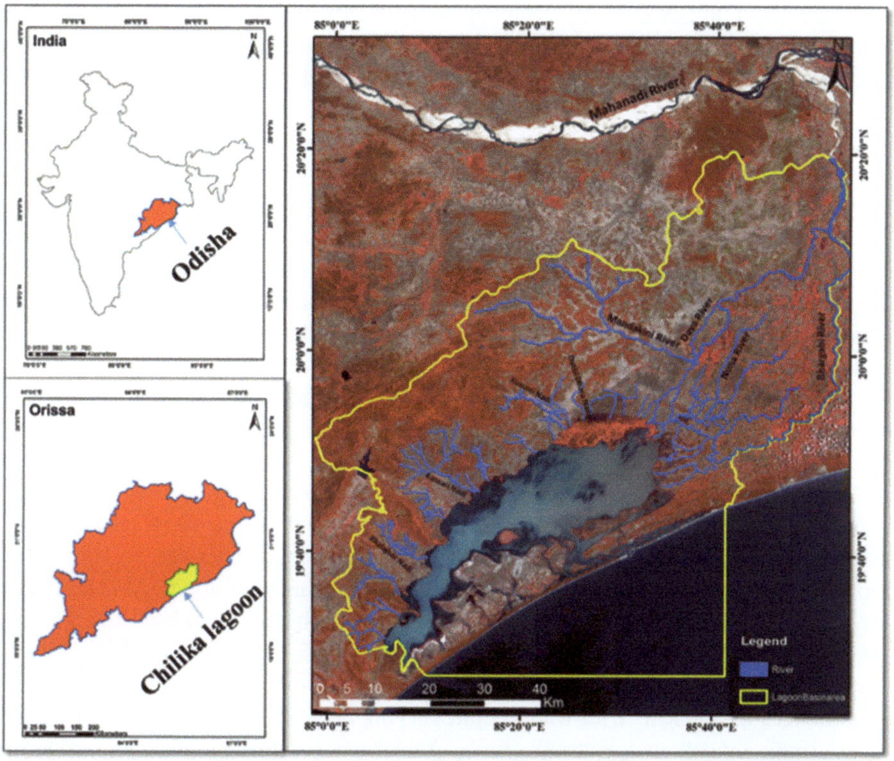

Fig. 10.1 Location map of Chilika Basin on the east coast of India. The Landsat image (False Color Composite) shows the various landscape elements in the region; blue color represent water in and around the Chilika Lake, red color shows the vegetation cover. Major drainage lines feeding the Chilika Lake are also drawn in blue

environment and integrated ecological assessment (Hudak and Wessman 1998; Fung and Ledrew 1987; Yeh and Li 1998; Long et al. 2008). Historical maps and remote sensing data have been used in a GIS framework to document these changes and to understand (a) the level of degradation of the Chilika during 1980s, and (b) the recovery of the Chilika after the implementation of a rehabilitation programme by the Chilika Development Authority (CDA) during late 1993.

10.2 Study Area

The Chilika lake runs parallel to the coastline of the Khordha and Ganjam districts of Odisha in northeast to southwest direction. The length and width of the lagoon are about 65 km and 20 km, respectively. To the north, west, and south, the lagoon is bordered by the Eastern Ghat hill ranges whereas to the east, a 60 km long sand bar separates the Chilika from the Bay of Bengal (Kumar and Pattnaik 2012). The lagoon opens into the Bay of Bengal through a 24 km long narrow outer channel that runs parallel to the coastline (Sarkar et al. 2012). Until September 2000, the Chilika had a major mouth (apart from others) opening in the Bay of Bengal near Arakhakuda. An artificial opening was created near Satapada connecting the Chilika with the Bay of Bengal (Sarkar et al. 2012). At the southern end of the lagoon, the ~18 km long "Palur Canal" connects the Rambha Bay to the mouth of the Rushikulya Estuary. The Chilika along with six other small streams between the Mahanadi and the Rushikulya River forms the part of the Mahanadi Basin. Chilika's catchment area spans 4406 km^2, 68% of which is drained by non-perennial rivers and streams from the western catchment and 32% by the distributaries of the Mahanadi River entering into the lagoon from the north-eastern end (Sarkar et al. 2012). On the basis of water salinity, depth, and ecological conditions, the Chilika has been broadly divided into four sectors, namely the northern, central, southern and outer channel.

The Chilika catchment experiences a tropical monsoonal climate with an average annual rainfall of 1300 mm and temperatures between 14 °C and 40 °C (Sarkar et al. 2012). The wind changes its direction under the influence of southwest and northeast monsoons with speed varying from 19 to 58 km/h (Sarkar et al. 2012). The water level in the lagoon fluctuates seasonally and during high and low tides. During rainy season, the Chilika expands to an area of over 1165 km^2 with the water depth varying from 1.8 to 3.7 m whereas in dry season (December–June), the waterspread area reduces to 950 km^2 with water depth varying between 0.9 and 2.6 m (Mohanty et al. 1996; Ghosh and Pattnaik 2005). In winter (November–January), a large portion of the Chilika acts as breeding and nesting grounds for millions of migratory bird species. Severe cyclonic storms and sometimes, super-cyclones frequently hit the region during pre-monsoon months of May–June and end of monsoon/post-monsoon months of October–November. During these storms, the wind speed mounts to more than 200 km/h. Torrential rains with storm often accompany these surges 10–12 m high. These storms pose a serious threat to the environment of the lagoon and the people living around by causing a dramatic change in the population of the biotic communities (e.g. phytoplankton, zooplankton, higher plants, fishes and birds) and by flooding extensive lowland areas.

10.3 Data and Methods

This study utilised topographic maps and satellite imageries. The topographic maps (73E – 1, 2, 3, 5, 6, 9, 12, 15 and 16) of 1:50000 scales obtained from Survey of India were scanned and converted into digital format. These maps were georeferenced with longitude and latitude using ArcGIS (10.1) to demarcate the Chilika Lagoon boundary. Georeferenced Landsat satellite imageries were downloaded from USGS website (http://earthexplorer.usgs.gov) for the years 1980, 1988 (spatial resolution: 79 m), 2000, 2005, 2010, and 2015 (spatial resolution: 30 m). The ERDAS Imagine (2014 version) software was used to create False Color Composites (FCCs) and

ArcGIS (10.1) for extracting the study area. For LULC classification, 90 polygons were identified for different land use/land cover patterns in each imagery in ERDAS Imagine (2014) for a supervised classification using the maximum likelihood algorithm (Wu and Shao 2002; McIver and Friedl 2002). We selected 300 points randomly for accuracy assessment of the LULC classification of the Landsat images corresponding to 1980, 1988, 2000, 2005, 2010 and 2015. Accuracy assessment represents the error matrix, along with the overall accuracy and the Kappa coefficient. The Kappa coefficient is used to infer the superiority of one map production over another, and a value of more than 0.8 indicates a strong agreement or accuracy between classification map and the ground reference map (Jensen 2005). In the accuracy assessment, topographic sheets and Google Earth maps were used for the ground verification for the years of 2015, 2010, 2005, 2000. The geomorphic landforms within the lagoon were visually interpreted and manually digitized utilizing the ArcGIS (10.1).

10.4 Geomorphology of the Chilika Basin and Lagoon Based on Remote Sensing Data

The various geomorphic units in the Chilika Basin are shown in Fig. 10.2. The northern and western parts of the Lagoon are surrounded by structural hills, residual hills and pediplain whereas deltaic, alluvial, and coastal plains are present in the north-eastern and eastern parts. There is very little information available on the geomorphic evolution of the Chilika, particularly its origin. Pascoe (1964) suggested that it was a shallow brackish water inshore lake connected to the Bay of Bengal during the later stages of the Pleistocene period. However, a more recent study by Khandelwal et al. (2008) recovered a 8 m long sediment core from Chilika for a detailed sedimentological and pollen studies and argued that the Chilika was a part of a river or a river delta with fresh water vegetation about 13,500 years BP when the sea level was lower. It was only after ~9500 years BP, with the rise in sea level, it became an estuary and mangrove vegetation dominated the area. The fall in sea level after about 2000 cal years BP changed the morphology of the coast by forming barrier spits and sand ridges. Consequently, a large lagoon was formed replacing the marine species by fresh water species (Khandelwal et al. 2008). The discharge of sediments by the tributaries of the Mahanadi River might have added to the formation of the spit bars (Krishnan 1968). The longshore drift and tidal currents and probably strong winds shifted sands to the shore resulting in the growth of the barrier spits and sand ridges. The barrier beach protected the Chilika from any direct influence of the Bay of Bengal and gradually transformed the lake into a shallow lagoon. The presence of a number of shoals, sand spits, sand bars and openings of shallow depth helps in maintaining lake's estuarine character by considerably reducing the tidal flow in and out of the lake.

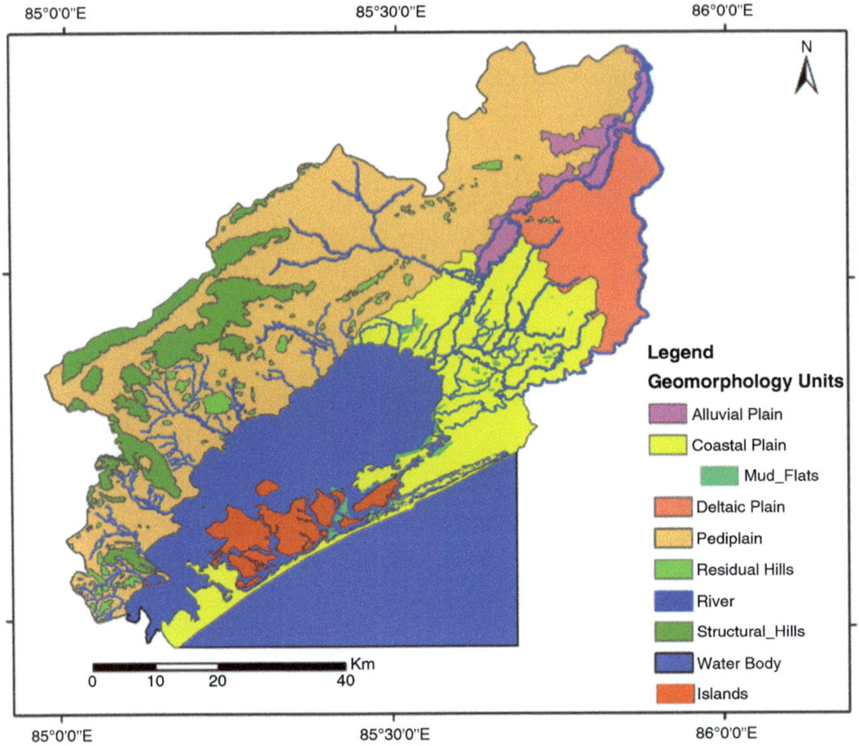

Fig. 10.2 Regional geomorphology of the Chilika Basin prepared from satellite images (this work)

10.4.1 Geomorphic Landform Changes

To assess the landform changes within and around the Chilika, we have prepared geomorphic maps of the lagoon using satellite images corresponding to the months of April and May of 1980, 1988, 2000, 2005, 2010 and 2015 enabling a systematic assessment of changes in the landforms. The emphasis of geomorphic mapping was on (a) lagoon boundary, (b) barrier islands, and (c) sand bars (Fig. 10.3). We discuss the landform changes for two different periods namely (a) 1980–2000 (pre-intervention period) and (b) 2000–2015 (post-intervention period) to highlight the impact of the restoration program, particularly the opening of a mouth adjoining Satapada.

Table 10.1 documents some of the significant changes in landforms within the lagoon for the period 1980–2000 and 2000–2015. The lagoon area increased from 865.42 km^2 to 932.62 during 1980–1988 and decreased to 909.28 during 1988–2000 and then increased to 876.10 km^2 between 2000 and 2015. On the other hand, both island area, as well as sand bars, decreased during the period 1980–2000. The period 2000–2015, however, shows a slightly different trend. During this period, the island area kept on declining, but sand bars increased drastically from 4.78 km^2 to

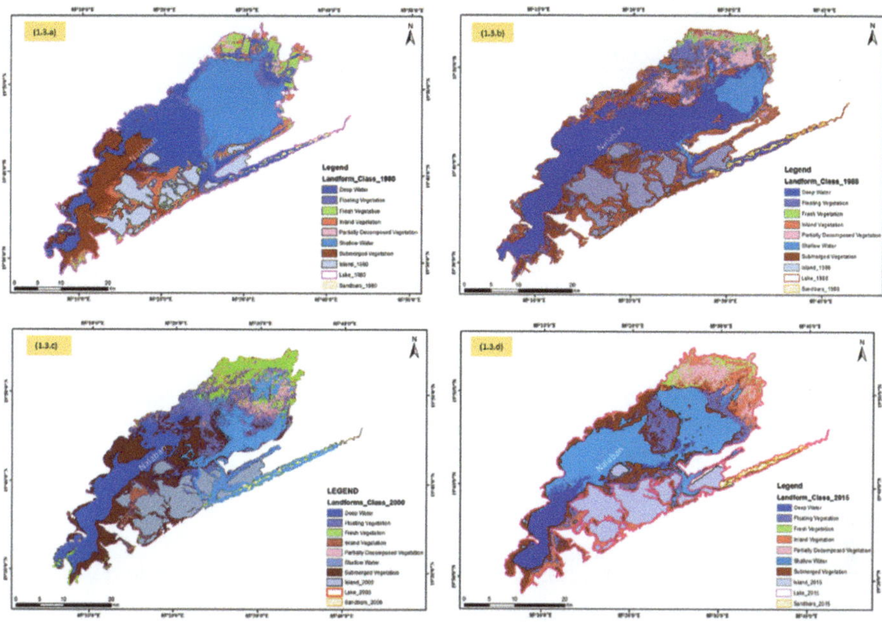

Fig. 10.3 Geomorphic maps of the Chilika for (**a**) 1980, (**b**) 1988, (**c**) 2000 and (**d**) 2015 documenting major landform changes during the period 1980–2015 prepared from repetitive satellite images (this work)

11.33 km^2. It appears that the elevations of the islands are only slightly higher than the lagoon resulting in their submergence as the water level increases. While some of the sand bars were also submerged during the pre-intervention period, a drastic increase in their area in the post-intervention period suggests a much higher sediment flux through the new opening in September 2000.

Probing further, some of the most remarkable changes within the lagoon include a significant variation in various geomorphic elements (Table 10.1). During the period 1980–2000, all vegetation related elements (viz. floating vegetation, fresh vegetation, inland vegetation, and partially decomposed vegetation) show an increase whereas the submerged vegetation decreased drastically (Figs. 10.3, 10.4, and 10.5). This change may be the manifestation of decreased salinity, depth and degradation of the overall ecological condition of the lagoon.

Interestingly, these trends seem to have reversed in the post-intervention period (November 2000–2015), and most vegetation related elements, except for partially decomposed vegetation, show a decrease. The area under submerged vegetation was observed to increase significantly during the latter period, which may be attributed to the revival of the conducive salinity regime. Although the deep water areas decreased drastically during this period, the shallow water areas showed a remarkable increase. We interpret this as a manifestation of increased sediment flux and siltation in the Chilika after the opening of the new mouth. An added factor is

Table 10.1 Landform changes within the Chilika lake between 1980 and 2015

Type of Landform	1980 Area (Km^2)	2000 Area (Km^2)	2015 Area (Km^2)	Remarks and Inference (after the implementation of programme)
Lagoon	865.42	871.80	876.10	Continuous increase; sedimentation rate increased, depth decreased causing submergence of island and increase in inundation
Island	175.49	167.44	158.90	
Sand bars	6.14	4.78	11.33	Decrease and then abrupt increase; passage to old mouth closed
Deep water	284.80	234.44	80.84	Deepwater continuously decreased but shallow water decreased and then abruptly increased; sedimentation rate increased, depth of the lake decreased
Shallow water	273.66	92.48	306.26	
Floating vegetation	144.69	245.37	112.94	Increase and then abrupt decrease; the salinity influx increased, fresh, floating and inland vegetation decreased
Fresh vegetation	65.39	79.71	23.12	
Inland vegetation	102.49	147.05	111.37	
Partially decomposed	11.04	45.06	55.08	Continuous increase; salinity influx increased, decomposed vegetation increased
Submerged vegetation	186.48	54.42	174.12	Decrease and then abrupt increase; salinity influx increased, submerged vegetation increased

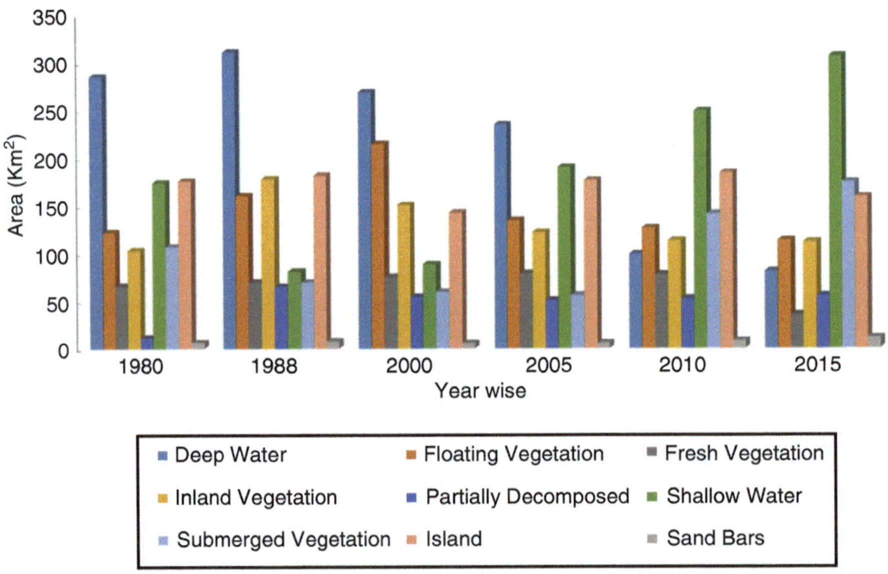

Fig. 10.4 Landform changes in Chilika Basin during the years 1980, 1988, 2000, 2005, 2010 and 2015

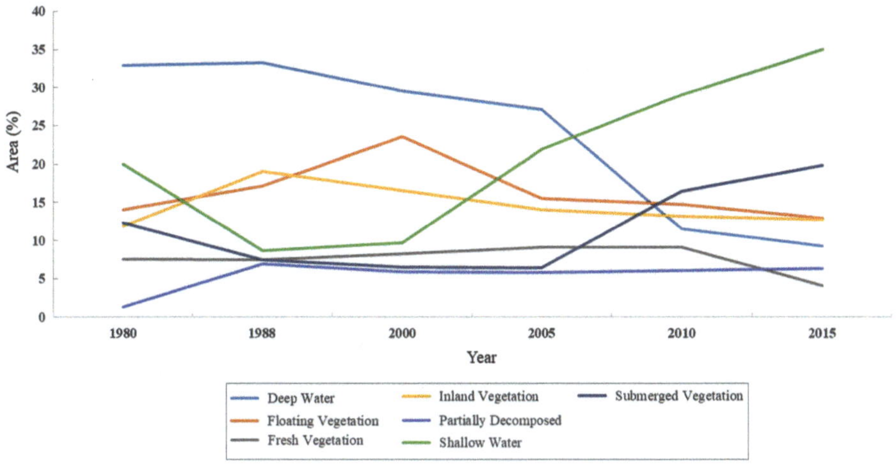

Fig. 10.5 Landform as well as aquatic vegetation changes in Chilika Basin during the years 1980, 1988, 2000, 2005, 2010 and 2015

Table 10.2 Tidal inlet changes in between 1980 and 2015

Type of landform	1980 (m)	2000 (m)	Mouth drift (1980–2000) (m)	2010 (m)	Mouth drift (2000–2010) (m)	2010 (m)	Mouth drift (2000–2015) (m)
Old mouth	264.50	160.00	7113.53	Closed	–	Closed	–
Artificial new mouth	–	232.00	–	320.00	1184.80	431.00	2269.46
Rate of mouth drifting/year			355.68		118.48		151.29

possibly increased flux from the tributaries of the Mahanadi River. Another notable change is the increase in the aerial extent of the Chilika lagoon, mainly due to a decrease in water depth because of high sediment input from the Mahanadi and its tributaries (Figs. 10.3, 10.4, and 10.5). We suggest that the storage capacity of the lagoon decreased due to sedimentation and therefore water spilled to the shore areas thereby increasing the lagoon area.

We have also investigated the morphological changes around the tidal inlets during the period of observation. The natural mouth of the lagoon was 264.50 m wide in 1980 and situated in the NE part of the lake (Table 10.2, Figs. 10.6a and 10.6c). But in the year 2000, the natural mouth opening reduced to a width of 160 m and further shifted in NE direction approximately by 7 km (Fig. 10.6f and g). The drifting rate calculated was around 355.68 m per year that was much higher than the rate observed in the last 35 years.

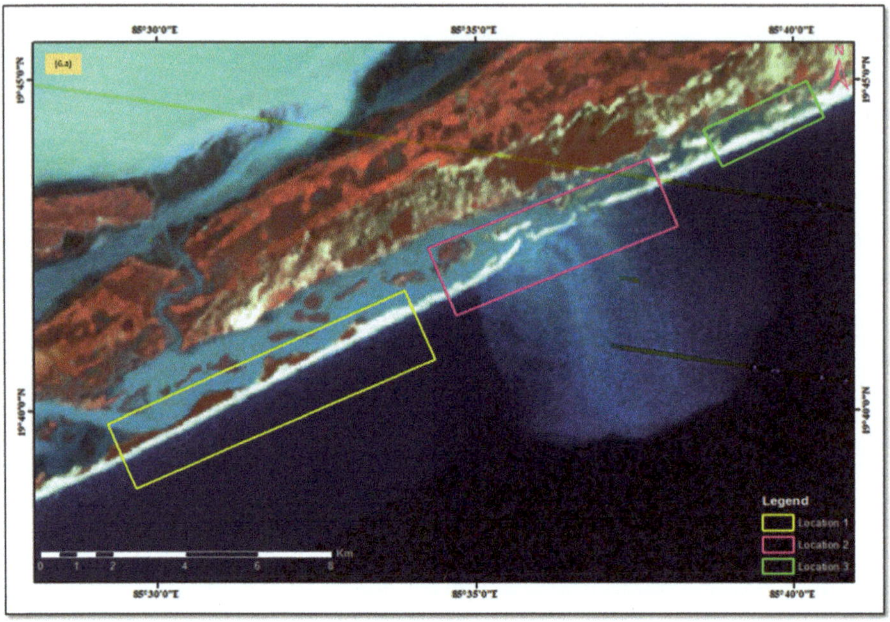

Fig. 10.6 (**a**) Landsat image of the area around the tidal inlet of Chilika. Rectangles mark the windows in which the detailed investigations were carried out

The reducing width of the old mouth severely threatened the Chilika due to problems such as shrinkage of water spread area, a decrease in depth due to siltation, salinity reduction, macrophyte infestation, eutrophication, and loss of biodiversity.

After a series of assessments on the wetland environment on changes in salinity, sedimentation rates, and freshwater macrophytes, the CDA decided to open a new mouth to improve the salinity and the biodiversity in and around the lagoon. In September 2000, a 232 m wide new artificial mouth was opened (Fig. 10.6e). The opening of new lagoon mouth and desiltation of the channel to the sea through the barrier beach at Satapada restored the natural flows of water and sediment by reducing the distance of outflow from the lake by 18 km. This opening vastly improved the lagoon-sea connection and helped to minimise the problems of sedimentation and invasive macrophytes (Ghosh et al. 2006).

The immediate positive impact was the stark and rapid recovery of the fishery along with the considerable improvement in salinity flux. Using the satellite measurements, we observed that by the year 2010, this artificial mouth shifted ~ 1 km towards the northeast with an opening of 320 m and a drift rate of 118.48 m year^{-1}. The natural mouth had closed by that time and further accumulation occurred around this by 2015 (Fig. 10.6i). The artificial mouth with a width of 431 m drifted further to a distance of 2269.46 m by 2015 (Fig. 10.6h) with a drift rate of 151.29 m year^{-1}. The reconstruction of spit growth and its dynamics indicate that as the width of the mouth increased, so did the rate of drift. The drift was mainly due to large

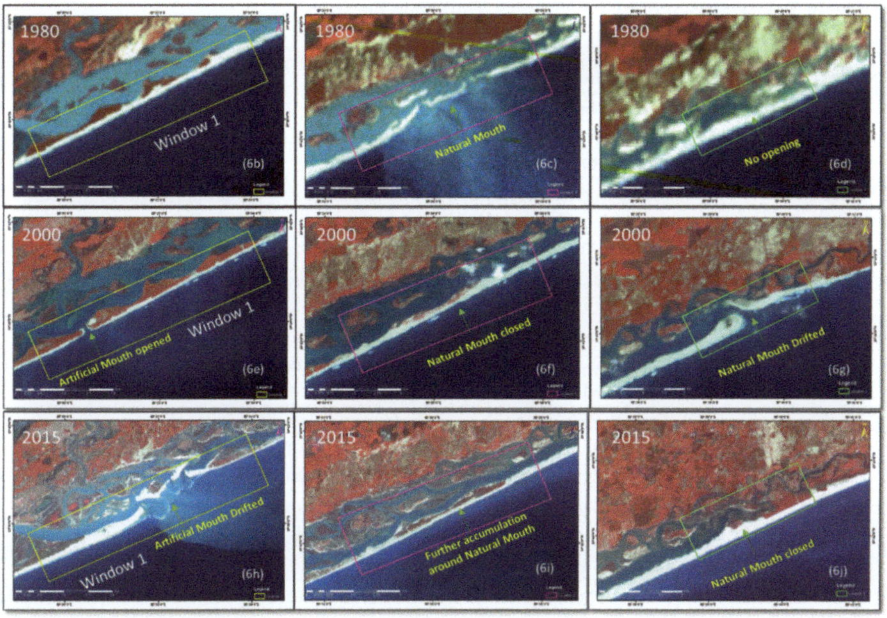

Fig. 10.6 (b–j) Changes in tidal inlets and opening of new artificial mouth in the three windows during the period 1980–2015

northerly littoral sediment transport that continuously shifted the inlet mouth towards the north, resulting in elongation of the channel and reduction in channel cross section (Ghosh et al. 2006).

10.5 LULC Classification

The term land cover refers to the kind and state of vegetation and subsequent usage such as human structures, soil type, biodiversity and water (Mayer et al. 1995). On the basis of National Remote Sensing Center (NRSC) land use classification system (NRSA 2008), there are nine LULC patterns that were identified in the study area (Table 10.3). These include: (1) settlements (2) agriculture, plantation (3) agriculture, fallow (4) forest, evergreen (5) forest, scrub forest (6) lagoon (7) inland wetland (8) barren/wasteland, and (9) water bodies. In the accuracy assessment, the overall accuracy of the classified image ranged from 80% to 84% and the Kappa coefficient was found to vary between 0.76 and 0.81 which is considered to be acceptable.

Figures 10.7 and 10.8 show the LULC maps of the study area for the period 1980–2015. Tabulations and area calculations provide a comprehensive data set in terms of the overall landform changes, the type, and extent of changes that occurred (Table 10.4). In the year between 1980 and 1988, the spatial pattern of LULC in the

Table 10.3 Description of LULC classes identified in the study area

Land use/Land cover types	Description
Settlement	Area of human habitation developed due to non-agricultural use and that has a cover of buildings, transport & communication, utilities associated with water, vegetation and vacant lands
Agriculture, plantation	Areas under agricultural tree crops adopting certainagricultural management techniques. It includes crops which are normally grown in the hilly regions (tea, coffee, rubber etc.,) and closely associated with forest cover
Agriculture, fallow	Lands taken up for cultivation but temporarily allowed to rest, uncropped for one or more seasons, but not less than 1 year
Forest, Evergreen	Areas that comprise of the thick and dense canopy of tall trees which predominately remain green throughout the year. It includes both coniferous and tropical leaved evergreen species
Forest, scrub forest	Forest areas where the crown density is less than 10% of the canopy cover, generally seen at the fringes of the dense forest cover and settlements where there is biotic and abiotic interference
Lake wetland	Water logged areas (seasonal and perennial) in and around the lake
Inland wetland	Areas that include ox bowlakes, cut off meanders, playas, swamp marsh etc.
Barren/wasteland	Areas where rock exposure of varying lithology, often barren and devoid of soil and vegetation cover is present. They occur amidst hill forests as opening or an isolated exposures on plateaus and plains
Water bodies	Areas with surface water either impounded in the form of ponds, lakes, reservoirs, or flowing streams, rivers, canals etc.

Source: National LULC mapping using multi-temporal AWiFs data, NRSA/LULC/1:250K/2008-3, 2008

Chilika Lake exhibits water, agriculture plantation, inland wetland and settlement as the most extensive types of LULC. This LULC pattern accounted for about 29–30%, 21–22%, 13–14%, and 10–11% respectively of the total study area (Table 10.4). But in the years between 2000 and 2015, the most extensive type of LULC constituted of water, agriculture plantation, settlement, and inland wetland in the study area and they represented about 26–32%, 17–20% 14–18% and 12–15% respectively of the total area (Table 10.4). The doubling of settlement area from a mere 9.74% in 1980 to 18.30% in 2015 reflects a significant human impact on the system during the last 3–4 decades.

The increase has been particularly steep during 1980–2000, and it slowed down remarkably in the post-2000 period. It also appears that agriculture (plantation) and scrub forest suffered the maximum loss as both these classes show a continuous decrease during the entire period of 1980–2000. We interpret that most of the agriculture and scrub forest classes have been converted to settlements over time. Interestingly, the evergreen forest shows some increase between 1980 and 2000 but then got stabilised in the post-2000 period. The lake area shows a decrease between 1980 and 2000 but then has been increasing thereafter with minor variations, a clear manifestation of the opening of the new mouth.

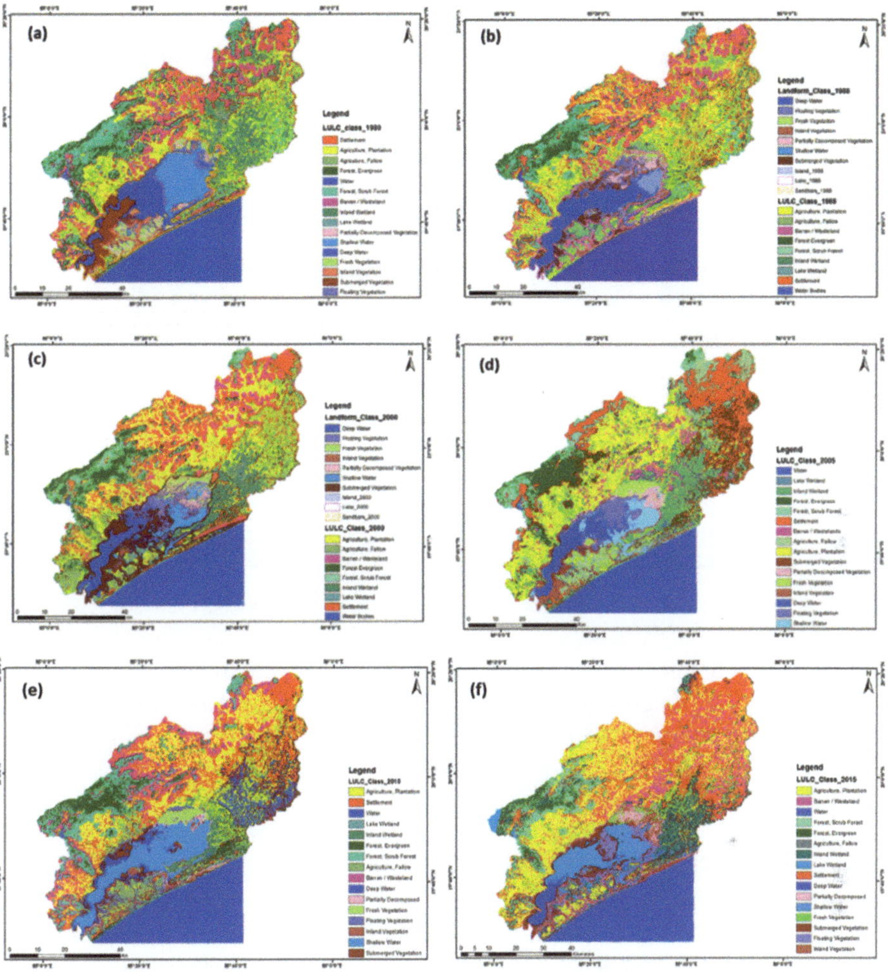

Fig. 10.7 Land use/land cover maps of Chilika Basin for (**a**) 1980, (**b**) 1988, (**c**) 2000, (**d**) 2005, (**e**) 2010 and (**f**) 2015 prepared from repetitive satellite images (this work)

During the period 1980–2015, significant changes were observed in the settlement, agriculture plantation and barren/wasteland areas while other classes do not show many differences. Between 1980 and 2015 (Table 10.4), the area of agriculture plantation (338.7 km^2) and barren landform (204.9 km^2) decreased and settlements (496.6 km^2) increased. A two-fold increase in the settlement class is observed between 1980–2000 and 2010–2015 due to the increase in urban population. Similarly, Lake Wetland (40.53 km^2) and water bodies (45.0 km^2) increased due to new mouth opening in the year 2000. The scrub forest areas (65.16 km^2) decreased and evergreen forest areas (62.57 km^2) increased. This change implies

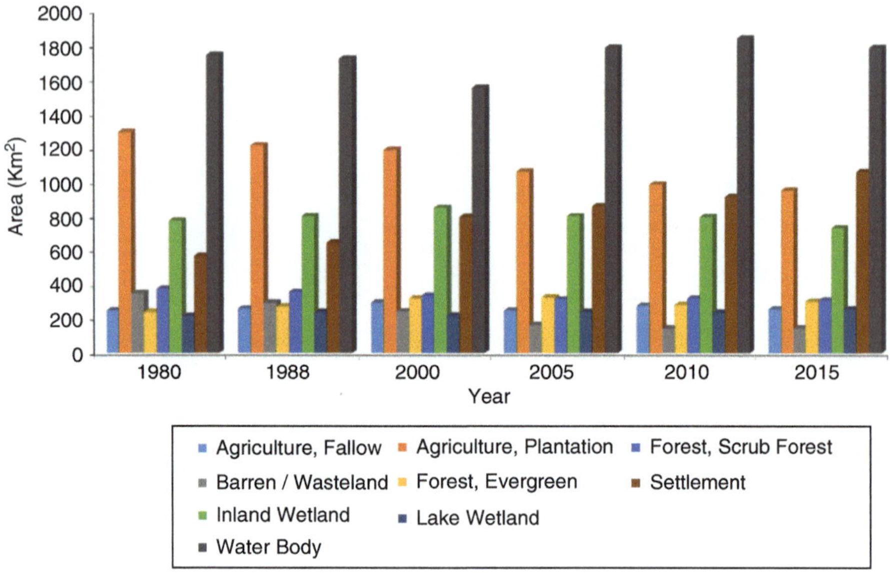

Fig. 10.8 Land use/land cover changes in Chilika Basin during the period 1980–2015

that these scrub forest areas converted into evergreen forest due to high dense vegetation cover. Interestingly, overall agriculture fallow land does not show much variation between 1980 and 2015 (Fig. 10.9).

10.6 Sedimentary Environment of Chilika

Three hydrological subsystems influence the hydrology of the Chilika Lagoon namely, (a) the Mahanadi River system on the northern side, (b) small river channels on the western side, and (c) the Bay of Bengal on the eastern side (Kumar and Pattnaik 2012). The distributaries of the Mahanadi River namely, Daya, Nuna, Bhargavi and Makara drain directly into the Chilika (Fig. 10.1). These distributaries of the Mahanadi though controlled by the Naraj barrage upstream contribute maximum silt load and fresh water into the Chilika (Pattnaik 2002). Ghosh et al. (2006) and Khandelwal et al. (2008) estimated that 1.5 million tons of silt load are discharged every year by the Mahanadi distributaries along with 51% of the total freshwater inflow into the lake. As many as 52 small rivers and rivulets from western catchment (e.g. Kansari, Kusumi, Janjira, Tarimi) account for 0.3 million tons of sediments supply annually along with the total freshwater input of 39% (Ghosh et al. 2006). The longshore sediment transport (littoral drift) of the Bay of Bengal also contribute about 0.1 million tons of sediments annually (Sarkar et al. 2012). The discharge by the rivers to the Chilika is high during peak flood seasons of the Southwest monsoons (June–September) and very low during rest of the year. The

Table 10.4 LULC classification for 1980, 2000, 2005, 2010 and 2015 images

Land cover type	1980		2000		2005		2010		2015		Remarks
	Area (Km^2)	Area (%)	Area (Km^2)	Area (%)	Area (Km^2)	Area (%)	Area (Km^2)	Area (%)	Area (Km^2)	Area (%)	
Settlement	564.38	9.74	812.78	14.02	858.84	14.82	913.50	15.76	1060.98	18.30	Continuous increase, significant
Agriculture, plantation	1292.52	22.30	1171.29	20.21	1062.76	18.34	987.97	17.05	953.82	16.46	Continuous decrease
Agriculture, fallow	246.35	4.25	264.22	4.56	246.60	4.25	276.57	4.77	255.44	4.41	No change
Forest, Evergreen	238.42	4.11	327.59	5.65	324.52	5.60	281.09	4.85	300.99	5.19	Some increase between 1980–2000 and then no change
Forest, scrub forest	372.95	6.43	328.07	5.66	312.38	5.39	318.71	5.49	307.79	5.31	Continuous decrease, minor
Lake wetland	214.74	3.71	174.36	3.01	238.67	4.12	235.59	4.06	255.27	4.40	Decrease between 1980–2000 and then increase
Inland wetland	774.54	13.36	708.09	12.22	801.68	13.83	796.40	13.74	729.51	12.59	No major change
Barren/ wasteland	348.27	6.01	211.69	3.65	162.46	2.80	142.44	2.46	143.37	2.47	Significant decrease throughout
Water bodies	1743.63	30.08	1797.72	31.02	1787.89	30.85	1843.53	31.81	1788.63	30.86	No change

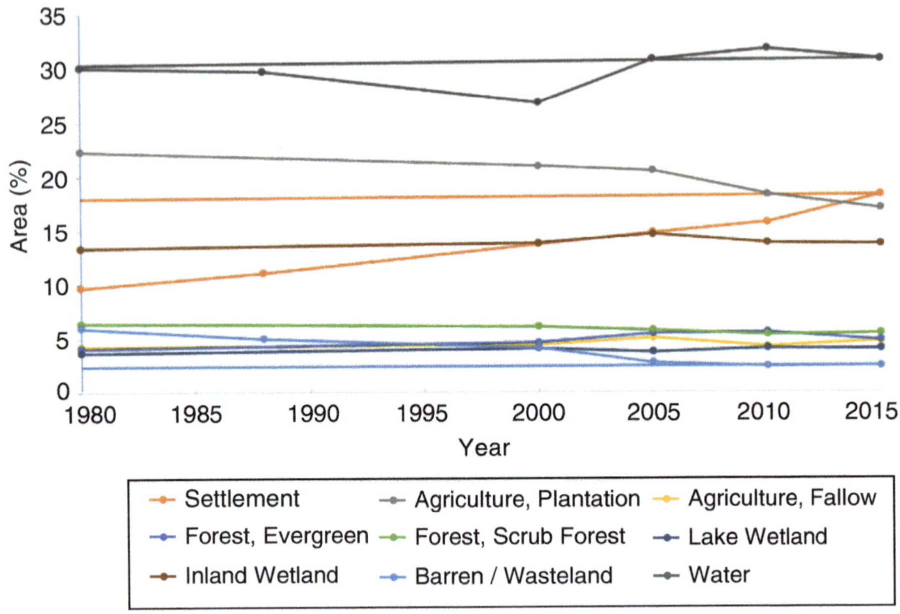

Fig. 10.9 Trend analysis of land use/land cover in Chilika between 1980 and 2015

lagoon receives 10% of freshwater input from direct precipitation (Khandelwal et al. 2008). Due to high sediment supply by the Mahanadi distributaries, the sedimentation rate is comparatively higher in northern (7.6 mm $year^{-1}$) and central sectors (8 mm $year^{-1}$) and low in the southern sector (2.8 mm $year^{-1}$) (Sarkar et al. 2012).

Sediments to the Chilika are mostly derived from the denudation of hills and plateaus of the Eastern Ghats exposed on the northern, western and southern borders of the lake. The sediments are mostly of silt-clay size with biogenic debris and occasional intercalated lenses of sand (Zachmann et al. 2009). Sand predominates on the eastern and western shores of the central and southern sectors, at the bottom of the outer channel and spits and bars. Because of long-distance transport, the Mahanadi distributaries supply fine-grained sediments of mud size whereas the small rivers/rivulets originating from the nearby hills mostly carry coarser sediments. A seaward decrease in mud has been observed with an increase in sand content towards the mouth of the outer channel (Rao et al. 2000). The sediments show a predominance of quartz along with minor admixtures of feldspar (microcline), muscovite, and clay minerals. The clay minerals are composed of hydro–muscovite, illite, and kaolinite (Zachmann et al. 2009). The heavy metals exhibit slightly decreasing concentrations from northeast to southwest, thus indicating the supply of contaminants from the Mahanadi River. The organic matters within the sediments of the lagoon are from the decomposition of macrophytes and supplies by the rivers. The organic matter content is highest in the northern sector (1.29%) that decreases seaward (Pattnaik 1987; Rao et al. 2000).

The monsoon rains strongly influence the hydrography of the lagoon. Sediment and water supplies from the catchments upstream as well as oceanic longshore transport are maximum during monsoon months, and the lake consequently experiences massive flooding during these months. Due to the influence of both tides and perennial freshwater inflow from the rivers, the lake salinity ranges from seawater strength to freshwater. The salinity decreases during the monsoon due to the substantial influx of freshwater from the northern and central sectors. The southern sector is the least affected even during monsoon and maintains its brackish-water conditions. During the summer, the water level of the lagoon is at its lowest, and this facilitates the influx of saline water from the outer channel into the lake and salinity of the lake increases. The winds also play an essential role in regulating salinity of the lake by mixing water from different sectors.

The Chilika lagoon environment is suitable for the survival of freshwater, brackish, and marine water organisms (Siddiqi and Rao 1995). The population of these organisms varies both spatially as well as seasonally where the salinity plays an important role. Biodiversity is severely affected by the spatial and temporal salinity gradients caused by freshwater inflows from the river systems, seawater intrusion under the tidal influence and weak circulation (Ghosh et al. 2006; Sahu et al. 2014). The outer channel is more prone to these changes where polyhaline and mesohaline organisms get gradually replaced by oligohaline organisms with a decrease in salinity. However, siltation of the lagoon is identified as a major source for salinity decrease and a threat to the ecosystem (Kumar and Pattnaik 2012).

Our work on the LULC changes suggests that in last few decades, the change in the land use pattern in the direct catchments of Chilika Lagoon has led to an overall decline of agriculture plantation (5.84%) and barren lands (3.54%) (Figs. 10.6 and Table 10.4). Probably increased rates of erosion upstream have ultimately increased the sediment load considerably although there are no measured rates of erosion available from this region. Moreover, the river inflows to the lagoon are being regulated upstream through a series of hydraulic structures influencing water availability for maintenance of ecosystem processes and functions of the wetland downstream (Ghosh et al. 2006). Since 2000, the increased tidal influence the lagoon probably acted as a sink for terrestrial material, and as a result, the average depth of the lake decreased (Fig. 10.3). However, this observation needs to be supplemented by detailed bathymetric surveys of the lake. The accumulation of sediments causes gradual choking of outer channel restricting the outflux of silt-laden freshwater. Siltation often results in shifting and blockage of the outer channel and mouth, preventing the smooth exchange of water and sediment between the lake and the sea (Chandramohan and Nayak 1994). Siltation of the outer channel and disturbed tidal flux has also resulted in shrinkage of the lagoon area and with a massive addition of suspended sediments transparency of the lake has been significantly affected. Though seasonal and episodic, storm surges, cyclones, monsoonal floods and longshore drift change coastal geomorphic features and add to the problems of salinity and siltation.

Apart from siltation, the Chilika environment is being degraded by several natural and human factors taking place within the Mahanadi River Basin as well as the coastal processes within the Bay of Bengal. The extensive habitation surrounding the lake and in the catchment basin is a serious threat to the lagoon because increased agricultural practices, disposal of untreated domestic garbage and wastewater has resulted in increased inputs of nutrients and pollutants to the lake through the various river and tributary systems. Nutrient enrichment has led to massive macrophyte growth, and their decomposition has created anoxia and hypoxia-like conditions in the lake (Sahu et al. 2014). Increase in freshwater weeds have considerably reduced the breeding of fish and led to a decrease in fishing activity.

10.7 Concluding Remarks and Future Perspectives

The land is a natural resource for supporting the life system. This chapter has highlighted that the water bodies, agricultural plantation, and settlement are the predominant LULC in the study area. These LULC classes form an essential ecosystem in Chilika. The lagoon is facing a significant problem of siltation mainly due to improper utilisation of LULC. Agriculture plantation and barren lands are more vulnerable due to urbanisation such as engineering construction, settlements, and transport. Changes in the island landforms within the lake are mainly due to the river as well as sea input into the lake. Sandbar area fluctuations reflect that depends upon the sedimentation input as well as the opening of the mouth. After the opening of the new mouth, there is an increase in the number of small sand bodies in the form of islands in the northern part of the lake. Sedimentation in the Chilika remains a serious concern, and there is no reasonable estimate of annual sediment flux into the basin. The tidal inlet was drifting in the NE direction due to the influence of longshore currents at a higher rate in the pre-hydrological invention, and the rate of drift has been quite significant in the artificial mouth as well after the hydrological invention. It is necessary to monitor the rate of tidal inlet changes because it is the primary factor for salinity changes in the lake. This chapter has also demonstrated the effective use of remote sensing data in a GIS framework to understand the LULC change and its causal factors. However, there is a strong need for generating high-resolution time series using satellite images of intermediate periods and detailed field investigations to confirm some of the changes to evolve a sustainable management strategy.

The hydrology of the Chilika Lagoon is influenced significantly by the river systems in the northern and western catchment and by the Bay of Bengal. The Mahanadi distributaries are responsible for significant freshwater, and silt load into the lagoon whereas the longshore sediment transport of the Bay of Bengal is mostly restricted to the coast and outer channel. The wetland environment is under serious threat because of the changes in salinity values controlled by freshwater inflow from the rivers and the tidal exchange between the lagoon and sea. Natural factors like storm surges, cyclones, monsoonal floods, sediment deposition by longshore drift (tidal inlet changes) and human-induced factors like change in land use pattern upstream to the lagoon, increased agricultural practices, disposal of untreated domestic garbage and wastewater are degrading the lagoon environment which is needed to be monitored and managed adequately.

A detailed work focusing on deep sediment cores from different parts of the lagoon may be extremely rewarding to reconstruct the temporal as well as spatial evolution of the Chilika. No geochemical data is available on the Chilika sediments to understand the nutrient cycling and geochemical evolution of the lagoon through geological time. In addition, detailed geochemistry of stratigraphically controlled core samples should answer several aspects of the lagoon evolution and the paleoclimate of the area. Lagoons are the appropriate sites for the reconstruction of relative sea level and paleosalinity changes. Sea level change with time and the relative contribution of the saline water into the lagoon can be studied by reconstructing the salinity variations using various proxies such as evaporate concentration, the $^{87}Sr/^{86}Sr$, Sr/Ca ratios along with $\delta^{18}O$ and $\delta^{13}C$ of the biogenic carbonates. These isotopic and elemental ratios are ideal proxies for paleosalinity because of the typically different values in the fresh and sea waters. Also, biogenic carbonates incorporate Sr isotopes in their crystal lattice during precipitation with no important effect, recording the $^{87}Sr/^{86}Sr$ ratio of the waters in which they grew. A consistent trend is expected between the above proxies as a result of mixing between fresh and seawater. Furthermore, the seawater contribution can be quantified with the help of available paleo salinity data from the Bay of Bengal (Indian Ocean).

References

Chandramohan P, Nayak BU (1994) A study of the improvement of the Chilika lake tidal inlet east coast of India. J Coast Res 10:909–918

Fung T, Ledrew E (1987) Application of principal components analysis to change detection. Photogramm Eng Remote Sens 53(12):1649–1658

Ghosh AK, Pattnaik AK (2005) Chilika Lagoon: Experience and lessons learned brief. In: Proceeding of the ILEC, managing lakes and their basins for sustainable use: a report for lake basin managers and stakeholders, International Lake Environment Committee Foundation, Kusatsu

Ghosh AK, Pattnaik AK, Ballatore TJ (2006) Chilika lagoon: restoring ecological balance and livelihoods through re-salinization. Lakes Reserv Res Manag 11:239–255

Hudak AT, Wessman CA (1998) Textural analysis of historical aerial photography to characterize woody plant encroachment in south African savanna. Remote Sens Environ 66(3):317–330

Jensen JR (2005) Thematic map accuracy assessment. In: Keith CC (ed) Introductory digital image processing–a remote sensing perspective, Prentice Hall series in geographic information science, 3rd edn. Prentice Hall, Saddle River, pp 495–515

Khandelwal A, Mohanti M, García-Rodríguez F, Scharf BW (2008) Vegetation history and sea level variations during the last 13,500 years inferred from a pollen record at Chilika Lake, Orissa, India. Veg Hist Archaeobotany 17(4):335–344

Kumar R, Pattnaik AK (2012) Chilika – an integrated management planning framework for conservation and wise use. Wetlands International South Asia, New Delhi. 162 pp

Krishnan MS (1968) Geology of India and Burma. Higginbothams (P) Ltd, Madras

Long H, Wu X, Wang W, Dong G (2008) Analysis of urban-rural land-use change during 1995–2006 and its policy dimensional driving forces in Chongqing, China. Sensors 8 (2):681–699

Marshall H (2016) World's lakes endangered over-use, water withdrawals and pollution puts one billion people at risk. Retrieved from http://www.hoffmanpr.com/press-release/worlds-lakes-endangered-over-use/

Mayer RC, Davis JH, Schoorman FD (1995) An integrative model of organizational trust. Acad Manag Rev 20:709–734

McIver DK, Friedl MA (2002) Using prior probabilities in decision tree classification of remotely sensed data. Remote Sens Environ 81:253–261

Mohanty PK, Dash SK, Mishra PK, Murty AS (1996) Heat and momentum fluxes over Chilika: a tropical lagoon. Indian J Mar Sci 25:184–188

NRSA (2008) National LULC mapping using multi-temporal AWiFs data, NRSA/LULC/1:250K/2008-3

Pascoe EH (1964) A manual of the geology of India and Burma, vol 3. Geological Survey of India, Delhi

Pattnaik DK (1987) Distribution of nutrients in the surfacial sediments of Chilika Lake. M. Phil. thesis, Berhampur University, Berhampur, Orissa

Pattnaik SS (2002) Estimation of sediment flow into Chilika Lake. In: The proceedings of international workshop on sustainable development of Chilika Lagoon. Chilika Development Authority, Bhubaneswar

Rao KK, Jayalakshmy KV, Venugopal P, Gopalakrishnan TC, Rajagopal MD (2000) Foraminifera from the Chilika Lake on the east coast of India. J Mar Biol Assoc India 42(1&2):47–61

Sahu BK, Pati P, Panigrahy RC (2014) Environmental conditions of Chilika Lake during pre and post hydrological intervention: an overview. J Coast Conserv 18(3):285–297

Sarkar SK, Bhattacharya A, Bhattacharya AK, Satpathy KK, Mohanty AK, Panigrahi S (2012) Chilika Lake. In: Bengtsson L, Herschy RW, Fairbridge RW (eds) Encyclopedia of lakes and reservoirs. Springer, Dordrecht, pp 148–156

Siddiqi SZ, Rao KVR (1995) Limnology. In: Ghosh AK (ed) Wetland ecosystem series 1: Fauna of Chilika Lake. Zoological Survey of India, New Alipore, pp 11–136

Valdez VC, Ruiz-Luna A, Berlanga-Robles CA (2016) Effects of land use changes on ecosystem services value provided by coastal wetlands: recent and future landscape scenarios. J Coast Zone Manag 19:1. https://doi.org/10.4172/jczm.1000418

Williams WD (2002) Environmental threats to salt lakes and the likely status of inland saline ecosystems in 2025. Environ Conserv 29(2):154–167

Wu W, Shao G (2002) Optimal combinations of data, classifiers, and sampling methods for accurate characterizations of deforestation. Can J Remote Sens 28:601–609

Yeh GO, Li X (1998) Principal component analysis of stacked multi-temporal images for the monitoring of rapid urban expansion in the Pearl River. Int J Remote Sens 19(8):1501–1518

Zachmann DW, Mohanti M, Treutler HC, Scharf B (2009) Assessment of element distribution and heavy metal contamination in Chilika Lake sediments (India). Lakes Reserv Res Manag 14:105–125

Zorrilla-Miras P, Palomo I, Gomez-Baggethun E, Martin-Lopez B, Lomas PL, Montes C (2014) Effects of land-use change on wetland ecosystem services: a case study in the Doñana marshes (SW Spain). Landsc Urban Plan 122:160–174

Chapter 11
Spatiotemporal Assessment of Phytoplankton Communities in the Chilika Lagoon

Suchismita Srichandan and Gurdeep Rastogi

Abstract Phytoplankton are the primary producers in aquatic ecosystem and play crucial role in the nutrient cycling, carbon fixation, and regulating the overall food-web dynamics. In addition to ensuring ecological services, phytoplankton species composition is also considered an efficient bio-indicator of the water quality. Thus, phytoplankton composition, diversity, and their distribution could be used as a biological proxy to assess the ecological health of a water body. Considering the ecological significance of phytoplankton, various studies have targeted them to understand their spatiotemporal variation and environmental drivers in the Chilika lagoon. Phytoplankton community structure of Chilika lagoon is influenced by several environmental factors (nutrients, light, and salinity) of which salinity predominantly determines the composition and distribution of phytoplankton communities. In Chilika lagoon, spatial variation in salinity regime provides a variety of habitats (e.g. oligohaline (0–5 ppt), mesohaline (5–18 ppt), and polyhaline (>18 ppt)) for the proliferation of freshwater, estuarine, and marine phytoplankton forms. Based on the published literature, a total of 739 phytoplankton species have been documented from the Chilika lagoon, which included a diverse assemblage of species spectrum represented by Bacillariophyta (270 species), Dinophyta (88 species), Cyanophyta (103 species), Chlorophyta (178 species), Euglenophyta (92 species), Chrysophyta (5 species) and Xanthophyta (3 species). Among these, Bacillariophyta has been shown to be the most diverse and abundant in the phytoplankton communities. The total inventory of 709 phytoplankton species during the post-restoration study (2000–2014) included 612 new records which were documented for the first time from Chilika lagoon. Long-term systemic monitoring of phytoplankton is essential to understand their intrinsic spatiotemporal variability and also to recover maximum species diversity in lagoon. Further, continuous and detailed observation of phytoplankton community is necessary to monitor the occurrence of toxic species and harmful algal blooms. In addition to the application of classical microscopy based taxonomic approach to document phytoplankton

S. Srichandan · G. Rastogi (✉)
Wetland Research and Training Centre, Chilika Development Authority, Balugaon, Odisha, India

C. M. Finlayson et al. (eds.), *Ecology, Conservation, and Restoration of Chilika Lagoon, India*, Wetlands: Ecology, Conservation and Management 6,
https://doi.org/10.1007/978-3-030-33424-6_11

species diversity, efforts should also be directed to integrate the molecular tools such as high-throughput DNA sequencing to understand the genetic diversity of smaller size nano-phytoplankton and pico- phytoplankton in the lagoon ecosystem.

Keywords Phytoplankton · Coastal Lagoon · Chilika · Brackish · Salinity

11.1 Background

Coastal lagoons are commonly distinguished as highly productive ecosystems due to their shallow depth and restricted water exchange with the sea (Bec et al. 2011). In addition, the productivity of coastal lagoons are influnced by development of strong boundaries and salinity gradients. Depth profiles also plays pivotal role by regulating solar insolation and controls benthic-pelagic coupling (Pérez-Ruzafa et al. 2011). Most of the coastal lagoons are simultaneously influenced by riverine and seawater influx resulting in brackish water salinity regime. The brackish water habitat hosts a wide array of biodiversity including major feeding and breeding grounds of fishes and birds. Lagoons also extends ecological services by providing food (largely fishes), stock of freshwater, maintainance of hydrobiology, climate regulation, flood protection, water purification, oxygen production, fertility, recreation and ecotourism (Newton et al. 2018). Hence, coastal lagoons have immense ecological, economic, and social values by supporting livelihoods of fisher folks and coastal communities (Newton et al. 2014). Coastal lagoons are typically characterized by shallow depth, bi-directional horizontal flows, and frequent mixing in the water column which results in a highly variable gradient in the physicochemical properties (Rakhesh et al. 2015). The variability in these properties also influences the phytoplankton community composition over the spatial and temporal scales in a lagoonal environment.

The challenge of deciphering the roles of phytoplankton community assembly remains a central problem of aquatic ecology (Cloern and Dufford 2005) and aquatic ecosystems are characterized by remarkable phytoplankton diversity (Goebel et al. 2013). Long-term changes in phytoplankton communities have been a major concern for global changes, which could be used to track ecosystem's response to the eutrophication and climate changes (Chen et al. 2010). Therefore, knowledge of phytoplankton community structure and associated variability at larger temporal scales covering multiple years is essential to understand underlying environmental changes, caused due to various drivers and pressures, which are accentuated by climate change. Krebs (1994) has opined that phytoplankton dynamics is influenced by bottom-up and/or top-down factors. Bottom-up factors (e.g., temperature, light intensity, salinity, nutrients, nitrogen, and phosphorus) control species growth, while top-down factors (e.g., predation and competition) control their biomass (Wehr and Descy 1998).

Phytoplankton are recognized as a major entity in global biogeochemical cycles (Myklestad 2000) and supply intermittently new, potentially labile dissolved organic carbon to aquatic systems (Sondergaard et al. 1985; Kirchman et al. 1991). Phytoplankton are also an important source of primary production and determine the potential productivity of entire aquatic food webs (Wissel and Fry 2005). Some species of phytoplankton are considered to be an important food source for pelagic and benthic species (e.g. fishes, and molluscs) (Pasquaud et al. 2010). For example, larval oysters feed on smaller phytoplankton cells (Olson and Olson 1989). Further, phytoplankton are generally considered as indicators of climate change due to their short lifespan and ability to produce resting stages (Guerrero and Rodriguez 1998; McQuoid et al. 2002). Studies have also shown that besides their role as bio-indicators of climate change, phytoplankton are also reliable indicators for assessing the pollution and eutrophication in aquatic ecosystems. For instance, spatiotemporal distribution of phytoplankton composition and biomass with emphasis on harmful algal blooms as indicators of eutrophication has been studied in the Cienfuegos Bay (Cuba) (Moreira-González et al. 2014). In another study, phytoplankton species composition was applied as a bio-diagnostic tool in relation to associated pollution in the Iyagbe lagoon (South-western Nigeria) (Onyema 2013).

Phytoplankton community represents an assemblage of heterogeneous microscopic algal forms, more or less dependent upon prevailing water currents (Kudela and Peterson 2009). Phytoplankton distribution in an estuarine lagoonal ecosystem is closely linked to several physicochemical and biological factors. Of these, salinity has been recognised as an important factor in determining community composition and their distribution (Huang et al. 2004; Lionard et al. 2005; Varona-Cordero et al. 2010; Lueangthuwapranit et al. 2011; Harris and Vinobaba 2012; Canini et al. 2013). Other environmental factors such as pH, temperature, light (influenced by turbidity) and nutrients also regulate the spatiotemporal distribution of these communities. For example in Bach Dang estuary (Vietnam) phytoplankton community structure was influenced by nutrient, turbidity, and heavy metals (Rochelle-Newall et al. 2011). In another study, temperature, salinity, silicate, and total phosphorus affected phytoplankton structure in Tagus estuary (Portugal) (Brogueira et al. 2007).

The Chilika lagoon represents a biologically diverse and ecologically unique ecosystem located along the east coast of India (19°28′–19°54′ N and 85°06′–85°35′ E). The lagoon is a shallow bar-built estuary with a surface area of 906 km^2 in summer and 1165 km^2 in monsoon (Mohanty et al. 2015). The lagoon experiences tropical monsoon-forced climate with average annual precipitation of 1238 mm (Gupta et al. 2008). Semi-diurnal tides facilitate seawater influx into the lagoon, mostly restricted to seawater inlet (Ganguly et al. 2015). Simultaneous mixing of river water and seawater makes the hydrological regime highly dynamic in lagoon. About 78% of the freshwater flows into the lagoon are from 12 major rivers located in the northern and western catchment of the lagoon (Srichandan et al. 2015a). Historically, based on the salinity gradient, the lagoon has been spatially delineated into four ecological sectors; northern sector, central sector, southern sector, and outer channel (Fig. 11.1).

Phytoplankton are frequently characterized in relation to discrete size classes–picoplankton (<2 μm), nanoplankton (2–20 μm), microplankton (20–200 μm) and macroplankton (>200 μm) (Brewin et al. 2010). Majority of studies on

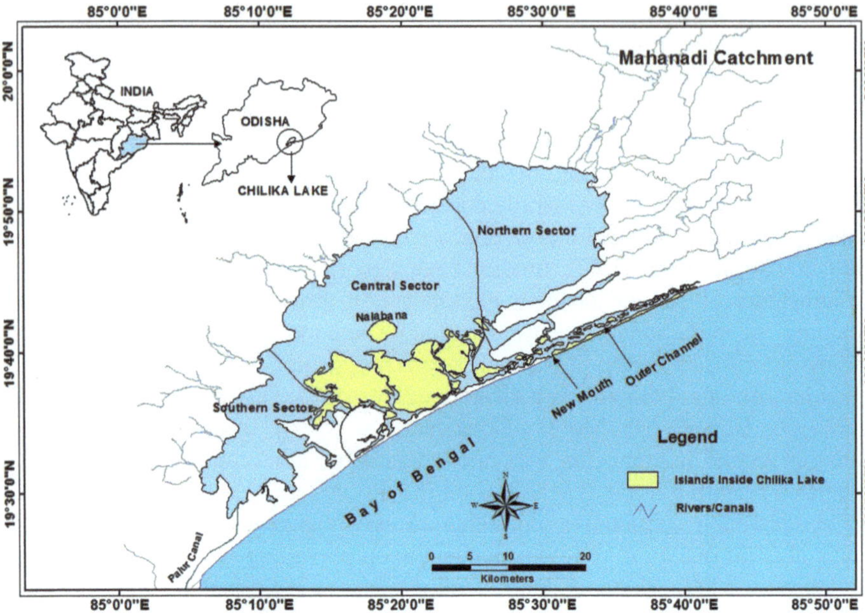

Fig. 11.1 Map of Chilika lagoon showing four sectors (northern sector, central sector, southern sector, outer channel) of the lagoon

phytoplankton diversity and distribution in Indian coastal ecosystems have relied on the classical morphological identification using light microscopy (Selvaraj et al. 2003; Srichandan et al. 2015a; b). Conventional light microscopy is only suited to discriminate large size (>10 μm) phytoplankton cells and has minimal use in assessing genetic diversity of smaller size (≤ 2 μm) phytoplankton. Although, electron microscopy generally allows assignment to taxonomic classes, but most of the picoplankton do not have enough ultra-structural features for the identification at lower taxonomic level. Recent studies on natural plankton assemblages have also employed flow cytometry and photosynthetic pigment analysis which provide information on the structure and dynamics of the phototrophic and/or autotrophic behaviour of the planktonic organism, but phylogenetic information supplied by these methods is limited (Diez et al. 2001).

Recently, advances in high-throughput DNA sequencing has allowed sequencing of hundreds of environmental samples in a very cost-effective manner, generating millions of sequence reads which can provide a realistic estimate on the true extent of the genetic diversity of picophytoplankton. For example, 454 pyrosequencing of 18S rRNA genes from the Pacific coastal waters, for the first time presented a comprehensive picture of the diversity of marine picoeukaryotes (Cheung et al. 2010). The 18S/16S rRNA genes are widely used in picophytoplankton diversity studies allowing discrimination of both heterotrophic and phototrophic picophytoplankton at different taxonomic levels (Xiao et al. 2014). In contrast, sequencing of functional genes, provide direct linkages to the essential functions in carbon biogeochemistry that picophytoplankton performs. Diversity, gene abundance, and gene expression studies based on the *rbcL* (Ribulose-1, 5-bisphosphate carboxylase oxygenase) gene, which encodes the large subunit of the CO_2-fixing enzyme Rubisco (Ribulose-1, 5-bisphosphate carboxylase oxygenase), have produced valuable insights into community composition and environmental patterns in the different aquatic ecosystem (Samanta and Bhadury 2014).

Phytoplankton communities in Chilika are mixed assemblages consisting freshwater, estuarine, and marine species (Panigrahi et al. 2009; Srichandan et al. 2015a). Most of these species have wide salinity tolerance and are eurytopic in nature (Srichandan et al. 2015b). Literature also suggests that Chilika experiences significant seasonal changes in nutrient and salinity regime during dry and wet seasons (Srichandan et al. 2015b). Excessive nutrient loading could lead to eutrophication and may promote the development of Cyanophyta blooms (Conley et al. 2009; Stal 2012). Given the ecological significance of phytoplankton, number of studies have explored the taxonomic diversity of phytoplankton communities in the Chilika lagoon (Biswas 1932; Devasundaram and Roy 1954; Patnaik 1973; Patnaik and Sarkar 1976; Raman et al. 1990; Adhikary and Sahu 1992; Rath and Adhikary 2008; Panigrahi et al. 2009; Jha et al. 2009; Mohanty and Adhikary 2013; Mukherjee et al. 2016). In addition, environmental factors (e.g. salinity, transparency, dissolved nutrients) which drive the spatiotemporal distribution of phytoplankton communities in the Chilika lagoon have been well studied (Srichandan et al. 2015a, b).

In view of the foregoing discussions on the importance of phytoplankton, the present chapter provides a detailed overview of the current knowledge on the species diversity, spatiotemporal distribution, and environmental drivers of phytoplankton communities in the Chilika lagoon. This article also highlights consideration for a future line of phytoplankton research to enable bridging the knowledge gaps, particularly related to smaller size pico and nanophytoplankton.

11.2 Floral Classification of Phytoplankton

Phytoplankton are ususally classed in two major groups, (i) non-motile, fast-growing Bacillariophyta (diatom) and (ii) motile Dinophyta (dinoflagellates), capable of vertical migration in the water column in response to photosynthetically available solar irradiance (Moreno-Díaz et al. 2015). Other groups *viz.* Cyanophyta (blue-green algae), Chlorophyta (green algae), Euglenophyta (euglenoids), Chrysophyta (silicoflagellates) and Xanthophyta (yellow-green algae) are also the members of phytoplankton communities and often predominates under certain favourable circumstances.

In total, 739 phytoplankton species represented by Bacillariophyta (270 species), Dinophyta (88 species), Cyanophyta (103 species), Chlorophyta (178 species), Euglenophyta (92 species), Chrysophyta (5 species) and Xanthophyta (3) have been documented so far from the Chilika lagoon (Tables 11.1, 11.2, 11.3, 11.4,

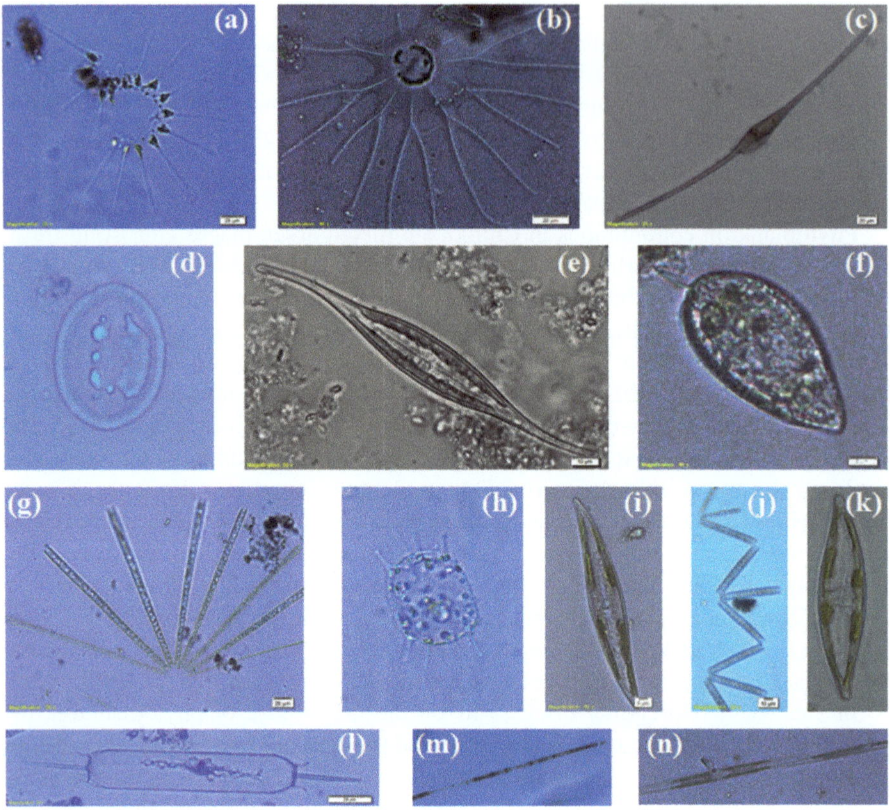

Plate 11.1 Pictures of some dominant phytoplankton species recorded in Chilika Lagoon (**a**) *Asterionellopsis glacialis*[1,2,3,4] (**b**) *Bacteriastrum* sp.[2]; (**c**) *Ceratium fusus*[6]; (**d**) *Cocconeis pediculus*[4]; (**e**) *Gyrosigma fasciola*[6]; (**f**) *Prorocentrum micans*[4,6]; (**g**) *Thalassiothrix frauenfeldii*[2]; (**h**) *Odontell mobiliensis*[1,3,4]; (**i**) *Pleurosigma normanii*[3,4,5,6]; (**j**) *Thalassionema nitzschioides*[3,6]; (**k**) *Navicula transitrans*[6]; (**l**) *Ditylum brightwelli*[2]; (**m**) *Trichodesmium erythraeum*[2,4]; (**n**) *Pseudonitzschia* sp.[5,6]. [1]Devasundaram and Roy (1954), [2]Patnaik (1973), [3]Rath and Adhikary (2008), [4]Panigrahi et al. (2009), [5]Srichandan et al. (2015a), [6]Srichandan et al. (2015b)

11.5, 11.6, and 11.7). The photomicrographs of some dominant phytoplankton species recorded in Chilika lagoon are depicted in Plate 11.1. However, it should be noted that different studies have used various techniques for collection, preservation, concentration, enumeration, and taxonomic identification of phytoplankton communities. Due to this reason, the data would not be directly comparable across different studies. For example, some studies have used plankton nets of 10–20 μm (Rath and Adhikary 2008; Srichandan et al. 2015a, b) and 45 μm (Mohanty and Adhikary 2013) mesh sizes, while others have used a gravity sedimentation method without plankton net (Panigrahi et al. 2009). The studies also differ in sampling frequency and sample size (in terms of the amount of water collected and number of sampling sites). Most of the earlier studies have employed seasonal sampling in the sense that water samples were collected only once during a given season. In recent studies, comprehensive monitoring of phytoplankton communities was carried out on a monthly basis over the period of 3 years (2011–2014) covering sampling locations which spanned all four sectors of the lagoon (Srichandan et al. 2015a, b). Such studies have highlighted the importance of conducting long-term systemic monitoring of phytoplankton communities and provided a detailed understanding of their variability caused either by the intrinsic environmental forces or by the extreme events such as a cyclone.

11.2.1 Bacillariophyta

Bacillariophyta represents unicellular and uni-nucleate algae with a size range of about 15 mm–400 mm in maximum dimension, although some smaller and a few considerably larger forms exist in the aquatic ecosystem. Bacillariophyta can be used as a suitable bioindicator for water quality assessments due to their short generation time and sensitivity to subtle environmental changes (Stevenson and Pan 1999; Goma et al. 2005). Bacillariophyta have been reported to constitute the bulk of the phytoplankton assemblages in many estuarine ecosystems. For instance, in Tagus (Portugal), Mahanadi (India), Batticaloa (Sri Lanka), and Bach Dang (Vietnam) estuaries, phytoplankton communities were dominated by Bacillariophyta (Cabeçadas 1999; Naik et al. 2009; Harris and Vinobaba 2012; Chu et al. 2014). In Chilika lagoon, Bacillariophyta has also been reported to dominate the phytoplankton community due to their eurythermal and euryhaline adaptations (Srichandan et al. 2015a). Literature also suggests that Bacillariophyta can tolerate a wide range of fluctuation in salinity and temperature (Sasamal et al. 2005). For instance, Aquino et al. (2015), while studying seasonal and spatial variation in phytoplankton community structure of Passos River estuary in Brazil remarked that Bacillariophyta are spatially affected by salinity and occurs in most estuaries in the world. However, nutrient availability and their stoichiometry regulates bacillariophytic metabolism and often results in a change in species composition in response to changing water quality (Lie et al. 2011). Molar ratios of available macronutrient concentration as dissolved silicate (16): dissolved inorganic nitrogen (16): dissolved inorganic phosphorus (1) is required for optimum growth of Bacillariophyta (Redfield et al. 1963). Apart from the nutrient availability in estuarine ecosystems, the growth and distribution of Bacillariophyta is also governed by water transparency that determines the availability of light in the water column (Resende et al. 2005; Masuda et al. 2011).

In Chilika lagoon, 270 Bacillariophyta species belonging to 95 genera have been reported (Table 11.1). Devasundaram and Roy (1954) investigated plankton community assemblages particularly in Balugaon, Kalupadaghat, Rambha, Satpara and Arkhakuda regions of Chilika lagoon and documented 31 species of Bacillariophyta, mostly marine in nature. Later, Patnaik (1973) identified 40 taxa of Bacillariophyta of which *Chaetoceros* sp., *Coscinodiscus* sp., *Asterionellopsis glacialis*, *Rhizosolenia* sp., *Bacteriastrum hyalinum*, *Grammatophora* sp. and *Nitzschia*

Table 11.1 List of Bacillariophyta species from Chilika

Phylum	Bacillariophyta
Class	Coscinodiscophyceae
Order	Coscinodiscales
Family	Heliopeltaceae
	Actinoptychus sp. Ehrenberg 1843[11, *, **]
Family	Coscinodiscaceae
	Coscinodiscopsis jonesiana (Greville) E.A.Sar & I.Sunesen in Sar, Sunesen & Hinz 2008[11,*,**], *Coscinodiscus centralis* Ehrenberg 1839[1,6,10,11,*], *Coscinodiscus curvatulus* Grunow in A.Schmidt 1878[10,*,**], *Coscinodiscus granii* L.F.Gough 1905[2], *Coscinodiscus marginatus* Ehrenberg 1843[6,7,9,*,**], *Coscinodiscus radiatus* Ehrenberg 1840[10,*,**], *Coscinodiscus gigas* Ehrenberg 1841[6,7,10,11,*,**], *Coscinodiscus* sp. Ehrenberg 1839[2,3,6,7,10,11,*], *Palmerina hardmaniana* (Greville) G.R.Hasle 1996[11,*,**]
Family	Hemidiscaceae
	Azpeitia neocrenulata (S.L.VanLandingham) G.Fryxell & T.P.Watkins[8,*,**], *Hemidiscus cuneiformis* Wallich 1860[11,*,**], *Hemidiscus kanayanus* Simonsen[8,*,**], *Hemidiscus* sp. Wallich 1860[10,11,*,**]
Order	Asterolamprales
Family	Asterolampraceae
	Asteromphalus flabellatus (Brébisson) Greville 1859[10,*,**], *Asteromphalus hookeri* Ehrenberg 1844[10,*,**], *Asteromphalus wyvillii* F.S.Castracane degli Antelminelli[10,*,**], *Asteromphalus* sp. Ehrenberg 1844[10,11,*,**]
Order	Aulacoseirales
Family	Aulacoseiraceae
	Aulacoseira granulata (Ehrenberg) Simonsen 1979[8,*,**], *Aulacoseira granulata var. angustissima* (Otto Müller) Simonsen 1979[8,*,**], *Aulacoseira italica* (Ehrenberg) Simonsen 1979[8,11,*,**], *Aulacoseira* sp. Thwaites, 1848[11,*,**]
Order	*Cocconeidales*
Family	Cocconeidaceae
	Cocconeis pediculus Ehrenberg 1838[6,7,9,11,*,**], *Cocconeis placentula* Ehrenberg 1838[7,8,10,11,*,**], *Cocconeis* sp. Ehrenberg, 1836[1,10,11,*]
Order	Corethrales
Family	*Corethraceae*
	Corethron hystrix Hensen 1887[2,10,*], *Corethron* sp. Castracane 1886[11,*,**]
Order	*Rhizosoleniales*
Family	Rhizosoleniaceae
	Dactyliosolen fragilissimus (Bergon) Hasle 1996[2,10,*], *Guinardia delicatula* (Cleve) Hasle 1997[1,2,7,8,*], *Guinardia striata* (Stolterfoth) Hasle 1996[1,2,8,10,11,*] *Guinardia flaccida* (Castracane) H.Peragallo 1892[6,8,10,*,**], *Pseudosolenia calcar-avis* (Schultze) B.G.Sundström 1986[1], *Rhizosolenia bergonii* H.Peragallo 1892[1], *Neocalyptrella robusta* (G.Norman ex Ralfs) Hernández-Becerril & Meave del Castillo 1997[1,10,11,*], *Rhizosolenia castracanei* H.Peragallo 1888[10,*,**], *Rhizosolenia crassispina* J.L.B. Schröder 1906[11,*,**], *Rhizosolenia imbricata* Brightwell 1858[1,10,11,*], *Rhizosolenia setigera* Brightwell 1858[1,2,6,8,10,11,*], *Rhizosolenia setigera f. pungens* (A.Cleve) Brunel 1962[10,*,**], *Rhizosolenia styliformis* T.Brightwell 1858[1,8,10,11,*], *Rhizosolenia* sp. Brightwell 1858[3,4,10,11,*]

(continued)

Table 11.1 (continued)

Family	Probosciaceae
	Proboscia alata (Brightwell) Sundström 1986[1,2,8,10,11,*]
Order	Melosirales
Family	Melosiraceae
	Melosira borreri Greville 18,336,[10,11,*,**], *Melosira decussata* (Ehrenberg) Kützing[9,*,**], *Melosira sulcata* (Ehrenberg) Kützing 1844[10,11,*,**], *Melosira* sp. C. Agardh 1824[10,11,*,**]
Family	Paraliaceae
	Paralia sulcata (Ehrenberg) Cleve 1873[10,11,*,**], *Paralia* sp. Heiberg 1863[11,*,**]
Order	*Stephanopyxales*
Family	*Stephanopyxidaceae*
	Stephanopyxis turris (Greville) Ralfs in Pritchard 1861[1,6,10,11,*]
Order	*Triceratiales*
Family	*Triceratiaceae*
	Triceratium sp. Ehrenberg 1839[4,10,11,*]
Class	Bacillariophyceae
Order	Cymbellales
Family	Anomoeoneidaceae
	Adlafia minuscula (Grunow) Lange-Bertalot in Lange-Bertalot & Genkal 1999[6,8,*,**]
Order	Rhaphoneidales
Family	Rhaphoneidaceae
	Adoneis pacifica G.W.Andrews & P.Rivera 1987[10,*,**], *Rhaphoneis amphiceros* (Ehrenberg) Ehrenberg 1844[10,*,**], *Asterionellopsis glacialis* (Castracane) Round 1990[1,2,3,6,7,10,11,*]
Order	Naviculales
Family	Amphipleuraceae
	Amphiprora gigantea Grunow 1860[6,*,**], *Amphiprora obtusa* W.Gregory[10, *,**], *Amphiprora* sp. Ehrenberg 1843[11,*,**]
Family	Diploneidaceae
	Diploneis elliptica (Kützing) Cleve 1894[8,10,11,*,**], *Diploneis oblongella* (Nägeli ex Kützing) Cleve-Euler 1922[8,10,11,*,**], *Diploneis puella* (Schumann) Cleve 1894[8,10,11,*,**], *Diploneis smithii* (Brébisson) Cleve 1894[10,11,*,**], *Diploneis robustus* R.Subrahmanyan[11,*,**], *Diploneis weissflogii* (A.W.F.Schmidt) Cleve 1894[10,11,*,**], *Diploneis* sp. Ehrenberg ex Cleve 1894[10,11,*,**]
Family	*Plagiotropidaceae*
	Ephemera sp. Paddock 1988[11,*,**], *Meuniera membranacea* (Cleve) P.C.Silva 1996[10,11,*,**], *Plagiotropis* sp. Pfitzer 1871[10,11,*,**], *Manguinea rigida* (M.Peragallo) Paddock 1988[11,*,**]
Family	Stauroneidaceae
	Craticula cuspidata (Kutzing) D.G.Mann 1990[6,*,**], *Stauroneis pusilla* Ehrenberg[6,*,**]
Family	Sellaphoraceae
	Fallacia pygmaea (Kützing) Stickle & D.G.Mann 1990[7,*,**]
Family	Naviculaceae
	Gyrosigma acuminatum (Kützing) Rabenhorst 1853[6,8,10,*,**], *Gyrosigma balticum* (Ehrenberg) Rabenhorst 1853[10,11,*,**], *Gyrosigma fasciola* (Ehrenberg) J.W.Griffith &

(continued)

Table 11.1 (continued)

	Henfrey 1856[11,*,**], *Gyrosigma* sp. Hassall 1845[10,11,*,**], *Navicula cryptocephala* Kützing 1844[8,*,**], *Navicula distans* (W.Smith) Ralfs 1861[11,*,**], *Navicula lanceolata* Ehrenberg 1838[6,8,10,*,**], *Navicula protracta* (Grunow) Cleve 1894[6,*,**], *Navicula rhynchocephala* Kützing 1844[8,*,**], *Navicula salinarum* Grunow 1880[6,*,**], *Navicula transitans* Cleve 1883[10,11,*,**], *Navicula veneta* Kützing 1844[8,*,**], *Navicula* sp. Bory 1822[6,10,11,*,**]
Family	Neidiaceae
	Neidium affine var. *amphirhynchus* (Ehrenberg) Cleve 1894[9,*,**]
Family	Pinnulariaceae
	Pinnularia alpina W.Smith 1853[6,7,10,*,**], *Pinnularia major* (Kützing) Rabenhorst 1853[9,*,**], *Pinnularia subsimilis* H.P.Gandhi[9,*,**], *Pinnularia nobilis* (Ehrenberg) Ehrenberg 1843[6,*,**], *Pinnularia nodosa* (Ehrenberg) W.Smith 1856[9,*,**], *Pinnularia viridis* (Nitzsch) Ehrenberg 1843[8,*,**], *Pinnularia* sp. Ehrenberg 1843[10,11,*,**]
Family	*Pleurosigmataceae*
	Pleurosigma angulatum (J.T.Quekett) W.Smith 1852[11,*,**], *Pleurosigma directum* Grunow 1880[10,11,*,**], *Pleurosigma elongat*um W.Smith 1852[2,3,5,7,10,11,*], *Pleurosigma naviculaceum* Brébisson 1854[9,*,**], *Pleurosigma normanii* Ralfs 1861[4,6,7,8,9,10,11,*], *Pleurosigma* sp. W.Smith 1852[2,3,10,11,*]
Order	Thalassiophysales
Family	Catenulaceae
	Amphora lineolata Ehrenberg 1838[10,*,**], *Amphora ostrearia* Brébisson ex Kützing 1849[10,11,*,**], *Amphora ovalis* (Kützing) Kützing 1844[6,10,*,**], *Amphora pediculus* (Kützing) Grunow ex A.Schmidt 1875[8,*,**], *Amphora* sp. Ehrenberg ex Kützing 1844[3,11,*]
Order	Tabellariales
Family	Tabellariaceae
	Asterionella formosa Hassall 1850[8,11,*,**], *Asterionella* sp. Hassall 1850[3,11,*], *Diatoma elongata* (Lyngbye) C.Agardh 1824[6,*,**], *Diatoma vulgaris* Bory 1824[8,*,**], *Diatoma* sp. Bory 1824[11,*,**], *Tabellaria fenestrat*a (Lyngbye) Kützing 1844[5,6,10,11,*], *Tabellaria flocculosa* (Roth) Kützing 1844[9,*,**], *Tabellaria* sp. Ehrenberg ex Kützing 1844[10,11,*,**]
Order	Bacillariales
Family	Bacillariaceae
	Bacillaria paxillifera (O.F.Müller) T.Marsson 1901[2,3,6,7,10,11,*], *Bacillaria* sp. J.F. Gmelin 1791[1,10,*], *Fragilariopsis oceanica* (Cleve) Hasle 1965[10,*,**] *Cylindrotheca closterium* (Ehrenberg) Reimann & J.C.Lewin 1964[2,4,6,7,8,10,11,*], *Hantzschia amphioxys* (Ehrenberg) Grunow 1880[9,*,**], *Nitzschia acicularis* (Kützing) W.Smith 1853[8,10,*,**], *Nitzschia clausii* Hantzsch 1860[8,10,*,**], *Nitzschia fibula-fissa* Lange-Bertalot 1980[8,*,**], *Nitzschia intermedia* Hantzsch 1880[8,*,**], *Nitzschia longissima* (Brébisson) Ralfs 1861[1,2,3,8,10,11,*], *Nitzschia obtusa* W.Smith 1853[6,7,11,*,**], *Nitzschia pacifica* Cupp 1943[10,*,**], *Nitzschia pungens* Grunow ex Cleve 1897[1,2,10,*], *Nitzschia recta* Hantzsch ex Rabenhorst 1862[8,*,**], *Nitzschia sigma* (Kützing) W.Smith 1853[6,10,11,*,**], *Nitzschia* sp. Hassall 1845[3,10,11,*], *Pseudo-nitzschia pungens* (Grunow ex Cleve) Hasle 1993[10,11,*,**], *Pseudo-nitzschia seriata* (Cleve) H.Peragallo 1899[2,7,10,*], *Pseudo-nitzschia seriata f. obtusa* (Hasle) Hasle 1993[8,*,**], *Pseudo-nitzschia* sp. H.Peragallo 1900[10,11,*,**], *Psammodictyon panduriforme* (W.Gregory) D. G.Mann 1990[6,10,*,**], *Tryblionella acuta* (Cleve) D.G.Mann 1990[9,*,**]

(continued)

Table 11.1 (continued)

Order	Surirellales
Family	Surirellaceae
	Campylodiscus clypeus (Ehrenberg) Ehrenberg ex Kützing 1844[8,*,**], *Campylodiscus horologium* W.C.Williamson 1848[2], *Campylodiscus* sp. Ehrenberg ex Kützing 1844[2,3,10,11,*], *Iconella tenera* (W.Gregory) Ruck & Nakov 2016[2], *Surirella elegans* Ehrenberg 1843[2,10,*], *Surirella birostrata* Hustedt ex Ant.Mayer 1917[8,*,**], *Surirella brebissonii* Krammer & Lange-Bertalot 1987[8,10,*,**], *Surirella eximia* Greville 1857[10,*,**], *Surirella fastuosa* (Ehrenberg) Ehrenberg 1843[10,11,*,**], *Surirella fluminensis* Grunow 1862[10,*,**], *Surirella minuta* Brébisson ex Kützing 1849[8,10,*,**], *Surirella robusta* Ehrenberg 1841[10,*,**], *Surirella* sp. Turpin 1828[10,11,*,**]
Family	Entomoneidaceae
	Entomoneis alata (Ehrenberg) Ehrenberg 1845[8,*,**], *Entomoneis paludosa* (W.Smith) Reimer 1975[10,*,**]
Order	Cymbellales
Family	Cymbellaceae
	Cymbella affinis Kützing 1844[9,*,**], *Cymbella aspera* (Ehrenberg) Cleve 1894[8,*,**], *Cymbella cistula* (Ehrenberg) O.Kirchner 1878[8,*,**], *Cymbella tumida* (Brébisson) Van Heurck 1880[8,10,*,**], *Cymbella* sp. C.Agardh 1830[3,6,10,11,*]
Order	Fragilariales
Family	Fragilariaceae
	Fragilaria acus (Kützing) Lange-Bertalot 2000[8,10,*,**], *Fragilaria capucina* Desmazières 1830[8,*,**], *Fragilaria crotonensis* Kitton 1869[6,7,8,9,*,**], *Fragilaria radians* (Kützing) D.M.Williams & Round 1987[9,*,**], *Fragilaria* sp. Lyngbye, 1819[6,10,11,*,**], *Synedra crystallina* (C.Agardh) Kützing 1844[9,*,**], *Synedra* sp. Ehrenberg 1830[3,10,11,*]
Order	Cymbellales
Family	Gomphonemataceae
	Gomphonema constrictum var. *capitatum* (Ehrenberg) Grunow 1880[8,*,**], *Gomphonema grunowii* R.M.Patrick & Reimer 1975[8,*,**], *Gomphonema intricatum* Kützing 1844[8,*,**], *Gomphonema micropus* Kützing 1844[9,*,**], *Gomphonema olivaceum* (Hornemann) Brébisson 1838[8,*,**], *Gomphonema truncatum* Ehrenberg 1832[8,*,**], *Gomphonema* sp. Ehrenberg 1832[10,11,*,**]
Order	Rhabdonematales
Family	Grammatophoraceae
	Grammatophora undulata Ehrenberg[3,6,*], *Grammatophora* sp. Ehrenberg 1840[2,3]
Order	Licmophorales
Family	Licmophoraceae
	Licmophora abbreviata C.Agardh 1831[6,10,11,*,**], *Licmophora* sp. C.Agardh 1827[10,11,*,**]
Family	Ulnariaceae
	Tabularia fasciculata (C.Agardh) D.M.Williams & Round 1986[4,9,10,*], *Ulnaria ulna* (Nitzsch) Compère 2001[3,6,7,8,9,10,*]
Order	Thalassionematales
Family	Thalassionemataceae
	Lioloma pacificum (Cupp) Hasle 1996[10*,**], *Thalassionema frauenfeldii* (Grunow) Tempère & Peragallo 1910[1,2,3,8,10,11,*], *Thalassionema nitzschioides* (Grunow) Mereschkowsky 1902[1,2,6,8,10,11,*], *Thalassionema* sp. Grunow ex Mereschkowsky

(continued)

Table 11.1 (continued)

	1902[3,11,*], *Thalassiothrix longissima* Cleve & Grunow 1880[1,10,11,*], *Thalassiothrix* sp. Cleve & Grunow 1880[3,10,11,*]
Order	Lyrellales
Family	Lyrellaceae
	Lyrella clavata (Gregory) D.G.Mann 1990[10,11,*,**], *Lyrella hennedyi* (W.Smith) Stickle & D.G.Mann 1990[10,11,*,**]
Order	Mastogloiales
Family	Mastogloiaceae
	Mastogloia elliptica (C.Agardh) Cleve 1893[9,*,**], *Mastogloia exigua* F.W.Lewis 1861[4], *Mastogloia exilis* Hustedt 1933[10,*,**], *Mastogloia* sp. Thwaites ex W.Smith 1856[11,*,**]
Order	Rhopalodiales
Family	Rhopalodiaceae
	Rhopalodia gibberula (Ehrenberg) Otto Müller 1895[9,*,**]
Class	Mediophyceae
Order	Toxariales
Family	Ardissoneaceae
	Ardissonea formosa (Hantzsch) Grunow[11,*,**]
Family	Climacospheniaceae
	Climacosphenia moniligera Ehrenberg 1843[6,*,**]
Family	Toxariaceae
	Toxarium undulatum J.W.Bailey 1854[10,*,**]
Order	Eupodiscales
Family	Eupodiscaceae
	Auliscus sculptus (W.Smith) Brightwell 1860[6,10,*,**], *Odontella aurita* (Lyngbye) C. Agardh 1832[10,11,*,**], *Odontella granulata* (Roper) R.Ross 1986[7,*,**], *Odontella litigiosa* (Van Heurck) Hoban 1980[9,*,**], *Odontella longicruris* (Greville) M.A.Hoban 1983[10,11,*,**], *Odontella* sp. C.Agardh 1832[3,6,10,11,*], *Odontella sinensis* (Greville) Grunow 1884[1,2,7,8,10,11,*]
Order	Chaetocerotales
Family	Chaetocerotaceae
	Bacteriastrum comosum Pavillard 1916[8,11,*,**], *Bacteriastrum delicatulum* Cleve 1897[8,11,*,**], *Bacteriastrum furcatum* Shadbolt 1854[6,8,11,*,**], *Bacteriastrum hyalinum* Lauder 1864[1,2,6,8,11,*], *Bacteriastrum varians* Lauder 1864[10,11,*,**], *Bacteriastrum* sp. Shadbolt 1854[3,10,11,*]
Order	Biddulphiales
Family	Biddulphiaceae
	Biddulphia biddulphiana (J.E.Smith) Boyer 1900[3,5], *Eucampia cornuta* (Cleve) Grunow 1883[8,*,**], *Eucampia zodiacus* Ehrenberg 1839[8,10,11,*,**], *Eucampia* sp. Ehrenberg 1839[3]
Order	Hemiaulales
Family	Hemiaulaceae
	Cerataulina pelagica (Cleve) Hendey 1937[10,11,*,**], *Hemiaulus hauckii* Grunow ex Van Heurck 1882[8,10,11,*,**], *Hemiaulus sinensis* Greville 1865[8,10,11,*,**], *Hemiaulus* sp. Heiberg 1863[3,11,*]

(continued)

Table 11.1 (continued)

Order	Eupodiscales
Family	Eupodiscaceae
	Cerataulus heteroceros (Grunow) P.A.Sims & J.Witkowski 2012[4,6,7,10,11,*], *Trieres mobiliensis* (J.W.Bailey) Ashworth & Theriot 2013[1,6,7,10,11,*]
Order	Chaetocerotales
Family	Chaetocerotaceae
	Chaetoceros aequatorialis Cleve 1901[11,*,**], *Chaetoceros affinis* Lauder 1864[1,2,3,5,6,8,11,*], *Chaetoceros atlanticus* Cleve 1873[8,*,**], *Chaetoceros borealis* Bailey 1854[8,*,**], *Chaetoceros brevis* F.Schütt 1895[3], *Chaetoceros coarctatus* Lauder 1864[2], *Chaetoceros compressus* Lauder 1864[1,2,3,7,8,10,11,*], *Chaetoceros curvisetus* Cleve 1889[1,2,6,10,11,*], *Chaetoceros decipiens* Cleve 1873[9,10,11,*,**], *Chaetoceros densus* (Cleve) Cleve 1899[7,*,**], *Chaetoceros diadema* (Ehrenberg) Gran 1897[8,11,*,**], *Chaetoceros didymus* Ehrenberg 1845[7,10,11,*,**], *Chaetoceros diversus* Cleve 1873[2,6,7,10,*], *Chaetoceros eibenii* Grunow1882[1,6,*], *Chaetoceros laciniosus* F.Schütt 1895[2], *Chaetoceros laevis* Leuduger-Fortmorel 1892[1,8,11,*], *Chaetoceros lauderi* Ralfs ex Lauder 1864[7,*,**], *Chaetoceros lorenzianus* Grunow 1863[1,2,6,7,8,10,11,*], *Chaetoceros messanensis* Castracane 1875[11,*,**], *Chaetoceros mitra* (Bailey) Cleve 1896[8,*,**], *Chaetoceros paradoxus* Cleve 1873[6,7,*,**], *Chaetoceros pendulus* Karsten 1905[2], *Chaetoceros perpusillus* Cleve 1897[2], *Chaetoceros peruvianus* Brightwell 1856[1,2,7,8,10,11,*], *Chaetoceros protuberans* Lauder 1864[8,*,**], *Chaetoceros seriacanthus* Gran 1897[8,*,**], *Chaetoceros subtilis* Cleve 1896[1,11,*], *Chaetoceros teres* Cleve 1896[8,*,**], *Chaetoceros wighamii* Brightwell 1856[8,*,**], *Chaetoceros* sp. Ehrenberg 1844[3,4,6,10,11,*]
Family	Leptocylindraceae
	Leptocylindrus danicus Cleve 1889[6,10,11,*,**], *Leptocylindrus minimus* Gran 1915[10,11,*,**], *Leptocylindrus* sp. Cleve 1889[2,10,*]
Order	Stephanodiscales
Family	Stephanodiscaceae
	Cyclotella maxima Kützing 1844[9,*,**], *Cyclotella meneghiniana* Kützing 1844[9,10,*,**], *Cyclotella striata* (Kützing) Grunow 1880[8,11,*,**], *Cyclotella* sp. (Kützing) Brébisson 1838[10,11,*,**], *Pantocsekiella kuetzingiana* (Thwaites) K.T.Kiss & E.Ács 2016[8,*,**], *Stephanodiscus* sp. Ehrenberg 1845[4]
Order	Lithodesmiales
Family	Lithodesmiaceae
	Ditylum brightwellii (T.West) Grunow 1885[2,6,10,11,*], *Ditylum sol* (A.Schmidt) Cleve 1901[1,10,11,*], *Ditylum* sp. J.W.Bailey ex L.W.Bailey 1861[3], *Lithodesmium undulatum* Ehrenberg 1839[10,11,*,**], *Lithodesmium* sp. Ehrenberg 1839[11,*,**]
Order	Briggerales
Family	Streptothecaceae
	Helicotheca sp. M.Ricard 1987[11,*,**], *Streptotheca* sp. Shrubsole 1890[11,*,**]
Order	Thalassiosirales
Family	Lauderiaceae
	Lauderia annulata Cleve 1873[2,6,7,10,11,*]
Family	Thalassiosiraceae
	Minidiscus sp. Hasle 1973[11,*,**], *Planktoniella* sp. F.Schütt 1892[2], *Thalassiosira eccentrica* (Ehrenberg) Cleve 1904[8,10,11,*,**], *Thalassiosira gravida* Cleve 1896[11,*,**], *Thalassiosira subtilis* (Ostenfeld) Gran 1900[5,6,10,*], *Thalassiosira* sp. Cleve 1873[3,11,*]

(continued)

Table 11.1 (continued)

Family	Skeletonemataceae
	Skeletonema costatum (Greville) Cleve 1873[2,5,6,8,*], *Skeletonema subsalsum* (Cleve-Euler) Bethge 1928[8,*,**], *Skeletonema* sp. Greville 1865[11,*,**]
Class	Bacillariophyta incertae sedis
Order	Bacillariophyta incertae sedis
Family	Bacillariophyta incertae sedis
	Mediopyxis helysia Kühn, Hargreaves & Halliger 2006[10,11,*,**], *Mediopyxis* sp. Medlin & Kühn 2006[10,11,*,**]

[1]Devasundaram and Roy (1954), [2]Patnaik (1973), [3]Patnaik and Sarkar (1976), [4]Raman et al. (1990), [5]Adhikary and Sahu (1992), [6]Rath and Adhikary (2008), [7]Panigrahi et al. (2009), [8]Jha et al. (2009), [9]Mohanty and Adhikary (2013), [10]Srichandan et al. (2015a), [11]Srichandan et al. (2015b)

*Inventorized during survey in post-restoration period (2000–2014)

**New records during survey in post-restoration period (reported for the first time from the lagoon)

sp. were found to be abundant across the outer channel of Chilika lagoon. The Bacillariophyta such as *Chaetoceros perpusillus*, *Chaetoceros peruvianus*, *Cheatoeceros lorenzianus*, *Asterionellopsis glacialis*, *Thalassionema frauenfeldii*, and *Ditylum brightwellii* were mostly represented in the central sector while *Chaetoceros affinis*, *Chaetoceros pendulus*, and *Chaetoceros* sp. were dominant in the northern sector. Patnaik and Sarkar (1976) documented 29 species of Bacillariophyta in Chilika lagoon out of which 18 species were new records. Raman et al. (1990) have mentioned the name of only 9 species (*Cerataulus heteroceros*, *Chaetoceros* sp., *Cylindrotheca closterium*, *Mastogloia exigua*, *Pleurosigma normanii*, *Rhizosolenia* sp., *Stephanodiscus* sp., *Synedra affinis*, and *Triceratium* sp.) of Bacillariophyta in their publication. Subsequently, Adhikary and Sahu (1992) did the sampling from stations spanning the entire lagoon and reported an additional two species (*Tabellaria fenestrata* and *Thalassiosira subtilis*) of Bacillariophyta. A study undertaken by Rath and Adhikary (2008) documented 57 species of Bacillariophyta in Chilika lagoon. Later, Panigrahi et al. (2009) recorded a total of 30 Bacillariophyta species during sampling between the years 2001–2003. The species such as *Chaetoceros paradoxus*, *Coscinodiscus gigas*, *C. marginatus*, *Coscinodiscus* sp., *Lauderia annulata*, and *Cylindrotheca closterium* were dominant in outer channel. Subsequently, the study of phytoplankton carried out in Chilika lagoon over the period 2003–2006 by Jha et al. (2009) led to documentation of 80 species of Bacillariophyta. Mohanty and Adhikary (2013) carried out an investigation on changes in the algal diversity subsequent to the opening of a new seawater inlet in the lagoon and recorded 20 more species (*Mastogloia elliptica*, *Odontella litigiosa*, *Chaetoceros decipiens*, *Cyclotella maxima*, *Cyclotella meneghiniana*, *Cymbella affinis*, *Rhopalodia gibberula*, *Gomphonema micropus*, *Hantzschia amphioxys*, *Melosira decussata*, *Neidium affine var. amphirhynchus*, *Pinnularia major*, *Tryblionella acuta*, *Pinnularia subsimilis*, *Pinnularia nodosa*, *Pleurosigma naviculaceum*, *Synedra crystallina*, *Fragilaria radians*, *Tabularia fasciculata* and *Tabellaria flocculosa*) to the existing Bacillariophyta species inventory of Chilika lagoon.

Recently, based on monthly sampling between year 2011–2012 from 13 stations of Chilika lagoon, 138 Bacillariophyta species belonging to 54 genera have been reported (Srichandan et al. 2015a). Out of these 138 Bacillariophyta, 53 species were new reports from Chilika lagoon. Among the encountered genera, *Chaetoceros* and *Surirella* have been found to be represented by the highest number of species (8 species). Bacillariophyta such as *Pleurosigma* sp., *P. normanii*, *Synedra* sp., *Thalassionema nitzschioides*, *Surirella* sp., *Chaetoceros* sp., *Coscinodiscus* sp., *Lithodesmium undulatum*, *Hemiaulas sinensis*, and *Paralia sulcata* have been found to be dominant throughout Chilika lagoon. The diversity of Bacillariophyta was higher during pre-monsoon season which has a higher salinity. In a recent study, 136 Bacillariophyta species have been registered in Chilika lagoon (Srichandan et al. 2015b). *Synedra* sp., *Nitzschia* sp., *Diploneis weissflogii*, *Surirella* sp., *Navicula* sp., *Pseudonitzschia* sp., *Thalassiosira* sp., *Coscinodiscus* sp., *Chaetoceros* sp. and *Cyclotella* sp. were dominant species and present in a wide range of salinity (oligohaline, mesohaline, and polyhaline). Srichandan et al. (2015b) have also reported the dominance of centric Bacillariophyta over pennate as also reported from other estuarine habitats, globally (Patil and Anil 2008; Canini et al. 2013). Thus, the post-restoration status of Bacillariophyta in Chilika lagoon stood at 252 species and a total of 188 Bacillariophyta were documented as new reports.

11.2.2 Dinophyta

Dinophyta are common to abundant in fresh, marine, and estuarine environments. In general, Dinophyta occupy the second position next to Bacillariophyta in the aquatic environment (Sahu et al. 2014). Dinophyta belongs to the diverse group of unicellular eukaryotes (Leander and Keeling 2004). Some Dinophyta are autotrophs and heterotrophs (Glibert and Legrand 2006; Gaines and Elbrächter 1987) while some lead mixotrophic mode of nutrition (Burkholder et al. 2008). Further, the heterotrophic and mixotrophic Dinophyta are able to feed on diverse prey items (e.g. bacteria, picoeukaryotes, nanoflagellates, bacillariophyta, other dinophyta, heterotrophic protists, and metazoans) due to their diverse feeding mechanisms (Jeong et al. 2010). In turn they are ingested by several kinds of predators. Thus, the role of the Dinophyta in food chain and food webs are very diverse. The variation in abundance, diversity, and composition of Dinophyta depend on changes in salinity, pH, nitrogen and phosphate (Yoo 1991; Cremer et al. 2007). For instance, a study from Chapora estuary (India) has reported that the major influential agents for Dinophyta distribution were temperature and salinity (Alkawri and Ramaiah 2010). In another study, salinity, nutrient, temperature and pH were the determining factors for the growth of Dinophyta in Santo Andre lagoon (Portugal) (Macedo et al. 2001).

In Chilika lagoon, Devasundaram and Roy (1954) documented 6 species of Dinophyta: *Dinophysis caudata*, *D. miles*, *Peridinium* sp., *Tripos furca*, *Ceratium trichoceros*, and *C. breve* (Table 11.2). Patnaik (1973) investigated seasonal fluctuations of plankton in the Chilika lagoon and recorded 7 species (*Tripos furca*, *Ceratium tripos*, *Tripos fusus*, *Tripos longipes*, *Noctiluca scintillans*, *Dinophysis*

Table 11.2 List of Dinophyta species from Chilika

Phylum	Miozoa (=Dinophyta)
Class	Dinophyceae
Order	Gymnodiniales
Family	Gymnodiniaceae
	Akashiwo sanguinea (K.Hirasaka) G.Hansen & Ø.Moestrup 2000[11,*,**], *Gymnodinium catenatum* H.W.Graham 1943[8,9,*,**], *Gymnodinium heterostriatum* Kofoid & Swezy 1921[6],[*,**], *Gymnodinium* sp. F.Stein 1878[10,11,*,**], *Gyrodinium* sp. Kofoid & Swezy 1921[10,*,**], *Polykrikos kofoidii* Chatton 1914[8,9,*,**]
Order	Gonyaulacales
Family	Goniodomataceae
	Alexandrium minutum Halim 1960[8,9,11,*,**], *Alexandrium monilatum* (J.F.Howell) Balech 1995[8,9,10,11,*,**], *Alexandrium* sp. Halim 19601[1,*,**], *Alexandrium ostenfeldii* (Paulsen) Balech & Tangen 1985[9,*,**], *Alexandrium tamarense* (Lebour) Balech 1995[8,9,*,**]
Family	Ceratiaceae
	Ceratium breve (Ostenfeld & Schmidt) Schröder 1906[1,9,*], *Ceratium gibberum* Gourret 1883[8,*,**], *Ceratium trichoceros* (Ehrenberg) Kofoid 1881[1,9,10,11,*], *Ceratium tripos* (O.F.Müller) Nitzsch 1817[2,3,6,7,8,9,10,11,*], *Ceratium symmetricum* Pavillard 1905[10,*,**], *Ceratium tripos var. atlanticum* Ostenfeld 1903[9,*,**], *Ceratium* sp. F.Schrank 1793[3,11,*], *Tripos brevis* (Ostenfeld & Johannes Schmidt) F.Gómez 2013[9,*,**], *Tripos contortus* (Gourret) F.Gómez 2013[8,11,*,**], *Tripos dens* (Ostenfeld & Johannes Schmidt) F.Gómez 2013[9,*,**], *Tripos extensus* (Gourret) F.Gómez 2013[10,*,**], *Tripos falcatus* (Kofoid) F.Gómez 2013[9,*,**], *Tripos furca* (Ehrenberg) F.Gómez 2013[1,2,3,5,7,8,9,10,11,*], *Tripos fusus* (Ehrenberg) F.Gómez 2013[2,3,8,9,10,11,*] *Tripos kofoidii* (Jörgenen) F.Gómez 2013[11,*,**], *Tripos lineatus* (Ehrenberg) F.Gómez 2013[6,7,10,11,*], *Tripos vultur* (Cleve) F.Gómez 2013[11,*,**], *Tripos longipes* (J.W.Bailey) F.Gómez 2013[2,6,9,*], *Tripos macroceros* (Ehrenberg) F.Gómez 2013[9,11,*,**], *Tripos minutus* (Jörgensen) F.Gómez 2013[3]
Family	Gonyaulacaceae
	Gonyaulax minima Matzenauer 1933[10,11,*,**], *Gonyaulax spinifera* (Claparède & Lachmann) Diesing 1866[8,9,10,*,**], *Gonyaulax scrippsae* Kofoid 1911[9,*,**], *Gonyaulax* sp. K.M.Diesing 1866[10,11,*,**], *Lingulodinium polyedra* (F.Stein) J.D. Dodge 1989[9,*,**]
Family	Protoceratiaceae
	Protoceratium reticulatum (Claparède & Lachmann) Bütschli 1885[9,10,11,*,**]
Family	Pyrophacaceae
	Pyrophacus horologium F.Stein 1883[8,9,10,11,*,**], *Pyrophacus steinii* (Schiller) Wall & Dale 1971[8,9,10,11,*,**], *Pyrophacus* sp. F.Stein 1883[10,11,*,**]
Order	Dinophysiales
Family	Amphisoleniaceae
	Amphisolenia astragalus Kofoid & Michener 1911[10,*,**]
Family	Dinophysaceae
	Dinophysis caudata Saville-Kent 1881[1,2,3,5,6,9,10,11,*], *Dinophysis fortii* Pavillard 1923[8,*,**], *Dinophysis miles* Cleve 1900[1,9,*], *Ornithocercus magnificus* Stein 1883[9,*,**]
Order	Peridiniales
Family	Glenodiniaceae
	Glenodinium pulvisculus (Ehrenberg) Stein 1883[8,10,11,*,**], *Glenodinium* sp. Ehrenberg 1836[10,11,*,**]

(continued)

Table 11.2 (continued)

	Peridiniopsis penardiformis (Lindemann) Bourrelly 1968[8,10,11,*,**], *Peridiniopsis quadridens* (Stein) Bourrelly 1968[8,*,**]
Family	Diplopsalidaceae
	Oblea Balech ex Loeblich Jr. & Loeblich III 1966[8,*,**], *Preperidinium meunieri* (Pavillard) Elbrächter 1993[11,*,**]
Family	Peridiniaceae
	Peridinium willei Huitfeldt-Kaas 1900[8,*,**], *Peridinium* sp. Ehrenberg 1830[1,3], *Parvodinium inconspicuum* (Lemmermann) S.Carty 2008[8,10,11,*,**]
Family	Peridiniales incertae sedis
	Peridiniella catenata (Levander) Balech 1977[8,*,**]
Family	Thoracosphaeraceae
	Scrippsiella acuminata (Ehrenberg) Kretschmann, Elbrächter, Zinssmeister, S. Soehner, Kirsch, Kusber & Gottschling 2015[8,*,**]
Order	Prorocentrales
Family	Prorocentraceae
	Prorocentrum arcuatum Issel 1928[11,*,**], *Prorocentrum balticum* (Lohmann) Loeblich 1970[10,11,*,**], *Prorocentrum gracile* Schütt 1895[10,11,*,**], *Prorocentrum belizeanum* M.A.Faust 1993[8,9,*,**], *Prorocentrum rostratum* Stein 1883[11,*,**], *Prorocentrum micans* Ehrenberg 1834[7,9,10,11,*,**], *Prorocentrum compressum* (J.W.Bailey) Abé ex J. D.Dodge 1975[11,*,**], *Prorocentrum cordatum* (Ostenfeld) J.D.Dodge 1975[8,9,10,11,*,**], *Prorocentrum lima* (Ehrenberg) F.Stein 1878[8,9,11,*,**], *Prorocentrum maximum* (Gourret) Schiller 1937[10,*,**], *Prorocentrum* sp. Ehrenberg 1834[10,11,*,**]
Family	Protoperidiniaceae
	Protoperidinium brevipes (Paulsen) Balech 1974[4,7,*], *Protoperidinium conicum* (Gran) Balech 1974[10,*,**], *Protoperidinium crassipes* (Kofoid) Balech 1974[11,*,**], *Protoperidinium depressum* (Bailey) Balech 1974[3,7,9,10,11,*], *Protoperidinium diabolus* (Cleve) Balech 1974[2,3], *Protoperidinium divergens* (Ehrenberg) Balech 1974[8,11,*,**], *Protoperidinium elegans* (Cleve) Balech 1974[9,*,**], *Protoperidinium leonis* (Pavillard) Balech 1974[8,10,11,*,**], *Protoperidinium minimum* A.J.Schilling 1891[9,*,**], *Protoperidinium oceanicum* (Vanhöffen) Balech 1974[8,9,10,11,*,**], *Protoperidinium ovatum* Pouchet 1883[8,*,**], *Protoperidinium pallidum* (Ostenfeld) Balech 1973[8,10,*,**], *Protoperidinium pedunculatum* (Schütt) Balech 1974[8,10,11,*,**], *Protoperidinium pellucidum* Bergh 1881[9,*,**], *Protoperidinium steinii* (Jørgensen) Balech 1974[9,10,11,*,**], *Protoperidinium* sp. R.S.Bergh, 1881[10,11,*,**]
Order	Pyrocystales
Family	Pyrocystaceae
	Pyrocystis lunula (Schütt) Schütt in Engler & Prantl 1896[10,11,*,**], *Pyrocystis* sp. Wyville-Thompson, 1876[10,11,*,**]
Class	Noctilucea
Order	Noctilucales
Family	Noctilucaceae
	Noctiluca scintillans (Macartney) Kofoid & Swezy 1921[2,5,9,11,*]
Family	Kofoidiniaceae
	Pomatodinium impatiens J.Cachon & Cachon-Enjumet 1966[8,*,**]

[1]Devasundaram and Roy (1954), [2]Patnaik (1973), [3]Patnaik and Sarkar (1976), [4]Raman et al. (1990), [5]Adhikary and Sahu (1992), [6]Rath and Adhikary (2008), [7]Panigrahi et al. (2009), [8]Jha et al. (2013), [9]Mukherjee et al. (2016), [10]Srichandan et al. (2015a), [11]Srichandan et al. (2015b)
*Inventorized during survey in post-restoration period (2000–2014)
**New records during survey in post-restoration period (reported for the first time from the lagoon)

caudata and *Protoperidinium diabolus*) of Dinophyta. Patnaik and Sarkar (1976) added two more species; *Tripos minutus* and *Protoperidinium depressum* to the previous list. Raman et al. (1990) reported *Protoperidinium brevipes* as the most dominant species in the lagoon. Subsequently, Adhikary and Sahu (1992) documented the occurrence of *Tripos furca*, *Noctiluca scintillans* and *Dinophysis caudata*. Rath and Adhikary (2008) reported the presence of five species, adding *Gymnodinium heterostriatum* and *Tripos lineatus* while Panigrahi et al. (2009) reported six species with *Prorocentrum micans* as new additions. A study undertaken by Jha et al. (2009) documented 33 species of Dinophyta in the Chilika lagoon, out of which 30 species were included in existing Dinophyta species list. Later, Mukherjee et al. (2016) has carried out investigation on Dinophyta diversity and distribution in Chilika lagoon with description of new records including toxic species namely *Alexandrium minutum*, *A. ostenfeldii*, *A. tamarense*, *A.monilatum*, *Lingulodinium polyedrum*, *Dinophysis caudata*, *D. fortii*, *Prorocentrum cordatum*, *P. micans*, *P. belizeanum*, *P. lima*, *Noctiluca scintillans*, and *Gymnodinium catenatum*. Among the reported 38 species, 12 species (*Alexandrium ostenfeldii*, *Lingulodinium polyedrum*, *Protoperidinium pellucidum*, *Protoperidinium elegans*, *Protoceratium reticulatum*, *Tripos brevis*, *Tripos dens*, *Tripos falcatus*, *Tripos macroceros*, *Ceratium tripos var. atlanticum*, *Amphisolenia astragalus* and *Ornithocercus magnificus*) were new records for Chilika lagoon which mostly prevailed in the outer channel followed by southern sector and central sector. Mukherjee et al. (2016) have also demonstrated salinity as the key factor that drives the Dinophyta distribution in the lagoon. For example, *Tripos macroceros*, and *Prorocentrum micans* were observed in a salinity range of 8.9–33.1 ppt where as, *A. ostenfeldii* was observed between 30.5 and 33.10 ppt (Mukherjee et al. 2016).

A survey undertaken for the period 2011–2012 on spatiotemporal distribution of phytoplankton assemblages reported 38 species of Dinophyta (Srichandan et al. 2015a). A recent study between 2012 and 2014 on interannual and cyclone-driven variability in phytoplankton communities recorded 47 species of Dinophyta (Srichandan et al. 2015b). *Protoperidinium* sp., *Gymnodinium* sp., *Prorocentrum cordatum*, *Tripos fusus*, *Prorocentrum micans*, *Protoperidinium oceanicum*, *Alexandrium* sp. and *Gonyaulax* sp. were the dominant species (Srichandan et al. 2015a, b).

An upsurge in the relative dominance of Dinophyta with concurrent increase in toxic dinophyte species namely; *Alexandrium* sp. (565 cells L^{-1}), *Gonyaulax* sp. (999 cells L^{-1}), and *Prorocentrum cordatum* (448 cells L^{-1}) was observed in southern sector after the passage of very severe cyclonic storm *Phailin* (Srichandan et al. 2015b). It was suggested that several physical and physiological processes could have contributed to higher Dinophyta abundance particularly in the southern sector of the lagoon. For example, riverine run-off may contribute to the formation of low saline and nutrient-rich freshwater layer at the surface where Dinophyta could concentrate their cells due to their swimming behaviour (Srichandan et al. 2015b). Till date, 88 Dinophyta species have been reported from the Chilika lagoon. Among these species, 84 Dinophyta have been reported in inventorisation survey during post-restoration phase and 71 species were new reports.

11.2.3 Cyanophyta

Cyanophyta are unicellular or filamentous organisms that are ubiquitous in nature and are found nearly in all aquatic environments. High diversity and abundance of Cyanophyta depend on high temperature and slightly alkaline conditions. Nutrient rich freshwater discharge and high turbidity due to suspended sediments further favor their growth (Harsha and Malammanavar 2004). For example, the growth of Cyanophyta due to high temperature and water column stability has also been reported in Neuse River estuary (North Carolina), York River estuary (Virginia), Florida Bay (Florida), and San Francisco Bay (California) (Phlips et al. 1999; Ning et al. 2000; Sin et al. 2000; Valdes-Weaver et al. 2006). In Na Thap River Estuary (Thailand), turbidity was the major factor responsible for variation in Cyanophyta diversity and abundance (Lueangthuwapranit et al. 2011). In Chilika lagoon, Cyanophyta have also been reported to dominate the phytoplankton community due to their specific adaptation to survive in highly turbid freshwater part (northern sector) of the lagoon (Srichandan et al. 2015b). Further, a combination of physical upward movement, nutrient, and favorable temperature conditions promoted benthic Cyanophyta growth in Chilika lagoon (Srichandan et al. 2015b).

Biswas (1932) initially documented 11 species of Cyanophyta from Chilika lagoon (Table 11.3). Patnaik (1973) have further added 4 Cyanophyta species to the existing species inventory and found that marine species *Trichodesmium erythraeum* was abundant in the outer channel while freshwater species such as *Anabaena* sp., *Nostoc* sp. were largely represented in the southern and northern sectors. Patnaik and Sarkar (1976) added another 3 Cyanophyta species (*Lyngbya* sp., *Microcystis* sp., and *Oscillatoria* sp.) to the existing list.

Raman et al. (1990) have investigated phytoplankton species composition in the Chilika lagoon before the opening of an artificial inlet in September 2000. It was observed that freshwater Cyanophyta *Pseudanabaena limnetica* dominated in the central sector. Another study, reported 8 Cyanophyta species representing freshwater and marine water forms (Adhikary and Sahu 1992). A seasonal survey of phytoplankton communities was conducted in the year 2000–2001 which reported a total of 12 species, of which 8 were the first report from Chilika lagoon (Rath and Adhikary 2008). Later, Panigrahi et al. (2009), Jha et al. (2009), Mohanty and Adhikary (2013), and Srichandan et al. (2015a) have documented 15, 39, 24 and 28 Cyanophyta species, respectively from the lagoon. A recent investigation on phytoplankton community in the lagoon has reported 39 Cyanophyta species (Srichandan et al. 2015b). Among these, three species (*Cylindrospermum* sp., *Phormidium* sp. and *Anabaena* sp.) were found to be most abundant and consistent in central and northern sectors. Thus, in total, updated documentation of Cyanophyta species stands at 103 of which 95 species were inventorized during post-restoration period which includes 81 new records collected for the first time.

Table 11.3 List of Cyanophyta species from Chilika

Phylum	Cyanophyta
Class	Cyanophyceae
Order	Nostocales
Family	Aphanizomenonaceae
	Anabaenopsis arnoldii Aptekar 1926[8,*,**], *Anabaenopsis elenkinii* V.V.Miller 1923[11,*,**], *Anabaenopsis* sp. V.V.Miller 1923[11,*,**],
	Aphanizomenon sp. A.Morren ex É.Bornet & C.Flahault, 1886 '1888'[10,11,*,**], *Dolichospermum spiroides* (Klebhan) Wacklin, L.Hoffmann & Komárek 2009[11,*,**], *Dolichospermum flosaquae* (Brébisson ex Bornet & Flahault) P.Wacklin, L.Hoffmann & J.Komárek 2009[6,7,10,11,*,**], *Sphaerospermopsis aphanizomenoides* (Forti) Zapomelová, Jezberová, Hrouzek, Hisem, Reháková & Komárková 2010[8,*,**]
Family	Chlorogloeopsidaceae
	Chlorogloeopsis fritschii (A.K.Mitra) A.K.Mitra & D.C.Pandey 1967[8,*,**]
Family	Hapalosiphonaceae
	Fischerella sp. (É.Bornet & C.Flahault) M.A.Gomont 1895[6,7,*,**]
Family	Nostocaceae
	Anabaena orientalis S.C.Dixit 1936[8,*,**], *Anabaena oscillarioides* Bory ex Bornet & Flahault 1886[9,*,**], *Anabaena torulosa* Lagerheim ex Bornet & Flahault 1886[1,4,6,10,*], *Anabaena* sp. Bory ex Bornet & Flahault 1886[2,3,5,10,11,*], *Cylindrospermum* sp. Kützing ex É.Bornet & C.Flahault 1886[10,11,*,**], *Nostoc linckia* Bornet ex Bornet & Flahault 1886[8,11,*,**], *Nostoc punctiforme* Hariot 1891[8,*,**], *Nostoc* sp. Vaucher ex Bornet & Flahault 1886[2,3,5,7,11,*], *Trichormus variabilis* (Kützing ex Bornet & Flahault) Komárek & Anagnostidis 1989[9,*,**], *Wollea* sp. É.Bornet & C.Flahault 1886[8,*,**]
Order	Chroococcales
Family	Aphanothecaceae
	Aphanothece sp. C.Nägeli 1849[11,*,**], *Gloeothece rupestris* (Lyngbye) Bornet 1880[8,10,*,**], *Gloeothece* sp. C.Nägeli 1849[11,*,**]
Family	Chroococcaceae
	Chroococcus dispersus (Keissler) Lemmermann 1904[8,11,*,**], *Chroococcus minimus* (Keissler) Lemmermann 1904[8,*,**], *Chroococcus minutus* (Kützing) Nägeli 1849[8,*,**], *Chroococcus turgidus* (Kützing) Nägeli 1849[6,7,9,11,*,**], *Chroococcus* sp. Nägeli 1849[10,11,*,**], *Cyanosarcina spectabilis* (Geitler) Kovácik 1988[8,*,**], *Dactylococcopsis fascicularis* Lemmermann 1898[8,*,**], *Dactylococcopsis raphidioides* Hansgirg 1888[8,*,**],
Family	Cyanobacteriaceae
	Cyanobacterium diachloros (Skuja) Komárek, Kopecky & Cepák 1990[9,*,**]
Family	Gomphosphaeriaceae
	Gomphosphaeria dubium[10,*,**], *Gomphosphaeria* sp.Kützing 1836[10,11,*,**]
Family	Microcystaceae
	Anacystis sp. Meneghini 1837[10,*,**], *Gloeocapsa alpina* Nägeli 1865[11,*,**], *Gloeocapsa coracina* Kützing 1843[8,*,**], *Gloeocapsa livida* (Carmichael) Kützing 1847[8,*,**], *Gloeocapsa* sp. Kützing 1843[5,10,11,*], *Microcystis aeruginosa* (Kützing) Kützing 1846[8,9,11,*,**], *Microcystis wesenbergii* (Komárek) Komárek ex Komárek 2006[9,*,**], *Microcystis flosaquae* (Wittrock) Kirchner 1898[11,*,**], *Microcystis smithii* Komárek & Anagnostidis 1995[8,*,**], *Microcystis* sp. Lemmermann 1907[3,7,10,11,*]
Order	Synechococcales
Family	Coelosphaeriaceae
	Coelosphaerium dubium Grunow 1865[8,*,**], *Snowella* sp. A.A.Elenkin 1938[11,*,**]

(continued)

Table 11.3 (continued)

Family	Leptolyngbyaceae
	Leptolyngbya tenuis (Gomont) Anagnostidis & Komárek 1988[8,10,11,*,**]
Family	Merismopediaceae
	Aphanocapsa grevillei (Berkeley) Rabenhorst 1865[11,*,**], *Aphanocapsa marina* Hansgirg in Foslie 1890[9,*,**], *Aphanocapsa* sp. C.Nägeli 1849[10,11,*,**], *Aphanocapsa rivularis* (Carmichael) Rabenhorst 1865[8,10,*,**], *Merismopedia convoluta* Brébisson ex Kützing 1849[8,*,**], *Merismopedia punctata* Meyen 1839[8,9,*,**], *Merismopedia elegans* A.Braun ex Kützing 1849[6,7,11,*,**], *Merismopedia warmingiana* (Lagerheim) Forti 1907[9,*,**], *Merismopedia tenuissima* Lemmermann 1898[4], *Merismopedia glauca* (Ehrenberg) Kützing 1845[6,7,8,9,11,*,**], *Merismopedia* sp. F.J.F.Meyen 1839[10,11,*,**], *Synechocystis aquatilis* Sauvageau 1892[4,6,7,*], *Synechocystis pevalekii* Ercegovic 1925[8,*,**], *Synechocystis* sp. C.Sauvageau 1892[11,*,**]
Family	Pseudanabaenaceae
	Jaaginema pseudogeminatum (G.Schmid) Anagnostidis & Komárek 1988[8,*,**], *Pseudanabaena limnetica* (Lemmermann) Komárek 1974[4,7,9,10,*], *Pseudanabaena minima* (G.S.An) Anagnostidis 2001[9,*,**]
Order	Oscillatoriales
Family	Coleofasciculaceae
	Coleofasciculus chthonoplastes (Thuret ex Gomont) M.Siegesmund, J.R.Johansen & T.Friedl 2008[1]
Family	Coleofasciculaceae
	Geitlerinema claricentrosum (N.L.Gardner) Anagnostidis 1989[8,*,**], *Geitlerinema earlei* (N.L.Gardner) Anagnostidis 1989[9,*,**]
Family	Microcoleaceae
	Arthrospira platensis Gomont 1892[6,7,*,**], *Arthrospira gigantea* (Schmidle) Anagnostidis 1998[8,*,**],*Planktothrix prolifica* (Gomont) Anagnostidis & Komárek 1988[8,11,*,**], *Kamptonema chlorinum* (Kützing ex Gomont) Strunecký, Komárek & J. Smarda 2014[8,10,*,**], *Kamptonema proteus* (Skuja) Strunecký, Komárek & J.Smarda 2014[9,*,**], *Kamptonema laetevirens* (H.M.Crouan & P.L.Crouan ex Gomont) Strunecký, Komárek & J.Smarda 2014[1,2,3,5], *Microcoleus paludosus* Gomont 1892[1], *Porphyrosiphon versicolor* (Gomont) Anagnostidis & Komárek 1988[8,10,11,*,**], *Johanseninema constrictum* (Szafer) Hasler, Dvorák & Poulícková 2014[8,*,**], *Trichodesmium erythraeum* Ehrenberg ex Gomont 1892[2,5,10,11,*], *Trichodesmium* sp. Ehrenberg ex Gomont 1892[7,*,**]
Family	Oscillatoriaceae
	Lyngbya aestuarii Liebman ex Gomont 1892[1,2,3,5,6,7,9,*], *Lyngbya confervoides* C. Agardh ex Gomont 1892[1,5], *Lyngbya majuscula* Harvey ex Gomont 1892[8,10,11,*,**], *Lyngbya anomala* (C.B.Rao) Umezaki & Watanabe 1994[8,*,**], *Lyngbya* sp. C.Agardh ex Gomont 1892[3,10,11,*], *Oscillatoria anguina* Bory ex Gomont 1892[8,*,**], *Oscillatoria chilkensis* Biswas 1932[1,3], *Oscillatoria curviceps* C.Agardh ex Gomont 1892[8,10,11,*,**], *Oscillatoria limosa* C.Agardh ex Gomont 1892[9,*,**], *Oscillatoria peromata* Skuja 1949[8,9,10,*,**], *Oscillatoria princeps* Vaucher ex Gomont 1892[6,7,8,9,10,11,*,**], *Oscillatoria sancta* Kützing ex Gomont 1892[9,*,**], *Oscillatoria simplicissima* Gomont 1892[9,*,**], *Oscillatoria tenuis* C.Agardh ex Gomont 1892[11,*,**], *Oscillatoria* sp. Vaucher ex Gomont 1892[3,10,11,*] *Phormidium corium* Gomont ex Gomont 1892[1,7,*], *Phormidium fragile* Gomont 1893[1,5], *Phormidium submembranaceum* Gomont 1892[1,6,7,*], *Phormidium aerugineo-caeruleum* (Gomont) Anagnostidis & Komárek 1988[1], *Phormidium ambiguum* Gomont 1892[9,*,**], *Phormidium* sp. Kützing ex Gomont 1892[2,10,11,*]
Order	Spirulinales
Family	Spirulinaceae

(continued)

Table 11.3 (continued)

	Spirulina labyrinthiformis Gomont 1892[9,*,**], *Spirulina major* Kützing ex Gomont 1892[9,*,**], *Spirulina subsalsa* Oersted ex Gomont 1892[8,*,**], *Spirulina subtilissima* Kützing ex Gomont 1892[6,9,*,**], *Spirulina* sp. Turpin ex Gomont 1892[10,11,*,**]

[1]Biswas (1932), [2]Patnaik (1973), [3]Patnaik and Sarkar (1976), [4]Raman et al. (1990), [5]Adhikary and Sahu (1992), [6]Rath and Adhikary (2008), [7]Panigrahi et al. (2009), [8]Jha et al. (2009), [9]Mohanty and Adhikary (2013), [10]Srichandan et al. (2015a), [11]Srichandan et al. (2015b)
*Inventorized during survey in post-restoration period (2000–2014)
**New records during survey in post-restoration period (reported for the first time from the lagoon)

11.2.4 Chlorophyta

Chlorophyta are green colored phytoplankton with chlorophyll a and b, xanthophylls, and carotenes as the dominant photosynthetic pigments (Dawson 1966). The Chlorophyta prevails in a wide range of environments ranging from freshwater to estuarine and marine conditions. In general, Chlorophyta occur preferably in freshwater upstream regions of estuaries. For example, in upper reaches of Tapi Estuary (India), dominance of chlorophytic phytoplankton communities such as *Ankistrodesmus falcatus*, *Chlorella vulgaris*, *Scenedesmus quadricauda*, *Spirogyra indica*, *Pediastrum* sp. and *Closterium acerosum* have been observed (George et al. 2012). Chlorophyta population was numerically more abundant in the freshwater region of Chilika lagoon (Srichandan et al. 2015b). Literature also suggests that eutrophic conditions further maximize the diversity and density of Chlorophyta population (Saify et al. 1986).

Devasundaram and Roy (1954) and Patnaik (1973) investigated the entire Chilika lagoon and recorded only one species of Chlorophyta are presented by *Spirogyra* sp. A survey conducted between the year 2000 and 2001 on the phytoplankton communities reported 14 species of Chlorophyta (Rath and Adhikary 2008). Further, Panigrahi et al. (2009) documented 10 species of Chlorophyta. Subsequently, a study on Chilika between 2003 and 2006 reported 114 species of Chlorophyta (Jha et al. 2009). Mohanty and Adhikary (2013) studied the algal diversity of Chilika lagoon extensively in different seasons and reported 14 species of Chlorophyta. A study on Chilika between the year 2011 and 2012 recorded 32 species belonging to 25 genera. The freshwater Chlorophyta *Eudorina* sp. were most abundant in the northern sector of the lagoon (Srichandan et al. 2015a). Another study during the year 2012–2014 has documented a total of 54 species of Chlorophyta (Srichandan et al. 2015b). To date, 178 Chlorophyta species have been reported from the Chilika lagoon (Table 11.4). Among the encountered 178 species, 173 species were all new records and inventorized during post-restoration period.

Table 11.4 List of Chlorophyta species from Chilika

Phylum	Chlorophyta
Class	Trebouxiophyceae
Order	Chlorellales
Family	Chlorellaceae
	Actinastrum hantzschii Lagerheim 18825,[10,*,**], *Actinastrum* sp. Lagerheim 1882[9,10,*,**], *Dictyosphaerium* sp. Nägeli 1849[9,10,*,**], *Geminella* sp. Turpin 1828[10,*,**]
Family	Oocystaceae
	Crucigeniella irregularis (Wille) P.M.Tsarenko & D.M.John 2002[10,*,**], *Eremosphaera eremosphaeria* (G.M.Smith) R.L.Smith & Bold 1966[7,*,**], *Eremosphaera viridis* De Bary 1858[7,*,**], *Glochiococcus aciculiferus* (Lagerheim) P. C.Silva 1996[7,*,**], *Oocystis* sp. Nägeli ex A.Braun 1855[10,*,**] *Trochiscia aspera* (Reinsch) Hansgirg 1888[7,*,**], *Trochiscia pachyderma* (Reinsch) Hansgirg[7,*,**], *Trochiscia reticularis* (Reinsch) Hansgirg 1888[7,*,**]
Order	Trebouxiales
Family	Botryococcaceae
	Botryococcus braunii Kützing 1849[7,*,**], *Botryococcus* sp. Kützing, 18499,[10,*,**]
Order	Trebouxiophyceae ordo incertae sedis
Family	Trebouxiophyceae incertae sedis
	Crucigenia sp. Morren 1830[10,*,**]
Class	Chlorophyceae
Order	Chlamydomonadales
Family	Actinochloridaceae
	Actinochloris sp. Korschikov 1953[10,*,**]
Family	Palmellopsidaceae
	Asterococcus superbus (Cienkowski) Scherffel 1908[7,*,**], *Asterococcus* sp. Scherffel 1908[10,*,**]
Family	Palmellopsidaceae
	Chlamydocapsa planctonica (West & G.S.West) Fott 1972[7,9,10,*,**]
Family	Chlamydomonadaceae
	Chlamydomonas microsphaera Pascher & Jahoda 1928[7,*,**], *Chlamydomonas sphagnicola* (F.E.Fritsch) F.E.Fritsch & H.Takeda 1916[7,*,**]
Family	Chlorococcaceae
	Chlorococcum infusionum (Schrank) Meneghini 1842[7,*,**]
Family	Hormotilaceae
	Dendrocystis raoi M.O.P.Iyengar 1962[7,*,**]
Family	Volvocaceae
	Eudorina elegans Ehrenberg 1832[5,6,7,9,10,*,**], *Eudorina* sp. Ehrenberg 1832[9,10,*,**], *Pandorina cylindricum* M.O.P.Iyengar 1981[7,*,**], *Pandorina morum* (O.F.Müller) Bory 1824[7,*,**], *Pandorina* sp. Bory 1824[9,10,*,**], *Pleodorina californica* W.R.Shaw 1894[7,*,**], *Pleodorina indica* (Iyengar) H.Nozaki 1989[7,*,**]
Family	Goniaceae
	Gonium compactum M.O.P.Iyengar in M.O.P.Iyengar & Desikachary 1981[7,*,**], *Gonium pectorale* O.F.Müller 1773[7,*,**], *Gonium* sp. O.F.Müller 1773[10,*,**]

(continued)

Table 11.4 (continued)

Family	Haematococcaceae
	Haematococcus sp. Flotow 1844[10,*,**]
Family	Sphaerocystidaceae
	Sphaerocystis schroeteri Chodat 1897[7,10,*,**], *Sphaerocystis* sp. Chodat 1897[9,*,**]
Family	Tetrabaenaceae
	Tetrabaena socialis (Dujardin) H.Nozaki & M.Itoh 1994[7,*,**]
Family	Treubariaceae
	Treubaria sp. C.Bernard 1908[10,*,**]
Order	Sphaeropleales
Family	Selenastraceae
	Ankistrodesmus falcatus (Corda) Ralfs 1848[7,9,*,**], *Ankistrodesmus falcatus var. radiatus* Lemmermann 1908[7,*,**], *Ankistrodesmus* sp. Corda 1838[9,10,*,**], *Kirchneriella lunaris* (Kirchner) Möbius 1894[7,*,**], *Messastrum gracile* (Reinsch) T.S. Garcia 2016[5,10,*,**], *Selenastrum* sp. Reinsch 1867[9,10,*,**], *Monoraphidium* sp. Komárková-Legnerová 1969[4,6,*]
Family	Scenedesmaceae
	Coelastrum astroideum De Notaris 1867[9,10,*,**], *Coelastrum cambricum* W.Archer 1868[5,*,**], *Coelastrum microporum* Nägeli 1855[10,*,**], *Coelastrum* sp. Nägeli 1849[10,*,**], *Desmodesmus protuberans* (F.E.Fritsch & M.F.Rich) E.Hegewald 2000[8,*,**], *Desmodesmus perforatus* (Lemmermann) E.Hegewald 2000[7,*,**], *Desmodesmus opoliensis* (P.G.Richter) E.Hegewald 2000[10,*,**], *Enallax costatus* (Schmidle) Pascher 1943[10,*,**], *Scenedesmus arcuatus* (Lemmermann) Lemmermann 1899[10,*,**], *Scenedesmus falcatus* Chodat 1926[10,*,**], *Scenedesmus calyptratus* Comas 1980[8,*,**], *Scenedesmus quadricauda* (Turpin) Brébisson 1835[5,6,7,10,*,**], *Scenedesmus obtusus* Meyen 1829[10,*,**], *Scenedesmus* sp. Meyen 1829[9,10,*,**], *Tetradesmus bernardii* (G.M.Smith) M.J.Wynne 2016[10,*,**], *Tetradesmus dimorphus* (Turpin) M.J.Wynne 2016[8,*,**], *Tetradesmus lagerheimii* M.J.Wynne & Guiry 2016[5,6,10,*,**], *Tetradesmus obliquus* (Turpin) M.J.Wynne 2016[7,8,9,10,*,**], *Willea apiculata* (Lemmermann) D.M.John, M.J.Wynne & P.M.Tsarenko 2014[7,*,**], *Willea rectangularis* (A.Braun) D.M.John, M.J.Wynne & P.M.Tsarenko 2014[7,*,**], *Westella botryoides* (West) De Wildeman 1897[7,10,*,**]
Family	Characiaceae
	Fernandinella sp. Chodat 1922[10,*,**]
Family	Microsporaceae
	Microspora stagnorum (Kützing) Lagerheim 1887[7,*,**], *Microspora willeana* Lagerheim 1887[8,*,**]
Family	Hydrodictyaceae
	Monactinus simplex (Meyen) Corda 1839[5,7,10,*,**], *Pediastrum duplex* Meyen 1829[5,6,10,*,**], *Pediastrum duplex var. rotundatum* Lucks 1907[7,*,**], *Pediastrum duplex* var. *subgranulatum* Raciborski 1889[5,*,**], *Pediastrum* sp. Meyen 1829[9,10,*,**], *Stauridium tetras* (Ehrenberg) E.Hegewald 2005[5,6,8,10,*,**], *Pseudopediastrum boryanum* (Turpin) E.Hegewald 2005[10,*,**], *Tetraëdron proteiforme* (W.B.Turner) Brunnthaler 1915[7,*,**], *Tetraëdron lobulatum* (Nägeli) Hansgirg 1888[7,*,**], *Tetraedron gracile* (Reinsch) Hansgirg 1889[5,7,*,**], *Tetraëdron trigonum* (Nägeli) Hansgirg 1888[5,7,10,*,**], *Tetraedron trigonum var. minus* (Reinsch) De Toni[7,*,**], *Tetraëdron* sp. Kützing 1845[9,10,*,**]
Family	Schizochlamydaceae
	Planktosphaeria gelatinosa G.M.Smith 1918[7,*,**], *Schizochlamys gelatinosa* A.Braun 1849[7,*,**]

(continued)

Table 11.4 (continued)

Order	Chaetophorales
Family	Chaetophoraceae
	Draparnaldia sp. Bory 1808[9,*,**]
Order	Oedogoniales
Family	Oedogoniaceae
	Oedogonium nanum Wittrock ex Hirn 1900[7,*,**], *Oedogonium* sp. Link ex Hirn 1900[3,6,9,10,*]
Class	Ulvophyceae
Order	Ulotrichales
Family	Ulotrichaceae
	Ulothrix aequalis Kützing 1845[7,*,**], *Ulothrix rorida* Thuret 1850[9,*,**], *Ulothrix tenerrima* (Kützing) Kützing 1843[7,*,**], *Ulothrix zonata* (F.Weber & Mohr) Kützing 1833[7,*,**], *Ulothrix* sp. Kützing 1833[9,*,**]
Class	Conjugatophyceae (Zygnematophyceae)
Order	Desmidiales
Family	Closteriaceae
	Closterium acerosum var. *elongatum* Brébisson 1856[7,*,**], *Closterium gracile var. tenue* (Lemmermann) West & West 1902[7,*,**], *Closterium acutum* Brébisson 1848[7,*,**], *Closterium lunula* Ehrenberg & Hemprich ex Ralfs 1848[7,9,*,**], *Closterium strigosum* Brébisson 1856[7,*,**], *Closterium macilentum* Brébisson 1856[7,*,**], *Closterium pygmaeum* Gutwinski 1890[7,*,**], *Closterium venus* Kützing ex Ralfs 1848[8,*,**] *Closterium* sp. Nitzsch ex Ralfs 1848[4,6,9,10,*]
Family	Desmidiaceae
	Cosmarium awadhense B.N.Prasad & R.K.Mehrotra 1977[8,*,**], *Cosmarium calcareum* Wittrock 1872[7,*,**], *Cosmarium costatum* Nordstedt 1875[7,*,**], *Cosmarium crenatum* Ralfs ex Ralfs 1848[7,*,**], *Cosmarium decoratum* West & G.S.West 1895[8,*,**], *Cosmarium geminatum* P.Lundell 1871[7,*,**], *Cosmarium impressulum* Elfving 1881[5,*,**], *Cosmarium indentatum* Grönblad 1920[7,*,**], *Cosmarium lundellii* Delponte 1877[8,*,**], *Cosmarium miscellum* Skuja 1964[8,*,**], *Cosmarium moniliforme* Ralfs 1848[7,*,**], *Cosmarium novae-semliae* Wille 1879[7,*,**], *Cosmarium pachydermum* var. *aethiopicum* (West & G.S.West) West & G.S.West 1905[7,*,**], *Cosmarium papilliferum* Schmidle[7,*,**], *Cosmarium pachydermum* var. *incrassatum* Scott & Grönblad 1957[7,*,**], *Cosmarium pachydermum* P.Lundell 1871[7,9,*,**], *Cosmarium phaseolus* Brébisson ex Ralfs 1848[7,*,**], *Cosmarium punctulatum* Brébisson 1856[8,*,**], *Cosmarium subspeciosum* Nordstedt 1875[7,*,**], *Cosmarium quadrifarium* f. *hexastichum* (P.Lundell) Nordstedt 1889[7,*,**], *Cosmarium quadrifarium* P.Lundell 1871[7,*,**], *Cosmarium ungerianum* (Nägeli) De Bary 1858[7,*,**], *Cosmarium* sp. Corda ex Ralfs 1848[4,9,10,*], *Desmidium swartzii* C.Agardh ex Ralfs 1848[7,*,**], *Euastrum dubium* Nägeli 1849[8,*,**], *Euastrum oblongum* Ralfs 1848[7,*,**], *Euastrum* sp. Ehrenberg ex Ralfs 1848[10,*,**], *Desmidium* sp. C.Agardh ex Ralfs 1848[6,9,*,**], *Micrasterias papillifera* Brébisson ex Ralfs 1848[7,*,**], *Micrasterias* sp. C.Agardh ex Ralfs 1848[9,10,*,**], *Spinocosmarium quadridens* (H.C.Wood) Prescott & A.M.Scott 1942[7,*,**], *Spondylosium* sp. Brébisson ex Kützing 1849[9,10,*,**], *Staurastrum alchora* West & West[7,*,**], *Staurastrum anatinum* Cooke & Wills 1881[7,9,*,**], *Staurastrum bicornatum* Johnson[7,*,**], *Staurastrum bieneanum* Rabenhorst 1862[7,*,**], *Staurastrum boreale* West & G.S.West 1905[7,*,**], *Staurastrum brevispinum var. inerme* Wille[7,*,**], *Staurastrum cingulum* (West & G.S.West) G.M.Smith 1922[7,*,**], *Staurastrum crenulatum* (Nägeli) Delponte 1877[7,*,**], *Staurastrum curviceps* Scott. & Groenblad[7,*,**], *Staurastrum cyclacanthem* West & West[7,*,**], *Staurastrum dilatatum* Ehrenberg ex Ralfs 1848[7,*,**], Staurastrum dilatatum var. productum Scott & Groenblad[7,*,**], *Staurastrum elongated forma* chilikensis[7,*,**], *Staurastrum floriferum* West & G.S.West 1896[7,*,**], *Staurastrum manfeldtii var. pseudosebaldi* (Wille) Coesel

(continued)

Table 11.4 (continued)

	& Meesters 2013[7,*,**], *Staurastrum leptacanthum* Nordstedt 1869[7,*,**], *Staurastrum margaritaceum* Meneghini ex Ralfs 1848[7,*,**], *Staurastrum quadrangular* (Breb.) Ralfs[7,*,**], *Staurastrum ophiura* var. *horridum* A.M.Scott[7,*,**], *Staurastrum polymorphum* Brébisson 1848[7,*,**], *Staurastrum setigerum* Cleve 1864[7,*,**], *Staurastrum proboscideum* (Brébisson) W.Archer 1861[7,*,**], *Staurastrum pseudosebaldi var. compactum* A.M.Scott & Grönblad 1957[7,*,**], *Staurastrum pseudosuecicum* Prescott & A.M.Scott[7,*,**], *Staurastrum punctulatum* Brébisson 1848[7,*,**], *Staurastrum turgescens* De Notaris 1867[7,*,**], *Staurastrum sexcostatum* Brébisson ex Ralfs 1848[7,*,**], *Staurastrum* sp. Meyen ex Ralfs 1848[7,9,10,*,**], *Staurodesmus lobatus* (Børgesen) Bourrelly 1966[7,*,**], *Staurodesmus convergens* (Ehrenberg ex Ralfs) S.Lillieroth 1950[7,*,**], *Staurodesmus* sp. Teiling 1948[7,9,*,**] *Xanthidium armatum* Brébisson ex Ralfs 1848[7,*,**], *Xanthidium pseudobengalicum* R. L.Grönblad[7,*,**], *Xanthidium* sp. Ehrenberg ex Ralfs 1848[7,*,**]
Family	Gonatozygaceae
	Genicularia kinhani (Archer) Rabenhorst[7,*,**], *Genicularia spirotaenia* (De Bary) De Bary 1858[7,*,**], *Genicularia* sp. De Bary 1858[9,*,**]
Order	Zygnematales
Family	Zygnemataceae
	Mougeotia sp. C.Agardh 1824[9,10,*,**], *Spirogyra hyalina* Cleve 1868[7,*,**], *Spirogyra puncticulata* C.C.Jao 1934[7,*,**], *Spirogyra subsalsa* Kützing 1845[7,*,**], *Spirogyra* sp. Link 1820[1,2,3,4,5,6,7,8,9,10,*] *Zygnema* sp. C.Agardh 1817[10,*,**], *Tamnogametum mayyanadense* Erady & Rajappan 1959[7,*,**]
Class	Coleochaetophyceae
Order	Coleochaetales
Family	Coleochaetaceae
	Coleochaete orbicularis Pringsheim 1860[7,*,**]

[1]Devasundaram and Roy (1954), [2]Patnaik (1973), [3]Patnaik and Sarkar (1976), [4]Raman et al. (1990), [5]Rath and Adhikary (2008), [6]Panigrahi et al. (2009), [7]Jha et al. (2009), [8]Mohanty and Adhikary (2013), [9]Srichandan et al. (2015a), [10]Srichandan et al. (2015b)
*Inventorized during survey in post-restoration period (2000–2014)
**New records during survey in post-restoration period (reported for the first time from the lagoon)

11.2.5 Euglenophyta

The Euglenophyta is a group of unicellular flagellates found in freshwater and marine environments. The class is distinguished by solitary unicells (only one colonial genus exists) with two anteriorly inserted flagella of which one is emergent, condensed chromosomes throughout the cell cycle, a paraxial rod associated with one or both flagella, a proteinaceous pellicle composed of individual strips each of which is lined by microtubules, and a beta-1, 3 glucan storage product known as paramylum. The diversity of Euglenophyceae members in aquatic environment can be attributed to high nutrient loading from various point and non-point sources indicating organic pollution in a water body (Kumar and Hosmani 2006; Laskar and Gupta 2009). In general, Euglenophyta are known to be dominant in freshwater regimes (preferably in upper reaches) of estuarine ecosystems in comparison to middle and lower reaches. For example, members of euglenophytic phytoplankton were observed to be dominated in upper reaches of Tapi estuary (India) (George et al. 2012). Similarly, a higher abundance of Euglenophyta has been observed in the freshwater zone i.e. northern sector of the Chilika lagoon (Srichandan et al. 2015a).

In Chilika lagoon, Euglenophyta was recorded for the first time by Jha et al. (2009) and documented 53 Euglenophyta species. Subsequently, a study on Euglenophyta diversity was carried out by Mohanty and Adhikary (2013) during 2010–2011.They encountered six species (*Lepocinclis acus*, *Euglena agilis*, *Euglenaria caudata*, *Lepocinclis playfairiana*, *Trachelomonas abrupta*, and *Trachelomonas hispida*) and found that their occurrence in northern and central sectors have attributed to increased eutrophication associated with anthropogenic discharge by human habitation around Chilika lagoon. Subsequently, Srichandan et al. (2015a) investigated phytoplankton community structure including Euglenophyta from the entire Chilika lagoon and added 4 more species to the existing Euglenophyta species list. This study has also revealed that Euglenophyta formed the most dominant group in the northern sector (freshwater zone) of the lagoon. The author has opined that this group occurs preferably in the nutrient-rich freshwater zone and serves as bio-indicator of organic pollution. Recently, a survey conducted between 2012 and 2014 have added 30 more species to the list (Srichandan et al. 2015b). Thus, the total number of Euglenophyta has increased to 92 species which were all inventorized new records during the post-restoration period (Table 11.5). It was also noticed that tropical cyclone *Phailin* which struck the lagoon on 12th October 2013 profoundly affected the Euglenophyta community composition in Chilika lagoon. After *Phailin*, the recovery of freshwater euglenophytes (e.g., *Strombomonas acuminata*, *Trachelomonas* sp.) was observed for the first time from the southern sector of the lagoon. In addition, the freshwater euglenophytes such as *Phacus circumflexus*, *Strombomonas acuminata*, *Trachelomonas granulata*, *Trachelomonas lefevrei*, *Trachelomonas manginii*, and *Lepocinclis acus* were recorded for the first time from the outer channel.

11.2.6 Chrysophyta

Chrysophyta (Golden-brown algae) are a group of marine pigmented heterokonts (Daugbjerg and Henriksen 2001) with a cosmopolitan distribution. They can be a major component in coastal and estuarine waters (e.g. Jochem and Babenerd 1989; Gómez and Gorsky 2003). They are generally autotrophs (Rigual-Hernández et al. 2010) while as opined out by Martini (1977) they have mixotrophic behavior. Further, Chrysophyta have been used as indicators of productivity (Takahashi et al. 2009), atmospheric and water mass variations (Onodera and Takahashi 2005). Chrysophyta are strongly influenced by environmental parameters, particularly by temperature and salinity (Henriksen et al. 1993). In Chilika lagoon, Chrysophyta were more numerous in brackish water salinity regime (Srichandan et al. 2015b).

In Chilika lagoon, only five taxa of Chrysophyta have been reported for the first time by Srichandan et al. (Srichandan et al. 2015a, b) and among them, three taxa

Table 11.5 List of Euglenophyta species from Chilika

Phylum	Euglenophyta (=Phylum Euglenozoa)
Class	Euglenophyceae
Order	Eutreptiales
Family	Astasiaceae
	Astasia klebsii Lemmermann 1910[1,3,4,*,**]
Order	Euglenophyceae incertae sedis
Family	Colaciaceae
	Colacium sp. Ehrenberg, 1834[4,*,**]
Order	Euglenales
Family	Euglenaceae
	Euglena acus var. *rigida* E.Hübner, 1886[1,*,**], *Euglena agilis* H.J.Carter 1856[2,*,**], *Euglena cantabrica* E.G.Pringsheim 1956[4,*,**], *Euglena chlamydomonas*[4,*,**], *Euglena deses* Ehrenberg 1834[4,*,**], *Euglena elastica* Prescott 1944[1,3,4,*,**],*Euglena schmitzii* Gojdics 1953[1,3,4,*,**], *Euglena elongata* W.Schewiakoff 1892[4,*,**], *Euglena gaumei* Allorge & Lefèvre 1931[1,*,**], *Euglena geniculata* F.Schmitz 1884[4,*,**], *Euglena granulata* (G.A.Klebs) F.Schmitz 18841[4,*,**], *Euglena repulsans* J.Schiller 1952[4,*,**], *Euglena sanguinea* Ehrenberg 1832[1,3,4,*,**], *Euglena sociabilis* P.A. Dangeard, 1902[3,4,*,**], *Euglena texta* (Dujardin) Hübner 1886[4,*,**], *Euglena van-goori* Deflandre 1928[1,*,**], *Euglena variabilis* G.A.Klebs 1883[1,3,4,*,**], *Euglena viridis* (O.F. Müller) Ehrenberg 1830[1,4,*,**], *Euglena wangii* S.P.Chu 1946[1,3,4,*,**], *Euglena* sp. Ehrenberg 1830[3,4,*,**], *Euglenaformis proxima* (Dangeard) M.S.Bennett & Triemer 2014[1,*,**], *Euglenaria anabaena* (Mainx) Karnkowska & E.W.Linton 2010[1,*,**], *Euglenaria clavata* (Skuja) Karnkowska & E.W.Linton 2010[1,4,*,**], *Euglenaria caudata* (E.F.W.Hubner) A.Karnowska-Ishikawa, E.Linton & J.Kwiatowski 2010[2,4,*,**], *Euglenopsis vorax* G.A.Klebs 1892[1,3,4,*,**], *Monomorphina nordstedtii* (Lemmermann) T.G.Popova 1955[1,*,**], *Monomorphina pyrum* (Ehrenberg) Mereschkowsky 1877[4,*,**], *Strombomonas acuminata* (Schmarda) Deflandre 1930[4,*,**], *Strombomonas eurystoma* (F.Stein) T.G.Popova 1966[4,*,**], *Strombomonas giardiana* (Playfair) Deflandre 1930[1,4,*,**], *Strombomonas tambowika* (Svirenko) Deflandre 1930[4,*,**], *Strombomonas* sp. Deflandre 1930[4,*,**], *Trachelomonas abrupta* Svirenko [Swirenko] 1914[2,*,**], *Trachelomonas bulla* F.Stein 1878[1,4,*,**], *Trachelomonas armata* (Ehrenberg) F.Stein 1878[4,*,**], *Trachelomonas crebae var. brevicollaris* prescott[1,*,**], *Trachelomonas cylindrica* Ehrenberg 1834[4,*,**], *Trachelomonas granulata* Svirenko 1914[4,*,**], *Trachelomonas hispida* (Perty) F.Stein 1878[2,3,4,*,**], *Trachelomonas hispida* var. *crenulatocollis* (Maskell) Lemmermann 1910[1,*,**], *Trachelomonas lefevrei*[4,*,**], *Trachelomonas manginii* Deflandre 1926[4,*,**], *Trachelomonas oblonga* Lemmermann 1899[4,*,**], *Trachelomonas planctonica* Svirenko 1914[4,*,**], *Trachelomonas similis* A.C.Stokes 1890[4,*,**], *Trachelomonas* sp. Ehrenberg 1835[3,4,*,**]
Family	Phacaceae
	Lepocinclis acus (O.F.Müller) B. Marin & Melkonian, 2003[1,2,3,4,*,**], *Lepocinclis acuta* Prescott 1938[1,3,4,*,**], *Lepocinclis caudata* (A.M. da Cunha) Pascher 1927[4,*,**], *Lepocinclis fusiformis* (H.J.Carter) Lemmermann 1901[1,4,*,**], *Lepocinclis fusiformis* var. *major* F.E.Fritsch & Rich 1930[1,*,**], *Lepocinclis ovum* (Ehrenberg) Lemmermann 1901[4,*,**], *Lepocinclis oxyuris* (Schmarda) B.Marin & Melkonian 2003[1,3,4,*,**], *Lepocinclis oxyuris var. minor* (Skvortzov) D.A.Kapustin 2011[1,*,**], *Lepocinclis playfairiana* (Deflandre) Deflandre 1932[1,2,3,4,*,**], *Lepocinclis repulsans*[4,*,**], *Lepocinclis sphagnophila* Lemmermann 1904[1,4,*,**], *Lepocinclis spirogyroides* B. Marin & Melkonian 2003[1,*,**], *Lepocinclis steinii* Lemmermann 1901[4,*,**],

(continued)

Table 11.5 (continued)

	Lepocinclis teres (F.Schmitz) Francé 1897[4,*,**], *Lepocinclis tripteris* (Dujardin) B. Marin & Melkonian 2003[1,*,**], *Lepocinclis* sp. Perty 1849[4,*,**], *Phacus anacoelus* A. C.Stokes 1885[1,*,**], *Phacus anacoelus* var. *undulata* Skvotzov 1928[1,*,**], *Phacus ankylonoton* Pochmann 1942[1,*,**], *Phacus bergi* Prescott 1944[1,4,*,**], *Phacus caudatus* Hübner 1886[1,4,*,**], *Phacus chloroplastes var. incisa* Prescott[1,*,**], *Phacus chloroplastes* Prescott 1944[1,*,**], *Phacus circumflexus* Pochmann 1942[4,*,**], *Phacus curvicauda* Svirenko 1915[1,*,**], *Phacus helikoides* Pochmann 1942[1,*,**], *Phacus lemmermannii* Svirenko 1915[1,*,**], *Phacus longicauda* (Ehrenberg) Dujardin 1841[1,3,4,*,**], *Phacus limnophilus* (Lemmermann) E.W.Linton & A.Karnkowska-Ishikawa, 2010[4,*,**], *Phacus monilatus* (Stokes) Lemmerman 1901[4,*,**], *Phacus orbicularis var. caudatus* Skvortzow[1,*,**], *Phacus orbicularis* K.Hübner 1886[1,3,4,*,**], *Phacus raciborskii* Drezepolski 1925[1,*,**], *Phacus pleuronectes* (O.F.Müller) Nitzsch ex Dujardin 1841[1,*,**], *Phacus pseudoswirewkoi* Prescott 1944[1,*,**], *Phacus segretii var. ovatum* Prescott[1,*,**], *Phacus segretii* Allorge & Lefèvre 1925[1,*,**], *Phacus spiralis* Allegre & T.L. Jahn 1943[4,*,**], *Phacus spirogyra var. maxima* Prescott[1,*,**], *Phacus suecicus* Lemmermann 1910[1,*,**], *Phacus swirenkoi* Skvortzov 1928[1,4,*,**], *Phacus tortus* (Lemmermann) Skvortzov 1928[1,*,**], *Phacus triqueter* (Ehrenberg) Perty 1852[1,*,**], *Phacus* sp. Dujardin 1841[3,4,*,**]

[1]Jha et al. (2009), [2]Mohanty and Adhikary (2013), [3]Srichandan et al. (2015a), [4]Srichandan et al. (2015b)
*Inventorized during survey in post-restoration period (2000–2014)
**New records during survey in post-restoration period (reported for the first time from the lagoon)

Table 11.6 List of Chrysophyta species from Chilika

Phylum	Ochrophyta
Class	Dictyochophyceae
Order	Dictyochales
Family	Dictyochaceae
	Dictyocha fibula Ehrenberg 1839[1,*,**], *Dictyocha* sp. Ehrenberg 1837[1,2,*,**], *Octactis octonaria* (Ehrenberg) Hovasse 1946[2,*,**]
Class	Chrysophyceae
Order	Chromulinales
Family	Dinobryaceae
	Dinobryon sertularia Ehrenberg 1834[1,*,**]
Family	Chromulinaceae
	Uroglena sp. Ehrenberg 1834[1,*,**]

[1]Srichandan et al. (2015a), [2]Srichandan et al. (2015b)
*Inventorized during survey in post-restoration period (2000–2014)
**New records during survey in post-restoration period (reported for the first time from the lagoon)

(*Dictyocha fibula*, *Dictyocha* sp., and *Octactis octonaria*) were representative of marine water environment (Table 11.6). However, this particular group of phytoplankton is largely understudied with respect to diversity and distribution in Chilika lagoon and possible application in long-term lagoonal environmental monitoring. The present work suggests more comprehensive study on the Chrysophyta taxa in Chilika lagoon in future.

Table 11.7 List of Xanthophyta species from Chilika

Phylum	Xanthophyta (=Ochrophyta)
Class	Xanthophyceae
Order	Mischococcales
Family	Gloeobotrydaceae
	Gloeobotrys limneticus (G.M.Smith) Pascher 1938[2,*,**]
Order	Tribonematales
Family	Tribonemataceae
	Tribonema bombycinum (C.Agardh) Derbès & Solier in Castagne 1851[1,2,*]
Order	Mischococcales
Family	Ophiocytiaceae
	Ophiocytium variable Bohl.[2,*,**]

[1]Adhikary and Sahu (1992), [2]Jha et al. (2009)
*Inventorized during survey in post-restoration period (2000–2014)
**New records during survey in post-restoration period (reported for the first time from the lagoon)

11.2.7 Xanthophyta

Xanthophyta are generally known as yellow green algae. These are non-motile, unicellular or colonial eukaryotic algae exhibits unique pigmentation which gives a yellow or fresh green appearance. This group of photosynthetic algae primarily occurs in freshwater, although a substantially found in marine environments. Literature suggests that mostly yellow green algae incline to be ecologically limited to small water bodies (Sahoo and Kumar 2015). This class characteristically possesses chlorophyll-a, β carotene and xanthophylls. The diversity of Xanthophyta in aquatic environment is large, but their biology, ecology, and biogeography are known for only a few of the more common taxa.

In Chilika lagoon, only three taxa (*Gloeobotrys limneticus*, *Tribonema bombycinum*, and *Ophiocytium variable*) of Xanthophyta have been reported (Adhikary and Sahu 1992; Jha et al. 2009) (Table 11.7). These species have been observed in freshwater zone i.e. northern sector and brackish water zone i.e. central sector. However, Xanthophyta is mostly understudied in Chilika lagoon. Further to clarify these data gaps and uncertainties, careful efforts, longer monthly studies, and the use of modern taxonomic keys are need to be implemented in phytoplankton monitoring programs. All the three species reported from Chilika were inventorized during post-restoration period.

11.3 Spatio-Temporal Distribution of Phytoplankton

Seasonal and spatial variability in phytoplankton communities of Chilika lagoon has been broadly described by Srichandan et al. (2015a, b) over an annual and inter-annual scale (Table 11.8). The survey conducted between 2011 and 2012, reported

Table 11.8 Spatio-temporal variations in some dominant species of phytoplankton recorded from Chilika lagoon

Season/ sector	Phytoplankton taxa 2011–2012 (Srichandan et al. 2015a)	2012–2013 (Srichandan et al. 2015b)	2013–2014 (Srichandan et al. 2015b)
Southern sector			
Monsoon	*Dictyocha* sp., *Thalassiothrix longissima*, *Gymnodinium* sp.	*Prorocentrum micans*, *Prorocentrum cordatum*, *Diploneis weissflogii*	*Diploneis* sp., *Alexandrium* sp., *Dictyocha* sp.
Post-monsoon	*Pleurosigma normanii*, *Pleurosigma* sp., *Gymnodinium* sp.	*Dictyocha* sp., *Diploneis weissflogii*, *Ceratium fusus*	*Gonyaulax* sp., *Alexandrium* sp., *Prorocentrum cordatum*
Pre-monsoon	*Dictyocha* sp., *Synedra* sp., *Synedra ulna*	*Dictyocha* sp., *Synedra* sp., *Pleurosigma* sp.	*Gloeocapsa alpina*, *Gonyaulax* sp., *Alexandrium* sp.
Central sector			
Monsoon	*Pleurosigma normanii*, *Dictyocha* sp., *Protoperidinium* sp.	*Prorocentrum micans*, *Protoperidinium oceanicum*, *Protoperidinium* sp.	*Cylindrospermum* sp., *Phormidium* sp., *Dictyocha* sp.
Post-monsoon	Anabaena sp., *Pleurosigma* sp., *Pleurosigma normanii*	*Dictyocha* sp., *Anabaena* sp., *Cocconeis placentula*	*Phormidium* sp., *Anabaena* sp., *Diploneis* sp.
Pre-monsoon	*Prorocentrum cordatum*, Synedra sp., *Diploneis elliptica*	*Synedra* sp., *Navicula* sp., *Nitzschia* sp.	*Phormidium* sp., *Alexandrium* sp., *Gonyaulax* sp.
Northern sector			
Monsoon	*Anabaena* sp., *Eudorina* sp., *Mougeotia* sp.	*Euglena* sp., *Actinastrum* sp., *Trachelomonas* sp.	*Trachelomonas* sp., *Phormidium* sp., *Anabaena* sp.
Post-monsoon	*Trachelomonas* sp., *Oscillatoria* sp., *Aphanocapsa* sp.	*Anabaena* sp., *Trachelomonas* sp., *Trachelomonas lefevrei*	*Cylindrospermum* sp., *Anabaena* sp., *Strombomonas tambowika*
Pre-monsoon	*Cylindrospermum* sp., *Aphanocapsa* sp., *Anabaena flos-aquae*	*Cylindrospermum* sp., *Anabaena* sp., *Trachelomonas* sp.	*Cylindrospermum* sp., *Gomphosphaeria* sp., *Anabaena* sp.
Outer Channel			
Monsoon	*Nitzschia* sp., Pleurosigma sp., *Aphanizomenon* sp.	*Thalassionema nitzschioides*, *Navicula transitrans*, *Pleurosigma normanii*	*Phormidium* sp., *Cylindrospermum* sp., *Alexandrium* sp.
Post-monsoon	*Chaetoceros* sp., *Surirella* sp., *Pleurosigma normanii*	*Surirella* sp., *Thalassiosira* sp., *Coscinodiscus* sp.	*Gyrosigma fasciola*, *Amphiprora* sp., *Spiro-gyra* sp.
Pre-monsoon	*Thalassiosira subtili*, *Pseudonitzschia pungens*, *Pleurosigma* sp.	*Nitzschia* sp., *Surirella* sp., *Amphora* sp.	*Pseudonitzschia* sp., *Alexandrium* sp., *Cyclotella* sp.

that Bacillariophyta such as *Nitzschia* sp., *Chaetoceros* sp., and *Thalassiosira subtilis* were ubiquitous during monsoon, post-monsoon and pre-monsoon, respectively in outer channel of the lagoon (Srichandan et al. 2015a). In contrast, Chrysophyta (*Dictyocha* sp.) dominated the phytoplankton communities during monsoon and pre-monsoon in southern sector. However, Bacillariophyta (*Pleurosigma normanii*) was predominant during post-monsoon. High abundance of freshwater Cyanophyta (*Anabaena* sp., *Cylindrospermum* sp.) and Euglenophyta (*Trachelomonas* sp.) was recorded in the northern sector of the lagoon. Central sector was dominated by species of Bacillariophyta (*Pleurosigma normanii*), Cyanophyta (*Anabaena* sp.) and Dinophyta (*Prorocentrum cordatum*) during monsoon, post-monsoon, and pre-monsoon seasons, respectively.

Subsequently, the survey conducted between 2012 and 2013 have shown that the phytoplankton communities in the central sector of the lagoon were dominated by *Prorocentrum micans*, *Dictyocha* sp. and *Synedra* sp. during monsoon, post-monsoon and pre-monsoon, respectively (Srichandan et al. 2015b). In the northern sector, the species such as *Euglena* sp., *Trachelomonas* sp. and *Actinastrum* sp. thrived well during monsoon period while *Anabaena* sp. and *Trachelomonas* sp. dominated during post-monsoon and pre-monsoon seasons. In the southern sector, *Prorocentrum micans* was predominant during monsoon while *Dictyocha* sp. during both post-monsoon and pre-monsoon seasons. However, phytoplankton flora of outer channel was mainly represented by *Thalassionema nitzschioides* in monsoon, *Surirella* sp. in post-monsoon and *Nitzschia* sp. in pre-monsoon.

Another survey undertaken during the period 2013–2014 have shown that in the central sector, freshwater Cyanophyta, i.e. *Phormidium* sp. was mostly represented in post-monsoon and pre-monsoon period while *Cylindrospermum* sp. was largely represented during monsoon (Srichandan et al. 2015b). In the northern sector, *Cylindrospermum* sp. was the significant species during post-monsoon and pre-monsoon period while *Trachelomonas* sp. was more abundant during monsoon season. In outer channel, the most abundant species encountered during monsoon, post-monsoon and pre-monsoon were *Phormidium* sp., *Gyrosigma fasciola*, and *Pseudonitzschia* sp. respectively. In southern sector, epipelic Bacillariophyta (*Diploneis* sp.) dominated during monsoon period where as toxic Dinophyta (*Gonyaulax* sp.) and Cyanophyta (*Gloeocapsa alpina*) dominated during post-monsoon and pre-monsoon season, respectively.

In addition to the general spatio-temporal trends with respect to physico-chemical forcing the phytoplankton community of the lagoon well responded to the extreme climatic events such as tropical cyclone *Phailin* (Srichandan et al. 2015b). An increase in freshwater Cyanophyta *Cylindrospermum* sp., have been observed in central sector and outer channel of Chilika lagoon during post-*Phailin* period. Further, it was also suggested that the enhanced growth of *Cylindrospermum* sp. was attributed to the sediment-resuspension along with physical upward movement. Tropical cyclone *Phailin* had a significant impact on the phytoplankton community composition of southern sector of the lagoon. Toxic dinophytes (e.g. *Alexandrium* sp., *Gonyaulax* sp., *Prorocentrum cordatum*) have been observed in considerably higher number during post-*Phailin* period.

11.4 Phytoplankton Population Density

Typically, in an estuarine ecosystem, phytoplankton abundance is highest during dry season (pre-monsoon) while lowest abundance is recorded during wet (monsoon) season (Perumal et al. 2009; Prabhahar et al. 2011). The pre-monsoon season is usually characterized with increase in salinity, enhanced temperature, sufficient solar irradiance and stable environmental conditions (Saravanakumar et al. 2008). In contrast, heavy rainfall, cloudy sky, river/terrestrial run-off, induced high turbidity limit the light availability in water column and reduce salinity causes reduction in phytoplankton density during monsoon (Perumal et al. 2009). Although in Chilika lagoon several studies deciphered the time scale phytoplankton community structure. However, a well marked spatial and temporal variations in phytoplankton population density has been reported by Srichandan et al. (2015a, b) (Fig. 11.2). In Chilika lagoon, overwhelming dominance of a benthic Bacillariophyta i.e. *Pleurosigma normanii* was observed during monsoon season (Srichandan et al. 2015a). It was suggested that disturbance of benthic habitat by wind and water current was the main factor for the occurrence of large number of this benthic pennate phytoplankton in the surface water. Other factors could be use of mechanized boat for fishing and dredging operations which also cause re-suspension of bottom sediments in water column.

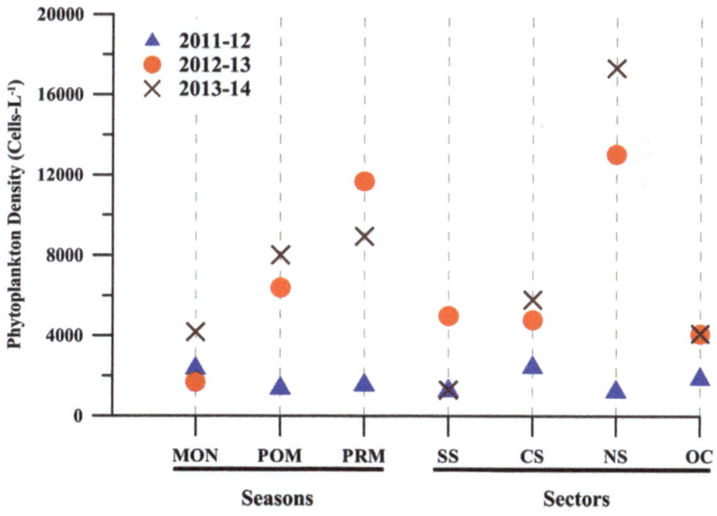

Fig. 11.2 Phytoplankton population density at spatio-temporal scale. *NS* Northern Sector, *CS* Central Sector, *SS* Southern Sector, *OC* Outer Channel, *MON* monsoon, *POM* post-monsoon, *PRM* pre-monsoon

11.5 Spatial and Seasonal Variation in Phytoplankton Abundance

Srichandan et al. (Srichandan et al. 2015a, b) have reported a clear spatial variation in phytoplankton density with respect to four ecological sectors of the Chilika lagoon (Fig. 11.3). Euglenophyta dominated the phytoplankton communities at lower salinity zone (i.e. northern sector), while Bacillariophyta were ubiquitous throughout the higher salinity zones (i.e., southern, central and outer channel) (Srichandan et al. 2015a). When tropical cyclone *Phialin* hit the lagoon in October 2013, it caused a drastic reduction in salinity (avg. 1.9 ppt) resulting proliferation of Cyanophyta in central sector besides northern sector (Srichandan et al. 2015b).

A marked temporal variation in phytoplankton density with respect to seasons has also been described by Srichandan et al. (Srichandan et al. 2015a, b) (Fig. 11.4). Seasonal changes in freshwater influx during monsoon appeared to be a controlling factor in determining the phytoplankton species composition and their abundances. The survey conducted between year 2011 and 2012 have shown that Bacillariophyta were the most dominant group in the lagoon irrespective of the season albeit with

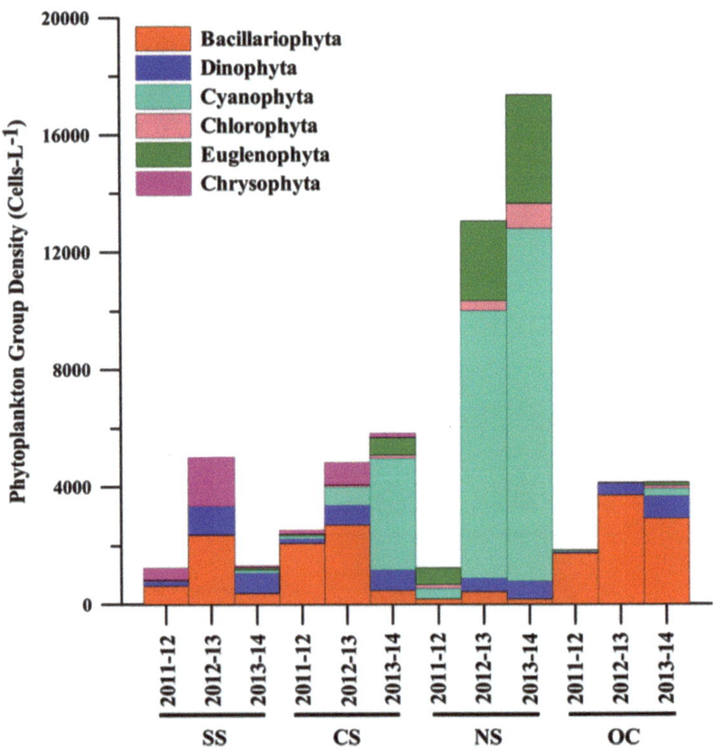

Fig. 11.3 Phytoplankton community composition of Chilika lagoon at spatial scale. *SS* Southern Sector, *CS* Central Sector, *NS* Northern Sector, *OC* Outer Channel

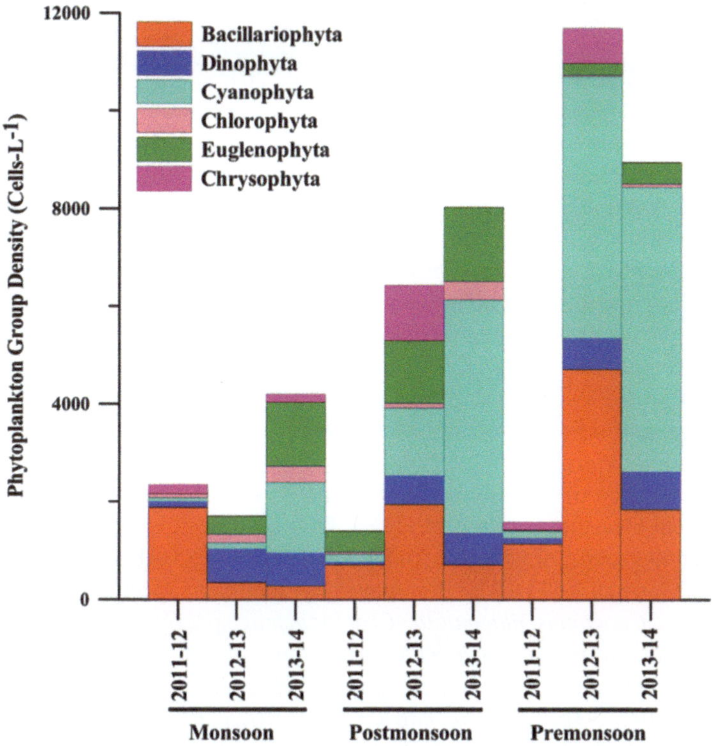

Fig. 11.4 Phytoplankton community composition of Chilika lagoon at temporal scale

varying cell densities (Srichandan et al. 2015a). Bacillariophyta were more abundant in monsoon season with mean cell density of 1879 cells L^{-1}, which subsequently decreased to 710 cells L^{-1} in post-monsoon season and further increased to 1134 cells L^{-1} in pre-monsoon season.

11.6 Phytoplankton and Environmental Variables

Phytoplankton communities in a lagoon are largely determined by a series of environmental parameters such as temperature, light, pH, salinity, dissolved oxygen, wind force, and tidal rhythm. In many estuaries, salinity has been considered as a key environmental variable for controlling the distribution and phytoplankton community composition. For example, in Schelde estuary in Belgium and Netherlands (Lionard et al. 2005), Suwannee River estuary in Florida (Quinlan and Phlips 2007), Bach Dang estuary in Vietnam (Rochelle-Newall et al. 2011), Pearl River Estuary in South China (Zhang et al. 2014) and Passos River estuary in Northeast Brazil (Aquino et al. 2015) salinity determined the spatial and temporal distribution of phytoplankton. In Chilika lagoon, salinity played a crucial role by governing the abundance and distribution of phytoplankton (Patnaik 1973; Patnaik and Sarkar 1976; Panigrahi et al. 2009; Srichandan et al. 2015a, b). Patnaik (1973) have determined that appearance and disappearance of freshwater, brackishwater and marine forms of phytoplankton mostly depended on the salinity conditions of the lagoon. Further, Raman et al. (1990) and Srichandan et al. (2015a) have determined that salinity was the predominant factor in controlling the distribution of phytoplankton in the Chilika lagoon. For instance, Srichandan et al. (2015a) have observed Dinophyta and Chrysophyta as the dominant phytoplankton groups in southern sector due to stable salinity regime. In outer channel, marine phytoplankton forms were prevalent due to higher salinity regime because of direct connectivity to the Bay of Bengal. Due to high freshwater discharge from rivers, northern sector was mostly represented by freshwater phytoplankton forms. Further, due to inter-mixing of freshwater and seawater, central sector was represented by both freshwater and marine phytoplankton taxa.

Depending upon the salinity preference and according to the biotic categories in the ecological classification, the phytoplankton communities have been classified into 3 different groups; oligohaline (0–5 ppt), mesohaline (5–18 ppt), and polyhaline (>18 ppt) (Marshall 1993). The dominant phytoplankton species *Amphiprora* sp., *Amphora* sp., *Cocconeis placentula*, *Coscinodiscus* sp., *Cyclotella* sp., *Diploneis* sp., *Diploneis weissflogii*, *Navicula transitans*, *Navicula* sp., *Nitzschia* sp., *Pleurosigma normanii*, *Pleurosigma* sp., *Surirella* sp., *Synedra* sp., *Alexandrium* sp., *Gonyaulax* sp., *Prorocentrum micans*, *Prorocentrum cordatum*, *Protoperidinium* sp., *Anabaena* sp., *Cylindrospermum* sp., *Phormidium* sp., *Spirogyra* sp., *Euglena* sp., *Trachelomonas* sp., and *Dictyocha* sp. had a wide salinity preference ranging from oligohaline, mesohaline and polyhaline in Chilika lagoon (Srichandan et al. 2015b). Few species such as *Gomphosphaeria* sp., *Actinastrum* sp., *Trachelomonas lefevrei*, and *Strombomonas tambowika* were found only at oligohaline regions while *Thalassiosira* sp. was restricted only to polyhaline regions. Some species (*Pseudonitzschia* sp., *Thalassionema nitzschioides*, *Tripos fusus*, and *Protoperidinium oceanicum*) were observed both at mesohaline and polyhaline regions but were entirely absent in oligohaline regions. Species viz. *Gyrosigma fasciola*, *Gloeocapsa alpina* were distributed only in the oligohaline and mesohaline regions of the Chilika lagoon.

Nitrate and phosphate has been considered limiting nutrient to algal growth (Mukhopadhyay et al. 2006; Gle et al. 2008). Chu et al. (2014) observed in a highly turbid estuary of Southeast Asia that inorganic nutrient concentrations and their respective ratios were found to be principal factors that structured phytoplankton diversity and influenced the emergence of potentially toxic species. In Chilika lagoon, maximum nitrate and phosphate concentrations were recorded during pre-monsoon season. It was also suggested that higher nitrogenous nutrient concentration during pre-monsoon could be related to higher residence time of the water in the lagoon during the low-flow period (pre-monsoon) (Srichandan et al. 2015b).

Besides seasonal variability, nitrate and phosphate concentrations also show distinct spatial variability. For instance, freshwater head of tropical estuaries such as Tagus (Portugal) and Bach Dang (Vietnam) estuaries are greatly influenced by direct riverine inputs which reflect higher nutrient loading (Brogueira et al. 2007; Chu et al. 2014). Similarly, Srichandan et al. (2015b) have observed that freshwater zone (i.e. northern sector) of the lagoon displayed the higher amount of nitrate and phosphate due to riverine inputs compared to other three sectors of the lagoon. In Chilika lagoon it has been shown that nitrate ($r = -0.295$, $p < 0.05$) and phosphate ($r = -0.284$, $p < 0.05$) has great influence on phytoplankton communities especially on the Dinophyta abundance and diversity (Srichandan et al. 2015b). Similar to nitrate and phosphate concentrations, marked spatio-temporal variation in silicate concentration has also been observed in estuarine ecosystems. For instance, persistently higher silicate concentration was reported in monsoon season in Zuari estuary (India), (Patil and Anil 2011). Similarly in Chilika lagoon, maximum silicate concentration has been observed during monsoon (Srichandan et al. 2015a, b). It was suggested that decreased silicate concentration in pre-monsoon could be due to the utilization of silicate by a large number of Bacillariophyta for the synthesis of their shells. This was evident from a strong negative correlation between chlorophyll (Chl-*a)* and silicate during pre-monsoon ($r = -0.331$, $p < 0.05$). The source of silicate in lagoon is mainly the heavy inflow of freshwater from riverine distributaries and land drainage of catchment area.

Turbidity has been frequently cited as a key factor controlling the distribution, abundance and diversity of phytoplankton in estuaries. For instance, in Dhamra River Estuary (India) and Na Thap River Estuary (Thailand), distributions and compositions of phytoplankton have been reported to have relationship with changes in turbidity (Palleyi et al. 2011; Lueangthuwapranit et al. 2011). In Chilika lagoon, several studies have mentioned turbidity as the major controlling factor of primary producer (Patnaik 1973; Srichandan et al. 2015a, b). It was also observed that passage of tropical cyclone *Phailin* increased the turbidity (221.4 NTU (nephelometric turbidity units)) via influx of exogenous material of terrestrial origin and restricted the development of phytoplankton bloom after *Phailin*. In addition, satellite remote sensing imagery has also revealed that the phytoplankton biomass did not change much due to high turbidity prevailing in the lagoon after *Phailin* (Srichandan et al. 2015b). Furthermore, many studies have shown that phytoplankton community structure is highly correlated with pH. For example, a positive correlation between Cyanophyta abundance and pH has been noted in the estuarine region of southeastern coast of Tamilnadu, India (Ramanathan et al. 2013). Similarly, a strong positive correlation between Cyanophyta abundance and pH ($r = 0.450$, $p < 0.01$) has been observed in Chilika lagoon (Srichandan et al. 2015b). Thus phytoplankton flora of Chilika lagoon is susceptible to change under the influence of mainly salinity, light availability, pH, and nutrients resulting heterogeneity in species composition, and population size of phytoplankton.

11.7 Future Directions

Compared to understanding on the microplankton (20–200 μm), there are significant knowledge gaps regarding the species composition of picoplankton and nanoplankton in Chilika lagoon. Detailed literature search indicated that the genetic diversity of picophytoplankton community of Chilika lagoon is completely unexplored and warrants a thorough investigation using high-throughput DNA sequencing. In fact, the molecular genetic diversity of picophytoplankton, as such, from any of Indian coastal ecosystems remains poorly understood. This necessitates the application of high-throughput DNA sequencing in the area of phytoplankton ecology to understand the diversity and distribution of smaller size phytoplankton. Further, intense monitoring is necessary to study the dynamics of phytoplankton population with respect to tidal and diurnal variation in Chilika lagoon. Further, climate change is recognized as a major threat to the survival of species and integrity of ecosystems world wide. In Chilika lagoon, rise in water temperatures by 0.39 °C in a decade have already been observed (Pandey 2015). Changes in the size-structure of phytoplankton communities in response to warming are now being documented across a range of ecosystem types and spatial scales. Therefore, further intensive studies on phytoplankton dynamics in Chilika lagoon in the context of climate change assumes greater importance. Apart from response to varying temperature, phytoplankton plays an important role in cloud formation by producing dimethyle sulfide which acts as cloud condensation nuclei. Hence, role of lagoon phytoplankton in such aspects need to be investigated. Since the lagoon is prone to anthropogenic pollution and deterioration of water quality, possibilities and scope of phytoremediation strategies should be explored. As the lagoon supports livelihood of millions of fisher folk who depend on the capture fisheries, the feeding habit of planktivore fishes should be explored for possible implementation of production enhancement strategies.

References

Adhikary SP, Sahu JK (1992) Distribution and seasonal abundance of algal forms in Chilika Lake, east coast of India. Jpn J Limnol 53:197–205

Alkawri AAS, Ramaiah N (2010) Spatio-temporal variability of dinoflagellate assemblages in different salinity regimes in the West Coast of India. Harmful Algae 9:153–162

Aquino EP, Figueiredo LGP, Borges GCP, Ferreira LC, Passavante JZDO, Gloria MD, Silva-Cunha GD (2015) Seasonal and spatial variation in phytoplankton community structure of an estuary in Northeastern Brazil. Trop Ecol 56(1):125–131

Bec B, Collos Y, Souchu P, Vaquer A, Lautier J, Fiandrino A, Benau L, Orsoni V, Laugier T (2011) Distribution of picophytoplankton and nanophytoplankton along an anthropogenic eutrophication gradient in French Mediterranean coastal lagoons. Aquat Microb Ecol 63:29–45. http://sci-hub.tw/10.3354/ame01480

Biswas K (1932) Algal flora of the Chilika lake. Mem Asiat Soc Bengal 11:65–198

Brewin RJW, Lavender SJ, Hardman-Mountford NJ, Hirata T (2010) A spectral response approach for detecting dominant phytoplankton size class from satellite remote sensing. Acta Oceanol Sin 29(2):14–32. http://sci-hub.tw/10.1007/s13131-010-0018-y

Brogueira MJ, Oliveira MDR, Cabeçadas G (2007) Phytoplankton community structure defined by key environmental variables in Tagus estuary, Portugal. Mar Environ Res 64(5):1–29. http://sci-hub.tw/10.1016/j.marenvres.2007.06.007

Burkholder JM, Glibert PM, Skelton HM (2008) Mixotrophy, a major mode of nutrition for harmful algal species in coastal waters. Harmful Algae 8:77–93

Cabeçadas L (1999) Phytoplankton production in Tagus estuary (Portugal). Oceanol Acta 22:51–65

Canini ND, Metillo EB, Azanza RV (2013) Monsoon influenced phytoplankton community structure in a Philippine mangrove estuary. Trop Ecol 54(3):331–343

Chen B, Xu Z, Zhou Q, Chen C, Gao Y, Yang S, Ji W (2010) Long-term changes of phytoplankton community in Xiagu waters of Xiamen, China. Acta Oceanol Sin 29(6):104–114. http://sci-hub.tw/10.1007/s13131-010-0081-4

Cheung MK, Au CH, Chu KH, Kwan HS, Wong CK (2010) Composition and genetic diversity of picoeukaryotes in subtropical coastal waters as revealed by 454 pyrosequencing. ISME J 4:1053–1059. http://sci-hub.tw/10.1038/ismej.2010.26

Chu TV, Torréton J-P, Mari X, Nguyen HMT, Pham KT, Pham TT, Bouvier T, Bettarel Y, Pringault O, Bouvier C, Rochelle-Newall E (2014) Nutrient ratios and the complex structure of phytoplankton communities in a highly turbid estuary of Southeast Asia. Environ Monit Assess 186(12):8555–8572. http://sci-hub.tw/10.1007/s10661-014-4024-y

Cloern JE, Dufford R (2005) Phytoplankton community ecology: principles applied in San Francisco Bay. Mar Ecol Prog Ser 285:11–28

Conley DJ, Paerl HW, Howarth RW, Boesch DF, Seitzinger SP, Havens KE, Lancelot C, Likens GE (2009) Controlling eutrophication: nitrogen and phosphorus. Science 323:1014–1015

Cremer H, Sangiorgi F, Wagner-Cremer F, Mcgee V, Lotter AF, Visscher H (2007) Diatoms (Bacillariophyceae) and Dinoflagellate Cysts (Dinophyceae) from Rookery Bay, Florida, USA. Caribb J Sci 43:23–58

Daugbjerg N, Henriksen P (2001) Pigment composition and rbcL sequence data from the silicoflagellate Dictyocha speculum: a heterokont alga with pigments similar to some haptophytes. J Phycol 37:1110–1120

Dawson EY (1966) Marine botany: an introduction. Holt, Rinehart and Winston, New York. 371p

Devasundaram MP, Roy JC (1954) A preliminary study of the plankton of the Chilka Lake for the year 1950–1951. In: Symposium on marine and fresh-water plankton in the Indo-Pacific, Indo-Pacific, Fish. Com. Publication, pp 1–7

Diez B, Pedros-Alio C, Massana R (2001) Study of genetic diversity of eukaryotic picoplankton in different oceanic regions by small-subunit rRNA gene cloning and sequencing. Appl Environ Microbiol 67(7):2932–2941

Gaines G, Elbrächter M (1987) Heterotrophic nutrition. In: Taylor FJR (ed) The biology of dinoflagellates. Botanical monograph, vol 21. Blackwell Scientific Publications, Oxford, pp 224–268

Ganguly D, Patra S, Muduli PR, Vishnu Vardhan K, Abhilas KR, Robin RS, Subramanian BR (2015) Influence of nutrient input on the trophic state of a tropical brackish water lagoon. J Earth Syst Sci 124(5):1005–1017

George B, Nirmal Kumar JI, Kumar RN (2012) Study on the influence of hydro-chemical parameters on phytoplankton distribution along Tapi estuarine area of Gulf of Khambhat, India. Egypt J Aqua Res 38:157–170

Gle G, Amo YD, Sautour B, Laborde P, Chardy P (2008) Variability of nutrients and phytoplankton primary production in a shallow macrotidal coastal ecosystem (Arcachon Bay, France). Estuar Coast Shelf Sci 76:642–656

Glibert PM, Legrand C (2006) The diverse nutrient strategies of harmful algae: focus on osmotrophy. In: Granéli E, Turner J (eds) The ecology of harmful algae. Springer, New York, pp 163–176

Goebel NL, Edwards CA, Zehr JP, Follows MJ, Morgan SG (2013) Modeled phytoplankton diversity and productivity in the California current system. Ecol Model 264:37–47. http://sci-hub.tw/10.1016/j.ecolmodel.2012.11.008

Goma J, Rimet F, Cambra J, Hoffmann L, Ector L (2005) Diatom communities and water quality assessment in Mountain Rivers of the upper Segre basin (La Cerdanya, Oriental Pyrenees). Hydrobiologia 551:209–225

Gómez F, Gorsky G (2003) Annual microplankton cycles in Villefranche Bay, Ligurian Sea, NW Mediterranean. J Plankton Res 25(4):323–339

Guerrero F, Rodriguez V (1998) Existence and significance of Acartia grani resting eggs (Copepoda: Calanoida) in sediments of a coastal station in the Alboran Sea (SE Spain). J Plankton Res 20:305–314

Gupta GVM, Sarma VVSS, Robin RS, Raman AV, Kumar M, Rakesh M, Subramanian BR (2008) Influence of net ecosystem metabolism in transferring riverine organic carbon to atmospheric CO_2 in a tropical coastal lagoon (Chilka Lake, India). Biogeochemistry 87(3):265–285

Harris JM, Vinobaba P (2012) Impact of water quality on species composition and seasonal fluctuation of plankton of Batticaloa lagoon, Sri Lanka. J Ecosyst Ecogr 2(4):1–6. http://sci-hub.tw/10.4172/2157-7625.1000117

Harsha TS, Malammanavar SG (2004) Assessment of phytoplankton density in relation to environmental variables in Gopalaswamy pond at Chitradurga, Karnataka. J Environ Biol 25:113–116

Henriksen P, Knipschildt F, Moestrup Ø, Thomsen HA (1993) Autoecology, life history and toxicology of the silicoflagellate Dictyocha speculum (Silicoflagellata, Dictyochophyceae). Phycologia 32:29–39

Huang L, Jian W, Song X, Huang X, Liu S, Qian P, Yin K, Wu M (2004) Species diversity and distribution for phytoplankton of the Pearl River estuary during rainy and dry seasons. Mar Pollut Bull 49:588–596

Jeong HJ, Yoo YD, Kim JS, Seong KA, Kang NS, Kim TH (2010) Growth, feeding and ecological roles of the mixotrophic and heterotrophic dinoflagellates in marine planktonic food webs. Ocean Sci J 45(2):65–91. http://sci-hub.tw/10.1007/s12601-010-0007-2

Jha BC, Vass KK, Panda S, Bhatta KS (2009) Algal biodiversity of Chilika lagoon. CIFRI, Barrackpore and CDA, Bhubaneswar, pp 1–144. ISSN 0970-616X

Jochem F, Babenerd B (1989) Naked Dictyocha speculum—a new type of phytoplankton blooms in the Western Baltic. Mar Biol 103:373–379

Kirchman DL, Suzuki Y, Garside C, Ducklow HW (1991) High turnover rates of dissolved organic carbon during a spring phytoplankton bloom. Nature, London 352:612–614

Krebs CJ (1994) Ecology: the experimental analysis of distribution and abundance, 4th edn. Harper Collins College Publishers, New York, p 801

Kudela RM, Peterson TD (2009) Influence of a buoyant river plume on phytoplankton nutrient dynamics: what controls standing stocks and productivity? J Geophys Res 114:15

Kumar NSV, Hosmani SP (2006) Algal biodiversity in fresh water and related physic chemical factors. J Nat Environ Pollut Technol 5:37–40

Laskar HS, Gupta S (2009) Phytoplankton diversity and dynamics of Chatla floodplain lake, Barak Valley, Assam, North East India – a seasonal study. J Environ Biol 30:1007–1012

Leander BS, Keeling PJ (2004) Early evolutionary history of dinoflagellates and apicomplexans (Alveolata) as inferred from HSP90 and actin phylogenies. J Phycol 40:341–350

Lie AAY, Wong CK, Lam JYC, Liu JH, Yung YK (2011) Changes in the nutrient ratios and phytoplankton community after declines in nutrient concentrations in a semi-enclosed bay in Hong Kong. Mar Environ Res 71(3):178–188. http://sci-hub.tw/10.1016/j.marenvres.2011.01.001

Lionard M, Muylaert K, Gansbeke DV, Vyverman W (2005) Influence of changes in salinity and light intensity on growth of phytoplankton communities from the Schelde river and estuary (Belgium/The Netherlands). Hydrobiologia 540:105–115. http://sci-hub.tw/10.1007/S10750-004-7123-X

Lueangthuwapranit C, Sampantarak U, Wongsai S (2011) Distribution and abundance of phytoplankton: influence of salinity and turbidity gradients in the Na Thap River, Songkhla Province, Thailand. J Coast Res 27(3):585–594. http://sci-hub.tw/10.2112/JCOASTRES-D-10-00123.1

Macedo MF, Duarte P, Mendes P, Ferreira JG (2001) Annual variation of environmental variables, phytoplankton species composition and photosynthetic parameters in a Costal lagoon. J Plankton Res 23(7):719–732

Marshall HG (1993) Monitoring phytoplankton populations in the Chesapeake Bay, USA. Idee Ekologii Tom 3, Ser Szkice nr 2:45–60

Martini E (1977) Systematics, distribution and stratigraphical application of silicoflagellates. In: Ramsey ATSE (ed) Oceanic micropaleontology. Academic, London, pp 1327–1343

Masuda LSM, Moser GAO, Barrera-Alba JJ (2011) Variação temporal do fitoplâncton no estuarino de Santos (SP) [Temporal variation of phytoplankton in the Santos estuary]. Braz J Aquat Sci Technol 15:79–93

McQuoid MR, Godhe A, Nordberg K (2002) Viability of phytoplankton resting stages in the sediments of a coastal Swedish fjord. Eur J Phycol 37:191–201

Mohanty D, Adhikary SP (2013) Assessment of changes in the algal diversity of Chilika lagoon after opening of New Mouth to Bay of Bengal. J Water Resour Prot 5:611–623

Mohanty SK, Mishra SS, Khan M, Mohanty RK, Mohapatra A, Pattnaik AK (2015) Ichthyofaunal diversity of Chilika Lake, Odisha, India: an inventory, assessment of biodiversity status and comprehensive systematic checklist (1916–2014). Check List 11(6):1–19. http://sci-hub.tw/10.15560/11.6.1817

Moreira-González A, Seisdedo-Losa M, Muñoz-Caravaca A, Comas-González A, Alonso-Hernández C (2014) Spatial and temporal distribution of phytoplankton as indicator of eutrophication status in the Cienfuegos Bay, Cuba. J Integr Coast Zone Manag 14(4):597–609

Moreno-Díaz G, Rojas-Herrera AA, Violante-González J, González-González J, Acevedo JLR, Ibáñez SG (2015) Temporal variation in composition and abundance of phytoplankton species during 2011 and 2012 in Acapulco Bay, Mexico. Open J Mar Sci 5:358–367

Mukherjee M, Karna SK, Suresh VR, Manna RK, Panda D, Sharma AP, Pati MK, Mandal S, Ali Y (2016) Dinoflagellate diversity and distribution in Chilika Lagoon with description of new records. Indian J Geo Mar Sci 45:999–1009

Mukhopadhyay SK, Biswas H, De TK, Jana TK (2006) Fluxes of nutrients from the tropical river Hooghly at the land-ocean boundary of Sundarbans, NE Coast of Bay of Bengal. Indian J Mar Sys 6:9–21

Myklestad SM (2000) Dissolved organic carbon from phytoplankton. In: Wangersky P (ed) The handbook of environmental chemistry [D], Marine chemistry. Springer, Berlin, pp 111–148

Naik S, Acharya BC, Mohapatra A (2009) Seasonal variation of phytoplankton in Mahanadi estuary, east coast of India. Indian J Mar Sci 38:184–190

Newton A, Icely JD, Cristina S, Brito A, Cardoso AC, Colijn F, Riva SD, Gertz F, Hansen JW, Holmer M, Ivanova K, Leppäkoski E, Canu DM, Mocenni C, Mudge S, Murray N, Pejrup M, Razinkovas A, Reizopoulou S, Pérez-Ruzafa A, Schernewski G, Schubert H, Carr L, Solidoro C, Viaroli P, Zaldívar JM (2014) An overview of ecological status, vulnerability and future perspectives of European large shallow, semi-enclosed coastal systems, lagoons and transitional waters. Estuar Coast Shelf Sci 140:95–122. http://sci-hub.tw/10.1016/j.ecss.2013.05.023

Newton A, Brito AC, Icely JD, Derolez V, Clara I, Angus S, Schernewski G, Inácio M, Lillebø AI, Sousa AI, Béjaoui B (2018) Assessing, quantifying and valuing the ecosystem services of coastal lagoons. J Nat Conserv 44:50–65

Ning X, Cloern JE, Cole BE (2000) Spatial and temporal variability of picocyanobacteria Synechococcus sp. in San Francisco Bay. Limnol Oceanogr 45:695–702

Olson RR, Olson MH (1989) Food limitation of planktotrophic marine invertebrate larvae: does it control recruitment success. Annu Rev Ecol Evol Syst 20:225–247

Onodera J, Takahashi K (2005) Silicoflagellate fluxes and environmental variations in the northwestern North Pacific during December 1997–May 2000. Deep-Sea Res I 52:371–388

Onyema IC (2013) Phytoplankton bio-indicators of water quality situations in the Iyagbe Lagoon, South-Western Nigeria. J Life Phy Sci 4(2):93–107
Palleyi S, Kar RN, Panda CR (2011) Influence of water quality on the biodiversity of phytoplankton in Dhamra River Estuary of Odisha Coast, Bay of Bengal. J Appl Sci Environ Manag 15 (1):69–74
Pandey K (2015) Lakes warming faster than oceans and India's Chilika is no exception: global study. http://www.downtoearth.org.in/news/lakes-warming-faster-than-oceans-and-india-s-chilika-is-no-exception-global-study-52243
Panigrahi S, Wikner J, Panigrahy RC, Satapathy KK, Acharya BC (2009) Variability of nutrients and phytoplankton biomass in a shallow brackish water ecosystem Chilika Lagoon, India. Limnology 10:73–85
Pasquaud S, David V, Lobry J, Girardin M, Sautour B, Elie P (2010) Exploitation of trophic resources by fish under stressful estuarine conditions. Mar Ecol Prog Ser 400:207–219
Patil JS, Anil AC (2008) Temporal variation of diatom benthic propagules in a monsoon influenced tropical estuary. Cont Shelf Res 28(17):2404–2416. http://sci-hub.tw/10.1016/j.csr.2008.06.001
Patil JS, Anil AC (2011) Variations in phytoplankton community in a monsoon influenced tropical estuary. Environ Monit Assess 182:291–300
Patnaik S (1973) Observation on the seasonal fluctuating of plankton in the Chilika lake. Ind J Fish 20:43–45
Patnaik S, Sarkar SK (1976) Observations on the distribution of phytoplankton in Chilika lake. J Inland Fish Soc India 8:38–48
Pérez-Ruzafa A, Marcos C, Pérez-Ruzafa IM, Pérez-Marcos M (2011) Coastal lagoons: "transitional ecosystems" between transitional and coastal waters. J Coast Conserv 15(3):369–392
Perumal NV, Rajkumar M, Perumal P, Rajasekar KT (2009) Seasonal variations of plankton diversity in the Kaduviyar estuary, Nagapattinam, southeast coast of India. J Environ Biol 30 (6):1035–1046
Phlips EJ, Badylak S, Lynch TC (1999) Blooms of the picoplanktonic cyanobacterium Synechococcus in Florida Bay, a subtropical inner-shelf lagoon. Limnol Oceanogr 44:1166–1175
Prabhahar C, Saleshrani K, Enbarasan R (2011) Studies on the ecology and distribution of phytoplankton biomass in Kadalur coastal zone, Tamilnadu, India. Curr Bot 2(3):26–30
Quinlan EL, Phlips EJ (2007) Phytoplankton assemblages across the marine to lowsalinity transition zone in a blackwater dominated estuary. J Plankton Res 29(5):401–416. http://sci-hub.tw/10.1093/plankt/fbm024
Rakhesh M, Madhavirani KSVKS, Charan Kumara B, Raman AV, Kalavati C, Prabhakara Rao Y, Rosamma S, Ranga Rao V, Gupta GVM, Subramanian BR (2015) Trophic–salinity gradients and environmental redundancy resolve mesozooplankton dynamics in a large tropical coastal lagoon. Reg Stud Mar Sci 1:72–84
Raman AV, Satyanarayana C, Adiseshasai K, Phani Prakash K (1990) Phytoplankton characteristics of Chilka lake, a brackish water lagoon along east coast of India. Ind J Mar Sci 19 (4):274–277
Ramanathan G, Sugumar R, Jeevarathinam A, Rajarathinam K (2013) Studies of Cyanobacterial distribution in estuary region of southeastern coast of Tamilnadu. Indian J Algal Biomass Util 4 (3):26–34
Rath J, Adhikary SP (2008) Biodiversity assessment of algae in Chilika Lake, East Coast of India. In: Monitoring and modelling lakes and coastal environments. Springer, Dordrecht, pp 22–33
Redfield AC, Ketchum BH, Richards FA (1963) The influence of organisms on the composition of sea-water. In: Hill MN (ed) The sea, vol 2. Wiley, New York, pp 26–77
Resende P, Azeiteiro U, Pereira MJ (2005) Diatom ecological preferences in a shallow temperate estuary (Ria de Aveiro, Western Portugal). Hydrobiologia 544:77–88
Rigual-Hernández AS, Bárcena MA, Sierro FJ, Flores JA, Hernnández-Almeida I, Sanchez-Vidal-A, Palanquez A, Heussner S (2010) Seasonal to interannual variability and geographic distribution of the silicoflagellate fluxes in the Western Mediterranean. Mar Micropaleontol 77:46–57

Rochelle-Newall EJ, Chu VT, Pringault O, Amouroux D, Arfi R, Bettarel Y, Bouvier T, Bouvier C, Got P, Nguyen TMH, Mari X, Navarro P, Duong TN, Cao TTT, Pham TT, Ouillon S, Torréton JP (2011) Phytoplankton distribution and productivity in a highly turbid, tropical coastal system (Bach Dang Estuary, Vietnam). Mar Pollut Bull 62:2317–2329. http://sci-hub.tw/10.1016/j.marpolbul.2011.08.044

Sahoo D, Kumar S (2015) Xanthophyceae, euglenophyceae and dinophyceae. In: The algae world, volume 26 of the series cellular origin, life in extreme habitats and astrobiology, pp 259–305. http://sci-hub.tw/10.1007/978-94-017-7321-8_9

Sahu G, Mohanty AK, Samantara MK, Satpathy KK (2014) Seasonality in the distribution of dinoflagellates with special reference to harmful algal species in tropical coastal environment, Bay of Bengal. Environ Monit Assess 186:6627–6644. http://sci-hub.tw/10.1007/s10661-014-3878-3

Saify T, Chaghtai SA, Parveen A, Durrani A (1986) Hydrology and periodicity of phytoplankton in the sewage fed Motia pond, Bhopal (India). Geobios 13(5):199–203

Samanta B, Bhadury P (2014) Analysis of diversity of chromophytic phytoplankton in a mangrove ecosystem using rbcL gene sequencing. J Phycol 50(2):328–340. http://sci-hub.tw/10.1111/jpy.12163

Saravanakumar A, Rajkumar M, Thivakaran GA, Serebiah JS (2008) Abundance and seasonal variations of phytoplankton in the creek waters of western mangrove of Kachchh-Gujarat. J Environ Biol 29:271–274

Sasamal SK, Panigrahy RC, Misra S (2005) Asterionella blooms in the northwestern Bay of Bengal during 2004. Int J Remote Sens 26(17):3853–3858

Selvaraj GSD, Thomas VJ, Khambadkar LR (2003) Seasonal variation of phytoplankton and productivity in the surf zone and backwater at Cochin. J Mar Biol Assoc India 45(1):9–19

Sin Y, Wetzel RL, Anderson IC (2000) Seasonal variations of size-fractionated phytoplankton along the salinity gradient in the York River Estuary, Virginia (USA). J Plankton Res 22:1945–1960

Sondergaard M, Riemann BJ, Orgensen NOG (1985) Extracellular organic carbon (EOC) released by phytoplankton and bacterial production. Oikos 45:323–332

Srichandan S, Kim JY, Bhadury P, Barik SK, Muduli PR, Samal RN, Pattnaik AK, Rastogi G (2015a) Spatiotemporal distribution and composition of phytoplankton assemblages in a coastal tropical lagoon: Chilika, India. Environ Monit Assess 187:47. http://sci-hub.tw/10.1007/s10661-014-4212-9

Srichandan S, Kim JY, Kumar A, Mishra DR, Bhadury P, Muduli PR, Rastogi G (2015b) Interannual and cyclone-driven variability in phytoplankton communities of a tropical coastal lagoon. Mar Pollut Bull 101(1):39–52

Stal LJ (2012) Cyanobacterial mats and stromatolites. In: Whitton BA (ed) The ecology of cyanobacteria. Springer, London, pp 65–125

Stevenson RJ, Pan Y (1999) Assessing environmental conditions in rivers and streams with diatoms. In: Stoermer EF, Smol JP (eds) The diatoms: applications for the environment and earth science. Cambridge University Press, Cambridge

Takahashi K, Onodera J, Katsuki K (2009) Significant populations of seven-sided Distephanus (Silicoflagellata) in the sea-ice covered environment of the central Arctic Ocean, summer 2004. Micropaleontology 55:313–325

Valdes-Weaver LM, Piehler MF, Pinckney JL, Howe KE, Rossignol K, Paerl HW (2006) Long-term temporal and spatial trends in phytoplankton biomass and class level composition in the hydrologically variable Neuse-Pimlico estuarine continuum, North Carolina, USA. Limnol Oceanogr 51(3):1410–1420

Varona-Cordero F, GutiErrez-mendieta FJ, Castillo MEMd (2010) Phytoplankton assemblages in two compartmentalized coastal tropical lagoons (Carretas-Pereyra and Chantuto-Panzacola, Mexico). J Plankton Res 32(9):1283–1299. http://sci-hub.tw/10.1093/plankt/fbq043

Wehr JD, Descy JP (1998) Use of phytoplankton in large river management. J Phycol 34:741–749

Wissel B, Fry B (2005) Tracing Mississippi River influences in estuarine food webs of coastal Louisiana. Oecology 144:659–672

Xiao X, Sogge H, Lagesen K, Tooming-Klunderud A, Jakobsen KS, Rohrlack T (2014) Use of high throughput sequencing and light microscopy show contrasting results in a study of phytoplankton occurrence in a freshwater environment. PLoS One 9(8):e106510

Yoo KI (1991) Population dynamics of dinoflagellate community in Masan bay with a note on the impact of environmental parameters. Mar Pollut Bull 23:185–188

Zhang X, Zhang J, Huang X, Huang L (2014) Phytoplankton assemblage structure shaped by key environmental variables in the Pearl River Estuary, South China (2014). J Ocean Univ China 13:73–82

Chapter 12
Fish and Fisheries of Chilika: Post-Restoration Scenario

Surya K. Mohanty and Debabrata Panda

Abstract This chapter presents the synthesis of research findings on finfish and shellfish fauna and their status, fish yield potential and fisheries output, fishing methods, issues and management needs for sustainable fisheries development of Chilika lagoon, Odisha, India. The hydrological intervention by opening a new connection (mouth) to the Bay of Bengal helped for eco-restoration of the brackish water lake during the year 2000. This intervention witnessed spectacular fisheries enhancement with 203.62% increase in fish landings during post restoration period as compared to pre restoration period. Chilika fish fauna are largely migratory inhabiting marine, brackish and freshwater environments. Out of 317 nos. of finfish species reported from Chilika, about 41% of species were documented only during post restoration period. More or less a stabilized yield level has been maintained during the last 14 years of post restoration period. However, about 48 threatened fish species under the category of critically endangered, endangered, vulnerable and near threatened were also reported from Chilika. Moreover, the mean annual fish landing (12,136 tonnes) during post restoration period is hovering close to the mean maximum sustainable yield (MSY) estimate (11,500 tonnes). In spite of significant enhancement in fish yield during post restoration period, the overall fishery scenario does not seem to be encouraging since majority of the commercially important fishes are being captured below their size at first maturity, indicating over exploitation of the resources. Even presently, the ecosystem is struggling with several challenges including maintaining suitable salinity gradient for enhanced fishery productivity in a climate change scenario and enforcement of stringent regulatory fishery management measures for holistic sustainable fishery development of the Chilika.

The correction to this chapter is available online at https://doi.org/10.1007/978-3-030-33424-6_17

S. K. Mohanty (✉)
Chilika Development Authority, Bhubaneswar, Odisha, India

D. Panda
ICAR-Central Institute of Freshwater Aquaculture, Kausalyaganga, Bhubaneswar, Odisha, India

C. M. Finlayson et al. (eds.), *Ecology, Conservation, and Restoration of Chilika Lagoon, India*, Wetlands: Ecology, Conservation and Management 6,
https://doi.org/10.1007/978-3-030-33424-6_12

Keywords Chilika lagoon · Post-restoration · Fish fauna · Conservation · Management

12.1 Introduction

Chilika (Fig. 12.1), the first Indian Ramsar Site (latitudes 19° 20′13.06″ N and 19° 54′47.02″ N and longitudes 085° 06′49.15″ E and 085° 35′32.87″ E) spanning 1100 km^2 area in monsoon and 906 km^2 in dry season is the largest brackish water lake (lagoon) in Asia (Mangla 1989; Dujovny 2009). Chilika is an assemblage of marine, brackish and freshwater ecosystems which is influenced by three hydrologic sub-systems namely, Mahanadi river distributaries, western catchment and Bay of Bengal (Mohapatra et al. 2007; Mohanty et al. 2008). Chilika is one of the region's finest repositories of aquatic biodiversity and a rich fishery resource supporting the livelihoods and nutritional security of more than 0.2 million local fishers (Mohanty et al. 2015). The unique and fragile ecosystem of Chilika gradually began to lose its ecological integrity due to coastal process, significant decrease in salinity regime and degraded drainage basin with associated anthropogenic impacts (Mohanty et al. 2009). Between 1950 and 2000, its fishery was in a rapid decline trend when the landing reached its lowest by the end of the 1990's. The fishery suffered serious setbacks since the later part of the 1980s with the salinity level sharply decreasing. The recruitment corridors (outer channel and Palur canal) gradually got silted up, adversely affecting the recruitment of fish and shellfish seed from the sea into the

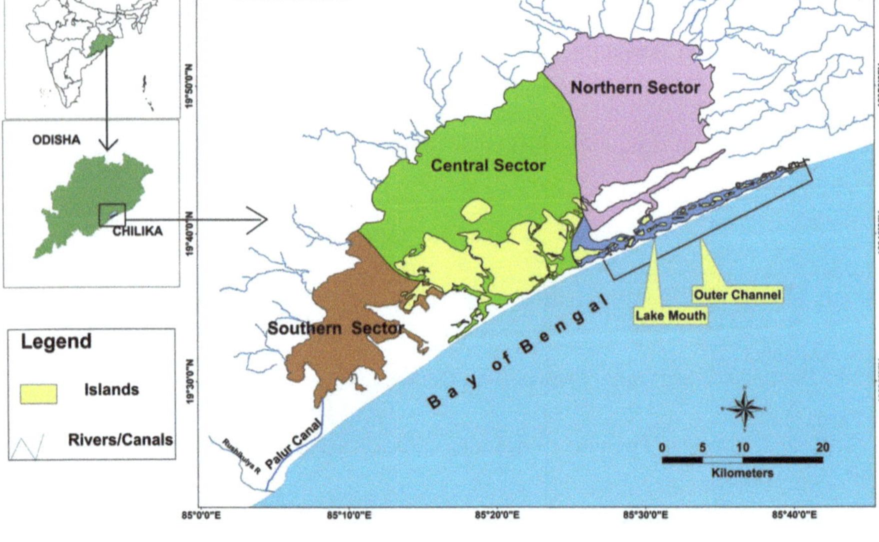

Fig. 12.1 Location map of Chilika Lagoon

Chilika, while silted up river mouth in the northern sector of the lagoon also affected freshwater seed recruitment from riverine sources (Mohanty et al. 2015). In the aftermath of the gradual reduction of seawater inflow due to closure of the old mouth and Palur canal, Chilika began transformation towards a freshwater ecosystem, causing substantial decrease in fish yield and ichthyofaunal composition. Continued degradation of the ecosystem, changes in ecological characteristics, overall loss of biodiversity and decline in productivity adversely affected the livelihoods of local communities. In 1993, Chilika was included in the Montreux Record (list of threatened Ramsar sites) (Pattnaik and Kumar 2016).

Restoration of the fragile ecosystem of Chilika and enhancement of its fisheries and bio-resources for the greater benefit of communities depending largely on fisheries for sustenance became imperative by the end of 1990's. The opening of the new mouth as a part of the hydrological intervention in September 2000 for wetland restoration rapidly showed a positive effect with dramatic enhancement in fisheries, overall ecology and biodiversity (Pattnaik and Kumar 2016). The annual fish catch from Chilika in 2001–2002 increased seven folds as compared to the catch in the previous year (1999–2000). Based on the recommendations of the Ramsar Advisory Mission, Chilika was delisted from the Montreux Record in 2002, and the restoration recognized with Ramsar Award and Evian Special Prize.

This chapter aims to enlighten about the fish and fishery enhancement of Chilika after hydrological intervention and challenges ahead. This presents a synthesis of monitoring and research information on Chilika ichthyofaunal and management issues. The first two sections contain discussion on the species richness and conservation status. Fishery characteristics, including gears and catch trends are discussed in the third section. The fourth section contains information on the biology and catch of nine commercially important species of Chilika. In the final section, management issues and recommendations for sustaining Chilika fisheries are discussed. A detailed review of the published literature on Chilika fisheries was worked out. Additionally, the unpublished data presented here were collected from the reports of the various research projects operated under the supervision of Chilika Development Authority (CDA), Govt. of Odisha, Bhubaneswar, India.

12.2 Fish and Shellfish Fauna of the Chilika

The first study of fish and shellfish fauna of Chilika was initiated by the Zoological Survey of India (ZSI) during 1914–1916, a century ago (Annandale and Kemp 1915; Chaudhuri 1916a, b) and the reports were published by 1923 (Chaudhuri 1917 and 1923; Hora 1923). Later, Jones and Sujansingani (1954) made further ichthyofaunal study in Chilika in the 1950s. Central Inland Fisheries Research Institute (CIFRI) conducted the 9 years organized ichthyofaunal inventory during 1957–1965 and contributed substantially to the ichthyofaunal record of the Chilika (Rajan et al. 1968). The first Chilika expedition was conducted during 1985–1987 by ZSI and the updated record of piscine fauna of Chilika was published in 1995 (Rama Rao 1995).

Prawn and crab fauna of the Chilika were studied by Reddy (1995) and Maya Deb (1995). Mohapatra et al. (2015) published the updated list of fish and shellfish fauna of Chilika. Mohanty et al. (2008) and Mohapatra et al. (2007) published the updated record of ichthyofaunal inventory from the Chilika. A comprehensive systematic checklist of 317 finfish fauna with updates on the taxonomy of fishes of Chilika for the period 1916–2014 was published by Mohanty et al. (2015), thus providing a baseline of the ichthyofaunal diversity study of the Chilika. However, the recent checklist of species including finfishes and shellfishes enlists 383 species (Mohapatra et al. 2015).

The ichthyofaunal of Chilika comprises 317 finfish species of 207 genera, in 88 families and 23 orders (Table 12.1); 31 species of prawns and lobsters of 16 genera in 11 families and single order (Table 12.2) and 35 species of brachyuran crabs of 27 genera in 15 families and single order (Table 12.3). The ichthyofaunal classification adopted mainly follows Eschmeyer and Fong (2014) and Nelson (2006), with genera and species arranged alphabetically. The classification of prawns and shrimps follows the scheme recommended by Radhakrishnan et al. (2012) and for brachyuran crabs as recommended by Ng et al. (2008) and Jeyabaskaran et al. (2002).

Of the total 317 finfish species, 271 (85.49%) and 46 (14.51%) were categorized as migratory and resident species respectively. The migratory species are either seasonal migrants or incidental visitors to the Chilika from the sea as well as the inflowing rivers (Mohanty et al. 2015). These include 14 catadromous and 13 anadromous species. The finfish diversity is represented by 35.65% marine, 43.85% brackish & 20.50% freshwater species (Mohanty et al. 2015). Similarly, shellfish fauna included 23 marine/brackish prawn species, 6 freshwater prawn species, 35 marine/brackish crab species and 2 species of marine/brackish lobster species. Fish fauna of Chilika includes 2 endemic gobiid species (*Acentrogobius griseus* and *Bathygobius ostreicola*) and one exotic cichlid species (*Oreochromis mossambicus*).

Although there are 129 commercially important fish species, 6 species (*Mugil cephalus, Planiliza macrolepis, Daysciaena albida, Eleutheronema tetradactylum, Lates calcarifer* & *Etroplus suratensis*) are considered as very high-value target species. The current ichthyofaunal analysis further documented for the first time 114 species of 48 families with ornamental value for home and public aquarium use, a resource with considerable economic potential (Mohanty et al. 2015). Further research is warranted to develop technologies for artificial propagation and rearing of important native ornamental fishes from Chilika. At least 56 species belonging to 29 families breed in the Chilika (Table 12.1). A complete list of ichthyofaunal characteristics of Chilika is presented in Table 12.4.

Of 29 prawn species, eight penaeid species & nine palaemonid species are considered as food prawns of Chilika. Similarly, out of 35 brachyuran crab species, eight portunid species have commercial importance. Two portunid species (*Scylla serrata* & *S. tranquebarica*) have high commercial value with high export potential. Two lobster species (*Panulirus ornatus* & *P. polyphagus*) do not form a commercial fishery in Chilika.

Table 12.1 Finfish fauna of Chilika (1916–2014). (Order-23, Family-88, Genus-207 & Species-317)

Family	Species	Environment	Conservation Status
Hemiscyllidae (Bamboo sharks)	*Chiloscyllium indicum* Gmelin (1789)[b]	M; F; B	NT[Δ]
Carcharhinidae (Requiem sharks)	*Carcharhinus leucas* Muller and Henle (1839)[b]	M; F; B	NT[Δ]
	Carcharhinus limbatus Muller and Henle (1839)	M; B	NT[Δ]
	Carcharhinus melanopterus Quoy and Gaimard (1824)	M; B	NT[Δ]
	Glyphis gangeticus Muller and Henle (1839)	M; F; B	Cr[Δ]
	Scoliodon laticaudus Muller and Henle (1838)[a]	M; B	NT[Δ]
Sphyrnidae (Hammerheaded shark)	*Eusphyra blochii* Cuvier (1816)[b]	M; B	NT[Δ]
	Sphyrna lewini Griffith and Smith (1834)[b]	M; B	EN[Δ]
Pristidae (Saw fish)	*Pristis clavata* Garman (1906)	M; B	EN [Δ]
Rhinobatidae (Guitar fishes)	*Rhynchobatus djiddensis* Forsskal (1775)[b]	M; B	Vu[Δ,B]
Dasyatidae (Stingrays)	*Himantura imbricata* Bloch and Schneider (1801)[aβ]	M; F; B	DD[Δ]
	Himantura marginata Blyth (1860)[b]	M; B	DD[Δ]
	Himantura uarnak Gmelin (1789)[a]	M; B	Vu[Δ]
	Himantura walga Muller and Henle (1841)[a]	M	NT[Δ]
	Pastinachus sephen Forsskal (1775) [aβ]	M; F; B	DD[Δ]
Myliobatidae (Eaglerays)	*Aetobatus flagellum* Bloch and Schneider (1801)[a]	M; B	EN[Δ]
	Aetobatus ocellatus Kuhl (1823)[a]	M	NE
	Aetomylaeus nichofii Bloch and Schneider (1801)[a]	M; B	Vu[Δ,B]
Notopteridae (Featherbacks)	*Chitala chitala* Hamilton (1822)[a]	F	EN[CAMP]
	Notopterus notopterus Pallas (1769)[aβ]	F; B	LC[Δ]
Elopidae (Tenpounders)	*Elops machnata* Forsskal (1775)[a]	M; B	LC[Δ]
Megalopidae (Tarpons)	*Megalops cyprinoides* Broussonet (1782)[a]	M; F; B	DD[Δ]

(continued)

Table 12.1 (continued)

Family	Species	Environment	Conservation Status
Anguillidae (Freshwater eels)	*Anguilla bengalensis* Gray (1831)[a]	M; F; B	NT[Δ]
	Anguilla bicolor McClelland (1844)[a]	M; F; B	NT[Δ]
Muraenidae (Moray eels)	*Strophidon sathete* Hamilton (1822)[a]	M; F; B	NE
Ophichthidae (Snake eels)	*Lamnostoma orientalis* McClelland (1844)	M; F; B	LC[Δ]
	Pisodonophis boro Hamilton (1822)[a]	M; F; B	LC[Δ]
	Pisodonophis cancrivorus Richardson (1848)	M; F; B	NE
Muraenesocidae (Pike congers)	*Congresox talabonoides* Bleeker (1853)[a]	M; B	VU[M,B]
	Muraenesox bagio Hamilton (1822)[b]	M; B	NE
	Muraenesox cinereus Forsskal (1775)[a]	M; F; B	VU[M,B]
Dussumieridae	*Dussumieria acuta* Valenciennes (1847)	M; F; B	NE
	Dussumieria elopsoides Bleeker (1849)[b]	M	NE
Clupeidae (Herrings and allies)	*Amblygaster leiogaster* Valenciennes (1847)[b]	M	NE
	Amblygaster sirm Walbaum (1792)	M	NE
	Anodontostoma chacunda Hamilton (1822)[a]	M; F; B	NE
	Corica soborna Hamilton (1822)[a]	M; F; B	LC[Δ]
	Ehirava fluviatilis Deraniyagala (1929)[b]	M; F; B	NE
	Escualosa thoracata Valenciennes (1847)[a]	M; F; B	NE
	Gonialosa manmina Hamilton (1822)[a]	F; B	VU[CAMP]
	Gudusia chapra Hamilton (1822)[a]	F; B	LC[Δ]
	Hilsa kelee Cuvier (1829) [aβ]	M; F; B	NE
	Nematalosa nasus Bloch (1795) [aβ]	M; F; B	LC[Δ]
	Sardinella fimbriata Valenciennes (1847)[b]	M; B	NE
	Sardinella longiceps Valenciennes (1847)[b]	M	LC[Δ]

(continued)

Table 12.1 (continued)

Family	Species	Environment	Conservation Status
	Sardinella melanura Cuvier (1829)	M	NE
	Tenualosa ilisha Hamilton (1822) [aβ]	M; F; B	VU[NBFGR,B]
	Tenualosa toli Valenciennes (1847)[b]	M; F; B	NE
Engraulidae (Anchovies)	*Setipinna phasa* Hamilton (1822)	F; B	LC[Δ]
	Stolephorus baganensis Hardenberg (1933)[a]	M; B	NE
	Stolephorus commersonnii Lacepede (1803)[a]	M; B	NE
	Stolephorus dubiosus Wongratana (1983)[a]	M; B	NE
	Stolephorus indicus Van Hasselt (1823)[aβ]	M; B	NE
	Thryssa gautamiensis Babu Rao (1971)[b]	M; B	DD[Δ]
	Thryssa hamiltonii Gray (1835)[a]	M; B	NE
	Thryssa kammalensoides Wongratana (1983)[β]	B	NE
	Thryssa malabarica Bloch (1795)[a]	M; B	NE
	Thryssa mystax Bloch and Schneider (1801)[a]	M; B	LC[Δ]
	Thryssa polybranchialis Wongratana (1983[aβ]	M	NE
	Thryssa purava Hamilton (1822) [aβ]	M; B	NE
	Thryssa setirostris Broussonet (1782)[b]	M; B	NE
	Thryssa vitrirostris Gilchrist and Thompson (1908)[b]	M; B	NE
Chirocentridae (Wolf herrings)	*Chirocentrus dorab* Forsskal (1775)	M; B	NE
Pristigasteridae (Pellonas)	*Ilisha elongata* Anonymous [Bennett] (1830)[b]	M; B	NE
	Ilisha megaloptera Swainson (1839)[a]	M; F; B	EN[NBFGR]
	Ilisha melastoma Bloch and Schneider (1801)	M; B	NE
	Opisthopterus tardoore Cuvier (1829)[b]	M; B	NE

(continued)

Table 12.1 (continued)

Family	Species	Environment	Conservation Status
Chanidae (Milkfish)	*Chanos chanos* Forsskal (1775)[a]	M; F; B	NE
Cyprinidae (Carps and minnows)	*Amblypharyngodon mola* Hamilton (1822)[a]	F	LC[Δ]
	Bangana ariza Hamilton (1807)	F	LC[Δ]
	Chela cachius Hamilton (1822)[a]	F; B	LC[Δ]
	Cirrhinus mrigala Hamilton (1822)[a]	F	LC[Δ]
	Cirrhinus reba Hamilton (1822)[a]	F	VU[CAMP]
	Crossocheilus latius Hamilton (1822)[a]	F; B	VU[NBFGR]
	Danio rerio Hamilton (1822) [β]	F	LC[Δ]
	Esomus danricus Hamilton (1822) [aβ]	F; B	LC[Δ]
	Catla catla Hamilton (1822)[a]	F; B	VU[CAMP]
	Labeo boga Hamilton (1822)[b]	F	LC[Δ]
	Labeo calbasu Hamilton (1822)[a]	F; B	LC[Δ]
	Labeo gonius Hamilton (1822)[b]	F	LC[Δ]
	Labeo rohita (Hamilton (1822)[a]	F; B	LC[Δ]
	Laubuka laubuca (Hamilton (1822)[a]	F; B	LC[Δ]
	Osteobrama peninsularis Silas (1952)[b]	F	DD[Δ]
	Osteobrama vigorsii Sykes (1839)	F; B	LC[Δ]
	Pethia ticto Hamilton (1822) [aβ]	F; B	LC[Δ]
	Puntius chola Hamilton (1822)[a]	F	LC[Δ]
	Puntius sophore (Hamilton (1822)[a]	F; B	LC[Δ]
	Puntius vittatus Day (1865)	F; B	LC[Δ]
	Rasbora daniconius Hamilton (1822)[a]	F; B	LC[Δ]
	Rasbora rasbora Hamilton (1822)[a]	F; B	LC[Δ]
	Salmostoma bacaila Hamilton (1822)[a]	F; B	LC[Δ]
	Systomus sarana Hamilton (1822)[a]	F; B	LC[Δ]

(continued)

Table 12.1 (continued)

Family	Species	Environment	Conservation Status
Cobitidae (Loaches)	*Lepidocephalichthys guntea* Hamilton (1822)	F; B	LC[Δ]
Bagridae (Bagrid catfishes)	*Mystus cavasius* Hamilton (1822)[a]	F; B	LC[Δ]
	Mystus gulio Hamilton (1822) [aβ]	F; B	LC[Δ]
	Mystus vittatus Bloch (1794) [aβ]	F; B	VU[CAMP,B]
	Sperata seenghala Sykes 1839)[a]	F; B	LC[Δ]
Siluridae (Eurasian catfishes)	*Ompok bimaculatus* Bloch (1794)[a]	F; B	EN[CAMP,B]
	Ompok pabda Hamilton (1822)[a]	F	VU[NBFGR]
	Wallago attu Bloch and Schneider (1801)[a]	F; B	NT[Δ]
Schilbeidae (Schilbid catfishes)	*Ailia coila* Hamilton (1822)[a]	F; B	VU[CAMP,B]
	Eutropiichthys vacha Hamilton (1822)	F; B	EN[CAMP,B]
	Silonia silondia Hamilton (1822)	F; B	VU[NBFGR]
Pangasiidae (Shark catfish)	*Pangasius pangasius* Hamilton (1822)[a]	F; B	VU[NBFGR,B]
Sisoridae (Sisorid catfish)	*Bagarius bagarius* Hamilton (1822)[a]	F; B	VU[NBFGR]
	Bagarius yarrelli Sykes (1839)[b]	F	EN[NBFGR]
Clariidae (Air-breathing catfish)	*Clarias magur* Hamilton (1822)[a]	F	EN[Δ]
Heteropneustidae (Airsac catfish)	*Heteropneustes fossilis* Bloch (1794)[a]	F; B	VU[NBFGR]
Ariidae (Sea catfish)	*Arius arius* Hamilton (1822)[a]	M; B	LC[Δ]
	Arius maculatus Thunberg (1792)	M; F; B	NE
	Nemapteryx caelata Valenciennes (1840) [aβ]	M; B	NE
	Osteogeneiosus militaris Linnaeus (1758) [aβ]	M; F; B	NE
	Plicofollis layardi Günther (1866)		NE
Plotosidae (Stinging catfishes)	*Plotosus canius* Hamilton (1822)[aβ]	M; F; B	NE
	Plotosus lineatus Thunberg (1787) [aβ]	M; B	NE
Synodontidae (Lizard fishes)	*Saurida tumbil* Bloch (1795)[b]	M	NE
	Synodus myops Forster (1801)[b]	M	NE

(continued)

Table 12.1 (continued)

Family	Species	Environment	Conservation Status
Mugilidae (Mullets)	*Ellochelon vaigiensis* Quoy and Gaimard (1825)	M; F; B	LC[Δ]
	Planiliza macrolepis Smith (1846)[a]	M; F; B	LC[Δ]
	Planiliza melinopterus Valenciennes (1836)[a]	M; F; B	LC[Δ]
	Chelon parsia Hamilton (1822)[a]	M; F; B	NE
	Planiliza subviridis Valenciennes (1836)[a]	M; F; B	NE
	Liza tade Bloch (1801)[a]	M; F; B	DD[Δ]
	Osteomugil cunnesius Valenciennes (1836) [aβ]	M; F; B	NE
	Crenimugil seheli Forsskal (1775)[a]	M; F; B	NE
	Valamugil speigleri Bleeker (1858)[a]	M; F; B	NE
	Mugil cephalus Linnaeus (1758)[a]	M; F; B	LC[Δ]
	Rhinomugil corsula Hamilton (1822) [aβ]	F; B	VU[NBFGR,B]
Atherinidae (Oldworld silversides)	*Atherinomorus duodecimalis* Valenciennes (1835)[b]	M; B	NE
	Atherinomorus lacunosus Forster (1801)[b]	M; F; B	NE
Aplocheilidae (Asian revulines)	*Aplocheilus panchax* Hamilton (1822) [aβ]	F; B	LC[Δ]
Belonidae (Needle fishes)	*Strongylura leiura* Bleeker (1850)[a]	M; B	NE
	Strongylura strongylura Van Hasselt (1823) [aβ]	M; B	NE
	Xenentodon cancila Hamilton (1822)[a]	M; F; B	LC[Δ]
Hemiramphidae (Halfbeaks)	*Hemiramphus far* Forsskal (1775)[b β]	M; B	NE
	Hyporhamphus limbatus Valenciennes (1847) [aβ]	M; F; B	LC[Δ]
Adrianichthydae (Adrianichthyids)	*Oryzias dancena* Hamilton (1822) [aβ]	F; B	LC[Δ]
Syngnathidae (Pipe fishes & Sea horses)	*Hippichthys cyanospilos* Bleeker (1854)[a aβ]	M; F; B	NE
	Hippocampus fuscus Ruppell (1838)[aβ]	M	VU[B]
	Ichthyocampus carce Hamilton (1822) [aβ]	M; F; B	LC[Δ]

(continued)

Table 12.1 (continued)

Family	Species	Environment	Conservation Status
Synbranchidae (Swamp eels)	*Ophisternon bengalense* McClelland (1844)[b]	F; B	LC[Δ]
Mastacembelidae (Spiny eels)	*Macrognathus aral* Bloch and Schneider (1801) [aβ]	F; B	LC[Δ]
	Macrognathus pancalus Hamilton (1822)[aβ]	F; B	LC[Δ]
	Mastacembelus armatus Lacepede (1800)[a]	F; B	LC[Δ]
Scorpaenidae (Scorpion fishes)	*Pterois radiata* Cuvier (1829)[a]	M	NE
Tetrarogidae (Waspfishes)	*Tetraroge niger* Cuvier (1829)[b]	M; F; B	LC[Δ]
Platycephalidae (Flatheads)	*Cociella crocodilus* Cuvier (1829)[b]	M; B	NE
	Kumococius rodericensis Cuvier (1829)[b]	M	NE
	Platycephalus indicus Linnaeus (1758)[a]	M; B	DD[Δ]
Ambassidae (Perchlets, glass fishes)	*Ambassis ambassis* Lacepede (1802)[a]	M; F; B	LC[Δ]
	Ambassis gymnocephalus Lacepede (1802) [aβ]	M; F; B	LC[Δ]
	Chanda nama Hamilton (1822)[a]	F; B	LC[Δ]
	Parambassis ranga Hamilton (1822) [aβ]	F; B	LC[Δ]
Latidae (Lates perches/Asian Seabass)	*Lates calcarifer* Bloch (1790)[a]	M; F; B	VU[P]
Serranidae (Groupers, Rock-cods)	*Epinephelus coioides* Hamilton (1822)[b]	M; B	NT[Δ]
	Epinephelus lanceolatus Bloch (1790)	M; B	Vu[Δ,B]
	Epinephelus malabaricus Bloch and Schneider (1801)[b]	M; B	NT[Δ]
	Epinephelus tauvina Forsskal (1775)[a]	M	DD[Δ]
Sillaginidae (Smealt whitings)	*Sillaginopsis panijus* Hamilton (1822)	M; F; B	NE
	Sillago sihama Forsskal (1775) [aβ]	M; B	NE
	Sillago vincenti Mc Kay (1980)[b]	M; B	NE
Lactariidae (False trevallies)	*Lactarius lactarius* Bloch and Schneider (1801)[b]	M; B	NE

(continued)

Table 12.1 (continued)

Family	Species	Environment	Conservation Status
Rachycentridae (Cobias)	*Rachycentron canadum* Linnaeus (1766)[a]	M; B	NE
Echeneidae (Sharksuckers, Discfishes)	*Echeneis naucrates* Linnaeus (1758)[a]	M; B	NE
Carangidae (Jacks, Trevallies, Pompanos &Scads)	*Alectis indica* Ruppell (1830)[a]	M; B	NE
	Alepes djedaba Forsskal (1775)[a]	M	NE
	Atule mate Cuvier (1833)[a]	M; B	NE
	Carangoides gymnostethus (Cuvier (1833)	M	NE
	Carangoides praeustus Anonymous [Bennett] (1830)[a]	M	NE
	Caranx ignobilis Forsskal (1775)	M; B	NE
	Caranx melampygus Cuvier (1833)	M; B	NE
	Caranx sexfasciatus Quoy and Gaimard (1825)[a]	M; F; B	LC$^{\Delta}$
	Megalaspis cordyla Linnaeus (1758)[a]	M; B	NE
	Parastromateus niger Bloch (1795)	M; B	NE
	Scomberoides commersonnianus Lacepede (1801)[b]	M; B	NE
	Scomberoides lysan Forsskal (1775)	M; B	NE
	Scomberoides tala Cuvier (1832)[a]	M	NE
	Scomberoides tol Cuvier (1832)[b]	M; B	NE
	Selar boops Cuvier (1833)[b]	M	NE
	Selar crumenophthalmus Bloch (1793)[b]	M	NE
	Selaroides leptolepis Cuvier (1833)[a]	M; B	NE
	Trachinotus blochii Lacepede (1801)	M; B	NE
	Trachinotus mookalee Cuvier (1832)[b]	M	NE
Leiognathidae (Pony fishes, Silverbellies)	*Aurigequulla fasciatus* Lacepede (1803)[b]	M; B	NE
	Deveximentum insidiator Bloch (1787)[a]	M	NE
	Deveximentum ruconius Hamilton (1822)[b]	M	NE

(continued)

Table 12.1 (continued)

Family	Species	Environment	Conservation Status
	Eubleekeria splendens Cuvier (1829)	M	LC[Δ]
	Gazza minuta Bloch (1795)[a]	M	LC[Δ]
	Karalla daura Cuvier (1829)	M	NE
	Karalla dussumieri Valenciennes (1835)[a]	M	NE
	Leiognathus equulus Forsskal (1775) [aβ]	M; B	LC[Δ]
	Nuchequula blochii Valenciennes (1835) [aβ]	M	NE
	Nuchequula gerreoides Bleeker (1851)[b]	M; B	NE
	Photopectoralis bindus Valenciennes (1835)[b]	M	NE
Lutjanidae (Snappers)	*Lutjanus argentimaculatus* Forsskal (1775)[a]	M; F; B	NE
	Lutjanus indicus Allen et al. (2013)[a]	M; B	NE
	Lutjanus johnii Bloch (1792)[a]	M; B	NE
	Lutjanus kasmira Forsskal (1775)[a]	M	NE
Datnioididae (Freshwater triple tails)	*Datnioides polota* Hamilton (1822) [aβ]	F; B	LC[Δ]
Gerreidae (Silver biddies)	*Gerres erythrourus* Bloch (1791)[b]	M; B	NE
	Gerres filamentosus Cuvier (1829)[a]	M; B	LC[Δ]
	Gerres limbatus Cuvier (1830)[a]	M; B	NE
	Gerres macracanthus Bleeker (1854)	M; B	NE
	Gerres oyena Forsskal (1775)[a]	M; B	NE
	Gerres phaiya Iwatsuki and Heemstra (2001)[a]	M; B	NE
	Gerres setifer Hamilton (1822) [aβ]	M; B	NE
Haemulidae (Grunts & Rubberlips)	*Plectorhinchus gibbosus* Lacepède (1802)	M	LC[Δ]
	Pomadasys argenteus Forsskal (1775)[a]	M; B	VU[B]
	Pomadasys kaakan Cuvier (1830)[b]	M; B	NE

(continued)

Table 12.1 (continued)

Family	Species	Environment	Conservation Status
	Pomadasys multimaculatus Playfair (1867)[b]	M; B	NE
Sparidae (Seabreams)	*Acanthopagrus berda* Forsskal (1775)[a]	M; B	NE
	Acanthopagrus longispinnis Valenciennes (1830)	M; B	NE
	Argyrops spinifer Forsskal (1775)	M	NE
	Crenidens crenidens Forsskal (1775)[a]	M	NE
	Rhabdosargus sarba Forsskal (1775)[a]	M; B	NE
Nemipteridae (Threadfin breams)	*Nemipterus japonicus* Bloch (1791)[b]	M	NE
Sciaenidae (croakers)	*Daysciaena albida* Cuvier (1830) [aβ]	M; B	NE
	Dendrophysa russelii Cuvier (1829) [aβ]	M; F; B	NE
	Johnius amblycephalus Bleeker (1855)	M; F; B	NE
	Johnius belangerii Cuvier (1830)[a]	M; B	NE
	Johnius carutta Bloch (1793)[b]	M; B	NE
	Johnius coitor Hamilton (1822)	M; B	LC[Δ]
	Johnius dussumieri Cuvier (1830)	M; B	NE
	Johnius macropterus Bleeker (1853)[a]	M	NE
	Nibea maculata Bloch and Schneider (1801)[b]	M	NE
	Otolithes ruber Bloch and Schneider (1801)[b]	M; B	NE
	Otolithoides biauritus Cantor (1849)	M	VU[M,B]
	Otolithoides pama Hamilton (1822)[a]	M; B	NE
	Paranibea semiluctuosa Cuvier (1830)[a]	M	NE
	Protonibea diacanthus Lacepede (1802)[a]	M; B	VU[M,B]
Polynemidae (Threadfin fishes)	*Eleutheronema tetradactylum* Shaw (1804) [aβ]	M; F; B	NE
	Leptomelanosoma indicum Shaw (1804)[a]	M; B	VU[M,B]
	Polydactylus plebeius Broussonet (1782)[b]	M; B	NE

(continued)

Table 12.1 (continued)

Family	Species	Environment	Conservation Status
	Polydactylus sextarius Bloch and Schneider (1801)[a]	M; B	NE
Mullidae (goatfishes)	*Upeneus sulphureus* Cuvier (1829)[b]	M; B	NE
Drepaneidae (Sicklefishes)	*Drepane punctata* Linnaeus (1758)[a]	M	NE
Monodactylidae (Moonies)	*Monodactylus argenteus* Linnaeus (1758)[a]	M; B	NE
	Monodactylus kottelati Pethiyagoda (1991)[b]	M; B	NE
Nandidae (leaf fishes)	*Nandus nandus* Hamilton (1822)[a]	F; B	LC[Δ]
Terapontidae (Terapon perches)	*Pelates quadrilineatus* Bloch (1790)	M; B	NE
	Terapon jarbua Forsskål (1775)[a]	M; B	LC[Δ]
	Terapon puta Cuvier (1829[aβ]	M; B	NE
	Terapon theraps Cuvier (1829[a]	M; B	LC[Δ]
Cichlidae (Cichlids)	*Etroplus suratensis* Bloch (1790) [aβ]	F; B	LC[Δ]
	Oreochromis mossambicus Peters (1852)[bβ]	F; B	NT[Δ]
Uranoscopidae (stargazers)	*Ichthyscopus lebeck* Bloch and Schneider (1801)[a]	M	NE
Blenniidae (blennies & allies)	*Omobranchus zebra* Bleeker (1868) [aβ]	M; B	NE
Eleotridae (gudgeons)	*Butis butis* Hamilton (1822)	M; B	LC[Δ]
	Eleotris fusca Forster (1801)	F; B	LC[Δ]
	Eleotris melanosoma Bleeker (1853)[bβ]	F; B	LC[Δ]
Gobiidae (gobies)	*Acentrogobius cyanomos* Bleeker (1849) [aβ]	M; B	NE
	Acentrogobius griseus Day (1876)	F; B	NE
	Acentrogobius masoni Day (1873)	F; B	NE
	Acentrogobius viridipunctatus Valenciennes (1837)	F; B	NE
	Acentrogobius madraspatensis Day (1868)	M; B	NE
	Bathygobius fuscus Ruppell (1830)	M; B	LC[Δ]
	Bathygobius ostreicola Chaudhuri (1916) [β]	B	DD[Δ]

(continued)

Table 12.1 (continued)

Family	Species	Environment	Conservation Status
	Brachygobius nunus Hamilton (1822)	F; B	NE
	Drombus globiceps Hora (1923)[a]	F; B	LC[Δ]
	Glossogobius giuris Hamilton (1822) [aβ]	F; B	LC[Δ]
	Gobiopterus chuno Hamilton (1822) [β]	F; B	DD[Δ]
	Oligolepis acutipennis Valenciennes (1837)	M; B	DD[Δ]
	Oligolepis cylindriceps Hora (1923) [aβ]	B	NE
	Oxyurichthys microlepis Bleeker (1849)[a]	M; B	NE
	Oxyurichthys tentacularis Valenciennes (1837)[a]	M; B	NE
	Parapocryptes rictuosus Valenciennes (1837)	M; B	NE
	Periophthalmus kalolo Lesson (1831)[a]	M; B	NE
	Psammogobius biocellatus Valenciennes (1837) [aβ]	M; B	LC[Δ]
	Pseudapocryptes elongatus Cuvier (1816)	F; B	LC[Δ]
	Pseudogobiopsis oligactis Bleeker (1875)	B	LC[Δ]
	Pseudogobius javanicus Bleeker (1856)	F; B	NE
	Stigmatogobius minima Hora (1923) [β]	B	NE
	Taenioides buchanani Day (1873)	F; B	NE
	Trypauchen vagina Bloch and Schneider (1801)[a]	M; B	NE
	Yongeichthys nebulosus Forsskal (1775)[b]	M; B	NE
Ephippidae (spadefishes)	*Ephippus orbis* Bloch (1787)[b]	M	NE
	Platax orbicularis Forsskal (1775)[b]	M; B	NE
Scatophagidae (scats)	*Scatophagus argus* Linnaeus (1766)[a]	F; B	LC[Δ]
Siganidae (Spinsfoots, Rabbitfishes)	*Siganus canaliculatus* Park (1797)[b]	M; B	NE
	Siganus javus Linnaeus (1766)[a]	M; B	NE
	Siganus vermiculatus Valenciennes (1835)[a]	M; B	LC[Δ]

(continued)

Table 12.1 (continued)

Family	Species	Environment	Conservation Status
Acanthuridae (surgeon fishes)	*Acanthurus mata* Cuvier (1829)[b]	M	LC[Δ]
	Acanthurus triostegus Linnaeus (1758)[b]	M	LC[Δ]
Sphyraenidae (barracudas)	*Sphyraena jello* Cuvier (1829)[b]	M	NE
	Sphyraena putnamae Jordan and Seale (1905)[b]	M	NE
Trichiuridae (Hairtail fishes)	*Eupleurogrammus glossodon* Bleeker (1860)[b]	M	NE
	Lepturacanthus savala Cuvier (1829)[b]	M; B	NE
	Trichiurus lepturus Linnaeus (1758)[b]	M; B	NE
Scombridae (mackerels, Seerfishes, tunas, albacores)	*Euthynnus affinis* Cantor (1849)[b]	M	LC[Δ]
	Rastrelliger kanagurta Cuvier (1816)[b]	M	DD[Δ]
	Scomberomorus lineolatus Cuvier (1829)[a]	M	LC[Δ]
Anabantidae (climbing perches)	*Anabas cobojius* Hamilton (1822)[a]	F	DD[Δ]
	Anabas testudineus Bloch (1792)[a]	F; B	DD[Δ]
Osphronemidae (Gouramies)	*Trichogaster fasciata* Bloch and Schneider (1801)[a]	F; B	LC[Δ]
	Trichogaster lalius Hamilton (1822)[a]	F	LC[Δ]
Channidae (snakeheads, Murrels)	*Channa gachua* Hamilton (1822)[b]	F	LC[Δ]
	Channa marulius Hamilton (1822)[b]	F	LC[Δ]
	Channa punctata Bloch (1793) [aβ]	F	LC[Δ]
	Channa striata Bloch (1793) [aβ]	F; B	LC[Δ]
Paralichthyidae (Lefteye flounders)	*Pseudorhombus arsius* Hamilton (1822)[a]	M; B	NE
	Pseudorhombus micrognathus Norman (1927)[b]	M	NE
	Pseudorhombus triocellatus Bloch and Schneider (1801)[b]	M	NE
Soleidae (soles)	*Brachirus orientalis* Bloch and Schneider (1801)[a]	M; F; B	NE
	Solea ovata Richardson (1846)[a]	M	NE

(continued)

Table 12.1 (continued)

Family	Species	Environment	Conservation Status
Cynoglossidae (tongue soles)	*Cynoglossus lida* Bleeker (1851)[b]	M	NE
	Cynoglossus lingua Hamilton (1822)[a]	M; B	NE
	Cynoglossus puncticeps Richardson (1846)[a]	M; B	NE
Triacanthidae (tripod fishes)	*Triacanthus biaculeatus* Bloch (1786) [aβ]	M; B	NE
Balistidae (triggerfishes)	*Abalistes stellaris* Bloch and Schneider (1801)[b]	M	NE
Tetraodontidae (puffers)	*Arothron reticularis* Bloch and Schneider (1801)	M; B	NE
	Arothron stellatus Anonymous (1798)	M; B	NE
	Chelonodontops patoca Hamilton (1822)[a]	M; B	NE
	Dichotomyctere fluviatilis Hamilton (1822)[a]	F; B	NE
	Gastrophysus oblongus Bloch (1786)[a]	M; B	NE
	Lagocephalus lunaris Bloch and Schneider (1801)	M; B	NE
	Leiodon cutcutia Hamilton (1822)[a]	F; B	LC[Δ]
Diodontidae (Porcupinefishes)	*Diodon hystrix* Linnaeus (1758)[b]	M	NE

[a]Post-restoration inventory; [β] Species breeding in the lagoon; M-Marine; F-Fresh; B-Brackish; Δ IUCN Red List Status; CAMP Report 1998; P Pooniah (1993); NBFGR National Bureau of Fish Genetic Resources (2010); CE/CR-Critically Endangered; VU-Vulnerable; NT-Near Threatened; LC-Least Concern; DD-Data Deficient; M-Menon (2004); B-Barman et al. (2007) NE-Not Evaluated; EN/E-Endangered.
[b]New records during post-restoration period;

12.3 Conservation Status

The conservation status of Chilika ichthyofauna was assessed based on IUCN red list database (IUCN 2014), www.fishbase.org (Froese and Pauly 2015), CAMP Report (Molur and Walker 1998), Zoological Survey of India (Barman et al. 2007) and other published literature (Ponniah 1993; Menon 2004 and Lakra et al. 2010). In total, 48 species in 28 families were classed as Critically Endangered (CR), Endangered (EN), Vulnerable (VU) and Near Threatened (NT). Further 88 species and 15 species were categorized respectively Least Concern (LC) and Data Deficient (DD). Number of species assessed for different categories of biodiversity status (Mohanty et al.

Table 12.2 Prawn & lobster fauna of Chilika (1915–2014). (Order-1, Family-11, Genus-16 & Species-31)

Family	Species	Habitat	Conservation Status
Prawn			
Penaeidae	*Metapenaeus affinis* H. milne Edwards (1837)[a]	Brackish; demersal	Not evaluated
	Metapenaeus dobsoni Miers (1878)[a]	Brackish; demersal	Not evaluated
	Metapenaeus ensis De Haan (1844)[b]	Brackish; demersal	Not evaluated
	Metapenaeus monoceros Fabricius (1798)[a]	Brackish; marine	Not evaluated
	Penaeus canaliculatus Oliver (1811)[b]	Marine	Not evaluated
	Fenneropenaeus indicus H. Milne Edwards (1837)[a]	Brackish; marine	Not evaluated
	Penaeus monodon Fabricius (1798)[a]	Brackish; demersal	Not evaluated
	Penaeus semisulcatus De Haan (1844)[a]	Brackish; marine	Not evaluated
Luciferidae	*Lucifer hanseni* Nobili (1905)	Marine	Not evaluated
Palaemonidae	*Cuapetes demani* Kemp (1915)[a]	Marine	Not evaluated
	Exopalaemon styliferus H. Milne Edwards (1840)[a]	Fresh; pelagic	Not evaluated
	Macrobrachium equidens Dana (1852)[b]	Fresh	Not evaluated
	Macrobrachium lamarrei lamarrei H. Milne Edwards (1837)[a]	Fresh	Not evaluated
	Macrobrachium malcolmsonii malcolmsonii H. Milne Edwards (1844)[a]	Fresh	Not evaluated
	Macrobrachium rosenbergii D. Man (1879)[b]	Fresh; benthic	Not evaluated
	Macrobrachium rude Heller (1862)[a]	Fresh; benthic	Not evaluated
	Macrobrachium scabriculum Heller (1862)	Fresh; benthic	Not evaluated
	Phycomenes indicus Kemp (1915)	Marine	Not evaluated
Alpheidae	*Alpheus lobidens* De Haan (1849	Marine; demersal	Not evaluated
	Alpheus malabaricus Fabricius (1775)	Brackish; demersal	Not evaluated
	Alpheus paludicola Kemp (1915)	Brackish; demersal	Not evaluated
	Athanas polymorphus Kemp (1915)	Brackish; benthic	Not evaluated
Ogyrididae	*Ogyrides striaticauda* Kemp (1915)	Marine; demersal	Not evaluated
Atyidae	*Caridina nilotica* Roux (1833)	Marine; fresh	Not evaluated
	Caridina propinqua De Mann (1908)[a]	Brackish	Not evaluated
Crangonidae	*Philocheras hendersoni* Kemp (1915)	Marine	Not evaluated
Pasiphaeidae	*Leptochela aculeocaudata* Paulson (1875)	Marine	Not evaluated

(continued)

Table 12.2 (continued)

Family	Species	Habitat	Conservation Status
Callianassidae	*Neocallichirus maxima* A. Milne-Edwards (1870)[a]	Marine	Not evaluated
Upogebiidae	*Wolffogebia heterocheir* Kemp (1915)[a]	Marine	Not evaluated
Lobster			
Palinuridae	*Panulirus ornatus* Fabricius (1798)[b]	Coastal, benthic	Least concern[Δ]
	Panulirus polyphagus Herbst (1793)[b]	Coastal/estuarine, benthic	Least concern[Δ]

[a]14 (Inventory of earlier recorded species)
[b]6 (New records during post-restoration period)

2015) is depicted in Fig. 12.2. The conservation status of 166 finfish species is globally unassessed. The conservation status of shellfishes indicated that except the two reported lobster species (least concern), the status is not evaluated for prawns and crabs globally.

Key threats to ichthyofaunal diversity of Chilika include siltation, encroachment of spawning grounds for illegal ghery (a type of pen enclosure) operation, year round obstruction by multiple gears in the two vital recruitment corridors (outer channel and Palur canal), large scale destructive fishing practices and unabated expansion of illegal large pen culture units (*gheries*).

12.4 Fishery Characteristics

12.4.1 Fishing Gears

Fishers in Chilika based on their nuanced understanding of ecosystem dynamics evolved a management regime based on partitioning of fishing grounds, use of diverse fishing gears and seasonal restrictions (Sekhar 2007). The traditional fishing gears of Chilika were described with an annotated list of 54 traditional nets with synonyms by Jhingran and Natarajan 1969. A paradigm shift in fishing gears used in Chilika was observed after introduction of polyamide and HDPE netting materials along with outboard boat engines. After the introduction of monofilament and multifilament twine-made nylon (PA-polyamide) nets and HDPE twines in the early eighties, the traditional gear materials like cotton and hemp and fishing methods such as *Jano, Dian, Uthapani, Split bamboo screen traps* etc. were gradually replaced by polyamide monofilament nets. The monofilament gill nets are passive gears made up of PA nylon monofilament and mainly operated for finfishes like, mullets, seabass, polynemids, catfishes, clupeiforms and sciaenids. The dimension of the net, mesh size, rigging and mode of operation of the gear varies based on the target species and ground conditions. Thus, there exist many types of gill nets in the

Table 12.3 Brachyuran crab fauna of Chilika (1915–2014). (Order-1, Family-15, Genus-27 & Species-35)

Family	Species	Habitat
Calappidae	*Ashtoret lunaris* Forsskal (1775)[b]	Marine; demersal
	Matuta planipes Fabricius (1798[a]	Coastal; brackish
Gecarcinucidae	*Cardisoma carnifex* Herbst (1796)	Mangrove swamp
Leucosiidae	*Philyra malefactrix* Kemp (1915)	Brackish; benthic
	Philyra alcocki Kemp (1915)[a]	Marine; brackish
Hymenosomatidae	*Elamena (Trigonoplax) cimex* Kemp (1915)	Marine
Epialtidae	*Doclea muricata* Herbst (1788)	Marine
Pilumnidae	*Benthopanope indica* de Man (1887)	Marine
Portunidae	*Charybdis callianassa* Herbst (1789)[b]	Coastal; marine
	Charybdis feriatus Linnaeus (1758)[b]	Coastal; brackish
	Podophthalmus vigil Fabricius (1798)[b]	Marine
	Portunus pelagicus Linnaeus (1758)[a]	Marine
	Portunus sanguinolentus Herbst (1783)[b]	Coastal; marine
	Scylla serrata Forsskal (1775)[a]	Coastal; marine
	Scylla tranquebarica Fabricius (1798)[b]	Coastal; brackish
	Thalamita crenata Milne-Edwards (1834)[a]	Coastal; brackish
Grapsidae	*Metopograpsus messor* Forsskal (1775)	Coastal;brackish
	Neosarmatium meinerti de Man (1887)	Marine
	Pachygrapsus propinquus de Man (1908	Marine
	Metopograpsus quadridentatus Stimpson (1858) [b]	Marine
Plagusiidae	*Plagusia squamosa* Herbst (1790)	Coastal benthic
Sesarmidae	*Nanosesarma batavicum* Moreira (1903)	Brackish; benthic
	Parasesarma plicatum Latreille (1806)	Marine
Varunidae	*Ptychognathus onyx* Alcock (1900)	Marine
	Varuna litterata Fabricius (1798)[a]	Marine; brackish
Camptandriidae	*Baruna socialis* Stebbing (1904)	Marine
	Camptandrium sexdentatum Stimpson (1858)	Marine; brackish
Dotillidae	*Dotilla pertinax* Kemp (1915)	Marine
	Dotilla intermedia de Man (1888)	Marine; brackish
	Dotilla myctiroides Milne-Edwards (1852)	Marine; brackish
Macrophthalmidae	*Euplax leptophthalmus* H. Milne Edwards (1852)	Brackish
Ocypodidae	*Ocypode ceratophthalmus* Pallas (1772)	Coastal; brackish
	Ocypode macrocera H. Milne Edwards (1852)[a]	Coastal; brackish
	Ocypode platytarsis H. Milne Edwards (1852)	Coastal; marine
	Uca annulipes Milne-Edwards (1837)	Brackish; benthic

[a]7 (Inventory of earlier recorded species)
[b]7 (New records during post-restoration period)

lagoon with different local names. A new type of three-walled gill net (trammel net) locally known as "*Dubi jaal*", a non-selective fishing gear with three different meshes, was introduced in Chilika since 1980s. The ***Khanda*** (barrier net with net-box traps) made of high-density polyethylene (HDPE) net has completely

Table 12.4 Characteristics of ichthyofauna of Chilika

Characteristics	Species (nos. or %)
Fin fish species by environment/habitats	Marine = 113 (35.65%) Brackish water = 139 (43.85%) Freshwater = 65 (20.50%)
Endemic species	2 (*Acentrogobius griseus* and *Bathygobius ostreicola*)
Exotic species (Non-native)	1 (*Oreochromis mossambicus*)
Migratory species (Marine & reverine migrants)	274 (85.89%)
Resident species	45(14.11%)
Catadromous species	15
Anadromous species	12
Commercially important species	129
High value economic species	6 (*Mugil cephalus, Planiliza macrolepis, Lates calcarifer, Daysciaena albida, Eleutheronema tetradactylum and Etroplus suratensis)*
Ornamental species	102
Food fish	287
Fishes breeding in the lake	56 (17.55%)
Species with 2 populations (1 in the lake and the other in the sea/river)	2 (*Nematanosa nasus* & *Rhinomugil corsula*)

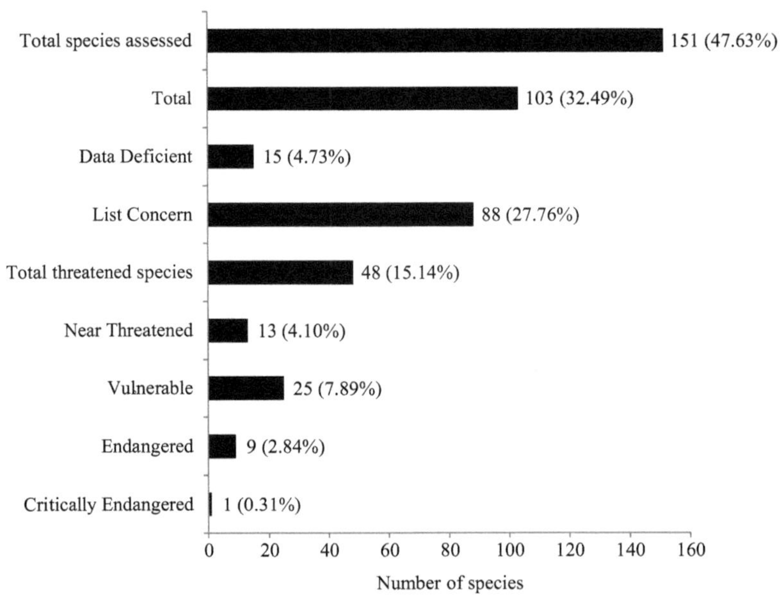

Fig. 12.2 Number of fish species assessed for different categories of biodiversity status. (Figure in parenthesis indicates percentage to total number of species)

replaced the traditional bamboo screen traps. The boat seine nets (*Patua* jaal & *Bhida jaal*) made of monofilament nylon nettings are also commonly used for catching schooling fishes like *Stolephorous & Thryssa sp.* and *Strongylura strongylura* (Spottail needlefish).

The use of fishing gears is influenced by target species, fishing area (fishing ground) and fish availability (Panda 2013). Gears used in the commercial fishery in the lagoon may be categorized into nine types (Fig. 12.3).

A study under the JICA-CDA technical cooperation project (2006–2009) concluded that the average catch share by long line, sieve nets, *Khanda* (barrier nets), gill nets & drag nets were 3%, 3%, 48%, 32% & 14% respectively.

Most of the gears operated in the lagoon at present are mostly non-selective. Use of destructive gear in the two fish migration pathways of the lagoon, namely the outer channel and Palur Canal is a matter of concern. *Khanda* fishing with small mesh sizes and long barrier netting walls (leader lines) are placed densely along the entire Outer Channel, even hardly leaving any space in the mid-stream for boats. The *Khanda* captures fishes and prawns of all sizes (juvenile to adult), thus resulting in large-scale capture of migratory fishes during the breeding season. The stationary bag nets (*Behundi* jaal) having largemouth and small meshed cod end-capture juveniles and small fishes and shellfishes in large quantity. The large boat seine (*Alimi* jaal) which is operated for 5–6 months during winter captures almost the entire immigrant fishes from the sea near the mouth in the Outer Channel. Destructive fishing gears of Chilika are indicated in Fig. 12.3.

12.4.2 Fish Catch

With a view to estimating fish yield from the monthly catch statistics are collected from 29 established fish landing centres (Fig. 12.4) out of which four major centres namely, Bhusandapur, Kalupada, Sorana and Balugaon on the western shore of the lagoon together share more than 79.50% of total landings.

With a view to compare the annual fisheries output during pre and post-restoration periods, the yearly fish catch data for 14 years of pre and post-restoration period (1986–1987 to 2014–2015) are depicted in Table 12.5 and Fig. 12.5. The opening of new artificial lagoon mouth during September 2000 (Hydrological intervention) witnessed a sudden leap of fisheries output during the year 2001–2002 (First year of eco-restoration) which showed 586.75% increase in comparison to the annual fishery yield in 1999–2000. The 14 years average annual fisheries output during the post-restoration period (2001–2002 to 2014–2015) showed a significant increase by 203.62% as compared to 14 years (1986–1987 to 1999–2000) average annual yield during the pre-restoration period. Yearly landings of fish, prawn, crab & total fishery output during the post-restoration period ranged from 6463.92–10286.34 t (mean 7757.79 t), 2347.78–6413.91 t (mean 4155.32 t), 111.07–358.26 t (mean 218.07 t) and 9955.83–14228.20 t (mean 12131.18 t)

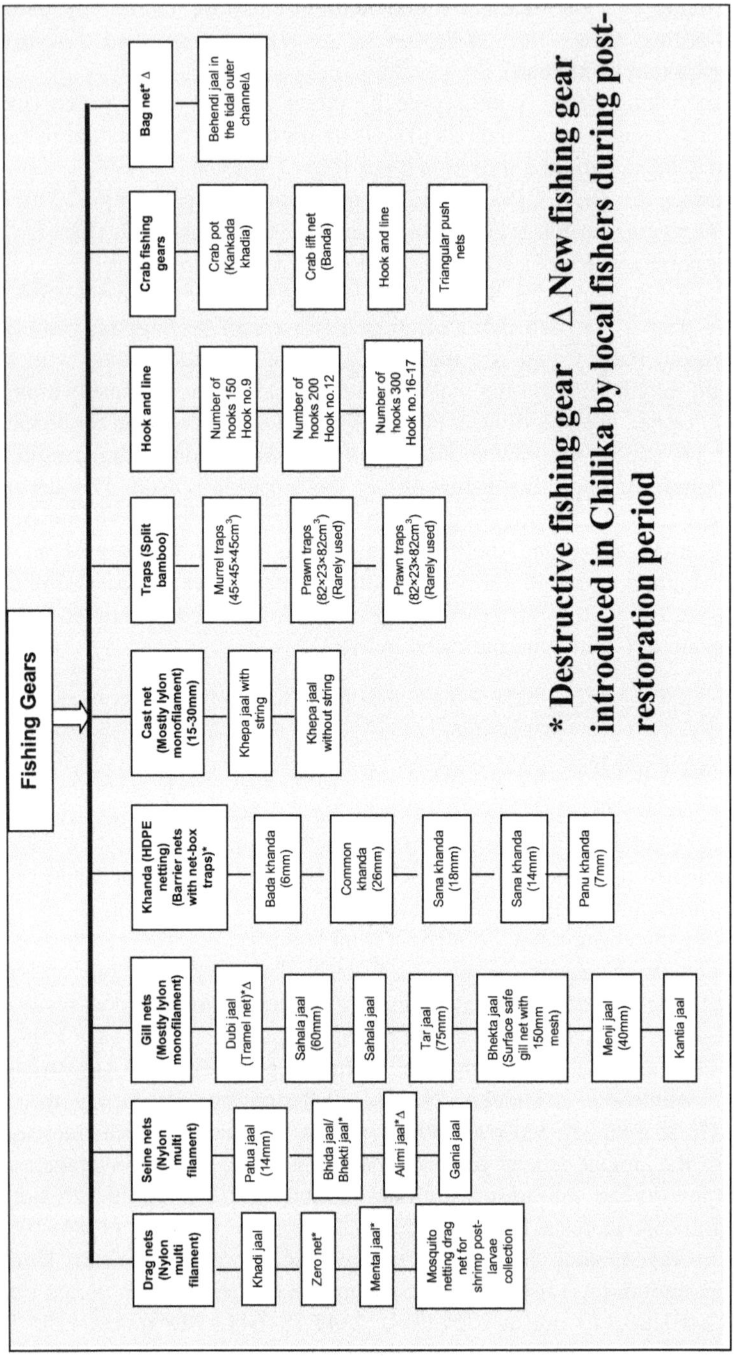

Fig. 12.3 Fishing gears used in Chilika since 1980s. *Destructive fishing gear Δ New fishing gear introduced in Chilika by local fishers during post-restoration period

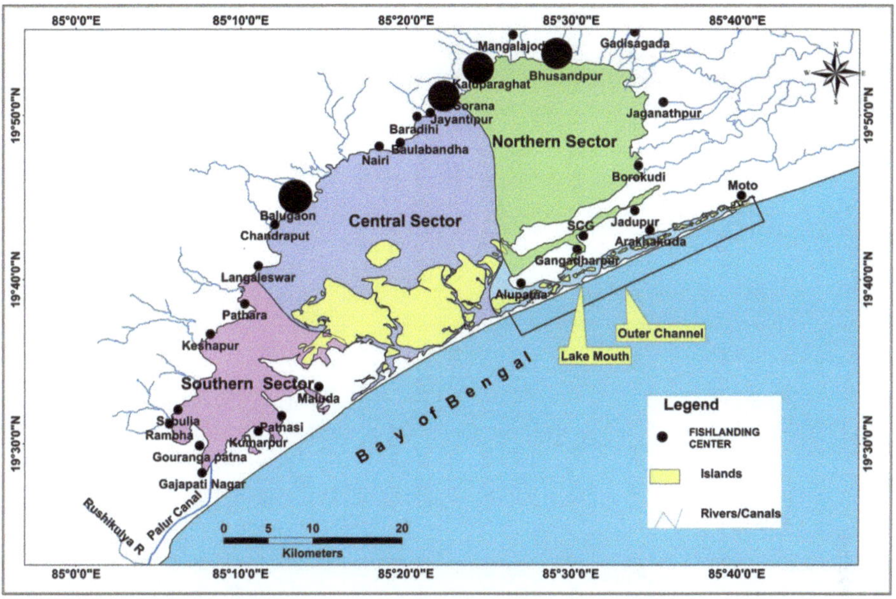

Fig. 12.4 Map of Chilika showing the locations of 29 established fish landing centres. *(Larger dots showing four major fish landing centers)*

respectively. Yearly average yield of fish, prawn and crab during the post-restoration period exhibited dramatic enhancement by 133.17%, 542.45% and 910.52% respectively as compared to pre-restoration period. Sarkar et al. (2012) also reported the spectacular enhancement of fisheries output from Chilika after hydrological intevention.

12.4.3 Fish Yield Potential and Maximum Sustainable Yield (MSY)

The ICAR- Central Inland Fisheries Research Institute (CIFRI), Barrackpore, India at the instance of CDA estimated the fish yield potential and Maximum Sustainable Yield (MSY) of Chilika during 2003 (Jha et al. 2005). The fish yield potential was estimated at 27,153 t, provided all the environmental variables operate to their optimum including the un-hindered recruitment of fish, prawn and crabs, both from marine and riverine sources (Jha et al. 2005). The MSY was estimated at 11,500 t per year, keeping in view the complex nature of fishery dynamics, dynamic estuarine wetland system, larger contribution to the commercial fishery by migratory species and complex nature of ecosystem functioning (Jha et al. 2005). The yearly fishery output during post-restoration period has been hovering close to the

Table 12.5 Year wise fisheries output and value during the post-restoration period 2001–2002 to 2014–2015 and comparison with mean estimates during pre-restoration (1986–1987 to 1999–2000). Qnty in tonnes & value in million INR

Year	Fish		Prawn		Fish & Prawn		Crab		Total Fisheries output	
	Quantity	Value	Quantity	Value	Quantity	Value	Quantity	Value	Quantity	Value
Mean estimates during pre-restoration (1986–87 to 1999–00)	**3327.09**	**39.40**	**646.79**	**22.90**	**3973.88**	**62.30**	**21.58**	**1.60**	**3995.46**	**63.90**
1999–00 (pre new lake mouth year)	1556.32	41.90	180.4	21.60	**1736.72**	**63.50**	9.03	1.00	**1745.75**	**64.50**
2000–01 (opening of new lake mouth)	3592.95	172.90	1296.26	157.20	**4889.21**	**330.10**	93.54	11.60	**4982.75**	**341.70**
2001–02	9530.03	344.70	2347.78	215.20	**11877.81**	**559.90**	111.07	11.70	**11988.88**	**571.60**
2002–03	8265.16	247.70	2478.82	216.90	**10743.98**	**464.60**	149.81	18.00	**10893.79**	**482.60**
2003–04	10286.34	401.20	3611.37	331.10	**13897.71**	**732.30**	155.51	19.40	**14053.22**	**751.70**
2004–05	8097.77	283.20	5000.71	463.80	**13098.48**	**747.00**	161.89	16.20	**13260.37**	**763.20**
2005–06	7774.81	265.70	4296.02	394.20	**12070.83**	**659.90**	154.08	15.40	**12224.91**	**675.30**
2006–07	6463.92	220.80	3368.97	310.60	**9832.89**	**531.40**	122.94	12.30	**9955.83**	**543.70**
2007–08	6610.23	261.30	3298.08	311.80	**9908.31**	**573.10**	143.05	15.70	**10051.36**	**588.80**
2008–09	6534.85	296.60	3929.68	381.60	**10464.53**	**678.20**	237.50	24.40	**10702.03**	**702.60**
2009–10	7892.97	379.00	3851.51	432.00	**11744.48**	**811.00**	210.89	29.90	**11955.37**	**840.90**
2010–11	7736.10	441.00	5043.62	685.20	**12779.72**	**1126.20**	285.90	34.30	**13065.62**	**1160.50**
2011–12	7456.03	491.30	6413.91	881.90	**13869.94**	**1373.20**	358.26	61.40	**14228.20**	**1434.60**
2012–13	7114.31	515.70	5034.05	832.10	**12148.36**	**1347.80**	318.58	33.10	**12466.94**	**1380.90**
2013–14	7699.71	652.00	4927.66	1141.20	**12627.37**	**1793.20**	308.97	44.90	**12936.34**	**1838.10**
2014–15	7146.77	634.40	4572.32	933.60	**11719.09**	**1568.00**	334.58	42.80	**12053.67**	**1610.80**
Variation (%) between pre and post-restoration	**+133.17**	**+885.25**	**+542.45**	**+2249.08**	**+199.78**	**+1386.56**	**+910.52**	**+1594.37**	**+203.62**	**+1391.77**

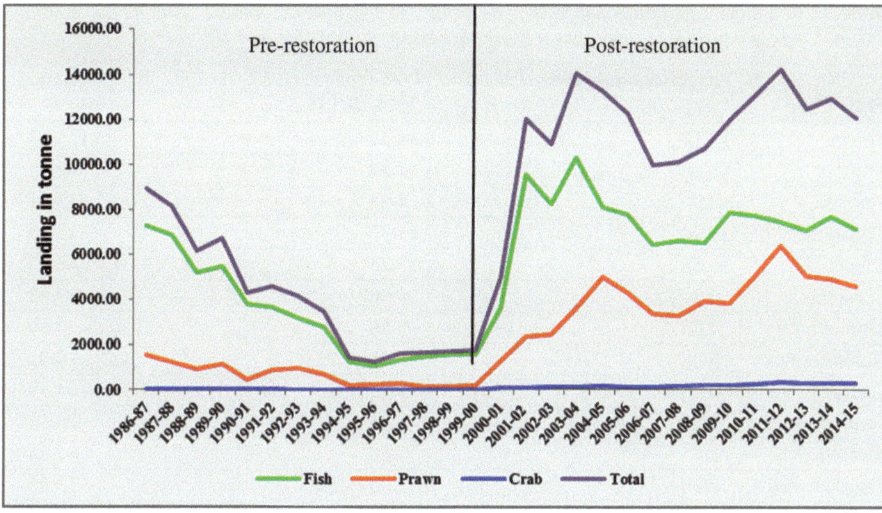

Fig. 12.5 Fish, prawn, crab & total yield (fisheries output) during pre (1986–1987 to 1999–2000) and post-restoration (2001–2002 to 2014–2015) period

estimated mean MSY level. Hence, more or less a stabilized yield level has been maintained during the last 14 years of post-restoration period.

12.4.4 Composition of Commercial Catch

Quality of commercial catch/landing is generally assessed from the percentage composition of commercially important fish groups/species to the commercial landing. Fish, prawn and crab contributed 63.95%, 34.25% and 1.80% respectively to the average yearly fisheries output during 14 years of the post-restoration period. As can be seen in Table 12.6, the average catch composition of prawn, clupeoids, sciaenids and crabs were higher during post-restoration period as compared to their composition during the pre-restoration period. Prawn yield from Chilika during post-restoration period has shown spectacular improvement with 542.45% increase in the percentage contribution to the average annual fisheries output in comparison to the pre-restoration period. Among penaeid prawns, *Penaeus monodon, Fenneropenaeus indicus, P. semisulcatus, Metapenaeus monoceros* and *M. dobsoni* registered 12.88%, 28.35%, 0.24%, 29.19% and 29.34% respectively to the total average yield of brackish water prawns during post-restoration period. The contribution of prawn fishery in terms of catch value at the average unit price of 2014–2015 to the average catch value of fisheries output in the post-restoration period worked out to 52.32%. Post-hydrological intervention period with 27.19% increase in salinity regime over pre-restoration period seems to have positively impacted the fisheries output in general and shellfish yield in particular. Average

Table 12.6 Percentage composition of fish groups in the average annual commercial landings (fisheries output) during pre and post-restoration period

Fish Group	Pre-restoration (1986–1987 to 1999–2000)	Post-restoration (2001–2002 to 2014–2015)	Variation (%)
Prawn	15.18	34.25	+125.63
Mullets	12.46	7.20	−42.22
Perches	10.00	4.63	−53.71
Catfishes	12.94	11.37	−12.13
Clupeoids	16.58	17.23	+3.92
Sciaenids	4.20	5.86	+39.52
Threadfins	4.35	2.58	−40.69
Beloniformes	3.28	3.25	−0.91
Cichlids	6.28	2.23	−64.49
Featherbacks	2.82	2.31	−18.08
Murrels	2.41	1.38	−42.74
Tripod fishes	2.20	1.95	−11.36
Carps & minnows	1.88	0.99	−47.34
Miscellaneous	5.42	2.97	−45.20
Crabs	0.54	1.80	+233.33

annual landing of shellfish (prawn and crab) during the post-restoration period contributed 36.08% to the average annual fisheries output and registered 555.07% increase as compared to average shellfish landing during pre-restoration period. The dramatic enhancement in shellfish landing after opening of the new lagoon mouth can be corroborated with salinity factor as has been indicated by higher significance value of correlation coefficient for prawn ($P < 0.01$) and crab ($P < 0.001$). However, the prawn and crab fisheries are influenced by their breeding and spawning success or failure in the adjacent coastal waters (Jhingran and Natarajan 1969) and their populations are more cyclical in nature, which is at least partly due to changes in coastal waters affecting spawning and recruitment to estuary/lagoon (Mohapatra et al. 2007).

12.5 Fishery with Biological Outlines of High-Value Target Species of Fish and Shellfish

Out of 129 commercially important fish species only 6 species (*Mugil cephalus, Planiliza macrolepis, Lates calcarifer, Daysciaena albida, Eleutheronema tetradactylum* and *Etroplus suratensis*) among finfishes and two penaeid prawns (*Penaeus monodon & Fenneropenaeus indicus*) and one portunid crab (*Scylla serrata*) are considered as high-value target species. The fishery and biological outline of these species are briefly discussed in the following paragraphs.

12.5.1 Mugil cephalus

Mugil cephalus (Linnaeus 1758) under Mugilidae family is locally known as "*Chilika Khainga*" (Fig. 12.6a) and is known to acquire largest size amongst all grey mullet species (https://en.wikipedia.org/wiki/List_of_largest_fish) and fetches a lucrative price (up to 300 INR/kg) in the market. The species migrate to the sea for spawning during the period September–January and November being the peak month for seaward migration. The species is considered isochronal spawning fish and spawns once in a year during the winter season and the juveniles enter into Chilika in large number. The adults of this species are very vulnerable to fishing in the Outer Channel and Palur Canal and in the inshore waters of Bay of Bengal adjoining the lagoon during their breeding migration. The food of ileophagus *M. cephalus* from Chilika mostly consists of benthic slime algae followed by decayed organic matters and fine particles of sand as well as mud (Jhingran and Natarajan 1969 and Panda 2013). The maximum size (total length) of the species recorded from Chilika was 712 mm (Jhingran and Natarajan 1969). The fish reaches approximately 201–350 mm in the I year, 351-500 mm in the II year and 501–625 mm in the III year of age (Jhingran and Natarajan 1969). The species

Fig. 12.6 Six high value targeted fish species (a: *Mugil cephalus*), (b: *Planiliza macrolepis*), (c: *Lates calcarifer*), (d: *Daysciaena albida*), (e: *Elutheronema tetradactylum*) & (f: *Etroplus suratensis*)

exhibits an isometric growth pattern and is a slow growing fish with a lifespan of about 5 years (Panda 2013).

The major fishing gears that capture this fish are gill nets, cast nets & barrier nets (*Khanda*).The fish contributed 13.06% to the average commercial landings from Chilika during the period 1957–1965 which was reduced to 2.99% during the post-restoration period (2001–2002 to 2014–2015). Wide fluctuation of the fishery was to a large extent attributable to recruitment vagaries. The landings of this grey mullet ranged between 251.37–681.39 t from 2001–2002 to 2014–2015 (Table 12.7). The estimated average annual yield for the post-restoration period was 362.13 t. The central sector of the lagoon registered highest yearly landing of the fish followed by the northern sector.The fishery is sustained by three-year classes of which the third is only of nominal interest (Jhingran and Natarajan 1969). The mean length of the species in commercial landings was recorded at 318 mm during 1957–1965 (Jhingran and Natarajan 1969, 1973) against 283.59 mm during post-restoration period (Panda 2013). The stock is highly exploited, and the spawning stock biomass (SSB) has declined by about 60% level from the virgin stock biomass (Suresh et al. 2015).

12.5.2 Planiliza macrolepis

This grey mullet species *Planiliza macrolepis* (Smith 1846) under Mugilidae family is locally known as "*Chilika Dangala*" (Fig. 12.6b). Being catadromous, this species breeds in the sea and undertakes seaward breeding migration during October–January, December being the peak month of migration. The migrating males are comparatively smaller in sizes than the females. Peak spawning of the species takes place in the coastal waters close to the lagoon mouth area (Jhingran and Natarajan 1969). The recruitment shows vagaries and varied appreciably form year to year during 2001–2002 to 2014–2015. The largest size of the species recorded from Chilika was 610 mm (Jhingran and Natarajan 1969). Like all ileophagus mullets *P. macrolepis* from Chilika basically a detritus feeder, the chief constituents of food being organic detritus and algal matters.

Gears like gill nets, barrier nets (*Khanda*) and cast nets are the principal fishing gears used to catch this species. Rod and line are also extensively used in winter months to capture the fish both in the main lagoon and the outer channel using *Spirogyra* (Green algae) as the main bait (Jhingran and Natarajan 1969). Bulk of the catches is made during August–November. The fish has been showing extreme fluctuations in the annual landings ranging from 12.01–99.51 t during post-restoration period, with annual average for the period at 46.68 t contributing 0.38% to the average annual yield of fish and shellfish. Yearly yield for the period 2001–2002 to 2014–2015 is presented in Table 12.7. Annual landing was the highest in the central sector followed by northern sector. Almost similar estimates of mean length in commercial landings were recorded as 276 mm during 1957–1965 (Jhingran and Natarajan 1973), while 275 mm during the post restoration period.

Table 12.7 Commercial landings (in tonnes) of high value target species of finfish and shellfish during 2001–2002 to 2014–2015

	Finfish						*Shellfish*		
Year	*M. cephalus*	*P. macrolepis*	*L. calcarifer*	*D. albida*	*E. tetradactylum*	*E. suratensis*	*P. monodon*	*F. indicus*	*S. serrata*
2001–02	346.40	70.71	181.37	1023.53	410.31	558.85	265.72	343.27	90.31
2002–03	359.68	32.37	149.70	858.18	468.35	102.05	337.19	317.95	114.04
2003–04	492.29	99.51	185.50	790.46	383.56	83.45	359.08	823.44	125.87
2004–05	681.39	87.19	112.02	852.31	295.38	69.60	503.45	1364.74	123.10
2005–06	507.81	41.27	142.13	806.64	356.59	121.90	376.82	1640.15	123.75
2006–07	260.47	24.23	99.21	583.88	226.45	22.83	345.37	825.62	100.95
2007–08	337.02	24.84	68.46	585.51	210.57	33.57	344.85	916.66	104.60
2008–09	297.79	12.01	73.08	428.54	306.69	102.97	426.09	845.87	196.85
2009–10	337.96	22.46	72.21	641.99	332.61	204.51	483.17	757.61	153.98
2010–11	340.18	28.06	79.58	816.76	357.81	264.60	694.78	1708.19	173.84
2011–12	251.37	55.03	115.24	555.83	252.69	402.40	847.99	1964.03	170..02
2012–13	289.99	47.63	85.75	541.69	238.64	380.80	602.87	1679.01	87.40
2013–14	291.68	46.30	129.19	553.93	234.37	254.41	828.72	1141.89	116.92
2014–15	275.77	61.96	116.37	548.89	247.50	112.09	571.99	1049.15	96.65
Mean	**362.13**	**46.68**	**114.99**	**684.87**	**308.68**	**193.86**	**499.15**	**1098.40**	**127.02**

12.5.3 Lates calcarifer

Lates calcarifer (Bloch 1790) under Latidae family locally known as "*Chilika Bhekti*" (Fig. 12.6c) is a very high value target species in Chilika; the average weighted unit price at current price level is 350 INR per kg. The fish is a protandrous and catadromous species growing to enormous sizes in Chilika; the largest recorded being 1210 mm (22.7 kg) in 2013. The fish attains a length of 400, 550, 688 and 800 mm in the first 4 years of its life (Jhingran and Natarajan 1966). Maturing and mature fishes have observed in the catches in Chilika Lagoon during April–July period indicating that the fish move out into the sea in June–July for spawning in the inshore waters of Bay of Bengal. The recruits noticeable in February are traceable to this spawning. There are also strong indications, on the basis of the appearance of another batch of recruits in July–August, that the fish spawns second time in January–March (Jhingran and Natarajan 1973). Recruitment to the fishery takes place in February at size 162 mm (modal value) and again in July–August at modal size 112-162 mm during 1958–1965 periods (Jhingran and Natarajan 1969). The fish feeds largely on fish (51.2%), prawns (39.3%), stomatopods (7.0%) and miscellaneous matter (2.5%) in Chilika (Jhingran and Natarajan 1966).

The species is captured mostly by gill nets and drag nets (*Bhekti bhida jaal*). The larger size fishes are mostly caught pre-dominantly by *Bhekti bhida jaal*. The fishery of this species showed wide fluctuation of 68.46–185.50 tonnes during the post-restoration period although the fishery fluctuated during 1957–1965 in the range of 55–749 tonnes (all-time high). The average annual yield during post-restoration period was 114.99 tonnes contributing 0.95% to the commercial catch. The central sector of the lagoon registered the highest landing of this species followed by the northern sector and southern sector. Yearly yield from 2001–2002 to 2014–2015 is presented in Table 12.7. The mean length in commercial catch was 405 mm during 1957–1965 (Jhingran and Natarajan 1969, 1973).

12.5.4 Daysciaena albida

Daysciaena albida (Cuvier 1830) under family Sciaenidae locally known as "*Chilika Borogo*" (Fig. 12.6d) is a commercially important high-value target species in Chilika, fetching average weighted unit price of 350 INR per kg in the fresh fish markets. The species feeds on prawns (35%), fish (20.7%), amphipods (15.1%), isopods (8.2%), stomatopods (6.1%), detritus (4.1%), higher plant matter (2.5%), algae (2.2%) and miscellaneous items (4%) (Jhingran and Natarajan 1966). Maturing specimens of this species have been observed in the lagoon during April–July period, May being the peak month. There are indications that this species breeds in the northern sector of the lagoon in summer months which does not, however, preclude its sea breeding (Jhingran and Natarajan 1973). Length frequency data analysis for this species indicated presence of two recruitment seasons, the most dominant in July–September at modal sizes 87-112 mm while the other, a weak one, in February at modal size 112 mm. This represents two peaks in spawning of this fish separated by a few months. The July–September recruits are most important from fisheries point of view. The fish attains a length of 238-263 mm in its first years, 363-388 mm in the second year, 463 mm in the third year, 563 mm in the fourth year and 638 mm in the fifth year. The maximum size of the species recorded from Chilika was 800 mm (Jhingran and Natarajan 1969).

The chief fishing gears for capturing this fish in Chilika are gill nets (*Borogo jaal*), drag nets (*Bhida jaal* and *Bhekti jaal*) and barrier nets (*Khanda*). The fishery of this species during post-restoration period fluctuated between 428.54–1023.53 tonnes (all time high during 2001–2002) averaging at 684.87 tonnes which contributed 5.65% to the commercial catch whereas the fluctuation of this fishery in the past (1957–1965) was 102.89–293.79 tonnes (Jhingran and Natarajan 1969). The mean length in commercial landings was 285 mm (Jhingran and Natarajan 1966, 1973). The northern sector of the lagoon registered the highest landing of the species followed by the central sector and southern sector. Details of yield for the period 2001–2002 to 2014–2015 is presented in Table 12.7. The higher yield at present indicated that the fishery of this species was improved during post-restoration period.

12.5.5 Eleutheronema tetradactylum

Eleutheronema tetradactylum (Shaw 1804) under family Polynemidae locally known as "*Chilika Sahala*" (Fig. 12.6e) is a commercially important high-value target species. This grows to a maximum length of 1000 mm in Chilika (Jhingran and Natarajan 1969), fetching an average weighted unit price of 320 INR per kg in the fresh fish markets. During the period 1957–1965 (pre-restoration) and 2001–2002 to 2014–2015 (Post-restoration period) it has been observed that the species breeds both in the sea and the lagoon and performs sea-lagoon and vice versa migrations. The recruitment for this species occurs twice in a year in Chilika during February–April and August–September (Jhingran and Natarajan 1973). The recruits observed during February–April in the lagoon are the most dominant and it is extremely likely that they are immigrants from the sea. These recruits are traceable to spawning of this fish in the sea during July–September period. Recruitment to the fishery at a modal size of 112 mm was observed during 1957–1965 (Jhingran and Natarajan 1969) while the same was observed to be 83 mm during post-restoration period (Panda 2013). The species reaches an average size of 217 mm in I year, 357 mm in II year, 537 mm in III year and 632 mm in IV year (Jhingran and Natarajan 1969). The largest size of the fish recorded from Chilika lagoon was 1000 mm (Jhingran and Natarajan 1969).

The species is mainly captured by gill nets, drag nets, barrier nets (*Khanda*) and hook & lines. Large sized fishes are generally caught by hook & lines, medium size fishes are caught by gill nets and drag nets while small sized fishes are caught by barrier nets. This fish is also susceptible for easy capture in the summer months in areas of Bay of Bengal adjoining the lagoon. Annual landing of this species during 1957–1965 fluctuated in the range of 122–364 tonnes averaging at 236 tonnes and contributing 6.44% to the average annual yield (Jhingran and Natarajan 1969). Maximum yield during 1957–1965 was from the northern sector and the minimum from the southern sector. Similar trend of sectoral landings were also noticed presently. Northern sector of the lagoon registered maximum landing of this species followed by central sector. The average yearly yield during the post-restoration period ranged from 210.57–468.35 tonnes, averaging at 308.68 tonnes which contributed 2.54% to the average annual landing. Details of yield for the period 2001–2002 to 2014–2015 is presented in Table 12.7. Although the quantum of average annual yield of this species from Chilika during post-restoration period showed 30.80% increase over the average annual landing during 1957–1965, the percentage contribution to the total annual landing decreased during the post-restoration period from 6.44–2.54% i.e. 60.56% decrease. The overall mean length of the fish in commercial catch over the entire period 1957–1965 was estimated at 216 mm which was reduced to 107.53 mm during the post-restoration period (Panda 2013). The species is under heavy fishing pressure and the spawning stock biomass has also reduced by 60% from the virgin stock. This situation warranted early enforcement of regulatory fishing regime in the lagoon (Panda 2013; Suresh et al. 2015).

12.5.6 Etroplus suratensis

Etroplus suratensis (Bloch 1790) under family Cichlidae locally known as "*Chilika Kundala*" (Fig. 12.6f) is a permanent resident of the lagoon. This species fetches average weighted price of INR 150 per kg at the landing centres in Chilika, which goes up to 350 INR per kg at fresh fish market in southern Indian states. The Chilika catch is mostly sent to Ernakulum fresh fish market from Chilika. This is the only native cichilid recorded in the lagoon so far. Maturing and mature specimens occur throughout the year in the lagoon showing its breeding all the year round with two breeding peaks, one during December–February and the other April–May. Recruitment to the fishery takes place in May–June and in October–December at modal sizes of 35-55 mm and 45-55 mm respectively (Jhingran and Natarajan 1969). The fishery, however is dominantly sustained by recruits from December–February spawning. The fish reaches an average length of 105 mm in the first year, 175 mm in the second year and 245 mm in third year. The maximum observed length of 243 during 2014–2015 (Suresh et al. 2015) is considerably less than the 335 mm recorded during 1957–1965 (Jhingran and Natarajan 1966, 1969, 1973).

This species is more abundant in monsoon and winter months in commercial catches. The major fishing gears used for capturing this species is gill nets including trammel nets, drag nets (*Patua jaal*, *bhida jaal* and *bhekti jaal*) and barrier nets (*Khanda*). Larger sized fishes (60-269 mm) are dominant in *bhekti jaal* (Jhingran and Natarajan 1969). The northern and central sectors largely sustain the fishery. During 2001–15, the average yearly landing ranged between 22.83 t (2006–2007) to 558.85 t (2001–2002), leading to an average of 193.86 t. Yearly landing during the post-restoration period of this fish is furnished in Table 12.7. However, Jones and Sujansingani (1954) recorded this species in their 'additional records' due to low contribution to the overall landings. The mean length of the fish in commercial catch 2001–2014 was estimated as 130.5 mm, which is lower than records of 1957–1965 (142 mm), indicating overexploitation of the stock (CIFRI 2015).

12.5.7 Penaeus monodon

Penaeus monodon Fabricius 1798 (Decapoda:Penaeidae) popularly known as Black tiger shrimp and locally called as "*Chilika Bagada Chingudi*" (Fig. 12.7a) is a high value exportable shrimp/prawn, majority of landing of which (up to 95%) is exported. At the landing centre, a kilogram was observed to fetch an average of 537 INR during 2001–2014. Black tiger shrimp is known to grow to the largest size among all penaeid prawns (growing up to 300 mm) and its life cycle is completed partly in sea and partly in lagoon. The species breeds in the sea and the post-larvae immigrate into lagoon environments for feeding and growth and they migrate back to the sea mostly at sub-adult stage. The species breeds round the year and post-larval incursion from sea to the lagoon also takes place throughout the year. Good catches are usually made between March and June but in some years it may commence in February and extend up to July. In certain years appreciable catches are also reported during September and November period. High or depressed landings are related to success or failure of "juvenile waves". February–July wave is dominant of the two, and accounts for a major fishery (Jhingran and Natarajan 1969). The post-larval abundance during the period October–January has a bearing on the success of this wave. There is a minor wave during August–December period.

Historically, the major fishing gears for prawn fishery in Chilika were split-bamboo made prawn traps locally known as "*Dhaudi & Baja*" and "*Thata*" as leader line. Since mid 1980s these traps have been replaced by barrier nets (*Khanda*) made of multi-filament HDPE nettings and net box traps (Remesan et al. 2011). Since 2000, the area occupied by *Khanda* nets have increased substantially. Recently, trammel nets (three walled gill nets) are also becoming a popular gear for this species. This fishery during 1957–1965 showed extreme fluctuations in the range 84–485 tonnes with an annual average for the period at 246 tonnes (Jhingran and Natarajan 1969), while during the post-restoration period the fishery fluctuated between 265.72–847.99 tonnes averaging at 499.15 tonnes and contributing to 4.11% to the average annual fisheries output during the period (Table 12.7).

12.5.8 Fenneropenaeus indicus

Fenneropenaeus indicus (Milne-Edwards 1837) (Decapoda:Penaeidae) popularly known as Indian white shrimp and locally called as "*Chilika Kantala Chingudi*" (Fig. 12.7b) is categorized as one of the high-value shellfish species. This penaeid prawn is a high value exportable shrimp next to *Penaeus monodon* fetching an average unit price of 236 INR per kg at landing centres during the post-restoration period. This prawn takes a dominant place among penaeids in Chilika though it shows extreme fluctuations from year to year.

The fishery of *F. indicus* is mainly attributable to two juveniles waves designated as 'April–July wave' and 'August–December wave' indicating their period of commencement and termination (Jhingran and Natarajan 1969). This was also noticed all the years during post-restoration period. The April–July wave is important which contributes about 40–60% to the total prawn catch. The overall average size of the prawn at detection and at departure was 91 mm and 108 mm respectively in the northern sector, 87 mm and 113 mm respectively in the central sector and 84 mm and 124 mm respectively in the southern sector (Jhingran and Natarajan 1969). The Kantala fishery during 1957–65 showed extreme fluctuations between 335.36–977.82 tonnes with an annual average for the period at 661.34 tonnes and contributed 18.05% to the average annual landings of fish and prawns (Jhingran and Natarajan 1969) while during the post-restoration period the fishery fluctuated between 317.95–1964.03 tonnes averaging at 1098.40 tonnes contributing 9.05% to the average annual fisheries output during the period (Table 12.7). Although the average annual landing of *F. indicus* during the post-restoration period was 1098.40 tonnes showing 66% increase over the pre-restoration period (1957–1965), the percentage contribution to the total landing decreased by 49.9% as compared to the contribution of *F. indicus* to total landings during 1957–1965. The maximum length recorded from Chilika was 140 mm during 1957–1965 (Jhingran and Natarajan 1969) while the largest size of 163 mm was recorded from the lagoon during 2002 (Jha et al. 2005). The common fishing gears for capturing *F. indicus* are barrier nets (*Khandas*), *khadi jaal* (drag net supported by bamboo sticks) and gill nets.

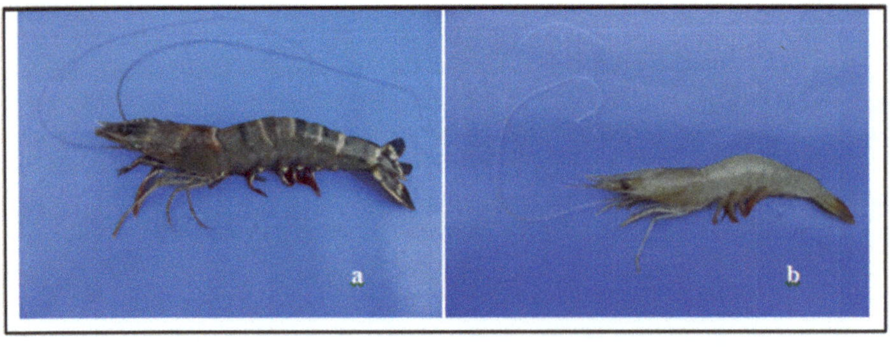

Fig. 12.7 Penaeid prawns (a: *Penaeus monodon*) & (b: *Fenneropenaeus indicus*)

12.5.9 Scylla serrata

Scylla serrata (Forsskål 1775) (Decapoda:Portunidae) popularly known as Mud crab (locally "*Chilika Kankada*") is a member of family Portunidae (Fig. 12.8). Two species under genus *Scylla* (*Scylla serrata & S. tranquebarica*) are found in Chilika. The crab fishery of Chilika was first reported by Jones and Sujansingani (1952). However, annual landing of mud crab from Chilika was first estimated in 1970s and time series data from 1971 onwards are available (Mohanty 1975). The coexistence of two mud crab species (*S. serrata & S. tranquebarica*) in Chilika was first reported by Mohanty et al. (2006). *S. serrata* migrate to the sea for breeding and the juveniles immigrate into the lagoon for feeding, growth and recruitment to the fishery.

Mud crab species are caught from Chilika Lagoon throughout the year and the active fishery season extends from February–October. The largest size of male and female *S. serrata* recorded from the lagoon were 159 mm carapace width (CW) and 680 g and 181 mm CW and 830 g respectively (Mohanty et al. 2006).The dominant size group of *S. serrata* in commercial landings was 61-90 mm CW (54.3%) for males and 91-120 mm CW (30.8%) for females during the post-restoration period. In general the juveniles and sub-adults constituted bulk of the commercial landings. Higher landings for *S. serrata* took place during February–May. *S. serrata* forms the major mud crab fishery, forming 61.2% of the total average annual mud crab landings during 2001–2002 to 2014–2015. Annual landing of *S. serrata* fluctuated between 3.00 tonnes (1994–1995) and 141.4 tonnes (1983–1984) during the period 1971–1972 to 1999–2000 (pre-restoration period) averaging at 51.21 tonnes. During post-restoration period annual landings of *S. serrata* fluctuated within the range 87.40 to 196.85 tonnes averaging at 127.02 tonnes (Table 12.7) showing an increase by 148.04% in comparison to the average annual landings during 1971–1972 to 1999–2000.

Fig. 12.8 Mud crab (*Scylla serrata*)

12.6 Conclusion

Although there has been dramatic enhancement in the landings during post-hydrological intervention, there are several indications of unsustainable regime as indicated below:

(a) Sustainability of hydrologic connectivity with the sea and rivers are regularly affected by natural siltation and coastal process which needs periodical maintenance dredging of inlet channels, Outer Channel, lead channels and lagoon mouth. Periodical renovation of the recruitment corridor Palur Canal is also imperative.
(b) Annual fish landing during post restoration period (9955.83–14,228.20 tonnes, averaging at 12,136.10 tonnes) is hovering around the maximum sustainable yield (11,500 tonnes). Several commercially important finfish species are being overfished and their spawning stock biomass (SSB) is alarmingly declining threatening the lagoon fishery. The data for commercially important species indicated that 65 to 88 percent of the catches comprised immature individuals, causing serious growth over-fishing and concern on the sustainability of fishery. Mesh size regulation and phasing out of destructive gears may help in attaining sustainability.
(c) Unregulated and large scale capture of immature and juvenile finfishes and shellfishes in the form of by-catch is also causing loss to biodiversity. Unchecked use of destructive fishing gears with small mesh sizes during post-restoration period causes serious biodiversity loss. These gears destroy larvae and juveniles of a number of commercial fish species, affecting their recruitment to fishery. This situation warrants immediate enforcement of regulatory fishing regime to reduce threats to the biodiversity.
(d) Seaward or lagoonward migration of brood fishes for breeding are also caught in large numbers in spawning season resulting in serious recruitment over-fishing. Regulated fishing and closed fishing period needs to be promulgated during the spawning season to avoid capture of brooders.
(e) Heavy concentration of barrier nets (*Khanda*) in the Outer Channel and Palur Canal causes severe hindrance in effective recruitment and breeding migration. Fishing in these areas needs to be regulated to facilitate seaward migration of mullets for breeding and safe return of juveniles into the lagoon.
(f) Almost 11.3% of the lagoon area is under illegal pen enclosure (*Ghery*) for aquaculture of prawns. This reduces the fishable area for fishers and obstructs free movement of migratory fishes and water circulation. Very stringent action need to be taken by management authority to remove these illegal *Gheries*.
(g) Proliferation of motorized boats and consequent use of fossil fuel is a potential threat to the lagoon due to oil pollution, especially in ecologically sensitive Outer Channel and Central Sector. Intensive movements of tourist motorized boats in the Outer Channel sector also have some negative impact on the ecosystem and especially to the movements of Chilika Dolphins.

Among the key challenges for future, prohibition of unauthorized fishing gears, regulation on number of motorized boats, mesh size regulation of fishing gears, restriction on operation of destructive gears, declaration and protection of nursery and spawning grounds, fishing restriction in ecologically sensitive areas, complete removal of encroachments in the form of illegal *Gheries* by enforcing stringent legislation with adequate punitive provisions seems to be important. Declaration of fishing holidays and providing alternative livelihood to the fishing community is also considered to be important. Climate change also possesses a key challenge for maintaining the suitable salinity gradient for fishery productivity. However, it is likely to be difficult to mitigate owing to several climate induced factors including increasing frequency of extreme climatic events like tropical cyclones, warming of lagoon surface water and high likelihood of freshwater flow reduction from Mahanadi River in future.

Acknowledgement It gives us great pleasure to place on record. Our sincere thanks to Chilika Development Authority for providing us secondary data on fisheries of Chilika Lake and to Dr. Ajit K. Pattnaik, Chief Executive,Chilika Development Authority for his constant encouragement and inspiration. We also gratefully extend our thanks to Dr. Ritesh Kumar, Director, Wetlands International-South Asia for critically reviewing the manuscript and providing valuable comments and suggestions to improve the quality of the manuscript. We thank Dr. Jajneseni Rout, Junior Research Fellow, Chilika Development Authority, Bhubaneswar for preparing GIS maps for the article.

Reference

Mohapatra A, Mohanty RK, Mohanty SK, Bhatta KS, Das NR (2007). Fisheries enhancement and biodiversity assessment of fish, prawn and mud crab in Chilika Lagoon through hydrological intervention. Wetlands Ecol Manag 15:229–251

Chapter 13
Avifauna of Chilika, Odisha: Assessment of Spatial and Temporal Changes

S. Balachandran, Ajit K. Pattnaik, P. Gangaiamaran, and Tuhina Katti

Abstract Waterbird populations in Chilika were systematically monitored since 2011 for assessment of status and trends and conservation needs. A gradual increase in the population of migratory species such as; Common Teal *Anas crecca*, Greylag Goose *Anser anser* Garganey *Spatula querquedula* and Painted Stork *Mycteria leucocephala* was observed over the last 4 years. On the contrary, the population trends show a general decline for two common dabbling ducks, Northern Pintail *Anas acuta* and Gadwall *Mareca strepera* and major diving ducks Tufted Duck *Aythya fuligula* and Common Pochard *Aythya ferina*. Decline in population of waders like Lesser Sand Plover *Charadrius mongolus*, Curlew Sandpiper *Calidris ferruginea*, Little Stint *Calidris minuta* and Ruff *Calidris pugnax* since 2011 corresponds with their global decline. The observed counts for 45 species recorded during the study exceeded their known 1% of the bio-geographical population.

The five species of ground-nesting birds at Nalabana Bird Sanctuary have abandoned the sites, and the breeding population of River Tern *Sterna aurantia* and Gull-billed Tern *Gelochelidon nilotica* has partially shifted to unprotected nesting areas like Panchakudi and Maldiguda in the central sector of the Chilika.

The neck collar studies indicated the partial shifting of the wintering populations to sites located further south of Chilika in Tamil Nadu and Karnataka. Re-sighting of neck-collared geese has also established high site fidelity to Chilika in four consecutive winters. Similarly, several colour-flagged waders, Black-tailed Godwit *Limosa*

S. Balachandran (✉)
Bombay Natural History Society, Mumbai, India

A. K. Pattnaik
Chilika Development Authority, Bhubaneswar, Odisha, India

P. Gangaiamaran
Wildlife Institute of India, Dehardun, India

T. Katti
Bombay Natural History Society, Mumbai, India

Bombay Natural History Society, Chilika Station, Odisha, India

C. M. Finlayson et al. (eds.), *Ecology, Conservation, and Restoration of Chilika Lagoon, India*, Wetlands: Ecology, Conservation and Management 6,
https://doi.org/10.1007/978-3-030-33424-6_13

limosa, Curlew Sandpiper, Little Stint and Redshank were sighted in the subsequent winters. Colour-flagged studies reveal the movement of some waders between Chilika and Point Calimere, and the connectivity with the Yellow Sea of China and South Korea. Though the Indian sub-continent is the primary wintering ground for the Central Asian Flyway population, the linkage of Chilika with the East Asian – Australasian Flyway was established during the current study. Ringing studies also further confirm movement and the existence of site-fidelity in several wader species. The study calls for urgent attention to securing the mudflats of Nalabana Island, especially for habitat specialist waders.

Keywords Chilika · Nalabana · Mangalajodi · Ducks · Waders · Terns · Gulls · Threatened species · Near-threatened species · Ground-nesting birds

13.1 Introduction

Migratory waterbirds use a network of wetlands during their long journeys. These wetlands play crucial roles by providing food and habitat for resting. Loss and degradation of wetlands stresses the populations of waterbirds (Hails 1997). Wetland habitats, therefore, need to be managed systematically to ensure that biodiversity and ecosystem services values of the site, especially those related to migratory waterbirds are retained and enhanced in the long-term (Ramsar 2007). Management of wetlands as waterbird habitats requires long term information on status and trends in waterbirds species and populations, habitat use and migration patterns (Holm and Clausen 2006).

The global importance of Chilika for individual waterbird species was recognised through the waterbird population monitoring study carried at Chilika from 2002 to 2016 by the Bombay Natural History Society in collaboration with the Chilika Development Authority (Balachandran et al. 2006a, b). The observed counts for 45 species recorded during the study exceeded their known 1% of the bio-geographical population (based on the datasets contained in Waterbird Population Estimate online database hosted by Wetlands International at wpe.wetlands.org).

Being located on the Indian east coast, significant populations of several waterbird species migrating along the East-Asia Australasian Flyway also winter in Chilika. This was validated based on ringing recoveries (e.g., Northern Shoveler *Spatula clypeata*, Gadwall, Curlew Sandpiper (Hussain et al. 1984), and satellite-tracking (e.g., Bar-headed Goose) studies (Balachandran and Sathiyaselvam (2010). Chilika also provides a stop-over site especially during the northward journey for the waders wintering further south of Chilika. The diverse bird habitats, ranging from beaches, mudflats, shallow brackish water zones with submerged vegetation and freshwater zones with floating and emergent vegetation, were observed to support atleast 225 species of birds of 50 families. Of these, 129 were waterbird species, and the rest wetland-dependent (Balachandran et al. 2006a, b). Chilika has also been recognised as a bottleneck site of the Central Asian Flyway.

Studies on waterbirds in Indian wetlands are still in the nascent stage. Of the 26 Ramsar Sites designated in India thus far, systematic studies on migratory waterbirds with bird ringing and bird population monitoring has only been carried in two sites, namely Point Calimere (Tamil Nadu) and Keoladeo National Park (Rajasthan). Short term bird ringing studies conducted by the BNHS during the1980s and 1990s at Chilika, and intensive studies carried out at Point Calimere and Pulicat indicated the three sites were being used as habitat, during different parts of the year, by a single population of Curlew Sandpiper (Balachandran 1998). Point Calimere and Chilika traditionally support large congregations of waterbirds, but the former predominantly supports waders and the latter ducks. The 15-year waterbird monitoring study (2000–2014) acquires significance for site management, as well as on the information on the health of the Central Asian Flyway. Key outcomes of the research have been reported in this paper (Table 13.1).

13.2 Study Area

Situated on the east coast of India, Chilika is an internationally significant waterbird reserve and one of the prime wintering grounds for migratory waterbirds in the Indian subcontinent. The congregation of large numbers of migratory ducks and waders during winter, northward and southward passages and nesting of resident species are the wetland's essential features. Based on rich biodiversity and socio-economic importance, Chilika was designated as a Ramsar Site in 1981.

Chilika is spread between 19°28′ and 19°54′N and 85°05′ and 85°28′E, within three districts of Odisha State, namely Khurda, Puri and Ganjam. The wetland is 64.5 km long, with a width varying between 5 and 18 km (Das and Samal 1988). Its water spread area ranges from 906 to 1165 km^2. The wetland has a variety of habitats including marshes, mudflats, open water of varying depths and salinity (some being fresh), and diverse aquatic vegetation within these habitats. For management purposes, the wetland has been zoned in four sectors based on hydrological regimes.

The Northern Sector is a shallow freshwater area with the diversified flora of both emergent and submerged types. The most dominant emergent macrophyte of this sector is a reed species *Phragmites karka.* The dominant submerged species are *Hydrilla verticillata, Vallisneria spiralis* and *Ceratophyllum demersum.* Some floatingforms, such as *Nymphaea, Utricularia, Polygonum, Ipomoea* and *Jussiea repens* are also present.

The more saline Southern Sector is comparatively deeper than the Northern sector. *Stuckenia pectinata, Potamogeton crispus, Najas minor, Najas graminea, Scirpus articulatus, Eragrostis japonica, Ceratophyllum demersum, Paspaldium geminatum* are the dominant species here. Well-established seagrass meadows are observed along the shoreline of this sector.

In the Central Sector, water depth and salinity are intermediate between the northern and southern sectors. The representative species of submerged macrophytes are *Stuckenia pectinata, Potamogeton crispus, Ruppia maritima, Najas foveolata, Najas graminea, Halophila ovalis, Halophila beccarii* and *Porteresia coarctata.*

Table 13.1 Population status of ducks and geese from 2001 to 2015

Common name	2001–2002	2002–2003	2003–2004	2004–2005	2011–2012	2012–2013	2013–2014	2014–2015	1% population threshold[a]
Gadwall	190,000	126,000	88,500	135,000	135,000	96,452	88,350	126,350	3000
Eurasian Wigeon	146,000	110,000	139,000	**230,000**	100,000	176,150	138,800	100,020	2500
Northern Pintail	**320,000**	280,000	260,000	178,000	142,078	91,200	150,500	190,000	20,000
Northern Shoveler	76,000	**95,000**	76,002	11,000	65,650	80,000	63,500	53,200	7100
Garganey	58,647	**70,000**	36,150	13,530	18,100	21,000	29,000	31,000	3500
Common Pochard	**90,000**	70,000	54,000	48,000	30,000	10,000	500	2010	800
Tufted Duck	**170,000**	110,000	27,500	23,000	41,661	20,000	26,000	20,000	1000
Red-crested Pochard	1600	2800	5300	1800	**2650**	3200	1000	2000	1000
Greylag Goose	64	40	120	180	72	300	351	**550**	250
Bar-headed Goose	1150	1560	650	1052	1500	**2013**	1474	1200	560

Figures in bold refer to maximum number recorded for the species

[a]As per WPE 5 (Wetlands International 2012)

Paspaldium flavidum and *Panicum repens* are the significant macrophytes of the Nalabana Bird Sanctuary.

The Outer Channel has higher salinity and tends to be marine, owing to linkages with Bay of Bengal. The dominant macrophyte species in the sector are *Stuckenia pectinata, Potamogeton crispus, Najas minor, Najas graminea, Scirpus articulates* and *Eragrostis japonica.*

Nalabana Island, situated within the Central Sector and covering an area of 15.52 km^2, is a major congregation area for a number of waterbirds. Acharya and Kar (1996) suggest that the island hosts as much as 75% of the total population of birds wintering in Chilika. The island is completely submerged during the monsoon, emerging gradually thereafter and drying up by end of April. The cyclical inundation helps in physiological and biological energy recycling as well regeneration of vegetation. Nalabana has been designated as a Wildlife Sanctuary since 1987 under the provisions of The Indian Wildlife (Protection) Act, 1972. Abundant aquatic flora and fauna, of micro and macro forms on the island serve as preferred food for the visiting waterbirds.

13.3 Study Method

Data was collected from sampling sites distributed in all sectors and major waterbird congregation sites of the wetland. The stretch between Kaluparaghat and Tinimuhani was monitored in the Northern Sector; Nalabana Island, Parikud, Kaluparaghat-Nairi and Barkul-Gangadharpur within the Central Sector; Rambha in the Southern Sector and stretch between Gurubai and Jahnikuda within the Outer Channel were the major monitoring sites (Fig. 13.1).

Bird count and ringing was conducted during 2001–2005 and during 2011–2015. Data for the period 2006–2010 was generated from a limited sampling frame, and thus has been excluded from analysis and reporting within this paper.

Fortnightly surveys were conducted throughout the migratory season (August–May) to monitor the number and population of waterbird species. Total count method through sighting by telescope, mounted on a boat or on foot, was used. Conspicuous species present in relatively small numbers or dispersed widely were counted individually. In case of large flocks, estimation was made notionally dividing the congregation into small groups of five to a thousand, depending on the size of the flock, and counting the number of groups. At least five groups of each flock were counted species-wise to calculate the average count for each species per group. The mean number of individuals per species was multiplied by the number of groups to arrive at the total number. In case of wide variation in the numbers counted for individual species within the groups of the same flock, a separate mean estimation was made for groups, having approximately similar species composition as well as similar numbers.

Ringing and colour flagging were carried out at Nalabana, Parikud, and Gurubai during November to April. Indigenous traps (mesh net, noose trap and claptrap) were used. Each trapped bird was marked with metal rings bearing serial numbers coded by Bombay Natural History Society and released after recording morphometric

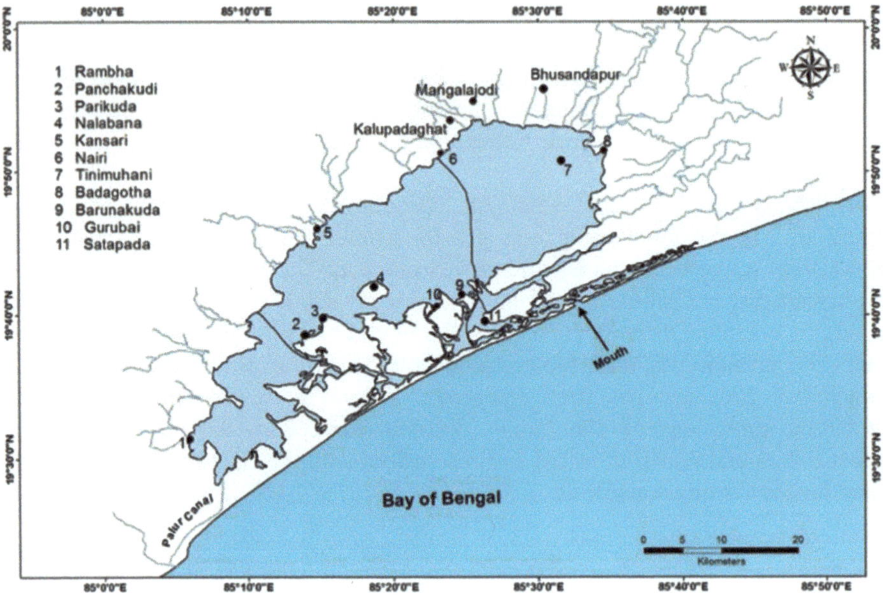

Fig. 13.1 Selected sampling sites in the Chilika

details and examining the stage of wing and tail moult. Since 2015, colour flagging of selected waterbirds has been initiated. Two white flags were placed on the right leg of the bird. The lower flag bears a unique alphanumeric code engraved in red on the outer side, and the inner side being entirely red.

Nest surveys were carried out during February to early May of 2001–2005 for the ground-nesting waders and terns. Nest searches were made in the known regular nesting sites namely Nalabana, Kansari, Parikud and Mangalajodi. As the ground-nesting birds had abandoned Nalabana Island in 2012, the search was extended to the Panchkudi Islet where River Tern had been observed to nest since 2011. The survey was carried out during early morning and late evening hours to minimise disturbance to the nesting and breeding birds.

13.4 Results

A total of 124 species of waterbirds and wetland-dependent birds (species which feed elsewhere and use wetlands for roosting) were recorded in Chilika. Sixty-six of these are migratory (Appendix). Amongst the 20 species of ducks and geese recorded, the Northern Pintail and Gadwall were the most abundant. The number of these two species usually exceeded 0.2 million, and never dropped below 0.1 million during the study period.

The counts of Eurasian Wigeon *Anas penelope* and Northern Shoveler were also observed to exceed 0.1 million during the study period. Common Pochard (numbers exceeding 30,000 in all years) was observed to be one of the abundant diving duck species recorded in Chilika. Similarly, the count of Common Coot *Fulica atra* exceeded 50,000 in all years. These counts signify the importance of Chilika in supporting migratory duck populations in the Central Asian Flyway, especially for Northern Pintail, Gadwall, Eurasian Wigeon, Northern Shoveler and Common Pochard.

Waders, with counts observed to range between 11% and 23% of the total, were the second dominant group of waterbirds in Chilika. The Black-tailed Godwit was the most abundant wader, having counts ranging between 33,000 and 60,000 during the study period. Notably, the counts of this species in Europe and Australia has declined in recent years (Birdlife Factsheet 2018). The counts of Lesser Sand Plover and Curlew Sandpiper exceeded 20,000 during 2002–2003.

13.4.1 Population Changes

13.4.1.1 Trends in Migratory Ducks and Geese

Changes observed in the population of most common migratory ducks and geese (all exceeding the 1% population threshold when assessed against the Waterbird Population Estimate 5) are summarised and discussed in this section.

1. Gadwall *Mareca strepera*

Gadwall is one of the abundant winter visitors that arrives at Chilika during September and October and stays up to mid-March. A few non-breeding birds can be seen up to May. This species was recorded in thousands in all sectors of the wetland, however, larger flocks were observed at Nalabana and the stretch between Kaluparaghat and Tinimuhani. In 2001–2002, their estimated population in Chilika (190,000 individuals) exceeded the known bio-geographical population by 60%. Benthall and Craven (1950) had recorded Gadwall as the most prevalent species of the lake in the 1950s. The population remained largely stable during the study period.

2. Eurasian Wigeon *Anas penelope*

The Eurasian Wigeon is a frequent and abundant winter visitor that arrives in October and leaves by mid-April. It was recorded in all four sectors. Wigeon prefers the area between Barkul and Kaluparaghat of the Central Sector. Craven (1949) had recorded several thousand Eurasian Wigeon at Nalabana Island in 1948. The population of this species was observed to decline from a high of 230,000 individuals in 2004–2005 to 100,020 by 2014–2015.

3. Northern Pintail *Anas acuta*

An abundant and widespread winter visitor, Northern Pintail arrives in Chilika by October and leaves by mid-April. A few individuals, probably first-year birds, stay up to mid-May. Mohapatra (1998) reported significant congregations of 100,000, Northern Pintails on the mudflats and shoreline of the lake. In this study, this species was observed to be well-spread in all sectors. The highest count of 320,000 was at Mangalajodi during December 2001. However, the population was observed to have declined conspicuously over the years.

4. Northern Shoveler *Anas clypeata*

This species is one of the familiar winter visitors, arriving in September and leaving by mid-April. It was observed in all sectors of the wetland, but higher counts were noted at Nalabana. Being a bottom feeder, the shallow water spread areas of Nalabana suits its feeding requirements. Their numbers at Nalabana ranged between 60,000–80,000 during 2001–2003. A dip in counts to 11,000 was noted during 2004–2005, which steadily recovered during the later periods.

5. Garganey *Spatula querquedula*

Garganey is a frequent visitor which arrives in September and leaves by late April. The highest estimate of 70,000 individuals was made in 2002–2003 after which the number was noted to decline to a low of 18,000 (Balachandran et al. 2006a, b, 2008). In the later periods, the counts ranged from 18,000 to 30,000.

6. Common Pochard *Aythya ferina*

Common Pochard is a common winter migrant that usually arrives in October–November and departs by mid-March. It was recorded from all parts of the wetland, barring the Outer Channel. The highest count was of 52,000 individuals recorded in December 2003. The counts have declined to 2010 in 2014–2015.

7. Tufted Duck *Aythya fuligula*

This duck is one of the most abundant winter visitors, arrives in October and leaves during mid-March. Counts were recorded in thousands in all parts of Chilika except Outer Channel. The peak count was of 170,000 individuals in the 2001–2002. A drastic decline in the population of this diving duck was observed.

8. Red-Crested Pochard *Netta rufina*

The Red-crested Pochard is a common winter visitor, arriving in November/December and leaving by mid-March. The count of this species has declined significantly from a high of 5000 recorded in 2003–2004 to around 1000 in 2013–2015.

9. Greylag Goose *Anser anser*

This goose is a regular but uncommon winter visitor, usually arriving in November and leaving by end January. Studies in 1940s have recorded the counts of Greylag Goose in Chilika to be higher than Bar-headed Goose (Benthall and

Craven 1950). Craven (1949) had recorded thousands of Greylag Goose in Nalabana Island. Ali and Ripley (1983) also commented on its abundance in Chilika in winters. During the early part of the present study, the population was less than 200 individuals. Since 2012–2013, there has been an increase in the counts to 550 individuals during 2014–2015.

10. Bar-headed Goose *Anser indicus*

A frequent winter visitor, Bar-headed Goose arrives in Chilika during mid-November and departs late March. Their overall population was noted to exceed 1% of its biogeographical population during the entire study period. Nalabana is its most favoured habitat. Bar-headed Geese were recorded to be familiar and plentiful in Chilika during 1945–1946 (Benthall 1947). Benthall and Craven (1950) recorded 1500 Bar-headed Geese at Satapada. A similar number of Bar-headed Geese were observed in Nalabana Island during 2002–2003. Notably, the counts of this species have also been observed to increase in Pong Dam (from 20,000 during 2002 to 35,000 during 2009 (Balachandran et al. 2006a) and Karaivetti Bird Sanctuary in Tamil Nadu (from 200 to 2000 in 1998) (Balachandran et al. 2004). The count of 1200 for this species in Nalabana during 2014–2015 was observed to be stable during the study period.

13.4.1.2 Trends in Waders and Flamingos

Trends in populations of waders and flamingos during the study period are summarized in Table 13.2 and discussed in the following paragraphs.

1. Curlew Sandpiper *Calidris ferruginea*

This species is one of the abundant winter visitors of Chilika, observed mostly in the shallow and brackish areas. Counts of this species during 2001–2005 exceeded 1% of its known biogeographical population. Notably, the count of 50,000 during the winter of 2003–2004 at Nalabana was equivalent to 54% of its biogeographical population. However, the counts have significantly declined to 21,000 individuals in 2011–2012 and further to 6500 individuals in 2014–2015.

2. Lesser Sandplover *Charadrius mongolus*

The Lesser Sand Plover is an abundant wintering wader, mostly confined to Nalabana, Parikud and Satapada. It arrives as early as August and departs during April. The population exceeded 1% of its biogeographic population during all study years. The highest number of 56,000 individuals was recorded during 2003–2004, equivalent to 56% of its biogeographic population. Some adults in partial or full breeding plumage and some first-year birds in non-breeding plumage also remained in Chilika during the summer. Population of Lesser Sandplover has undergone a significant decline in the recent years. The population never exceeded 20,000 individuals during 2011–2015, and the lowest population of 8800 individuals was recorded during 2014–2015.

Table 13.2 Population status of waders and flamingo from 2001 to 2015

Common name	2001–2002	2002–2003	2003–2004	2004–2005	2006–2011 (N.A)	2011–2012	2012–2013	2013–2014	2014–2015	1% biogeographic population
Curlew Sandpiper	25,000	27,000	54,000	21,000	–	21,000	15,000	12,000	6500	2400
Lesser Sand Plover	40,000	45,800	56,000	34,000	–	20,500	15,000	15,850	8803	1200
Little Stint	23,000	20,000	27,000	16,624	–	20,000	10,000	12,930	8805	2400
Common Redshank	7500	9130	15,500	3178	–	2000	2200	1350	1080	
Black-tailed Godwit	42,000	50,000	55,000	61,000		60,165	48,472	43,804	61,900	1500
Spotted Redshank	1405	900	850	1910	–	40	800	584	800	250
Grey Plover	1580	2300	1192	982	–	33	300	342	150	
Pacific Golden Plover	16,500	11,750	8680	3870	–	3730	17,700	11,500	5052	730
Marsh Sandpiper	3547	4280	21,250	5330	–	350	5750	3514	1650	1000
Great Knot	12	1	6	35	–	2	11	120	0	50
Red Knot	1	16	8	12	–	0	0	4	3	
Asian Dowitcher	60	28	56	39	–	3	400	71	150	150
Sanderling	0	0	0	0	–	0	50	0	50	
Greater Flamingo	5000	5000	142	1050	–	120	2000	800	2500	2400
Lesser Flamingo	152	190	52	72	–	0	30	50	0	3900

3. Little Stint *Calidris minuta*

Little Stint arrives in August/September and leaves in March/April. In all seasons during the study period, the population exceeded 1% of its biogeographical population. It is one of the abundant wintering waders in all sectors of Chilika except the Northern Sector (Ali and Ripley 1983). Nalabana harbours the largest population of Little Stint numbering around 25,000. In the last 3 years of the current study period, a decline has been noticed with the highest count of 8805 in the year 2014–2015.

4. Common Redshank *Tringa totanus*

Common Redshank is a regular winter visitor arriving in August/September and departing between late April and early May. The population exceeded the 1% threshold of its biogeographical population in all years of the study. The highest number of 11,000 individuals was estimated in the Outer Channel Sector during 2003–2004. This species has undergone a significant decline in population in the recent years. The population ranged from 1000 to 2200 individuals during the 2011–2015 seasons.

5. Spotted Redshank *Tringa erythropus*

The Spotted Redshank is a regular winter visitor, but less common than the Common Redshank. It arrives in September and leaves in April. It was recorded in all sectors but was usually most abundant at Nalabana. The highest number counted was 1910 in 2004–2005. Though there was a dip in the population in the 2011–2012 season, it has recovered during the last three seasons.

6. Grey Plover *Pluvialis squatarola*

This species is a typical and regular winter visitor arriving in September, and departing during April. A few adults in partial and full breeding plumages and some first-year birds in non-breeding plumage also remained here during the summer months. The highest number of around 2000 was recorded during February 2003 at Nalabana. In the recent years, the population has gone through a major decline and its number ranging between 700 and 1000.

7. Pacific Golden Plover *Pluvialis fulva*

The Golden Plover is a regular winter visitor which arrives towards the first week of September and remains up to April. Some individuals in breeding plumage also stay back during the summer months. The population was more than its 1% biogeographical population in all the years. It is one of the dominant waders for which the number exceeded 5000 in the Central and Northern sectors.

8. Marsh Sandpiper *Tringa stagnatilis*

A frequent winter visitor, Marsh Sandpiper arrives in thousands by August and leaves by mid-April. The population exceeded 1% biogeographical population in all winters of the study period. The highest estimate at Nalabana was of 12,000 individuals during 2003–2004. Their population in other wader habitats in the

Northern and Central sectors never to be exceeded over 2000. This species faced a major decline during 2013–2014, and a meagre 350 individuals were counted.

9. Great Knot *Calidris tenuirostris*

Great Knot is listed as endangered in the IUCN Red List of threatened species. An individual was caught at Parikud in 2002–2003, and a few individuals were sighted during 2004–2005. This species was thought to be a rare winter visitor, with a few records in Point Calimere (Ali and Hussain 1981), Pulicat (Mohapatra and Rao 1993) and Kolkata (Ali and Ripley 1983). Balachandran (1997) had reported Great Knot as a regular winter visitor to the south-east coast of India. In the year 2013–2014, 120 Great Knot were recorded at Nalabana. These records helped in bridging the understanding of the current distribution of this species in Indian wintering grounds, especially along the east coast.

10. Red Knot *Calidris canutus*

Though reported as a rare vagrant to India (Ali and Ripley 1983), Red Knot was found to be a regular winter visitor to south-east coast of India (Balachandran 1990; Mohapatra and Rao 1993). During the study period, only infrequent sightings were made from Chilika although three individuals were caught at Nalabana during 2002–2003 for ringing.

11. Asian Dowitcher *Limnodromus semipalmatus*

Ali and Ripley (1983) had observed the species to be common in winter in Chilika. During the early 2000 study, the presence of this species was confirmed, and 20 individuals were ringed. During the current study period, 70 to 400 number of this near threatened species has been recorded at Nalabana.

12. Sanderling *Calidris alba*

Around 50 individuals were recorded during the period 2012–2015 only at the new sea mouth, close to the beach area. There has been no previous record of this species from Chilika, although it has been recorded at other locations along the east coast (Ali and Ripley 1983).

13. Black-tailed Godwit *Limosa limosa*

Black-tailed Godwit is the most abundant of all the waders in Chilika. The wetland's significance as a wintering site for this Near-threatened species is indicated by the fact that the recorded population formed nearly one-fifth of bio-geographical population during the entire study period. The population counted each season was between 40,000 and 60,000. The increase in the population after 2010 discontinued during 2013–2014 migratory season. According to Davidson (2003), Black-tailed Godwit is one of the four species whose population is increasing along the Central Asian flyway.

14. Greater Flamingo *Phoenicopterus roseus*

Greater Flamingo is a typical seasonal visitor to Nalabana Island, arriving in August and leaving in May. Maximum number (5000) was recorded during

2001–2003, and in recent years the population has been fluctuating. This species was not observed to breed in Chilika.

15. Lesser Flamingo *Phoenicopterus minor*

Chilika forms the north-eastern range for Lesser Flamingo, with no evidence of breeding here. Throughout the study period, the counts were low (maximum 190 during 2001–2005), limited to Nalabana and for a short period each year. In 2013–2014, 50 individuals were recorded, with no sightings in the subsequent year.

16. Far Eastern Curlew *Numenius madagascariensis*

Far Eastern Curlew is a rare vagrant to India (Ali and Ripley 1983). Presence of this globally endangered species in Chilika could be affirmed on the basis of ringing of one individual during 2008.

13.4.1.3 Trends in Population of Resident Species

Trends in population of Spot-billed Pelican *Pelecanus philippensis*, Painted Stork *Mycteria leucocephala*, Asian Openbill *Anastomus oscitans* and Glossy Ibis *Plegadis falcinellus*. The four species resident in Chilika are summarized in Table 13.3.

Spot-Billed Pelican, a Near-threatened species was mostly sighted at Nalabana Island, mostly in low numbers. A count of 600 recorded during 2014–2015 is the highest thus far. Sighting of upto 5000 Painted Storks at Nalabana in May 2002, when the island was submerged, remains the highest observed in the study. During 2011–2015, the counts were restricted between 300 and 1200. The Asian Openbill, 23,000 of which were counted during 2012–2013, was observed to have a stable population in Chilika.

Glossy Ibis was regularly recorded during last 4 years (2011–2015) of the study period though in a small number. In 2008 only eight birds were sighted (Balachandran et al. 2008). They have been frequenting Nalabana Island for roosting in the night, sighted in the daytime only in January 2015. In 2014–2015, 6064 individuals were recorded.

13.4.1.4 Trends in Gulls

Brown-headed Gull *Larus brunnicephalus* and Palla's Gull *Larus ichthyaetus* are the commonly observed gull species in Chilika. Changes observed in their population during the study period are summarised in Table 13.4.

Brown-headed Gull is one of the frequent winter visitors to Chilika. It occurs in large congregations at Nalabana during daytime and night roosting. They arrive at the end of October and stay until early May. The highest count of 22,000 was recorded during 2012–2013.

Table 13.3 Population status of common resident waterbird species from 2001 to 2015

Common name	2001–2002	2002–2003	2003–2004	2004–2005	2011–2012	2012–2013	2013–2014	2014–2015	1% biogeographic population
Spot-billed Pelican	185	472	218	12	180	140	200	600	250
Painted Stork	5000	1160	254	82	300	600	800	800	
Asian Openbill-	15,800	12,070	13,512	19,300	5350	23,000	15,850	19,200	3000
Glossy Ibis					52	3000	3000	6064	

Table 13.4 Population statuses of gulls from 2001 to 2015

Common name	2001–2002	2002–2003	2003–2004	2004–2005	2006–2011 (N.A)	2011–2012	2012–2013	2013–2014	2014–2015	1% biogeographical population
Brown-headed Gull	16,560	12,786	18,238	7060	–	20,500	22,000	11,530	7090	1400
Pallas's Gull	125	87	56	215	–	2	3000	800	300	1000

The Pallas's Gull was observed regularly in Chilika during winter. It arrives in late October and leaves in March, preferring the beach region as a habitat. In the early 2000s, the counts never exceeded 300 at Nalabana and Sea mouth. But during the 2011–2015, the counts exceeded to 3000 individuals, peak recorded in 2012–2013. A large congregation of about 1500 individuals is largest sighting at Nalabana to date.

13.4.1.5 Trends in Ground Nesting Birds

Trends in counts of River Tern *Sterna aurantia* and Gull-Billed Tern *Gelochelidon nilotica*, two ground nesting birds are presented in Table 13.5.

River Tern, a threatened and restricted-range species, was recorded in all the sectors throughout the year. During the period 2001–2005, the species was observed to breed at Nalabana, with a notable increase in counts during breeding season (March–April). The highest count recorded for this species is 1200 in April 2003. A colony with 540 nests recorded in 2005 at Nalabana Island is the recorded maximum during the study period (Balachandran et al. 2005). During the later periods, the River Tern was observed to abandon Nalabana for nesting and shifted to Panchakudi islet. The size of the nesting colony of at Panchakudi was observed to gradually increasing.

Similarly, Gull-billed Tern used to breed at Nalabana Island in several hundred during early 2000 (Balachandran et al. 2005). Though seen sporadically through-out the year, the population increases from February due to the immigration of birds from elsewhere to the island for breeding. The highest number of 800 individuals was recorded during April 2003. A total of 326 nests recorded at Nalabana is one of the largest known breeding colonies for this species in India. In the year 2014–2015, only two nests were observed at Malathikuda, interspersed with River Tern nests, and the overall population reduced to just 180 individuals.

Till 2008, Nalabana was observed to be utilized by three species of terns (River Tern, Gull-billed Tern and Little Tern) and three species of waders (Black-winged Stilt *Himantopus himantopus*, Kentish Plover *Charadrius alexandrinus* and Oriental Pratincole *Glareola maldivarum*) for nesting. The nesting was not successful as flooding washed away the eggs and chicks before hatching and fledging. The number of nests was more than 200 except for the Black-winged Stilt that was observed with around 100 nests. The mound made to protect these birds were partially successful as some pairs successfully nested, however completely abandoned the site in few years and shifted to another islet Panchkudhi. However, Oriental Pratincole was observed to successfully nest in several parts of Chilika especially along the shorelines free from human habitation.

Table 13.5 Population status of ground-nesting Terns from 2001 to 2015

Common name	2001–2002	2002–2003	2003–2004	2004–2005	2011–2012	2012–2013	2013–2014	2014–2015	1% biogeographic population
River Tern	915	1380	1242	794	650	400	390	180	1380
Gull-billed Tern	954	1131	1133	844	655	450	790	180	1133

Table 13.6 List of Threatened birds in Chilika Lake and their population status in 2014–2015

Common name	IUCN status	Maximum count in Chilika
Pallas's Fish-eagle	Endangered	4
Great Knot	Endangered	150
Indian Skimmer	Vulnerable	1 (single record)
Asian Dowitcher	Near threatened	600
Spot-billed Pelican	Near threatened	350
Oriental Darter	Near threatened	10
Oriental White Ibis	Near threatened	2000
Eurasian Curlew	Near threatened	250
Far Eastern Curlew	Endangered	1 (single record)
Black-tailed Godwit	Near threatened	80,000
Painted Stork	Near threatened	5000
Lesser Flamingo	Near threatened	100
Ferruginous Duck	Near threatened	200
River Tern	Near threatened	700
Common Pochard	Vulnerable	50,000

13.4.2 Population of Threatened Birds

Fourteen species of birds observed at Chilika have been enlisted under Threatened and Near threatened categories as per the IUCN Red List of Threatened Species. The Near-threatened River Tern was observed to breed in Chilika. One pair of the migratory Pallas's Fish-eagle *Haliaeetus leucoryphus* was recorded on the Nalabana Island during all years of the study. A single Indian Skimmer *Rynchops albicollis* was sighted twice in the Outer Channel Sector in 2005–2006 and again in 2014–2015. As these migratory species have been observed to breed along the Mahanadi River, additional records may be forthcoming in the future. Table 13.6 summarizes the population status for the Threatened birds of Chilika, based on data from 2014–2015.

13.4.3 Results of Bird Ringing Satellite Tracking, Colour Flag, and Collaring Studies

During 2001–15, 13,327 birds of 88 species were marked with metal rings. A total of 627 birds (including 22 waders) were tagged with colour flags. Forty-three Bar-headed geese, six Northern Shoveler, four Northern Pintail, were fitted with neck collars. Eighty birds (ducks and geese) of eight species were tracked through satellite transmitters mounted on them.

On the basis of recapture of few individuals, waders were observed to move between Parikud, Nalabana Island and Gurubai. After the second week of December,

Table 13.7 Details of birds ringed at Parikud and recaptured at Nalabana

Species name	Date of ringing at Parikud	Date of recapture at Nalabana	Ring no.
Curlew Sandpiper	25.11.02	5.01.03	AB150070
Lesser Sandplover	26.11.02	16.03.03	AB150423
Grey Plover	09.12.02	9.04.03	B65579
Curlew Sandpiper	20.12.03	12.02.04	AB153215

Table 13.8 Details of birds ringed at Nalabana and recaptured at Satapada

Common name	Date of ringing at Nalabana	Date of recapture at Satapada	Ring no.
Curlew Sandpiper	10.03.03	24.03.03	AB150142
-do-	28.02.03	14.04.03	AB150666
Little Stint	03.04.02	14.04.03	A247069
Curlew Sandpiper	27.01.03	26.03.04	AB150730

by which time the shoreline of Parikud dries up, birds ringed during November were recaptured at Nalabana. Similarly, during the second week of March, by when Nalabana Island is exposed, the birds ringed in Nalabana between December and February were recaptured at Satapada during March–April.

The recaptures (Tables 13.7 and 13.8) at Nalabana and Satpada during the initial study period (2001–2005) were correlated with emergence of mudflats after prolonged submergence, coupled with the boom of benthic organisms which are a food source for a large number of waders. The occurrence of waders in large numbers for relatively short periods (2–3 months) at the three sites in succession–Parikud, Nalabana, and Satapada may indicate that availability and access to food sources persisted for a shorter period. Waders were observed to prefer tidal areas which have higher density of benthic organisms.

13.4.3.1 Local Movement of Ducks in Chilika Lake

Based on the outcome of the study the sector-wise bird population was documented. However, the overall bird density for the entire wetland could not be determined due to the high mobility of waterbirds from sector to sector. Northern Pintail and Gadwall were recorded in almost equal numbers in the Central and Northern sectors at different times of the day, and it is not clear whether they are the same population and move between these two sectors. During dusk hours at Nalabana, the numbers of birds of these two species increase threefold in comparison to the daytime population.

The congregation of diving ducks in the open water of the Northern Sector was observed to be stable until dusk. During early morning hours, the open waters were devoid of birds. Diving and dabbling ducks started congregating in the open waters of Northern Sector from around 09:00 h and reaching the peak between 10:00 and 12:00 h. The feeding population, observed during the daytime between 08:00 to

17:00 h from Kaluparaghat to Barkul move after 17:00 h. The population congregation on Nalabana Island commences from 17:00 h. It is presumed that these birds come to Nalabana for roosting from the Barkul-Kaluparaghat area after foraging. On the contrary, the birds of the open water area (Kaluparaghat, Nairy, Tinimuhani) were observed to be resting throughout the day.

The observations indicate that the dabbling ducks of the Northern Sector use Nalabana as a roosting habitat at dusk. The diving ducks, being nocturnal, might be going elsewhere for feeding at the night. Movement patterns of ducks could not be discerned owing to difficulties in capturing them.

13.4.3.2 Migratory Pattern of Species Ringed at Chilika

A total of 2764 birds were ringed during the last 4 years of the study period (2011–2012 [778]; 2012–2013 [762]; 2013–2014 [563]; 2014–2015 [669]). Curlew Sandpiper and Black-tailed Godwit constituted a significant proportion of those ringed. In 2014–2015, 17 individuals of seven species were recovered mostly at stopover sites between their breeding and wintering sites. The migratory patterns emerging from the study are presented in Fig. 13.2.

Two ducks (one Northern Shoveler and one Eurasian Wigeon) marked in Chilika were recovered in Russia after few months of ringing at Chilika. A male Northern Shoveler ringed on 27 February 2013 at Nalabana (Ring Number F-67519) was recovered on 12 September 2014 at Lake Krevankul (50.09 N, 80.43 E) in

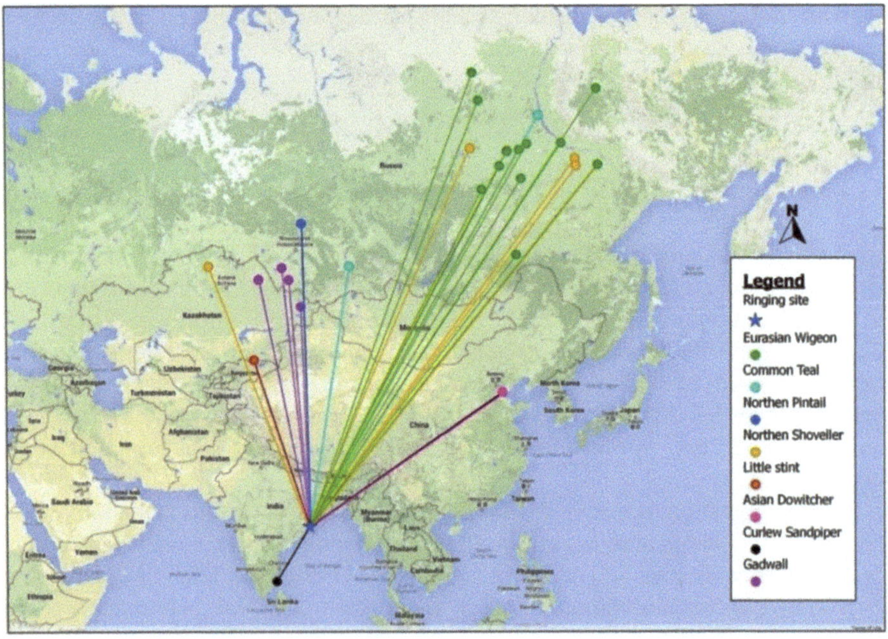

Fig. 13.2 Recoveries of birds ringed in Chilika

Vostochno – Kazakhstan. A female Eurasian Wigeon ringed on 23 March 2012 at Nalabana (Ring Number F-70145) was recovered on 15 May 2015 at Asyma (62.21 N, 126.38 E), Gornyy District, Yakutia, Russia.

One Curlew Sandpiper marked on 16 January 2015 from Nalabana, was recaptured further south along the east coast at Point Calimere (Tamil Nadu) in July 2015. Two Curlew Sandpipers and one Asian Dowitcher marked with colour flags were sighted at the Bohai Sea in China.

13.4.3.3 Re-Sighting of Bar-headed Geese Marked in Chilika Lake

A goose marked in Nalabana in February 2010 was re-sighted here after 4 years in January 2014 and again in December 2014. This may indicate wintering site fidelity of this species. Another individual marked in February 2010 at Nalabana was reported 2 years later in Koonthankulam (Tamil Nadu) in December 2012. Two Bar-headed Goose marked with BNHS neck collar H25 and H83 on 07, and 21 February 2014 from Nalabana were sighted once again near the mouth of River Rushikulya in the second and last week of January 2016.

13.4.3.4 Movement Pattern of Satellite Marked Birds

Of the 15 PTT fitted Bar-headed Geese at Chilika, seven (46.67%) reached the breeding grounds in Tibet, China and Mongolia. One Bar-headed Goose spent the summer at Qinghai Lake in China. Similarly, one Bar-headed Goose marked with PTT returned to Chilika and Rushikulya River during the subsequent season where it spent the winter again. Another marked bird went to Karnataka in the subsequent year. One colour-flagged individual from Mongolia was sighted at Nalabana during December 2007.The tracking studies indicate that the Bar-headed Geese wintering at Chilika mostly migrate from China and Mongolia (Fig. 13.3).

13.5 Discussion

The current study confirms that Chilika regularly supports over 0.5 million birds during the entire peak migratory season (December–March). Shallow zones, salinity gradient, extensive mudflats, seashore, abundant food and other microhabitats throughout the migratory season, provide the migratory birds with ample scope to congregate and utilise the food resources efficiently. Chilika also has the distinction of supporting a large congregation of migratory ducks (over 0.5 million), especially habitat specialists such as Gadwall and Tufted Duck which feed mostly on abundant macrophytes and thus seldom raid crops in the areas buffering Chilika. The wetland is also used as roosting and feeding site for the most abundant and wide spread migratory duck species, the Northern Pintail, which partially depends on adjoining crop fields for feeding.

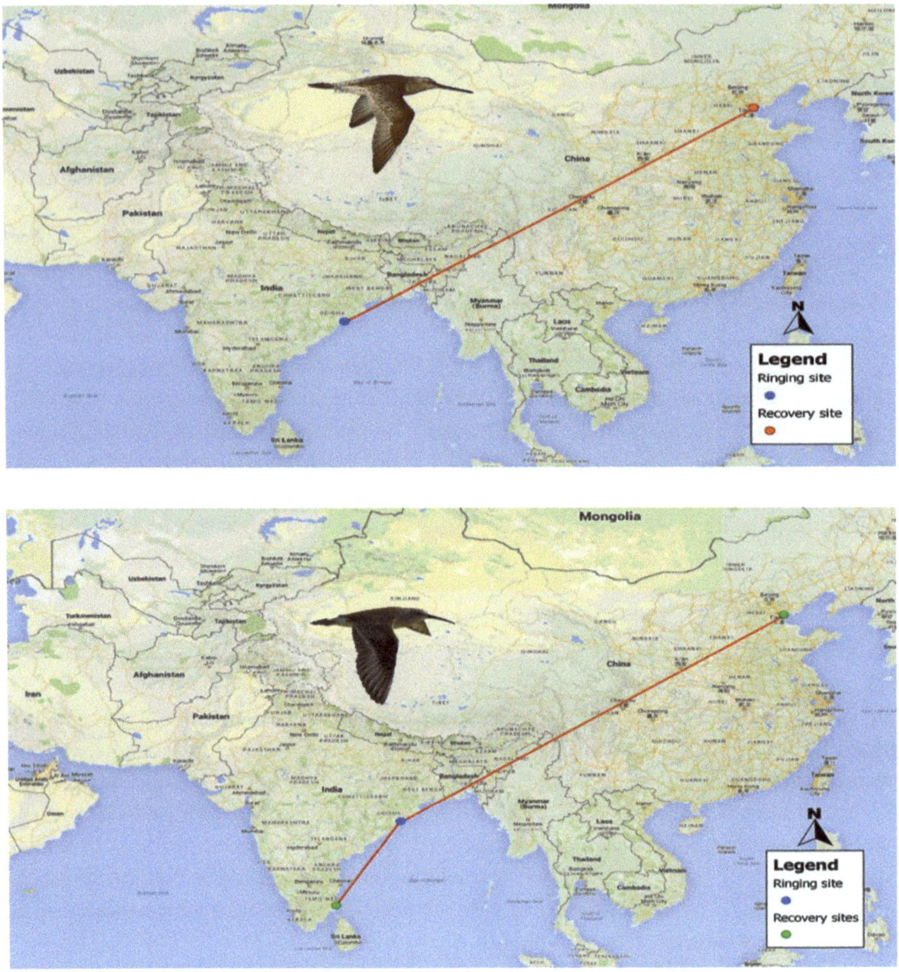

Fig. 13.3 Reports of colour-flagged waders from Chilika in China. Above: Asian Dowitcher *Limnodromus semipalmatu*. Below: Curlew Sandpiper *Calidris ferruginea*

The study highlights the ornithological significance of Nalabana, wherein several species were observed to congregate for feeding, resting and roosting. Abundance of food (macrophytes, macrobenthic organisms, and free swimming organisms inclusive of fish), and their accessibility to the birds, exposed mudflats and shorelines for roosting which are well protected from human and other disturbances are the plausible conducive factors. The heterogeneity in vegetation found at the island creates microhabitats providing the require niche for birds. Nalabana not only supports the highest diversity and density of waterbirds but is also more hospitable to the northern migrants for an extended period, as compared to other parts of the wetland.

The shallow open-water zones were ideal for large congregation of dabbling ducks such as Eurasian Wigeon and Gadwall which feed on the fresh shoots and seeds of *Potamogeton*. The freshwater marshes of Mangalajodi and Bhusandpur were favoured habitats for several resident species (such as jacanas, moorhens, whistling ducks, crakes and other rallids) and long migratory species Northern Pintail and Black-tailed Godwit.

The study indicated a gradual decline in the population of all the dominant diving ducks such as Tufted Duck (from 170,000 to 40,000 during 2001–2011) and Common Pochard (from 90,000 to 2010 during 2001–2014). Based on the declining population of Common Pochard in Chilika, the global status of this species was changed in 2016 from "Least concern" to "Vulnerable". On the other hand, the population of common dabbling ducks, such as Eurasian Wigeon and Northern Shoveler were observed to be stable. Counts of Greylag Goose, a species not common in Chilika, were noted to gradually increase. The population of Common Teal and Garganey observed to increase during early parts of the study are stabilising.

The cloud formation of waders noticed in the early years of the study has not been seen recently except for Black-tailed Godwit and on few occasions for Pacific Golden Plover. Population of waders has continued to decline, partially due to the loss of mudflats within Chilika. During 2011–2015, the population of Lesser Sand Plover, Curlew Sandpiper, Little Stint, Ruff and River Tern have gradually declined. The population of Near Threatened Black-tailed Godwit, observed to be increase till 2013, has declined thereafter. The threatened and rare waders such as Great Knot, Asian Dowitcher and Red Knot were observed in higher numbers towards the later parts of the study.

The formation of grass thickets at Nalabana was observed to attract freshwater preferring species such as Lesser Whistling Duck *Dendrocygna javanica*, Indian Spot-billed Duck *Anas poecilorhyncha*, and Comb Duck *Sarkidiornis melanotos* which never frequented this area till 2010. Common Shelduck *Tadorna tadorna* has been sighted during the recent years (2011–2012 and 2013–2014) at Parikud and was photographed for the first time from Nalabana. At Mangalajodi, the Glossy Ibis population has increased to 3000, which was recorded for the first time in a few numbers in 2008. The numbers of the Greater Flamingo is declining, primarily as their preferred habitats, the mudflats have shrunk.

Ringing and counting assessments affirm that Chilika is a significant habitat for waders. Of the 84 species of waders recorded in India (Ali and Ripley 1983), 46 species were recorded in Chilika during the study. Amongst these, the Lesser Sand Plover, Black-tailed Godwit, Curlew Sandpiper, Little Stint and Pacific Golden Plover were the most prevalent, accounting for 72–83% of all waders counts recorded from the wetland. The waders complete the later part of their primary moult before leaving for their breeding ground.

Among the three small common waders, namely Lesser Sand Plover, Curlew Sandpiper and Little Stint, the population of Lesser Sand Plover appears to be constant from autumn to spring. The absence of spring (March–April) or northward passage peak, and the almost stable population of Lesser Sand Plover throughout the season indicate that they can utilise the same migratory route for both the journeys. The recapture data also suggested that Lesser Sand Plover stay for a relatively more extended period than the two typical arctic breeders, i.e. Little Stint and Curlew Sandpiper, their stay duration at Chilika is still shorter than at the south Indian wintering grounds where they spend over 8 months. Moreover, the relatively lesser recapture rate of Lesser Sand Plover within a season suggests that a portion of transit population might also be occurring along with the wintering population in comparison to other sites. On the other hand, the absence of autumn peak (September and October) or southward passage, the lesser recapture rates within a season, and the occurrence of peak only in late winter and spring for the Little Stint and the Curlew Sandpiper indicate that majority numbers of these two species utilize the wetland for their return journey. Earlier bird migration studies also established this through recoveries and recaptures. The recovery of a Point Calimere ringed Curlew Sandpiper obtained at Chilika within the same season further confirms the migratory pattern of these arctic breeding waders.

13.6 Conclusion

This long-term study, one of first of its kind on Indian waterbirds, reaffirms the important role of Chilika for migratory waterbirds. This is significant considering the global decline in populations for several species. Habitat specialists such as waders which exclusively depend on the mudflats are the most affected. In Chilika most of the wader congregations from November are confined to Nalabana as it is the only site found with extensive pristine mudflats. However, these mudflats are being invaded with grass thickets and *Salicornia,* making the habitats less suited. This needs to be urgently addressed within the wetland management planning.

Waterbird populations in the Central Asia Flyway are under high pressure from hunting (Boere and Piersma 2012). Conservation of sites such as Chilika thus acquires high significance, especially for shorebirds. Wetland management also needs to be cognizant of intensifying human presence all along Chilika's shoreline and the consequence of the site as a habitat for waterbirds.

Appendix: List of Waterbirds and Water-Dependent Birds Recorded in Chilika During 2001–2015 (Status: *R* Resident, *M* Migrant, *SM* Seasonal Migrant, *B* Breeding)

Sl no.	Common name	Species name	Migratory status
1	Little Grebe	*Tachybaptus ruficollis*	R/B
2	Great Crested Grebe	*Podiceps cristatus*	M
3	Spot-billed Pelican	*Pelecanus philippensis*	R/SM/

(continued)

Sl no.	Common name	Species name	Migratory status
4	Little Cormorant	*Phalacrocorax niger*	R/B
5	Indian Shag	*Phalacrocorax fuscicollis*	R
6	Great Cormorant	*Phalacrocorax carbo*	R/SM
7	Darter	*Anhinga melanogaster*	R
8	Little Egret	*Egretta garzetta*	R/B
9	Grey Heron	*Ardea cinerea*	R/SM
10	Purple Heron	*Ardea purpurea*	R/B
11	Large Egret	*Casmerodius albus*	R
12	Median Egret	*Mesophoyx intermedia*	R
13	Cattle Egret	*Bubulcus ibis*	R
14	Indian Pond-Heron	*Ardeola grayii*	R
15	Black-crowned Night-Heron	*Nycticorax nycticorax*	R/B
16	Yellow Bittern	*Ixobrychus sinensis*	R/B
17	Chestnut Bittern	*Ixobrychus cinnamomeus*	R/B
18	Black Bittern	*Dupetor flavicollis*	R
19	Painted Stork	*Mycteria leucocephala*	R
20	Asian Openbill-Stork	*Anastomus oscitans*	R
21	Oriental White Ibis	*Threskiornis melanocephalus*	R
22	Eurasian Spoonbill	*Platalea leucorodia*	R
23	Greater Flamingo	*Phoenicopterus ruber*	M/M
24	Lesser Flamingo	*Phoenicopterus minor*	M
25	LargeWhistling-Duck	*Dendrocygna bicolor*	R/M
26	Lesser Whistling-Duck	*Dendrocygna javanica*	R
27	Greylag Goose	*Anser anser*	M
28	Bar-headed Goose	*Anser indicus*	M
29	Lesser White-fronted Goose	*Anser erythropus*	M
30	Brahminy Shelduck	*Tadorna ferruginea*	M
31	Common Shelduck	*Tadorna tadorna*	M
32	Comb Duck	*Sarkidiornis melanotos*	SM
33	Cotton Teal	*Nettapus coromandelianus*	R/B
34	Gadwall	*Anas strepera*	M
35	Eurasian Wigeon	*Anas penelope*	M
36	Spot-billed Duck	*Anas poecilorhyncha*	R/SM/B
37	Northern Shoveller	*Anas clypeata*	M
38	Northern Pintail	*Anas acuta*	M
39	Garganey	*Anas querquedula*	M
40	Common Teal	*Anas crecca*	M
41	Red-crested Pochard	*Rhodonessa rufina*	M
42	Common Pochard	*Aythya ferina*	M
43	Ferruginous Pochard	*Aythya nyroca*	M
44	Tufted Pochard	*Aythya fuligula*	M
45	Brahminy Kite	*Haliastur indus*	R/B
46	White-bellied Sea-Eagle	*Haliaeetus leucogaster*	R/SM/B

(continued)

Sl no.	Common name	Species name	Migratory status
47	Pallas's Fish-Eagle	*Haliaeetus leucoryphus*	R/M
48	Western Marsh-Harrier	*Circus aeruginosus*	M
49	Osprey	*Pandion haliaetus*	M
50	Water Rail	*Rallus aquaticus*	R/SM
51	Brown Crake	*Amaurornis akool*	R/SM
52	White-breasted Waterhen	*Amaurornis phoenicurus*	R
53	Baillon's Crake	*Porzana pusilla*	R/SM
54	Ruddy-breasted Crake	*Porzana fusca*	R/SM
55	Watercock	*Gallicrex cinerea*	R/SM/B
56	Purple Moorhen	*Porphyrio porphyrio*	R/B
57	Common Moorhen	*Gallinula chloropus*	M/B
58	Common Coot	*Fulica atra*	R/M/B
59	Pheasant-tailed Jacana	*Hydrophasianus chirurgus*	R/B
60	Bronze-winged Jacana	*Metopidius indicus*	R/B
61	Greater Painted-Snipe	*Rostratula benghalensis*	R
62	Pacific Golden-Plover	*Pluvialis fulva*	M
63	Grey Plover	*Pluvialis squatarola*	M
64	Little Ringed Plover	*Charadrius dubius*	M
65	Kentish Plover	*Charadrius alexandrinus*	R/M/B
66	Lesser Sand Plover	*Charadrius mongolus*	M
67	Yellow-wattled Lapwing	*Vanellus malabaricus*	R
68	Grey-headed Lapwing	*Vanellus cinereus*	M
69	Red-wattled Lapwing	*Vanellus indicus*	R
70	Pintail Snipe	*Gallinago stenura*	M
71	Common Snipe	*Gallinago gallinago*	M
72	Black-tailed Godwit	*Limosa limosa*	M
73	Whimbrel	*Numenius phaeopus*	M
74	Eurasian Curlew	*Numenius arquata*	M
75	Spotted Redshank	*Tringa erythropus*	M
76	Common Redshank	*Tringa totanus*	M
77	Marsh Sandpiper	*Tringa stagnatilis*	M
78	Common Greenshank	*Tringa nebularia*	M
79	Wood Sandpiper	*Tringa glareola*	M
80	Terek Sandpiper	*Xenus cinereus*	M
81	Common Sandpiper	*Actitis hypoleucos*	M
82	Ruddy Turnstone	*Arenaria interpres*	M
83	Asian Dowitcher	*Limnodromus semipalmatus*	M
84	Great Knot	*Calidris tenuirostris*	M
85	Red Knot	*Calidris canutus*	M
86	Sanderling	*Calidris alba*	M

(continued)

Sl no.	Common name	Species name	Migratory status
87	Little Stint	*Calidris minuta*	M
88	Rufous-necked Stint	*Calidris ruficollis*	M
89	Temminck's Stint	*Calidris temminckii*	M
90	Long-toed Stint	*Calidris subminuta*	M
91	Dunlin	*Calidris alpina*	M
92	Curlew Sandpiper	*Calidris ferruginea*	M
93	Broad-billed Sandpiper	*Limicola falcinellus*	M
94	Ruff	*Philomachus pugnax*	M
95	Black-winged Stilt	*Himantopus himantopus*	R/M/B
96	Pied Avocet	*Recurvirostra avosetta*	M
97	Great Stone-Plover	*Esacus recurvirostris*	R
98	Collared Pratincole	*Glareola pratincola*	R/M
99	Oriental Pratincole	*Glareola maldivarum*	R/B
100	Small Pratincole	*Glareola lactea*	R/M
101	Heuglin's Gull	*Larus heuglini*	M
102	Pallas's Gull	*Larus ichthyaetus*	M
103	Brown-headed Gull	*Larus brunnicephalus*	M
104	Black-headed Gull	*Larus ridibundus*	M
105	Gull-billed Tern	*Gelochelidon nilotica*	R/M/B
106	Caspian Tern	*Sterna caspia*	M
107	River Tern	*Sterna aurantia*	R/SM/B
108	Lesser Crested Tern	*Sterna bengalensis*	M
109	Large Crested Tern	*Sterna bergii*	M
110	Common Tern	*Sterna hirundo*	M
111	Little Tern	*Sterna albifrons*	R/M/B
112	Saunders's Tern	*Sterna saundersi*	M
113	Whiskered Tern	*Chlidonias hybridus*	M
114	White-winged Black Tern	*Chlidonias leucopterus*	M
115	Black Tern	*Chlidonias niger*	M
116	Indian Skimmer	*Rynchops albicollis*	SM
117	Small Blue Kingfisher	*Alcedo atthis*	R
118	White-breasted Kingfisher	*Halcyon smyrnensis*	R
119	Black-capped Kingfisher	*Halcyon pileata*	R
120	Lesser Pied Kingfisher	*Ceryle rudis*	R
121	Large Pied Wagtail	*Motacilla maderaspatensis*	R
122	Citrine Wagtail	*Motacilla citreola*	M
123	Yellow Wagtail	*Motacilla flava*	M
124	Grey Wagtail	*Motacilla cinerea*	M

Name reference: Ramsar (2007) Hand book 1 4th edition. Wise Use of Wetlands

References

Acharya S, Kar SK (1996) Checklist of waders (Charadriiformes) in Chilika lake, Orissa. Newsl Birdwat 36(5):89–90

Ali S, Hussain SA (1981) Studies on the movement and population structure of Indian avi-fauna. Annual report 1. Bombay Natural History Society, Mumbai

Ali S, Ripley SD (1983) Handbook of the birds of India and Pakistan. Compact edition, Oxford University Press, New Delhi

Balachandran S (1990) Studies on the coastal birds of Mandapam and the neighbouring islands (Peninsular India). Thesis, Annamalai University, India

Balachandran S (1997) Population, status, moult and measurement of Great Knot Calidris tenuirostris wintering in South India. Stilt 30:3–6

Balachandran (1998) Bird migration studies in India. Final report. Bombay Natural History Society, Mumbai

Balachandran, Sathiyaselvam (2010) Elucidating international migration routes of priority water-bird species in India and the Central Asian Flyway to evaluate their potential to transmit highly pathogenic avian influenza. Final report. Report submitted to the Food & Agriculture Organisation of the United States

Balachandran SA, Rahmani R, Sathiyaselvam P (2004) Habitat evaluation of Chilika Lake with special reference to birds as bio-indicators- II Annual report. Bombay Natural History Society, Mumbai

Balachandran SA, Rahmani R, Sathiyaselvam P (2005) Habitat evaluation of Chilika Lake with special reference to birds as bioindicators. Final report. Bombay Natural History Society, Mumbai

Balachandran S, Merwyn F, Sethi D (2006a) Habitat improvement measures for the birds of Pong Wetland. Final report. Bombay Natural History Society, Mumbai

Balachandran SA, Rahmani R, Sathiyaselvam P, Sirajudin H (2006b) Population monitoring of migratory waterbirds in Chilika and Bhitarkanika with special reference to Avian Influenza. Final report. Bombay Natural History Society, Mumbai

Balachandran SA, Rahmani R, Sathiyaselvam P, Gangaiamaran P (2007) An assessment of the impact of hydrological intervention on the bird population of Chilika. Interim report (2006–2007). Bombay Natural History Society, Mumbai

Balachandran SA, Rahmani R, Sathiyaselvam P, Gangaiamaran P (2008) An assessment of the impact of hydrological intervention on the bird population of Chilika. Final report (2006 to 2008). Bombay Natural History Society, Mumbai

Benthall EC (1947) Periodic abundance of Bar-headed Goose Anser indicus (Latham) on Chilika Lake. J Bombay Nat Hist Soc 47(1):168

Benthall EC, Craven LA (1950) Geese and duck on the Chilika Lake. Orissa J Bombay Nat Hist Soc 49(2):312–313

BirdLife International (2018) Species factsheet: Limosa limosa. Downloaded from http://www.birdlife.org on 30/07/2018

Boere GC, Piersma T (2012) Flyway protection and the predicament of our migrant birds: a critical look at international conservation policies and the Dutch Wadden Sea. Ocean Coast Manag 68:157e168

Craven LA (1949) Occurrence of the White-fronted Goose Anser albinofrons in Orissa. J Bombay Nat Hist Soc 48(2):365–366

Das NK, Samal RC (1988) Environmental survey of Chilika. In: Patro SN (ed) Chilika, The pride of our wetland heritage. Orissa Environmental Society, Bhubaneshwar, pp 96–103

Davidson N (2003) Status of wader populations on the Central/South Asian flyway. In: Declining populations workshop – abstracts of talks. International Wader Study Group (WSG) annual conference, Cadiz, September 2003

Hails AJ (1997) Wetlands, biodiversity and the Ramsar convention. Ramsar Convention Bureau, Gland

Holm TE, Clausen P (2006) Effects of water level management on autumn staging waterbird and macrophyte diversity in three Danish coastal lagoons. Biodivers Conserv 15:4399–4423
Hussain SA, Mohapatra KK, Ali S (1984) Avifaunal profile of Chilika Lake: a case for conservation. Technical report (4). Journal of the Bombay Natural History Society
Mohapatra KK (1998) Food and feeding habits of important aquatic birds of Chilika Lake Orissa. Final report. Bombay Natural History Society, Mumbai
Mohapatra KK, Rao P (1993) Some waders record from coastal Andhra Pradesh. J Bombay Nat Hist Soc 89:250–251

Chapter 14
Biodiversity of Benthic Fauna in Chilika Lagoon

Punyasloke Bhadury, Kapuli Gani Mohamed Thameemul Ansari, Areen Sen, and Vandana Kumari Gupta

Abstract Benthic communities represents the major component of aquatic sedimentary biodiversity and play important roles in major ecosystem processes beside serving as excellent proxy for tracking environmental and anthropogenically induced changes. Chilika lagoon, the largest brackish water lagoon of Asia, is a hot spot for biodiversity and harbors rich aquatic flora and fauna. Numerous studies have been undertaken to date with a focus towards unraveling assemblage structure and diversity of benthic macrofauna and meiofauna from Chilika lagoon. Among benthic macrofauna, Gastropods, Bivalves and Polychaetes are major players in terms of abundance and diversity. In case of meiobenthos, Free-Living Marine Nematodes and Foraminifera constitute major components in terms of abundance and diversity in Chilika lagoon. Lesser known groups of meiobenthos have not been fully explored from this ecosystem. The baseline level information obtained from cataloguing biodiversity of benthic fauna in Chilika lagoon could ultimately form the basis for its long-term ecological monitoring as well as the role of these groups towards sustaining rich fisheries in this unique lagoonal environment.

P. Bhadury (✉)
Integrative Taxonomy and Microbial Ecology Research Group, Department of Biological Sciences, Indian Institute of Science Education and Research Kolkata (IISERK), Nadia, West Bengal, India

Centre for Climate and Environmental Studies, Indian Institute of Science Education and Research Kolkata (IISERK), Nadia, West Bengal, India
e-mail: pbhadury@iiserkol.ac.in

K. G. M. T. Ansari · A. Sen
Integrative Taxonomy and Microbial Ecology Research Group, Department of Biological Sciences, Indian Institute of Science Education and Research Kolkata (IISERK), Nadia, West Bengal, India

V. K. Gupta
Integrative Taxonomy and Microbial Ecology Research Group, Department of Biological Sciences, Indian Institute of Science Education and Research Kolkata (IISERK), Nadia, West Bengal, India

Department of Marine Science, University of Calcutta, Kolkata, India

C. M. Finlayson et al. (eds.), *Ecology, Conservation, and Restoration of Chilika Lagoon, India*, Wetlands: Ecology, Conservation and Management 6,
https://doi.org/10.1007/978-3-030-33424-6_14

Keywords Benthic assemblage · Macrofauna · Meiofauna · Polychaetes · Gastropods · Bivalves · Nematodes · Foraminifers · Abiotic interaction · Ecological importance · Chilika lagoon, India

14.1 Background

The coastal zone is of fundamental importance to the development and sustainability of human population and advancement of society (Gray 1997). It forms the boundary between terrestrial and marine biomes and encompasses huge habitat variability ranging from coastal lagoons, estuaries, mangroves to fjords. Coastal lagoons represent transitional aquatic ecosystem interfacing fresh- and saline water environment (Kjerfve and Magill 1989). This type of ecosystem is dynamic and influenced by wave and tidal actions, variable freshwater flow and sediment load along with natural (e.g. cyclone, storm surge) and anthropogenic forcings (e.g. nutrient and organic load, eutrophication) (e.g Anthony et al. 2009). Due to resulting variability in abiotic conditions, coastal lagoons are characteristically unbalanced i.e. the thought of short or long term variation is somewhat virtual for these ecosystems compared to other coastal environments (Kjerfve 1986; Marzano et al. 2003). Due to their ecotonal setting, coastal lagoons are often characterized by unique biotic assemblages both in water column as well as in sediment substratum.

Chilika, the largest coastal lagoon of Asia, characterized by a shallow depth regime (Gupta et al. 2008), serves a pivotal role in local economy as it supports the livelihood of approximately 300,000 fisherfolk in the region (Dujovny 2009). The benthic fauna in the lagoon show variations based on monsoonal influenced precipitation (south-west monsoon) during July to October resulting in massive inflow of freshwater (5.09×10^9 m^3; Panda and Mohanty 2008) into the system from 52 rivers and rivulets. The lagoon can be broadly divided into four sectors based on its hydrology, biodiversity and fisheries yield, namely (i) Southern (ii) Central (iii) and Northern sectors; and (iv) Outer Channel. In Chilika lagoon, studies on benthic fauna have been undertaken to get a clearer understanding of their assemblage structure in relation to prevailing environmental conditions and role of abiotic factors in shaping observed assemblage patterns. The present chapter provides a detailed overview of studies undertaken exclusively on benthic macrofauna and meiofauna of Chilika lagoon to date, in addition with a focus on lesser known benthic foraminifera. Moreover, knowledge gaps for some faunal groups as well as overall importance of benthic faunal groups in understanding lagoon processes including long-term ecological monitoring have been highlighted.

A variety of sediment dwelling organisms utilize diverse floral assemblages that characterize such shallow water bodies (McGlathery et al. 2007). Derived from the Greek word βένθος, the term benthos was first coined by Ernst Haeckel in 1890 to describe "life forms of the deep ocean and sea floor". The benthic assemblage is composed of a wide range of plants, animals and microbes, thus forming a key functional component of complex aquatic food webs. The term 'phytobenthos' is used when referring to the plant members (i.e., various algae and aquatic plants), whereas 'zoobenthos' is applied in reference to all consumers (i.e., benthic protozoans and metazoans). ‘Benthic microflora’ (bacteria, fungi and many protozoans) constitute the decomposer assemblage and are involved in recycling of essential nutrients, trapping and fixation. Apart from their ecological roles different benthic faunal groups are widely used as indicators for tracking effects of environmental and anthropogenic changes in various ecosystems including coastal lagoons (Semprucci et al. 2015; 2016; Zeppilli et al. 2015). In coastal habitats environmental variables including salinity can collectively influence structure and functioning of benthic faunal assemblages (Eyre and Balls 1999; Semprucci et al. 2014a, b).

14.2 Benthic Faunal Classification

Generally benthic organisms are classified based on their functional attributes such as living habitat and body size (Phole and Thomas 2001). Benthic invertebrates can be differentiated by the position they occupy ‘above’ or ‘on’ or ‘in’ bottom sediments: hyperfauna – living just above the sediment surface (e.g. demersal fishes), epifauna – living on the sediment substrate (e.g. gastropods and copepods) and infauna – living inside or between sediment particles (e.g. polychaetes). Apart from these, organisms such as free-living marine nematodes and foraminifers can also inhibit the interstitial space among sediment particles. Based on body size, benthic organisms can be classified into four groups such as (1) megafauna – organisms more than 10 mm in size (e.g. most of the epifaunal organisms – crabs), (2) macrofauna – organisms in the size range of 0.5–10 mm (e.g. most of the infaunal organisms – polychaetes, molluscs), (3) meiofauna – organisms in the size range between 0.045 and 0.5 mm (e.g. most of the interstitial fauna – free-living marine nematodes) and (4) microfauna – organisms found less than 0.045 mm size (e.g. most of the microbial assemblage inhabiting sediment column) (Tagliapietra and Sigovini 2010).

14.2.1 Epifauna

Epifauna play important role in marine ecosystem processes. Epifaunal organisms aid in decomposition, breakdown, incorporation and turnover of organic matter in sediments and thereby help to recycle nutrients in the overlying water column. Hence they are an important link in coastal and marine food webs and found frequently in the diets of larger, more mobile, predators, some of which have commercial importance (e.g. demersal fisheries) (Khan et al. 2010). Investigations on the distribution and diversity of epifaunal communities have been widely undertaken across the Indian coasts (Khan et al. 2010 reference therein). Compared to other groups of bottom dwellers in Chilika, the epifaunal component appears to be understudied, though majority of the economically important species come under this category such as the brachyuran mud crab species like *Scylla serrata* and *Scylla tranquebarica* along with some penaeid shrimps (Mohapatra et al. 2007). In a recent study undertaken in the Outer Channel of Chilika lagoon four species belonging to Porifera namely,

Spongilla alba, Haliclona indistincta, Pione vastifica and *Protosuberites lacustris* have been documented living on the surface sediment (Mahapatro et al. 2015).

14.2.2 *Infauna*

14.2.2.1 Macrofauna

Macrofauna (Greek word: *macro* – larger) consist of organisms that live in or on sediment substrate or even some are attached to hard substrate within an aquatic environment and visible to the naked eyes. Several animal phyla have benthic macrofaunal representatives and found in various habitats ranging from estuaries, lagoons to deep sea environments and even freshwater environments. However, in case of freshwater environments, taxonomic composition of macrofauna may differ significantly from coastal or marine counterparts. Benthic macrofauna are found on different types of sediment substrate (e.g. rocky, sandy and muddy) encountered in the above habitats. These organisms usually represent size which is more than 0.5 mm (Higgins and Thiel 1988).

The first described investigation on Fauna of Chilika lagoon was undertaken in early twentieth century by Annandale and Kemp and was documented in Memories of Indian Museum in 1915. Studies were subsequently undertaken by various investigators in different periods (Mahapatro et al. 2015). In total, 253 macrofaunal species have been recorded from the Chilika lagoon (Fig. 14.1) and represented by polychaetes

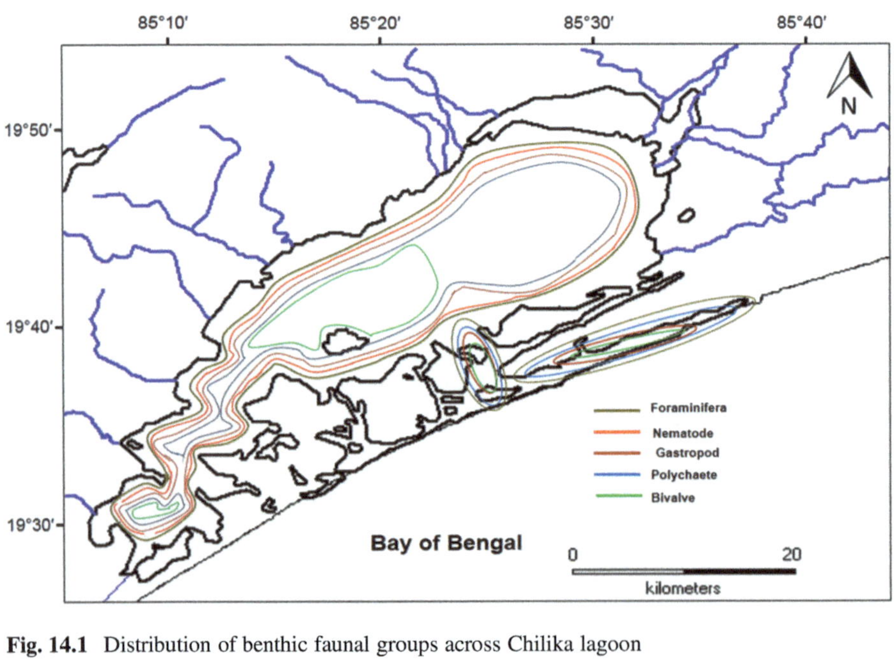

Fig. 14.1 Distribution of benthic faunal groups across Chilika lagoon

(70 species – including 4 species nomenclature not as per WoRMS Database), gastropods (74 species – including 16 species nomenclature not as per WoRMS database), bivalves (69 species – including 13 species nomenclature not as per WoRMS database), amphipods (28 species), isopods (6 species) and tanaids (3 species).

Polychaetes

Polychaetes are one of the most frequent, abundant and species-rich group among macrofauna with a wide geographical distribution and span from coastal habitats to deep sea floors (Fauchald and Jones 1979; Cosson-Sarradin et al. 1998; Carr 2012). The distribution and diversity of polychaetes depend on environmental factors such as nature of sediment, organic load, salinity, water depth and also latitudinal scales (Etter and Grassle 1992). Based on various estimates, it is thought that polychaete species richness can range from 25,000 to 30,000 (Snelgrove et al. 1997); however, to date, more than 10,000 species have been described (Minelli 1993; WoRMS Database). The ratio of described to undescribed species varies according to habitat and biogeographical region while intertidal and shallow subtidal regions from Europe (Fauvel 1923, 1927; Hartmann-Schroder 1971; Bastrop et al. 1998), North America (Hartman 1968, 1969; Blagoon et al. 1996) and Southern part of Africa (Day 1967; Simon et al. 2010) are comparatively well studied. This has led to many species being categorized as 'cosmopolitan' or at least being assumed to have wide geographical distribution. However, there are many regions around the world, where our knowledge of diversity of polychaetes remain largely unclear particularly from the Indian coastlines (e.g. Bhadury and Annapurna 2011).

An online portal (http://www.biosearch.in) is available which provide information on biota from the Indian coastlines including distribution and diversity of polychaetes. According to the most recent update, only 1142 polychaete species (valid) have been reported from the Indian coastlines. Out of the reported species, ca. 63% are from the Arabian Sea and remaining from the Bay of Bengal representing biotopes including mangroves, lagoons and seagrass beds (Ganesh and Raman 2007; Jayaraj et al. 2008; Joydas and Damodaran 2009; Manokaran et al. 2015). The recent reports on the distribution and diversity of polychaetes suggest that approximately 60% of species have been reported from the open ocean environment and the rest are from marginal environments (Sivaleela and Venkataraman 2012). The genera such as *Nereis* (43 species), *Eunice* (39 species), *Perinereis* (32 species), *Glycera* (24 species), *Lumbrinereis* (23 species) and *Nephtys* (19 species) have been found to be most diverse with broad distribution in Indian coastal ecosystems (http://www.biosearch.in/). In total, 1142 species have been reported from India which is comparatively higher than other biogeographic regions which are considered to be rich in polychaetes such as the Adriatic Sea (764 species) and Arabian Peninsula (807 species) including Red Sea, Gulf of Aden, Arabian Sea, Gulf of Oman and Arabian Gulf (Wehe and Fiege 2002; Mikae 2015).

In Chilika lagoon, Annandale and Kemp (1915) documented 23 species of polychaetes; however this study was restricted only in the Outer channel region. Southern

(1921) investigated benthic fauna of Chilika lagoon in Southern and Outer channel regions and recorded 28 species of polychaetes. Among them, 8 species (*Neanthes glandicincta* (Southern, 1921), *Dendronereides heteropoda* Southern, 1921, *Namalycastis indica* (Southern, 1921), *Sigambra constricta* (Southern, 1921), *Potamilla leptochaeta* Southern, 1921, *Pseudopolydora kempi* (Southern, 1921), *Barantolla sculpta* Southern, 1921 and *Mastobranchus indicus* Southern, 1921) were new distributional records from Chilika lagoon. Furthermore, continuous studies on documenting the distribution and diversity of polychaetes in this lagoon led to an improvement in overall information on polychaetes. However, most of the investigations have been restricted to a particular sector or sectors of the lagoon. For example, Das (2004) investigated only the Southern sector of the lagoon and recorded three species including putative species, namely *Nephtys* sp.1, *Capitella capitata* (Fabricius, 1780) and *Glycera* sp.1, while Mahapatro et al. (2009, 2015) carried out benthic faunal investigation with a focus on polychaetes throughout the Outer channel and recorded 28 more species adding to the existing polychaete species list of Chilika. Mahapatro et al. (2012) also documented the relationship between observed polychaete distribution and aquatic macrophyte density; they found that polychaete diversity was higher in Central and Northern sectors of Chilika lagoon since these regions had higher growth of aquatic macrophytes. Such trend has been also reported in published literature from other locations globally (e.g. Danovaro et al. 2002) Recently, Bhadury et al. (2015) covered entire lagoon except the Outer channel region and reported additional nine species of polychaete species out of encountered 15 species. To date, 70 polychaete species have been reported from the Chilika lagoon (Table 14.1).

The polychaete diversity observed in Chilika lagoon indicated that most of the species encountered are of marine origin; however, these species have largely adapted to the variable salinity gradients prevalent within the lagoon (e.g. Etter and Grassle 1992; Snelgrove et al. 1997). Among abiotic variables, salinity, nutrients and sediment texture play important roles in influencing the observed polychaete assemblage structure across this lagoon. For example most of the suspension feeding polychaetes (e.g. genera such as *Prionospio and Nereis*) has been reported from Central sector and Outer channel region whereas deposit feeders (e.g. genera such as *Neanthes,and Aglaophamus*) were dominant in Southern and Central sectors and surface deposit feeders and carnivores (e.g. *Nephtys, Glycera* genera) were found in the Central and Northern sectors. The suspension feeding polychaetes are generally dominant in marine ecosystems which have less turbidity and proper mixing zones with sandy nature of sediments whereas deposit feeders are found in silt/clay sediments along with high organic load (e.g. Fauchald and Jumars 1979; Manokaran et al. 2013). In Chilika lagoon, more than 40% of the area is inhabited by aquatic macrophytes, particularly in the Northern sector and it extends up to Central sector and thus phytal polychaete species abundance is high in these sectors (e.g. genus such as *Diopatra*) (Bell and Coen 1982). However, polychaete species diversity in Chilika lagoon is yet to be extensively inventorized at fine scale and therefore warrants further investigation as part of future studies.

Table 14.1 List of Polychaete species reported from Chilika lagoon

Polychaete species	References	Polychaete species	References
Aglaophamus dibranchis (Grube, 1877)	1	*Myriochele picta* Southern, 1921	1,2,7
Aglaophamus lyrocheata (Fauvel, 1902)	1	*Mysta picta* (Quatrefages, 1866)	7
Amphictene auricoma (O.F. Müller, 1776)	7	*Namalycastis indica* (Southern, 1921)	2,7
Aricidae (Acmira) lopezi Berkeley & Berkeley, 1956	7	*Neanthes agulhana* (Day, 1963)	1
Axiothella obockensis (Gravier, 1905)	7	*Neanthes chilkaensis* (Southern, 1921)	1,2,7
Barantolla sculpta Southern, 1921	2	*Neanthes glandicincta* (Southern, 1921)	2,7
Bipalponephtys cornuta (Berkeley & Berkeley, 1945)	7	*Neanthes indica brunnea* (Day, 1957)	1
Capitella capitata (Fabricius, 1780)	3,4,5,7,8	*Nephtys polybranchia* Southern, 1921	1,2,6,7,8
Chone fauveli McIntosh, 1916	7	*Nephtys* sp.1	3,4,5,8
Cossura coasta Kitamori, 1960	7	*Nephtys* sp.2	8
Dendronereis aestuarina Southern, 1921	2	*Nereis reducta* Hartmann-Schröder, 1960	1,2,7
Dendronereides heteropoda Southern, 1921	2	*Nereis* sp.1	4,5,8
Diopatra neapolitana Delle Chiaje, 1841	7	*Nereis* sp.3	8
Diopatra variabilis Southern, 1921	1,2	*Notomastus* sp.	4
Euclymene annandalei Southern, 1921	1,2,7	*Onuphis eremita* Audouin & Milne Edwards, 1833	7
Eulalia viridis (Linnaeus, 1767)	7	*Owenia fusiformis* Delle Chiaje, 1844	7
Eteone picta Quatrefages, 1866	7	*Oxydromus flexuosus* (Delle Chiaje, 1827)*	7
Fabriciola spongicola (Southern, 1921)	1,2,7	*Paraprionospio pinnata* (Ehlers, 1901)	7
Ficopomatus macrodon Southern, 1921	2	*Perinereis marionii* (Audouin & Milne Edwards, 1833)	1
Glycera alba (O.F. Müller, 1776)	1,2,7	*Perinereis nigropunctata* (Horst, 1889)	7
Glycera tridactyla Schmarda, 1861	8	*Pisione remota* (Southern, 1914)	7
Glycera sp.1	3,4,8	*Polydora hornelli* Willey, 1905	1,7
Glycinde oligodon Southern, 1921	1,2,7	*Polydora* sp.	4
Goniadopsis longicirrata (Arwidsson, 1899)	7	*Pomatoceros caeruleus* (Schmarda, 1861)	7
Hesione picta Müller in Grube, 1858	7	*Potamilla leptochaeta* Southern, 1921	2

(continued)

Table 14.1 (continued)

Polychaete species	References	Polychaete species	References
Heteromastus filiformis (Claparède, 1864)	5,7	*Prionospio cirrifera* Wirén, 1883*	7
Heteromastus similis Southern, 1921	1,2	*Prionospio dubia* Day, 1961	8
Hydroides elegans (Haswell, 1883)	7	*Prionospio polybranchiata* Fauvel, 1929	8
Laonome indica Southern, 1921	1,2	*Prionospio* sp. 2	4,8
Lumbrineris polydesma Southern, 1921	1,2,7	*Pseudopolydora kempi* (Southern, 1921)	2
Lumbrineris simplicis Hartman, 1959*	1,2	*Scoloplos (Scoloplos) marsupialis* (Southern, 1921)	1,2,7
Marphysa gravelyi Southern, 1921	1,2,7	*Sigambra constricta* (Southern, 1921)	2,7
Mastobranchus indicus Southern, 1921	2	*Spiophanes bombyx* (Claparède, 1870)	7
Mediomastus sp.	7	*Sternaspis costata* Marenzeller, 1879	1
Micronephthys oligobranchia (Southern, 1921)	1,2	*Tylonereis fauveli* Southern, 1921	1,2

References: 1- Annandale and Kemp 1915; 2 – Southern 1921; 3- Das 2004; 4 – Mahapatro et al. 2009; 5- Mahapatro et al. 2012; 6 – Mishra et al. 2013; 7 – Mahapatro et al. 2015; 8 – Bhadury 2015; * – nomenclature not as per WoRMS Database

Gastropods

Gastropods represent one of the most successful metazoan groups among Animalia, occupying all three major habitats such as marine, freshwater and terrestrial domains (e.g. Poppe and Tagaro 2006). They are by far the largest group of molluscs, and they comprise approximately 80% of total living molluscs (Poppe and Tagaro 2006). According to an estimate of described species of gastropods including living and fossil records, approximately 30,000 species are from marine (including estuaries, back waters, lagoons) and terrestrial ecosystems and remaining 5000 species are from freshwater ecosystems (Strong et al. 2007). However, most updated estimate report that total number of extant species range from 40,000 to over 150,000 species (Régnier et al. 2009). There are about 13,000 recognized genera for both recent and paleontological records explaining long and rich fossil documentations since early Cambrian era thus reflecting periodic extinctions of gastropods (Landing et al. 2002). In general, marine gastropods are most diverse in terms of species composition and even hundreds of species can be found in any coastal or marine ecosystem (e.g. estuary, lagoon and deep sea).

To date from India, a total of 2321 marine gastropod species have been reported (http://www.biosearch.in/); nevertheless appropriate information on freshwater and terrestrial gastropod diversity have some level of uncertainty (e.g. Annandale and Rao 1925). About 70% of gastropod species recorded from Indian coasts represent largely marginal habitats such as estuaries, backwaters, lagoons, mangroves, seagrass beds, seaweed stretches, rocky shores and sandy beaches. However, gastropod species diversity reported from inshore to deep sea environments are relatively very limited (Subba Rao 2003). The gastropod orders such as Neogastropoda (901 species) followed by Nudibranchia (115 species), Cephalaspidea (35 species) and Thecosomata (31 species) have been reported from the Indian coasts (http://www.biosearch.in/). For example, gastropod species diversity from Indian water is six times higher than from Norwegian coastal waters (365 species: Høisaeter 2009), while it is lesser than rest of Europe (http://www.somali.asso.fr/clemam/index.clemam.html) although Indian coastal waters are yet to be elaborately explored. Only 74 gastropod species have been documented in Chilika (Table 14.2); among them 56% species have been reported only from Outer channel region and remaining 44% species from inner part of Chilika lagoon. Annandale and Kemp (1915) and Preston (1915) initially reported 28 species of gastropods from the Outer channel region of Chilika and subsequently Mahapatro et al. (2009, 2012) recorded 4 more species from the same region. However, inner parts of the Chilika lagoon (Southern, Central and Northern) have not been thoroughly documented in terms of gastropod diversity and their distribution. Sahu et al. (2007) reported six species of gastropods namely, *Nassarius stolatus* (Gmelin, 1791), *Pirenella cingulata* (Gmelin, 1791), *Indothais lacera* (Born, 1778), *Indoplanorbis exustus* (Deshayes, 1834), *Thiara* sp. and *Notocochlis tigrina* (Röding, 1798) in Nalabana island (Central sector) of Chilika lagoon.

In another study, four gastropod species representing freshwater (*Indoplanorbis* sp. and *Lymnaea* sp.) and brackish water species (*Hydrobia* sp. and *Nassarius stolatus*) have been reported from the continental region of this lagoon (Mahapatro et al. 2012). Subsequently, Mishra et al. (2013) recorded six gastropod species in Central sector out of which three species (*Nassarius foveolatus* (Dunker, 1847), *Pseudanachis basedowi* (Hedley, 1918) and *Pericola ventricosa* (Swainson, 1822) were first report from Chilika lagoon. In the most recent and comprehensive study, Bhadury et al. (2015) encountered 27 gastropod species within the Chilika lagoon. Out of 27, 18 species were found to have wide-spread distribution throughout the lagoon while 7 species were found to have brackish water distribution and remaining 2 species were exclusively freshwater species. In general, calcified gastropods can mostly tolerate varying environmental gradients including physical factors such as tidal and wave actions, water current patterns and sediment texture can strongly influence their assemblage patterns across coastal ecosystems. Chilika lagoon is dominated by sandy sediment that along with macrophytes provide habitat for small gastropod species representing genera such as *Nassarius and Hydrobia* which are slightly burrowing organisms (Etter and Grassle 1992). However, it is worthwhile to mention that some of the encountered gastropod specimens have been identified to generic level and species level confirmation based on integrative taxonomy approaches is being presently adopted in Chilika lagoon (Bhadury 2015).

Table 14.2 List of Gastropod species reported from Chilika lagoon

Gastropod species	References	Gastropod species	References
Aceteocina estriata (Preston, 1914)	1	*Nassarius marratii* (E. A. Smith, 1876)	1
Aplysia sp.	6	*Nassarius orissaensis* (Preston, 1914)	1
Bufonaria echinata (Link, 1807)	6	*Nassarius* sp.3	7
Bullia melanoides (Deshayes, 1832)	7	*Nassarius stolatus* (Gmelin, 1791)	2,3,4,5,6,7
Bullia rhodostoma Reeve, 1847	7	*Nassarius versicolor* (C. B. Adams, 1852)	7
Bullia turrita Gray, 1839	7	*Nassarius vittatus* (A. Adams, 1853)	7
Bullia vittata (Linnaeus, 1767)	6	*Nerita balteata* Reeve, 1855	6
Chilkaia imitatrix Preston, 1915	1	*Neritina smithii* W. Wood, 1828	6
Chrysallida erucella A. Adams, 1863	1	*Notocochlis tigrina* (Röding, 1798)	2,6
Chrysallida nadiensis Preston, 1915∗	1	*Odostomia chilkaensis* Preston, 1914	1
Coliracemata innocens (Preston, 1915)	1	*Oliva oliva (Linnaeus, 1758)*	6
	6	*Oliva* sp.1	7
Conus hyaena Hwass in Bruguière, 1792	6	*Oliva* sp.2	7
Cyllene pulchella Adams & Reeve, 1850	7	*Pericola ventricosa* Swainson, 1822∗	5
Cyllene sulcata G. B. Sowerby II, 1859	7	*Peringia ulvae* (Pennant, 1777)	7
Drupella rugosa (Born, 1778)	2	*Phalium areola* (Linnaeus, 1758)	6
Epitonium clathrus (Linnaeus, 1758)	6	*Pirenella cingulata* (Gmelin, 1791)	2,3,5,6,7
Epitonium hamatulae Preston, 1915	1	*Pirenella conica* (Blainville, 1829)	1
Gangetia miliacea (G. Nevill, 1880)	1	*Pseudanachis basedowi* (Hedley, 1918)	5
Haminoea crocata Pease, 1860	1	*Quirella humilis* (Preston, 1905)	1
Hebra subspinosa (Lamarck, 1822)	1	*Smaragdia souverbiana* (Montrouzier in Souverbie & Montrouzier, 1863)	1
Hydrobia aurita Neumayr in Neumayr & Paul, 1875	7	*Solariella obscura* (Couthouy, 1838)	6
Hydrobia sp.	4	*Solariella satparaensis* Preston, 1914∗	1
Ilyanassa obsoleta (Say, 1822)	6	*Stenothyra blanfordiana* Nevill, 1880∗	1
Indoplanorbis exustus (Deshayes, 1834)	2,4,7	*Stenothyra chilkaensis* Preston, 1914∗	1

(continued)

Table 14.2 (continued)

Gastropod species	References	Gastropod species	References
Indothais lacera (Born, 1778)	4,7	*Stenothyra minima* (G. B. Sowerby I, 1837)	1
Litiopa (Alaba) copiosa Preston, 1915*	1	*Stenothyra obesula* Preston 1915*	1
Litiopa (Alaba) kempi Preston, 1914*	1	*Stenothyra orissaensis* Preston, 1914*	1
Littorina littorea (Linnaeus, 1758)	6	*Stenothyra trigona* Preston, 1915*	1
Lymnaea sp.	4	*Terebralia palustris* (Linnaeus, 1767)	7
Nassa denegabilis Preston, 1914*	1	*Terebralia* sp.1	7
Nassarius acuticostus (Montrouzier in Souverbie & Montrouzier, 1864)	7	*Terebralia sulcata* (Born, 1778)	7
Nassarius burchardi (Dunker in Philippi, 1849)	1	*Thiara scabra* (O. F. Müller, 1774)	6
Nassarius comptus (A. Adams, 1852)	7	*Thiara* sp.	2,5
Nassarius conoidalis (Deshayes, 1832)	7	*Tinostoma variegatum* Preston, 1914*	1
Nassarius dorsatus (Röding, 1798)	7	*Umbonium vestiarium* (Linnaeus, 1758)	1,6
Nassarius foveolatus (Dunker, 1847)	5	*Vanesia rambhaensis* Preston, 1914*	1

References: 1- Annandale and Kemp 1915; 2 – Sahu et al. 2007; 3 – Mahapatro et al. 2009; 4- Mahapatro et al. 2012; 5 – Mishra et al. 2013; 6 – Mahapatro et al. 2015; 7 – Bhadury 2015; * nomenclature not as per WoRMS Database

Bivalves

Class Bivalvia, the second largest class among molluscs is formally known as Lamellibranchia or Pelecypoda which includes mussels, oysters, scallops and clams. They are widespread in various type of aquatic environments (including freshwater and marine) such as mud flats of streams, sandy beaches, deep sea as well as rocky substrates (Purchon 1987; Bieler and Mikkelsen 2006). Class Bivalvia was recently classified by Carter et al. (2011) based on living and paleontological records (from early Cambrian), reported 10,000 described species including 2000 species which are found in freshwater environment. Among described species, approximately 87% inhabit marine ecosystems including brackish water, estuarine, coastal and deep sea environments.

The class Bivalvia contribute to 33.6% of total molluscan diversity in the Indian subcontinent (Appukuttan 1996) and approximately 920 species are of marine origin (Venkatesan and Mohamed 2015; http://www.biosearch.in/). About 89% of the bivalve species have been documented from sandy beaches and intertidal regions representing both the East and West coasts of India and rest from offshore environments (Venkatesan and Mohamed 2015). Under Bivalvia, orders such as Veneroida (407 species), Myoida (96 speceis), Arcoida (67 species), Mytiolida (64 species) and Pterioida (42 species) have been recorded in the Indian coast lines (http://www.biosearch.in/). The bivalve species recorded from Indian water is higher compared to Queensland in Australia (350 species: Healy and Potter 2010) or Rodrigues in Mauritius (109 species: Oliver et al. 2004). To date, 70 bivalve species (~8% of total diversity) have been recorded in Chilika lagoon (Table 14.3) which includes oysters, mussels and scallops. Annandale and Kemp (1915) documented first hand bivalve diversity (40 species) in the Outer channel region of Chilika lagoon. In the same year, Preston (1915) proposed a new genus and species (*Chilkaia imitatrix* Preston, 1915) from Chilika lagoon. In early 2000s, Das (2004) added three more bivalve species to the existing species inventory from Southern sector; further Sahu et al. (2007) investigated around the Nalaban Island and reported other four bivalve species to the existing list. Later, Mahapatro et al. (2012) and Mishra et al. (2013) investigated the continental region as well as Central and Northern sectors of the lagoon and recorded nine bivalve species.

A study undertaken by Mahapatro et al. (2015) documented 21 species of bivalves in the Outer Channel region of Chilika lagoon, out of which 10 species were new reports with respect to this lagoonal ecosystem. Recent investigation on benthic faunal assemblage inside the lagoon documented 11 bivalve species, among them three species (*Donax cuneatus* Linnaeus, 1758, *Mya arenaria* Linnaeus, 1758 and *Mya* sp.) were included in existing bivalve species list for Chilika (Bhadury 2015). Most of the bivalve species recorded in the Chilika lagoon inhabit sandy sediments and ca. 62% were reported from Outer channel region suggesting that salinity and sediment nature are the important factors influencing bivalve assemblage structure and distribution. In this lagoon, bivalve species diversity has been found to be higher in Central and Northern sectors. On the other hand in Southern sector which has relatively higher salinity exhibited lower diversity of bivalves and this could be attributed to the nature of sediment (silt/clay) along with water depth. In the Northern sector, bivalve diversity dominated mostly by spats (e.g. *Modiolus, Donax*) was frequently found to be attached with sea grasses and macrophytes. An important bivalve species which inhabits only in the Chilika lagoon, *Theora opalina* (Hinds, 1843), a transparent brackish water species, has been also reported in most of the studies (see Table 14.3).

Minor Groups of Benthic Macrofauna

Other than major macrofaunal groups (polychaetes, gastropods and bivalves), numerous minor macrofaunal groups also inhabit sediment-water interface in any aquatic ecosystem including coastal lagoons. Among them smaller crustacean groups constituted by amphipods, isopods, tanaids and smaller decapods represent minor benthic macrofaunal groups and their distribution and diversity are mainly

Table 14.3 List of Bivalve species reported from Chilika lagoon

Bivalve species	References	Bivalve species	References
Anomia achaeus Gray, 1850	8	*Meretrix casta* (Gmelin, 1791)	1,3,4,5,8,9
Bankia carinata (J.E. Gray, 1827)	1	*Meretrix meretrix* (Linnaeus, 1758)	1,8
Brachidontes modiolus (Linnaeus, 1767)	8	*Meretrix ovum* Lamarck, 1799	1
Brachidontes striatulus (Hanley, 1843)	1,8	*Modiolus modiolus* (Linnaeus, 1758)	3,8,9
Brachidontes subramosus (Hanley, 1843)	1,6,7,8	*Modiolus modulaides* (Röding, 1798)	4,9
Brachidontes undulatus (Dunker, 1857)	1,8	*Modiolus* sp.3	5,9
Chilkaia imitatrix Preston, 1915	2	*Mya arenaria* Linnaeus, 1758	9
Clementia papyracea (Gmelin, 1791)	1,4,5	*Mya* sp.	9
Confusella confusa (Preston, 1914)	1	*Neotrapezium sublaevigatum* (Lamarck, 1819)	1
Corbicula fluminea (O. F. Müller, 1774)	8	*Parreysia* sp.	7
Corbicula sp.	7	*Perna perna* (Linnaeus, 1758)	8
Crassostrea belcheri (G. B. Sowerby II, 1871)	5,8	*Perna viridis* (Linnaeus, 1758)	1,8
Crassostrea cuttackensis (Newton & E. A. Smith, 1912)	8	*Placuna placenta* (Linnaeus, 1758)	8
Crassostrea virginica (Gmelin, 1791)	1	*Potamocorbula chilkaensis* (Preston, 1911)	1
Cumingia hinduorum Preston, 1915∗	1	*Psammobia mahosaensis* Preston, 1915∗	1
Cuspidaria annandalei Preston, 1915	1,2	*Saccella commutata* (Philippi, 1844)	7
Diplodonta annandalei Preston, 1914	1	*Saccostrea cucullata* (Born, 1778)	8
Diplodonta barhampurensis Preston, 1915∗	1	*Scintilla chilkaensis* Preston 1915∗	1
Diplodonta chilkaensis Preston, 1914∗	1	*Solen annandalei* Preston, 1915	1,8
Diplodonta ovalis Preston, 1914	1	*Solen fonesii* Dunker, 1862	1
Donax incarnatus Gmelin, 1791	8	*Solen kempi* Preston, 1915	1
Donax sp.2	3,8	*Solen truncatus* Wood, 1815	1
Donax cuneatus Linnaeus, 1758	9	*Standella pellucida* (Gmelin, 1791)	1
Fulvia aperta (Bruguière, 1789)	1	*Striarca lactea* (Linnaeus, 1758)	8
Gluconoma sculpta G. B. Sowerby I, 1833∗	4	*Sunetta scripta* (Linnaeus, 1758)	8

(continued)

Table 14.3 (continued)

Bivalve species	References	Bivalve species	References
Hyotissa numisma (Lamarck, 1819)	1	*Tapes pinguis* Chemn., Romer, 1872∗	1
Kellya chilkaensis Preston, 1915∗	1	*Tegillarca granosa* (Linnaeus, 1758)	1,8,9
Kellya mahosaensis Preston, 1915∗	1	*Tellina aequistriata* Say, 1824	1
Laternula navicula (Reeve, 1863)	1	*Tellina chinensis* Hanley, 1845	1
		Tellina tenuis da Costa, 1778	8
Macoma sp.	4,5,7,9	*Theora opalina* (Hinds, 1843)	1,3,4,6,9
Mactra grandis Gmelin, 1791	7	*Thracia septentrionalis* Jeffreys, 1872	7
Mactra stultorum (Linnaeus, 1758)	8	*Timoclea imbricata* (G. B. Sowerby II, 1853)	7
Marcia opima (Gmelin, 1791)	1	*Tivela dillwyni* (Deshayes, 1853)∗	1
Martesia striata (Linnaeus, 1758)	1	*Transkeia satparaensis* (Preston, 1915)	1

References: 1- Annandale and Kemp 1915; 2 – Preston 1915; 3 – Das 2004; 4 – Sahu et al. 2007; 5 – Mahapatro et al. 2009; 6- Mahapatro et al. 2012; 7 – Mishra et al. 2013; 8 – Mahapatro et al. 2015; 9 – Bhadury 2015; ∗ – nomenclature not as per WoRMS Database

influenced by nature of sediment, water depth and types of predators that feed on these organisms (e.g. Joydas and Damodaran 2009).

Amphipods are common and wide spread in all type of aquatic habitats ranging from freshwater to deep sea; some have been reported in terrestrial environment (e.g. *Talitrus saltator*; Montagu, 1808). The Order Amphipoda contains 9791 valid species including about 1900 freshwater species and remaining being marine species (Barnard and Karaman 1980; Horton et al. 2016). Studies dealing with amphipod diversity in Indian coastal water are relatively limited compared to other benthic macrofaunal groups. Venkataraman and Wafar (2005) summarized and documented about 139 species from Indian coastlines and some of these species were previously included in another report by Surya Rao (1974). From the Indian coastal waters, 270 amphipod species have been reported representing both the East and West coasts (http://www.biosearch.in/). Out of these, 30 amphipod species have been reported from Chilika lagoon and among them, 23 species are from Outer Channel region (Annandale and Kemp 1915; Chilton 1924; Das 2004; Mahapatro et al. 2009, 2012, 2015; Bhadury 2015). Amphipod genera such as *Gammarus, Grandidierella* and *Orchestia* were found throughout the lagoon including the Outer channel (Table 14.4). Amphipods are most sensitive organisms to any kind of disturbance within an aquatic ecosystem and so from anthropogenic disturbances. In Chilika, the low diversity of amphipod observed may be due to high freshwater inflow and fishing activity which results in higher turbidity and thereby affecting amphipod assemblages.

Table 14.4 Minor groups of macrofauna reported from Chilika lagoon

Macrofaunal crustacean species	References	Macrofaunal crustacean species	References
Amphipods		*Perioculodes longimanus* (Bate & Westwood, 1868)	1
Americorophium triaeonyx (Stebbing, 1904)	1,6	*Photis longicaudata* (Bate & Westwood, 1862)	1,6
Ampelisca pusilla Sars, 1895	5,6	*Platorchestia platensis* (Krøyer, 1845)	1,5
Amphilochus brunneus Della Valle, 1893	1	*Quadrimaera incerta* (Chilton, 1883)	6
Ampithoe ramondi Audouin, 1826	6	*Quadrivisio bengalensis* Stebbing, 1907	1
Ceradomaera plumosa Ledoyer, 1973	1	*Synchelidium haplocheles* (Grube, 1864)	1
Eriopisa chilkensis (Chilton, 1921)	6	*Talorchestia martensii* (Weber, 1892)	1
Gammarus annandalei (Monod, 1924)	6	*Urothoe platydactyla* Rabindranath, 1971	6
Gammarus sp.1	4,5,7	*Victoriopisa chilkensis* (Chilton, 1921)	1,6
Gammarus sp.2	7	**Isopods**	
Grandidierella taihuensis Morino & Dai, 1990	3,6	*Calathura* sp.	2,3
Grandidierella gilesi Chilton, 1921	1	*Cirolana fluviatilis* Stebbing, 1904	2,4,5,6
Grandidierella megnae (Giles, 1888)	1	*Eurydice* sp.	7
Idunella chilkensis (Chilton, 1921)	1	*Idotea granulosa* Rathke, 1843	7
Indischnopus herdmani (Walker, 1904)	6	*Idotea* sp.	4,7
Melita festiva Chilton, 1885	6	*Sphaeroma* sp.	2,6
Melita inaequistylis Dana, 1852	1	**Tanaids**	
Niphargus chilkensis Chilton, 1921	6	*Apseudes* sp.1	2,4,5,7
Orchestia aestuarensis Wildish, 1987	6	*Apseudes* sp.2	7
Paracalliope fluviatilis (Thomson, 1879)	1	*Ctenapseudes chilkensis* (Chilton, 1924)	2,5,6
Parhyale hawaiensis (Dana, 1853)	1		

References: 1- Annandale and Kemp 1915; 2 – Chilton 1924; 3 – Das 2004; 4 – Mahapatro et al. 2009; 5- Mahapatro et al. 2012; 6 – Mahapatro et al. 2015; 7 – Bhadury 2015; ∗ – nomenclature not as per WoRMS Database

Order Isopoda represents a group of small crustaceans with high species diversity and commonly found in marine, freshwater and land environments. Isopods contain approximately 10,000 species, among them more than 50% are land based species while around 45% are marine with remaining being freshwater species (Brusca and Wilson 1991; Brusca 1997). In India only 122 species of marine isopods have been documented across the East and West coasts of India, while Chilika has only six species of isopods recorded to date (Table 14.4). Among them four species were recorded in Southern sector and remaining from Outer Channel (Bhadury 2015; Mahapatro et al. 2015). Most of the isopods are parasites and some are attached with aquatic plants while free-living (sediment dwelling) isopods are least abundant. The reported isopod species in the Chilika lagoon are mainly salinity dependent species, which are found only where the salinity values vary from mesohaline to polyhaline nature. Other than that, isopods are also sensitive to human disturbances (e.g. fishing, tourism), which are visible in certain sectors of the the Chilika lagoon and large number of predatory fish diversity also affect their assemblage structure in this ecosystem.

The Order Tanaidacea, represented by small marine-dwelling crustaceans is found mainly in marine sedimentary environment. To date, 940 species under this Order has been recorded and some species are planktonic (Bird 2015). The abundance of tanaids is found to be higher in inshore water compared to other marine habitats; some species are also found in brackish water environments (Larsen 2002). To date, only eight species have been documented from the Indian coastlines (http://www.biosearch.in/), out of which three species (*Apseudes* sp.1, *Apseudes* sp.2 and *Ctenapseudes chilkensis* (Chilton 1924) – Table 14.4) have been reported from Chilika lagoon (Chilton 1924; Mahapatro et al. 2009, 2012, 2015; Bhadury 2015). Tanaidacean diversity is not fully explored and this is particularly true for marginal ecosystems such as coastal lagoons as evident from the study undertaken in Chilika (Bhadury 2015). Thus our understanding of this group from the Indian coastal environments is still in its infancy and requires more sustained efforts to gain a better estimation of their diversity and document regional scale biogeographic patterns.

14.2.2.2 Meiofauna

The word meiofauna (Greek word: *meio* – smaller) was originally coined by M.F. Mare in 1942, which means benthic metazoan component between macrofauna and microfauna based on their intermediate sizes. However, body size boundaries range from 0.032 mm to 1.0 mm (Danovaro 2009). Additionally, some members of protozoa such as foraminifers and ciliates are also considered as meiobenthos based on their size and ecological attributes (Giere 2009). Meiofauna occurs in all types of sediments and occupies wide variety of habitats. The composition and functional importance of these communities vary enormously depending upon sediment characteristics, in addition to other abiotic factors (Heip et al. 1985). Meiofauna abundance are usually very high and can range from 10^5 and 10^6 ind./m^2, whereas the highest abundance has been recorded in intertidal estuarine habitats (Giere 2009) and also in deep sea sediments (Vanreusel et al. 2010). In terms of biomass, meiofaunal communities are the major contributor (1–2 g dw/m^2); however, these values (abundance and biomass) show variability across different types of biotopes (Balsamo et al. 2010).

The benthic meiofauna is represented by 24 phyla (Giere 2009) and organisms representing these phyla live their whole life (permanent meiofauna – nematode) or part of the life (temporary meiofauna – juveniles of macrofauna) in sediment. Generally, nematodes are the most abundant and diverse taxon among meiofaunal groups, followed by harpacticoid copepods as compared to other metazoan components (e.g. ostracods, gastrotrichs, turbularians) (Ansari et al. 2012a). Foraminifera also represent a highly abundant and diverse meiobenthos group in coastal sedimentary habitats (Balsamo et al. 2010). However, meiofaunal investigations undertaken in Chilika lagoon are much less compared to other coastal habitats. Only a handful of benthic meiofaunal groups such as nematodes (Ansari et al. 2015a) and foraminifera (Rao et al. 2000; Jayalakshmi and Rao 2001; Kumar et al. 2013, 2015) have been studied in Chilika lagoon albeit these studies were restricted to certain sectors of the lagoon or undertaken as part of short-term studies (Fig. 14.1).

Free-Living Marine Nematodes

Free-living marine nematodes (referred to as nematodes in this chapter), which dominate in terms of abundance and diversity, often represent more than 60–90% of benthic meiofauna particularly for abundance (Zeppilli et al. 2015). It is thought that the phylum Nematoda is hyperdiverse and (Appeltans et al. 2012) and may represent more than a million species (Lambshead 2004). However, approximately 9000 marine nematode species have been documented (Mokievsky and Azovsky 2002; Vanaverbeke et al. 2015) to date out of estimated 27,000 species (Hugot et al. 2001) and it has been recently proposed that more than 81% of nematode species are yet to be described (Semprucci and Balsamo 2012).

To date, 335 species belonging to 160 genera of nematodes have been reported from the Indian coasts (e.g. Chinnadurai and Fernando 2007; Sajan and Damodaran 2007; Ansari et al. 2012a, b, 2015b; Bhadury et al. 2015; http://www.biosearch.in/). The reported species diversity is comparable to other ecoregions globally, e.g. in Italian coastal water 445 species have been reported (Balsamo et al. 2010) while from the British Isle and rest of Europe (other than Italy) more than 1625 species of nematodes have been reported (Giere 2009). The continental shelf region of both East and West coasts of India has rich nematode species diversity (191 species – Ansari et al. 2012a, b, 2015b and 152 species – Sajan and Damodaran 2007; Sajan et al. 2010 respectively) comparable to other ecosystems such as deep sea (110 species – Singh and Ingole 2016), estuarine (78 species – Ansari and Parulekar 1998) and mangrove environments (56 species – Chinnadurai and Fernando 2007; Ansari et al. 2014). The nematode genera such as *Sabatieria* (15 species), *Desmodora* (13 species), *Daptonema* (12 species), *Halalaimus* (11 species) and *Theristus* (10 species) have been found to be most diverse with wide distribution across Indian coastal ecosystems.

From Chilika lagoon, 76 nematode species have been reported belonging to 32 genera and 14 families (Table 14.5). Among them, Order Chromadorida is represented by 7 families, followed by Monhysterida (4 families) and Enoplida (3 families). Among the encountered families, Comesomatidae has been found to be represented by highest number of species (11 species), followed by Oncholaimidae (10 species) and Linhomoeidae (7 species). Nematode genera such as *Viscosia, Sabatieria, Terschellingia, Daptonema* and *Metalinhomoeus* have been found to be dominant throughout Chilika lagoon (Ansari et al. 2015a). In this lagoon, salinity is an important abiotic factor and the ecosystem can be divided into oligohaline (<0.5) to mesohaline (<18) sectors inside the lagoon whereas the Outer Channel is polyhaline to euryhaline (>18) in nature. Majority of the identified nematodes from Chilika lagoon (98.95%) represents free-living marine form while the rest are exclusively fresh water in nature (e.g. Dorylaimida). Abundance and diversity of nematodes were found to be higher in Central sector, which showed fluctuating salinity, followed by Southern sector (high salinity) and lowest in the Northern sector of Chilika lagoon. Besides other abiotic factors including nature of sediment and freshwater flow affected nematode distribution and diversity in Chilika lagoon (Ansari et al. 2015a; Bhadury 2015) as also reported in other marine habitats globally (e.g. Vanaverbeke et al. 2002; Semprucci et al. 2013, 2014a; b; Fonseca et al. 2014). The reported species richness and diversity for nematodes is restricted within the Chilika lagoon and it is expected that diversity may increase further when studies are undertaken in the Outer Channel region as part of future efforts.

Harpacticoid Copepods

The Order Harpacticoida is one of the diverse groups within the subclass Copepoda under Class Maxillopoda. This is the only order among Copepoda found to inhabit sediment-water interface and at times also found to be attached with phytal fauna (Dussart and Defaye 2001). The Harpacticoid copepods represent the second most dominant meiofaunal component in benthos and they inhabit wide range of coastal and marine ecosystems. This group of organisms is more sensitive to any kind of disturbance that may occur in coastal ecosystems, in particular anthropogenic disturbances such as urban sewage release and oil spill (De Troch et al. 2013). Out of estimated species diversity of 30,000, almost 21,000 species of harpacticoids have been validly described (e.g. Dussart and Defaye 2001; Boxshall and Defaye 2008). From Indian coastlines, only 292 species have been described to date (http://www.biosearch.in/) and most of them are from inshore water (Altaff et al. 2004; Mantha et al. 2012) and continental shelf regions (Sajan and Damodaran 2007; Ansari et al. 2013). Among them, genera such as *Stenhelia* (19 species), *Laophonte* (12 species), *Paramesochra* (9 species) and *Peltidium* (8 species) have been reported across Indian coastal waters (http://www.biosearch.in/). In Chilika lagoon, only seven taxa of harpacticoid copepods (*Ameira* sp., *Microsettella* sp., *Macrosettela* sp., *Harpacticus* sp., *Laophonte* sp., *Stenhelia* sp. and *Diathrodes* sp.) have been reported and among them, three taxa (*Macrosettela* sp., *Laophonte* sp. and

Table 14.5 List of Free-living marine nematode species reported from Chilika lagoon (from Ansari et al. 2015a)

Adoncholaimuis sp.	Odontophora sp.1
Anoplostoma sp.	*Oncholaimellus calvadosicus* de Man, 1890
Aponema sp.	*Oncholaimellus* sp.
Axonolaimus paraspinosus Schuurmans Stekhoven & Adam, 1931	*Oncholaimus oxyuris* Ditlevsen, 1911
Axonolaimus sp.	*Oncholaimus* sp.
Calomaicrolaimus sp.	*Oxystomina elongata* Bütschli, 1874
Daptonema normandicum (de Man, 1890)	*Oxystomina* sp.
Daptonema oxycerca (de Man, 1888)	*Paracomesoma dubium* (Filipjev, 1918)
Daptonema setosum Bütschli, 1874	*Paracomesoma* sp.
Daptonema sp.	*Paralongicyatholaimus* sp.
Deontolaimus tardus (de Man, 1889) Holovachov & Boström, 2015	*Pomponema* sp.
Deontolaimus sp.	*Ptycholaimellus ponticus* (Filipjev, 1922) Gerlach, 1955
Desmodora (Desmodorella) sanguinea (Southern, 1914)	*Ptycholaimellus* sp.
Desmodora scaldensis de Man, 1889	*Rhynchonema* sp.
Desmodora (Desmodorella) schulzi (Gerlach, 1950)	*Sabatieria celtica* Southern, 1914
Desmodora sp.	*Sabatieria praedatrix* de Man, 1907
Dorylaimopsis punctata Ditlevsen, 1918	*Sabatieria pulchra* (Schneider, 1906)
Dorylaimopsis sp.	*Sabatieria punctata* (Kreis, 1924)
Halalaimus gracilis de Man, 1888	*Sabatieria* sp.
Halalaimus longicaudatus (Filipjev, 1927)	*Sabatieria* sp.
Halalaimus sp.	*Sphaerolaimus balticus* Schneider, 1906
Hopperia sp.	*Sphaerolaimus gracilis* de Man, 1876
Linhomoeus sp.	*Sphaerolaimus macrocirculus* Filipjev, 1918
Metachromadora remanei Gerlach, 1951	*Sphaerolaimus* sp.
Metachromadora sp.	*Spilophorella candida* Gerlach, 1951
Metachromadora suecica (Allgén, 1929)	*Spilophorella* sp.
Metalinhomoeus filiformis (de Man, 1907)	*Spirinia* sp.
Metalinhomoeus longiseta Kreis, 1929	*Terschellingia communis* de Man, 1888
Metalinhomoeus sp.	*Terschellingia longicaudata* de Man, 1907
Metoncholaimus sp.	*Terschellingia* sp.
Microlaimus conothelis (Lorenzen, 1973) Jensen, 1978	*Viscosia abyssorum* (Allgén, 1933) Warwick & Buchanon, 1970
Microlaimus sp.	*Viscosia elegans* Filipjev, 1922
Neochromadora poecilosomoides (Filipjev, 1918)	*Viscosia* sp.
Odontophora longisetosa (Allgén, 1928)	*Viscosia viscosa* (Bastian, 1865) de Man, 1890

Diathrodes sp.) are representative of freshwater-brackish water environment. However, this particular group of meiofauna is largely under-studied from the context of diversity and assemblage structure with respect to Chilika lagoon and yet to be tested as a bioproxy for long-term environmental monitoring in this lagoon.

Foraminifera

Phylum Foraminifera comprises of heterotrophic protists under the supergroup Rhizaria that are ubiquitous across marine (Pawlowski et al. 2011), freshwater (Holzmann and Pawlowski 2002) and terrestrial (Lejzerowicz et al. 2010) environments. There are two forms (benthic and planktonic) of foraminifera, amongst, benthic foraminifers are widely studied as they constitute 90% of deep sea biomass and are ideally suited for paleoecological studies (Murray 2006). Although, this group do not fall under meiofauna but given their importance in benthic sediments in terms of functioning, it represent a key taxonomic group in coastal sediments including lagoonal sediment globally. The distribution and diversity of benthic foraminifera is always associated with surrounding environmental variables, mainly dissolved oxygen, salinity, sediment nature and total organic carbon (Murray 2006). The total number of identified living benthic foraminifera species from previous records were found to be ~ 2140 out of which 602 were agglutinated, 341 porcelaneous and 1197 hyaline species (Murray 2007). Nevertheless, the number of total known extant species of foraminifera recorded to be 9106 species till date of which hyaline test overwhelmingly dominant (Pawlowski and Holzmann 2014). The number of recorded living benthic foraminifera species show huge variation across marginal marine environments (701 species listed from ~ 1.5 million individuals), shelf regions (989 species listed from ~ 0.6 million individuals) and deep sea (831 species listed from ~ 0.3 million individuals) (Murray 2007). By comparing different habitats globally it has been observed that species pool is comparatively higher in estuaries/lagoons (688 species) and relatively low in marshes (137 species) (Murray 2007).

To date, 559 foraminifera species have been recorded from Indian coastal environments (http://www.biosearch.in/). Among them, the genus *Quinqueloculina* has been highly diverse in terms of species composition (51 species), followed by *Bolivina* (33 species), *Spiroloculina* (27 species), *Globorotalia* (21 species), *Virgulinella* and *Nonion* (18 species each) (http://www.biosearch.in/). Previous investigation undertaken by Rao et al. (2000) from Chilika has documented a maximum of 69 species represented by 19 families under foraminifera. Amongst them, three foraminiferal species [*Globigerinita glutinata* Egger, 1893, *Globigerinoides rubra* (d'Orbigny, 1839) and *Globigerinoides sacculifer* (Brady, 1877)] were reported to be planktonic. Rao et al. (2000) however limited themselves to the Central, Northern and Outer Channel of the lagoon and concluded that foraminiferal species composition and richness were higher towards the marine source i.e. Outer Channel of Chilika. More recently Kumar et al. (2013) investigated benthic foraminiferal biofacies at Outer Channel region and reported the occurrence of 13 species of which five were of stenohaline in nature [*Asterorotalia inflata* (unaccepted species)], *Asterorotalia pulchella* (d'Orbigny, 1839), *Cibicides refulgens* Montfort, 1808 *Nonionoides grateloupii* (d'Orbigny, 1839) and *Pararotalia nipponica* Asano, 1936 while the remaining eight were considered to be endemic to Outer Channel region only, based on their reported tolerance to higher temperature and salinity regimes. The diversity of agglutinated foraminiferal species across the entire lagoon has been investigated by Kumar et al. (2015) and they reported 15 species of which *Miliammina fusca* (Brady, 1870), *M. fusca* var. and *Ammobaculites exiguous* Cushman & Brönnimann, 1948 were found to be relatively more abundant. Among the rarer forms observed by Kumar et al. (2015), living specimens of *Ammobaculites agglutinans* (d'Orbigny, 1846) and *Textularia earlandi* Parker, 1952 were observed from Outer Channel and Northern sector of the lagoon while members of the genus *Trochammina* were found to be living in Northern and Southern sectors. The study also reported *M. fusca* (Brady, 1870) to be the dominant species at Outer Channel which displayed a population increase in pre-monsoon and showed an opposite relationship with the occurrence of *M. fusca* var. with the later species being more sensitive to changes in prevailing conditions.

Bhadury et al. (2015) added 11 more species to the existing foraminiferal species list (*Ammomarginulina* sp., *Ammodiscus tenuis* Brady, 1881, *Ammotium salsum* (Cushman & Brönnimann, 1948), *Trochammina inflata* (Montagu, 1808), *Ammonia* sp.1, *Ammonia* sp.2., *Ammonia* sp.3., *Miliammina* spp., *Nonionella* sp. and *Quinqueloculina* sp.) based on a study undertaking over a period of 21 months. Thus the total number of valid foraminiferal species has increased to 87 species as of today from the entire Chilika lagoon (Table 14.6). The investigation revealed *Ammonia* spp. to be the most dominant foraminiferal genus across Chilika. Benthic foraminiferal abundance was found to be remarkably higher in the Central sector as compared to remainder of the lagoon. Agglutinated specimens of *Ammotium salsum* (Cushman & Brönnimann, 1948), *Ammomarginulina* sp., *Ammodiscus* sp. and *Miliammina* spp. constituted majority of the assemblages in the Southern sector. The dominant form *Ammonia* displayed a significant negative correlation with silicate based on which the investigators concluded the dominance of the group may increase under scenarios with limited primary production. Dominance of *Ammonia* in benthic assemblage has been previously reported from Venice lagoon, Italy (Donicci et al. 1997); and from Ria de Aveiro lagoon, Portugal (Martins et al. 2013) which are highly influenced by tidal influx alongside Araruama lagoon, Brazil (Debenay et al. 2001); Lake Varano lagoon, Italy (Frontalini et al. 2011, 2014) and Aegean coastal lagoons (Koukousioura et al. 2012; Dimiza et al. 2016) that have limited tidal influence similar to Chilika. The dominance observed thus appears to be independent of the microtidal nature of the lagoon.

Table 14.6 List of Foraminifera species reported from Chilika lagoon

Foraminifera species	References	Foraminifera species	References
Adelosina longirostra (d'Orbigny, 1826)	1	*Globigerinoides sacculifera* (Brady, 1877)	1
Ammonia sp.1	4	*Hanzawaia concentrica* (Cushman, 1918)	1
Ammonia sp.2	4	*Hanzawaia nitidula* (Bandy, 1953)	1
Ammonia sp.3	4	*Haplophragmoides canariensis* (d'Orbigny, 1839)	1
Ammonia beccarii (Linnaeus, 1758)	1,2	*Haynesina depressula* (Walker & Jacob, 1798)	2
Ammonia parkinsoniana (d'Orbigny, 1839)	1	*Hopkinsina pacifica* Cushman, 1933	1
Ammonia pauciloculata (Phleger & Parker, 1951)	1	*Lobatula lobatula* (Walker & Jacob, 1798)	2
Ammonia sobrina (Shupack, 1934)	1	*Lepidodeuterammina ochracea* (Williamson, 1858)	1,3
Ammonia tepida (Cushman, 1926)	1,2	*Miliammina fusca* (Brady, 1870)	1,2,3
Ammomarginulina sp.	4		
Ammobaculites agglutinans (d'Orbigny, 1846)	3,4	*Miliammina* spp.	4
Ammobaculites exiguus Cushman & Brönnimann, 1948	1,2,3	*Miliolinella subrotunda* (Montagu, 1803)	1
Ammodiscus tenuis Brady, 1884	4	*Nodulina dentaliniformis* (Brady, 1881)	3
Ammotium directum (Cushman & Brönnimann, 1948)	1,3	*Nonionella* sp.	4
Ammotium fragile Warren, 1957	1,3	*Nonionellina labradorica* (Dawson, 1860)	1
Ammotium salsum (Cushman & Brönnimann, 1948)	4	*Nonionoides grateloupii* (d'Orbigny, 1839)	2
Amphistegina radiata (Fichtel & Moll, 1798)	1	*Pararotalia nipponica* Asano, 1936	2
Asterorotalia inflata∗	2	*Planulina bassensis* Collins, 1974	1
Asterorotalia pulchella (d'Orbigny, 1839)	1,2	*Protelphidium schmitti* (Cushman & Wickenden, 1929)	1
Bolivina pseudoplicata Heron-Allen & Earland, 1930	1	*Quinqueloculina* sp.	4
Bolivina striatula Cushman, 1922	1	*Quinqueloculina agglutinans* d'Orbigny, 1839	1
Caronia exilis (Cushman & Brönnimann, 1948)	3	*Quinqueloculina dimidiata* Terquem, 1876	1
Cibicides refulgens de Montfort, 1808	1	*Quinqueloculina durandi* Cushman & Wickenden, 1929	1
Cribroelphidium excavatum var. *clavatum* (Cushman, 1930)	1	*Quinqueloculina lamarckiana* d'Orbigny, 1839	1

(continued)

Table 14.6 (continued)

Foraminifera species	References	Foraminifera species	References
Cribroelphidium incertum Williamson, 1858	1	*Quinqueloculina lata* Terquem, 1876	1
Cribroelphidium poeyanum (d'Orbigny, 1826)	1	*Quinqueloculina seminula* (Linnaeus, 1758)	1,2
Cribroelphidium excavatum (Terquem, 1875)	1	*Quinqueloculina tenagos* Parker, 1942	1
Elphidium advenum (Cushman, 1922)	1	*Quinqueloculina vulgaris* d'Orbigny, 1826	1
Elphidium alvarezianum (d'Orbigny, 1839)	1	*Remaneica kellettae* (Thalmann, 1932)	1,3
Elphidium articulatum (d'Orbigny, 1839)	1	*Riminopsis asterizans* (Fichtel & Moll, 1798)	1
Elphidium craticulatum (Fichtel & Moll, 1798)	1,2	*Rosalina globularis* d'Orbigny, 1826	1
Elphidium crispum (Linnaeus, 1758)	1	*Rosalina leei* Hedley & Wakefield, 1967	1
Elphidium discoidale (d'Orbigny, 1839)	1	*Siphonaperta agglutinata* (Cushman, 1917)	1
Elphidium galvestonense Kornfeld, 1931	1	*Strebloides advenus* (Cushman, 1922)	1
Elphidium gunteri Cole, 1931	1	*Textularia earlandi* (Parker, 1952)	1,3
Elphidium hispidulum Cushman, 1936	1	*Textularia* spp.	4
Elphidium mexicanum Kornfeld, 1931	1	*Triloculina trigonula* (Lamarck, 1804)	1
Elphidium rugulosum Cushman & Wickenden, 1929	1	*Trochammina advena* Cushman, 1922	1,3
Entzia macrescens (Brady, 1870)	1,3,4	*Trochammina hadai* Uchio, 1962	1,3
Globigerinita glutinata (Egger, 1893)	1	*Trochammina inflata* (Montagu, 1808)	4
Globigerinoides rubra (d'Orbigny, 1839)	1		

References: 1- Rao et al. 2000; 2- Kumar et al. 2013; 3- Kumar et al. 2015; 4- Bhadury 2015; * – nomenclature not as per WoRMS Database

14.3 Ecological Importance of Benthic Macro and Meiofauna

In recent years major focus has been on the use of benthic fauna for long-term ecological monitoring with a particular emphasis on anthropogenic disturbances. In particular, greater emphasis has been laid to understand dynamic processes associated with benthic communities, whereas, spatial patterns have received much lesser attention (Ambrogi et al. 1990). Therefore, knowledge of benthic faunal assemblage structure and associated temporal variability is important to track changes caused by human interference, in addition to natural disturbances. Many reports have successfully used different benthic faunal groups as biological proxy to measure the degree of pollution in coastal environments (e.g. Zeppilli et al. 2015 and references therein). The benthic fauna of Chilika lagoon appears to be composed of characteristic shallow coastal assemblages and considered to be major players in shaping sedimentary microhabitats as well as regulating ecosystem processes such as elemental cycling of carbon and nitrogen. Apart from being active part of detrital cycle, benthic fauna are also known to be active agents of ecosystem engineering. Kristensen (1988) has documented that benthic fauna mediate mechanisms such as (a) downward transport of organic matter within the sediment, (b) maintenance of redox potential by ion transport, (c) concentration of organic matter into faecal matter, (d) breakdown of aggregates owing to grazing activities and thereby render coastal ecosystems extremely productive.

Physical disturbance by different members of benthic fauna also contribute to ecological processes. For example, polychaetes (*Nereis* spp.) physically rework the sediment by their burrowing activities. Such activities increase ventilation within the sediment facilitating gaseous and nutrient transport to greater depth of the sediment. Bioturbation by macrofaunal groups also leads to redistribution of toxic elements within the sediment (Kristensen and Kostka 2005). Sediment burrows formed mostly by polychaetes and to a lesser extent by nematodes, stimulate several biogeochemical processes along the burrow wall (Mermillod-Blondin et al. 2004). Calcifying benthic organisms like molluscs and foraminifers act as major players of carbon sequestration into sediments and play a major role in global carbon cycle (Sanders 1968). These organisms almost permanently mineralize CO_2 in the sediment especially in shallow water habitats like Chilika that lie above the carbon compensation depth. Thus their ecological importance in Chilika lagoon is of utmost significance and also contributes to rich fisheries in this region.

Benthic faunal organisms are usually considered to be closely integrated into a 'detrital trophic complex' (Pinckney and Sandulli 1990; Moens et al. 2013) and they have various connections with other biotic compartments in benthic domain. Meiofauna together with microfauna (protozoans), microphytobenthos, prokaryotes and detritus, are part of the 'small food web' in varied ecosystems including lagoon. Other than that, meiofauna have direct relationship with macrofauna, occupying a pivotal position in benthic food web. In fact, small size is indeed coupled with high metabolic activity and rapid turnover results in high ratio of production to biomass (P:B). The benthic P:B is, however, taxon specific and can vary from species to species or season to season in an ecosystem. Therefore, it is difficult to assign a single value that could be generally valid; however it may vary due to local climatic conditions (Danovaro 2009).

Food web energy flow from lower benthic faunal assemblages to higher trophic levels can be illustrated using different models. In particular, downstream processes within the detrital cycle are strongly influenced by meiofaunal groups such as nematodes and foraminifera and thus they play a crucial role in breakdown and processing of organic matter respectively (Findlay and Tenore 1982; Alkemade et al. 1992; Papaspyrou et al. 2013). However, these aspects are not fully understood as part of food web energy flow with respect to Chilika lagoon and thus need to be explored as part of future studies.

In Chilika lagoon, increased disturbances, in particular from human intervention can have long-term ecological implications. Therefore, benthic macro and meiofaunal communities could be effectively used to track ecological changes if any, in this lagoon on a long-term basis. Some studies have shown that Chilika lagoon ecosystem is susceptible to seasonal eutrophication (Sahoo et al. 2014; Srichandran et al. 2015) and benthic meiofauna such as nematodes could be effectively used to track eutrophication induced ecological changes as been undertaken in other coastal ecosystems globally (Warwick and Robinson 2000; Wetzel et al. 2002; Armenteros et al. 2009).

14.4 Future Directions

While several studies have been undertaken on various aspects of benthic fauna in the Chilika lagoon, knowledge gaps exist with respect to lesser known groups including Gastrotricha, Kinorhyncha and Ostracoda. Additionally, aspects of food web model and P:B ratio for benthic fauna have not been thoroughly attempted in Chilika lagoon. One of the drawbacks has been limited application of benthic fauna as biological proxies to track natural and anthropogenic induced changes in this ecosystem. In particular, knowledge gap in terms of anthropogenic activities such as excessive fishing, increased tourism activities and continuous dredging on benthic faunal communities exist with respect to Chilika lagoon. At the same time our understanding of major threats posed to benthic faunal biodiversity is yet to be comprehensively understood in Chilika. Anthropogenic disturbances such as fishing activities, sewage discharge and tourism pressure are going to alter environmental quality in this lagoon. This may have cascading effect on benthic fauna and ultimately on larval recruitment process in Chilika lagoon. Such threats to benthic faunal diversity could ultimately have a deleterious effect on fisheries of Chilika lagoon. Therefore it is important to take a holistic approach for long-term ecological monitoring since Chilika lagoon supports rich fisheries as well as supports livelihood of thousands of fisher folks living along the coastal Bay of Bengal. Overall, further detailed biological inventory of benthic faunal assemblages can ultimately help towards understanding of major ecological processes including biogeochemical cycling as well as sustainable management of health of Chilika lagoonal ecosystem.

References

Alkemade R, Wielemaker A, De Jong SA, Sandee AJJ (1992) *Diplolaimella dievengatensis* in stimulating the mineralization of *Spartina anglica* detritus. Mar Ecol Prog Ser 90:149–155

Altaff K, Naveed MS, Sugumaran J (2004) Meiofauna of Chennai coast of Bay of Bengal with special reference to sampling and extraction methods. Convergence 6:61–90

Ambrogi R, Bedulli D, Zurijni G (1990) Spatial and temporal patterns in structure of macrobenthic assemblages. A three-year study in the northern Adriatic Sea in front of the Po River Delta. Mar Ecol 11:25–41

Annandale TN, Kemp S (1915) Introduction to the fauna of the Chilika lagoon. Mem Indian Mus 5:1–20

Annandale TN, Rao HS (1925) Materials for a revision of the recent Indian Limnaeidae (Mollusca, Pulmonata). Rec Indian Mus 27:1–137

Ansari ZA, Parulekar AH (1998) Assemblage structure of meiobenthos from a tropical estuary. Indian J Mar Sci 27:362–366

Ansari KGMT, Manokaran S, Raja S, Lyla PS, Ajmal Khan S (2012a) Checklist of nematodes (Nematoda: Adenophorea) from southeast continental shelf of India. Checklist 8:414–420

Ansari KGMT, Lyla PS, Ajmal Khan S (2012b) Faunal composition of metazoan meiofauna from the south-east continental shelf of India. Indian J Geo-Mar Sci 41:457–467

Ansari KGMT, Lyla PS, Ajmal Khan S, Manokaran S, Raja S (2013) Assemblage structure of benthic harpacticoid copepods from the southeast continental shelf of India. Proc Int Acad Ecol Environ Sci 3:87–100

Ansari KGMT, Manokaran S, Raja S, Lyla PS, Ajmal Khan S (2014) Interaction of free-living marine nematodes in the artificial mangrove environment (southeast coast of India). Environ Monit Assess 186:293–305

Ansari KGMT, Pattnaik AK, Rastogi G, Bhadury P (2015a) An inventory of free-living marine nematodes from Asia's largest coastal lagoon, Chilika, India. Wetlands Ecol Manag. http://sci-hub.tw/10.1007/s11273-015-9426-2

Ansari KGMT, Lyla PS, Ajmal Khan S, Bhadury P (2015b) Diversity patterns of free-living marine nematodes in the southwest continental shelf of Bay of Bengal and their link to abiotic variables. Mar Ecol. http://sci-hub.tw/10.1111/maec.12332

Anthony A, Atwood J, August P, Byron C et al (2009) Coastal lagoons and climate change: ecological and social ramifications in U.S. Atlantic and Gulf coast ecosystems. Ecol Soc 14:8

Appeltans W, Ahyong ST, Anderson G, Angel MV, Artois T, Bailly N, Bamber R, Barber A, Bartsch I, Berta A, Błażewicz–Paszkowycz M, Bock P, Boxshall G, Boyko CB, Nunes Brandão S, Bray RA, Bruce NL, Cairns SD, Chan T-Y, Cheng L, Collins AG, Cribb T, Curini–Galletti M, Dahdouh–Guebas F, Davie PJF, Dawson MN, De Clerck O, Decock W, De Grave S, de Voogd NJ, Domning DP, Emig CC, Erséus C, Eschmeyer W, Fauchald K, Fautin DG, Feist SW, Fransen CHJM, Furuya H, Garcia–Alvarez O, Gerken S, Gibson D, Gittenberger A, Gofas S, Gómez–Daglio L, Gordon DP, Guiry MD, Hernandez F, Hoeksema BW, Hopcroft RR, Jaume D, Kirk P, Koedam N, Koenemann S, Kolb JB, Kristensen RM, Kroh A, Lambert G, Lazarus DB, Lemaitre R, Longshaw M, Lowry J, Macpherson E, Madi LP, Mah C, Mapstone G, McLaughlin PA, Mees J, Meland K, Messing CG, Mills CE, Molodtsova TN, Mooi R, Neuhaus B, Ng PKL, Nielsen C, Norenburg J, Opresko DM, Osawa M, Paulay G, Perrin W, Pilger JF, Poore GCB, Pugh P, Read GB, Reimer JD, Rius M, Rocha RM, Saiz–Salinas JI, Scarabino, Schierwater B, Schmidt–Rhaesa A, Schnabe KE, Schotte M, Schuchert P, Schwabe E, Segers H, Self–Sullivan C, Shenkar N, Siegel V, Sterrer W, Stöhr S, Swalla B, Tasker ML, Thuesen EV, Timm T, Todaro MA, Turon X, Tyler S, Uetz P, van der Land J, Vanhoorne B, van Ofwegen LP, van Soest RWM, Vanaverbeke J, Walker–Smith G, Chad Walter T, Warren A, Williams GC, Wilson SP, Costello MJ (2012) The magnitude of global marine species diversity. Curr Biol 22:2189–2202

Appukuttan KK (1996) Marine molluscs and their conservation. In: Marine biodiversity conservation and management. Central Marine Fisheries Institute, Cochin, p 15

Armenteros M, Ruiz-Abierno A, Fernández-Garcés R, Pérez-García JA, Díaz-Ascencio L, Vincx M, Decraemer W (2009) Bio-diversity patterns of free-living marine nematodes in a tropical bay: Cienfuegos, Caribbean Sea. Est Coast Shelf Sci 85:179–189

Balsamo M, Albertelli G, Ceccherelli VU, Coccioni R, Colangelo MA, Curini-Galletti M, Danovaro R, D'Addabbo R, De Leonardis C, Fabiano M, Gallo M, Gambi C, Guidi L, Moreno M, Pusceddu A, Sandulli R, Semprucci F, Todaro MA, Tongiorgi P (2010) Meiofauna of the Adriatic Sea: present knowledge and future perspectives. Chem Ecol 26:45–63

Barnard JL, Karaman GS (1980) Classification of gammarid amphipoda. Crustaceana 6:5–16

Bastrop R, Jürss K, Sturmbauer C (1998) Cryptic species in marine polychete and their independent introduction from North America to Europe. Mol Biol Evol 15:97–103

Bell SS, Coen LD (1982) Investigations on epibenthic meiofauna II. Influence of microhabitat and macroalgae on abundance of small invertebrates on *Diopatra cuprea* (Bosc) (Polychaeta: Onuphidae) tube-caps in Virginia. J Exp Mar Biol Ecol 61:175–188

Bhadury P (2015) Inventorizing benthic faunal assemblages in the Chilika lagoon. ICZMP-CDA final report, pp 103

Bhadury P, Annapurna C (2011) Marine barcoding- how will it help Indian marine benthic studies. Ind J Geo-Mar Sci 40:645–647

Bhadury P, Mondal N, Ansari KGMT, Philip P, Pitale R, Prasade A, Nagale P, Apte D (2015) Checklist of free-living marine nematodes from intertidal sites along the central west coast of India. Checklist 11:1605. http://sci-hub.tw/10.15560/11.2.1605

Bieler R, Mikkelsen PM (2006) Bivalvia – a look at the branches. Zool J Linnean Soc 148:223–235

Bird GJ (2015) Tanaidacea (Crustacea: Peracarida) of the northeast Atlantic: *Chauliopleona* Dojiri and Sieg, 1997 and *Saurotipleona* n. gen. from the Atlantic Margin. J Nat Hist 49:25–26

Blagoon JA, Hilbig B, Scott PH (1996) The Annelida Part III- Polychaeta: Orbiniidae to Cossuridae. In taxonomic atlas of the benthic fauna of the Santa Maria Basin and the western Santa Barbara channel. Santa Barbara Museum of Natural History. CAL 6:1–418

Boxshall GA, Defaye D (2008) Global diversity of copepods (Crustacea: Copepoda) in freshwater. Hydrobiologia 595:195–207

Brusca R (1997) Isopoda. http://tolweb.org/Isopoda/6320/1997.08.06. Accessed at 6th Aug 1997

Brusca RC, Wilson GDF (1991) A phylogenetic analysis of the Isopoda with some classificatory recommendations. Mem Queensland Mus 31:143–204

Carr CM (2012) Polychaete diversity and distribution patterns in Canadian marine waters. Mar Biodivers 42:93–107

Carter JG, Altaba CR, Anderson LC, Araujo R, Biakov AS, Bogan AE, Campbell DC, Campbell M, Chen J, Cope JCW, Delvene G, Dijkstra HH, Fang Z, Gardner RN, Gavrilova VA, Goncharova IA, Harries PJ, Hartman JH, Hautmann M, Hoeh WR, Hylleberg J, Jiang B, Johnston P, Kirkendale L, Kleemann K, Koppka J, Kříž J, Machado D, Malchus N, MárquezAliaga A, Masse JP, McRoberts CA, Middelfart PU, Mitchell S, Nevesskaja LA, Özer S, Pojeta J Jr, Polubotko IV, Pons JM, Popov S, Sánchez T, Sartori AF, Scott RW, Sey II, Signorelli JH, Silantiev VV, Skelton PW, Steuber T, Waterhouse JB, Wingard GL, Yancey T (2011) A synoptical classification of the Bivalvia (Mollusca). Paleontol Contrib 4:1–47

Chilton C (1924) Fauna of the Chilka Lagoon. Tanaidacea and Isopoda. Mem Indian Mus 5:875–895

Chinnadurai G, Fernando OJ (2007) Meiofauna of mangroves of the southeast coast of India with special reference to the free-living marine nematode assemblage. Est Coast Shelf Sci 72:329–336

Cosson-Sarradin N, Sibuet M, Paterson GLJ, Vangriesheim A (1998) Polychaete diversity at tropical Atlantic deep-sea sites: environmental effects. Mar Ecol Prog Ser 165:173–185

Danovaro R (2009) Methods for the study of Deep-Sea sediments, their functioning and biodiversity. CRC Press, Boca Raton. 428pp

Danovaro R, Gambi C, Mirto S (2002) Meiofaunal production and energy transfer efficiency in a seagrass Posidonia oceanica bed in the western Mediterranean. Mar Ecol Prog Ser 234:95–104

Das S (2004) Effect of substratum on the assemblage structure of macrobenthic fauna from southern sector of Chilika lagoon. Indian J Fish 51:375–380

Day JH (1967) A monograph of the polychaeta of southern Africa. British Museum (Nat. Hist.) Publication, London. 878pp

De Troch M, Roelofs M, Riedel B, Grego M (2013) Structural and functional responses of harpacticoid copepods to anoxia in the Northern Adriatic: an experimental approach. Biogeosciences 10:4259–4272

Debenay JP, Tsakridis E, Southard R, Grossel H (2001) Factors determining the distribution of foraminiferal assemblages in Port Joinville Harbor (Ile d'Yeu, France): the influence of pollution. J Mar Micropaleontol 43:75–118

Dimiza MD, Triantaphyllou MV, Koukousioura O, Hallock P, Simboura N, Karageorgis AP, Papathanasiou E (2016) The Foram Stress Index: a new tool for environmental assessment of soft-bottom environments using benthic foraminifera. A case study from the Saronikos Gulf, Greece, Eastern Mediterranean. Ecol Indic 60:611–621

Donicci S, Barbero RS, Taroni G (1997) Live benthic foraminifera in the lagoon of Venice (Italy): population dynamics and its significance. Micropalaeontology 43:440–454

Dujovny E (2009) The deepest cut: political ecology in the dredging of a new sea mouth in Chilika Lagoon, Orissa, India. Conserv Soc 7:192–204

Dussart BH, Defaye D (2001) Introduction to the Copepoda. In: Guide to the identification of the microinvertebrates of the continental waters of the world, No. 16, 2nd edn. Backhuys Publishers, Leiden. 344pp

Etter RJ, Grassle JF (1992) Pattern of species diversity in the deep-sea as a function of sediment particle size diversity. Nature 360:609–610

Eyre BD, Balls PW (1999) A comparative study of nutrient processes along the salinity gradient of tropical and temperate estuaries. Estuar Coasts 22:313–326

Fauchald K, Jones GF (1979) A survey of five additional Southern California study sites (Report 18, Series 2). In Southern California Outer Continental Shelf Environmental Baseline Study, 1976/1977 (2nd year) Benthic Program. Principal investigators reports, vol II. Science Applications, Inc, La Jolla, pp 1–73

Fauchald K, Jumars PA (1979) The diet of worms: a study of polychaete feeding guilds. Oceanogr Mar Biol Ann Rev 17:193–284

Fauvel P (1923) Polychaetes errantes, vol 5. Faune de France, Paris, pp 1–488

Fauvel P (1927) Polychaetes sedentaires, vol 16. Faune de France, Paris, pp 1–494

Findlay S, Tenore KR (1982) Effect of a free-living marine nematode (*Diplolaimella chitwoodi*) on detrital carbon mineralization. Mar Ecol Prog Ser 88:161–166

Fonseca G, Maria TF, Kandratavicius N, Venekey V, Gheller PF, Galllucci F (2014) Testing for nematode-granulometry relationships. Mar Biodivers 44:435–442

Frontalini F, Margaritelli G, Francescangeli F, Rettori R, Armynot du Châtelet E, Coccioni R (2011) Benthic foraminiferal assemblages and biotopes in a coastal lake: the case study of Lake Varano (Southern Italy). Acta Protozool 52:147–160

Frontalini F, Semprucci F, Armynot du Châtelet E, Francescangeli F, Margaritelli G, Rettori R, Spagnoli F, Balsamo M, Coccioni R (2014) Biodiversity trends of the meiofauna and foraminifera assemblages of Lake Varano (southern Italy). Proc Biol Soc Wash 127:7–22

Ganesh T, Raman AV (2007) Macrobenthic assemblage structure of the northeast Indian shelf, Bay of Bengal. Mar Ecol Prog Ser 341:59–73

Giere O (2009) Meiobenthology: the microscopic motile fauna of aquatic sediments, 2nd edn. Springer, Berlin Heidelberg. 527pp

Gray JS (1997) Marine biodiversity: patterns, threats and conservation needs. Biodivers Conserv 6:153–175

Gupta GVM, Sarma VVSS, Robin RS, Raman AV, Jai Kumar M, Rakesh M, Subramanian BR (2008) Influence of net ecosystem metabolism in transferring riverine organic carbon to atmospheric CO2 in a tropical coastal lagoon (Chilka Lagoon, India). Biogeochemistry 87:265–285. http://sci-hub.tw/10.1007/s10533-008-9183-x

Hartman O (1968) Atlas of the errantiate polychaetous annelids from California. Allan Hancock Foundation, University of Southern California, Los Angeles. 828 pp

Hartman O (1969) Atlas of the sedentariate polychaetous annelids from California. Allan Hancock Foundation, University of Southern California, Los Angeles. 812 pp

Hartmann-Schroder G (1971) Annelida, Borstenwurmer, Polychaeta. In: Tierwelt Deutschlands, vol 58. Gustav Fischer Verlag, Jena, pp 1–594

Healy JM, Potter DG (2010) A preliminary checklist of the marine bivalves (Mollusca:Bivalvia) of Moreton bay, Queensland. In: Davie PJF, Phillips JA (eds) Proceedings of the Thirteen international marine biological workshop: The marine fauna and flora of Moreton Bay, Queensland, vol 54. Memories of Queensland Museum-Nature, Queensland, pp 235–252

Heip C, Vincx M, Vranken G (1985) The ecology of marine nematodes. Oceanograph Mar Biol Ann Rev 23:399–489

Higgins RP, Thiel H (1988) Introduction to the study of Meiofauna. Smithsonian Institute Press, Washington DC. 488 pp

Høisaeter T (2009) Distribution of marine, benthic, shell bearing gastropods along the Norwegian coast. Fauna Norvegica 28:5–106

Holzmann M, Pawlowski J (2002) Freshwater foraminiferans from Lagoon Geneva: past and present. J Foramin Res 32:344–350

Horton T, Lowry J, De Broyer C, Bellan-Santini D, Coleman CO, Daneliya M, Dauvin J-C, Fišer C, Gasca R, Grabowski M, Guerra-García JM, Hendrycks E, Holsinger J, Hughes L, Jazdzewski K, Just J, Kamaltynov RM, Kim Y-H, King R, Krapp-Schickel T, LeCroy S, Lörz A-N, Senna AR, Serejo C, Sket B, Tandberg AH, Thomas J, Thurston M, Vader W, Väinölä R, Vonk R, White K, Zeidler W (2016) World amphipoda database. Accessed at http://www.marinespecies.org/amphipoda on 2016-02-18

Hugot JP, Baujardand P, Morand S (2001) Biodiversity in helminths and nematodes as a field of study: an overview. Nematology 3:199–208

Jayalakshmi KV, Rao KK (2001) Frequency distribution of Foraminifera in the Chilka lake. J Mar Biol Assoc India 43:65–80

Jayaraj KA, Sheeba P, Jacob J, Revichandran C, Arun PK, Praseeda KS, Nisha PA, Rasheed KA (2008) Response of infaunal macrobenthos to the sediment granulometry in a tropical continental marginesouthwest coast of India. Est Coast Shelf Sci 77:743–754

Joydas TV, Damodaran R (2009) Infaunal macrobenthos along the shelf waters of the west coast of India, Arabian Sea. Indian J Mar Sci 38:191–204

Khan A, Manokaran S, Lyla S, Nazeer Z (2010) Biodiversity of epibenthic assemblage in the inshore waters of southeast coast of India. Biologia 65:704–713

Kjerfve B (1986) Comparative oceanography of coastal lagoons. In: Estuarine variability. Academic Press Inc, San Diego, pp 63–81

Kjerfve B, Magill KE (1989) Geographic and hydrodynamic characteristics of shallow coastal lagoons. Mar Geol 88:187–199

Koukousioura O, Triantaphyllou MV, Dimiza MD, Pavlopoulos K, Syrides G, Vouvalidis K (2012) Benthic foraminiferal evidence and paleoenvironmental evolution of Holocene coastal plains in the Aegean Sea (Greece). Quat Int 261:105–117

Kristensen E (1988) Benthic fauna and biogeochemical processes in marine sediments: microbial activities and fluxes. In: Blackburn TH, Sorensen J (eds) Nitrogen cycling in coastal marine environments. John Wiley and Sons Ltd, New York, pp 275–299

Kristensen E, Kostka JE (2005) Macrofaunal burrows and irrigation in marine sediment: microbiological and biogeochemical interactions. In: Kristensen E, Haese RR, Kostka JE (eds) Interactions between macro- and microorganisms in marine sediments. American Geophysical Union, Washington DC, pp 125–157

Kumar RA, Pallavi A, Ravisekhar C, Rao PG, Kaladhar R (2013) Biofacies of abundant foraminifera in relation to bottom sediments in Outer channel of Chilka lagoon, Odisha, East Coast of India. Gond Geol Mag 28:69–75

Kumar RA, Naidu KB, Kaladhar R (2015) Agglutinated foraminifera from the Chilika lagoon, Orissa, East Coast of India. Indian J Geo-Mar Sci 44:1–10

Lambshead PJD (2004) Marine nematode biodiversity. In: Chen ZX, Chen Y, Chen SY, Dickson DW (eds) Nematology: advances and perspectives. Volume 1: Nematode morphology, physiology and ecology. CABI Publishing, Wallingford. 436pp

Landing E, Geyer G, Bartowski KE (2002) Latest Early Cambrian small shelly fossils, trilobites and Hatch Hill dysaerobic interval on the Québec continental slope. J Paleontol 76:287–305

Larsen K (2002) 'Tanaidacea: Families.' Accessed at http://crustacea.net. 2nd October 2002

Lejzerowicz F, Pawlowski J, Fraissinet-Tachet L, Marmeisse R (2010) Molecular evidence for widespread occurrence of Foraminifera in soils. Environ Microbiol 12:2518–2526

Mahapatro D, Panigrahy RC, Panda S, Mishra RK (2009) Influence of monsoon on macrobenthic assemblage in outer channel area of Chilika lagoon, East coast of India. J Wetland Ecol 3:56–67

Mahapatro D, Mishra RK, Samal RN, Patanaik AK (2012) Study of macrobenthos on relation to eutrophication at Chilika lagoon, east coast of India. Mar Sci 2:139–148

Mahapatro D, Panigrahy RC, Panda S, Mishra RK (2015) Checklist of intertidal macrofauna of a brackish water coastal lagoon on east coast of India: the Chilika lagoon. Int J Mar Sci 5:1–13

Manokaran S, Ajmal Khan S, Lyla PS, Raja S, Ansari KGMT (2013) Feeding guild composition of shelf macrobenthic polychaetes of southeast coast of India. Trop Zool 26:120–139

Manokaran S, Ajmal Khan S, Lyla PS (2015) Macrobenthic composition of the southeast continental shelf of India. Mar Ecol 36:1–15

Mantha G, Moorthy MSN, Altaff K, Dahms HU (2012) Assemblage structure of the Hapacticoida (Crustacea: Copepoda) on the coast of Chennai, India. Zool Stud 51:463–475

Martins VA, Frontalini F, Tramonte KM, Figueira RCL, Miranda P, Sequeira C, Fernández S, Dias JA, Yamashita C, Renó R, Laut LLM, Silva FS, Antonieta da C, Rodrigues M, Bernardes C, Nagai R, Sousa SHM, Mahiques M, Rubio B, Bernabeu A, Rey D, Rocha F (2013) Assessment of the health quality of ria de Aveiro (Portugal): heavy metal sand benthic foraminifera. Mar Pollut Bull 70:18–33

Marzano CN, Liaci LS, Fianchini A, Gravina F, Mercurio M, Corriero G (2003) Distribution, persistence and change in the macrobenthos of the lagoon of Lesina (Apulia, southern Adriatic Sea). Oceanol Acta 26:57–66

McGlathery KJ, Sundbäck K, Anderson IC (2007) Eutrophication in shallow coastal bays and lagoons: the role of plants in the coastal filter. Mar Ecol Prog Ser 348:1–18

Mermillod-Blondin F, Rosenberg R, François-Carcaillet F, Norling K, Mauclaire L (2004) Influence of bioturbation by three benthic infaunal species on microbial communities and biogeochemical processes in marine sediment. Aquat Microbial Ecol 36:271–284

Mikae B (2015) A sea of worm: polychaete checklist of Adriatic Sea. Zootaxa 3943:1–172

Minelli A (1993) Biological systematics: the state of the art. Chapman & Hall, London. 387pp

Mishra S, Razauddin M, Das AK, Jha BC (2013) Quantitative and qualitative abundance of macrobenthos in central sector of Chilika lagoon, Odisha. J Chem Biol Phys Sci 3:1236–1246

Moens T, Braeckman U, Derycke S, Fonseca G, Gallucci F, Gingold R, Guilini K, Ingels J, Leduc D, Vanaverbeke J, Van Colen C, Vanreusel A, Vincx M (2013) Ecology of free-living marine nematodes. Handbook of zoology. De Gruyter, Berlin, pp 109–152

Mohapatra A, Rautray TR, Vijayan V, Mohanty RK, Dey SK (2007) Trace elemental characterization of some food crustacean tissue samples by EDXRF technique. Aquaculture 270:552–558

Mokievsky V, Azovsky A (2002) Re-evaluation of species diversity patterns of free-living marine nematodes. Mar Ecol Prog Ser 238:101–108

Murray JW (2006) Ecology and applications of benthic foraminifera. Cambridge University Press, Cambridge. 426pp

Murray JW (2007) Biodiversity of living benthic foraminifera: how many species are there? Mar Micropaleontol 64:163–176

Oliver PG, Holmes AM, Killeen IJ, Light JM, Wood H (2004) Annotated checklist of the marine Bivalvia of Rodrigues. J Nat Hist 38:3229–3272
Panda US, Mohanty PK (2008) Monitoring and modeling of Chilika environment using remote sensing data. Proc TAAL 2007:617–638
Papaspyrou S, Diz P, García-Robledo E, Corzo A, Jimenez-Arias JL (2013) Benthic foraminiferal assemblage changes and their relationship to environmental dynamics in intertidal muddy sediments (Bay of Cádiz, SW Spain). Mar Ecol Prog Ser 490:121–135
Pawlowski J, Holzmann M (2014) A plea for DNA barcoding of foraminifera. J Foramin Res 44:62–67
Pawlowski J, Christen R, Lecroq B, Bachar D, Shahbazkia HR, Amaral-Zettler L, Guillou L (2011) Eukaryotic richness in the abyss: insights from pyrotag sequencing. PLoS One 6:e18169
Phole WG, Thomas LHM (2001) Monitoring protocol for marine benthos: intertidal and sub-tidal macrofauna. A report by the marine bio-diversity monitoring committee (Atlantic Maritime Ecological Science Co-operative, Huntsman Marine Science Centre) to Ecological Monitoring and Assessment Network of Environment Canada. www.cciw.ca/emantem/research/preotocol/benthos/intro.html
Pinckney J, Sandulli R (1990) Spatial autocorrelation analysis of meiofaunal and microalgal populations on an intertidal sandflat: scale linkage between consumers and resources. Est Coast Shelf Sci 30:341–353
Poppe GT, Tagaro SP (2006) The new classification of gastropods according to Bouchet and Rocroi, 2005. Visaya, février. 10pp
Preston HB (1915) A further report on Mollusca from Chilka Lagoon on the East coast of India. Rec Indian Mus 11:291–310
Purchon RD (1987) Classification and evolution of the Bivalvia: an analytical study. Philos Trans R Soc Lond 316:277–302
Rao KK, Jayalakshmy KV, Venugopal P, Gopalakrishnan TC, Rajagopal MD (2000) Foraminifera from the Chilka Lagoon on the east coast of India. J Mar Biol Assoc India 42:47–61
Régnier C, Fontaine BT, Bouchet P (2009) Not knowing, not recording, not listing: numerous unnoticed Mollusk extinctions. Conserv Biol 23:1214–1221
Sahoo BK, Pati P, Panigrahy RC (2014) Environmental condition off Chilika lagoon during pre and post hydrological intervention: an overview. J Coast Conserv 18:285–297
Sahu P, Sethy PGS, Siddiqi SZ (2007) Distribution and abundance of benthic macrofauna in Nalaban Island, Chilika, Orissa. Indian J Fish 54:251–258
Sajan S, Damodaran R (2007) Faunal composition of meiobenthos from the shelf region off west coast of India. J Mar Biol Assoc India 49:19–26
Sajan S, Joydas TV, Damodaran R (2010) Meiofauna of the western continental shelf of India, Arabian Sea. Est Coast Shelf Sci 86:665–674
Sanders HL (1968) Marine benthic diversity: a comparative study. Am Nat 102:243–282
Semprucci F, Balsamo M (2012) Key role of free-living nematodes in the marine ecosystem. In: Boeri F, Jordan AC (eds) Nematodes: morphology, functions and management strategies. NOVA Science Publishers, Inc, Hauppauge, New York, pp 109–132
Semprucci F, Moreno M, Sbrocca S, Rocchi M, Albertelli G, Balsamo M (2013) The nematode assemblage as a tool for the assessment of marine ecological quality status: a case-study in the Central Adriatic Sea. Mediterr Mar Sci 14:48–57
Semprucci F, Balsamo M, Frontalini F (2014a) The nematode assemblage of a coastal lagoon (Lake Varano, Southern Italy): ecology and biodiversity patterns. Sci Mar 78:579–588
Semprucci F, Colantoni P, Sbrocca C, Baldelli G, Balsamo M (2014b) Spatial patterns of distribution of meiofaunal and nematode assemblages in the Huvadhoo lagoon (Maldives, Indian Ocean). J Mar Biol Assoc UK 94:1377–1385
Semprucci F, Losi V, Moreno M (2015) A review of Italian research on free-living marine nematodes and the future perspectives on their use as Ecological Indicators (EcoInds). Mediterr Mar Sci 16:352–365

Semprucci F, Balsamo M, Sandulli R (2016) Assessment of the Ecological quality (EcoQ) of the Venice lagoon using the structure and biodiversity of the meiofaunal assemblages. Ecol Indic 67:451–457

Simon CA, Ludford A, Wynne S (2010) Spionid polychaetes infesting cultured abalone Haliotis midae in South Africa. Afr J Mar Sci 28:161–171

Singh R, Ingole BS (2016) Structure and function of nematode communities across the Indian western continental margin and its oxygen minimum zone. Biogeosciences 13:191–209

Sivaleela G, Venkataraman K (2012) Distribution of marine polychaetes of India. Rec Zool Surv India 112:113–126

Snelgrove PVH, Blackburn TH, Hutchings PA, Alongi DM, Grassle JF, Hummel H, King G, Koike I, Lamshead PJD, Ramsing NB, Solis-Weiss V (1997) The importance of marine sediment biodiversity in ecosystem processes. Ambio 26:578–583

Southern R (1921) Polychaeta of the Chilka Lagoon and also of fresh and brackish waters in other parts of India. Mem Indian Mus 5:563–659

Srichandran S, Kim JY, Kumar A, Mishra DR, Bhadury P, Muduli PR, Rastogi G (2015) Interannual and cyclone-driven variability in phytoplankton communities of a tropical coastal lagoon. Mar Pollut Bull 101:39–52

Strong EE, Gargominy O, Ponder WF, Bouchet P (2007) Global diversity of gastropods (Gastropoda; Mollusca) in freshwater. Hydrobiologia 595:149

Subba Rao (2003) Indian sea shells (Part-1): polyplacophora and Gastropoda. Rec Zool Sur India – Occ Paper Ser 192:1–416

Surya Rao KU (1974) Intertidal amphipods from the Indian coast. Proc Indian Acad Sci 38B:190–205

Tagliapietra D, Sigovini M (2010) Benthic fauna: collection and identification of macrobenthic invertebrates. Terre et Environ 88:253–261

Vanaverbeke J, Gheskiere T, Steyaert M, Vincx M (2002) Nematode assemblages from subtidal sandbanks in the Southern Bight of the North Sea: effect of small sedimentological differences. J Sea Res 48:197–207

Vanaverbeke J, Bezerra TN, Braeckman U, De Groote A, De Meester N, Deprez T, Derycke S, Gilarte P, Guilini K, Hauquier F, Lins L, Maria T, Moens T, Pape E, Smol N, Taheri M, Van Campenhout J, Vanreusel A, Wu X, Vincx M (2015) NeMys: world database of free-living marine nematodes at http://www.nemys.ugent.be/aphia.php?p=taxadetails&id=2132. Added on 02.03.2015

Vanreusel A, Fonseca G, Danovaro R, Da Silva MC, Esteves AM, Ferrero T, Gad G, Galtsova V, Gambi C, Genevois VF, Ingels J, Ingole B, Lampadariou N, Merckx B, Miljutin D, Miljutina M, Muthumbi A, Netto S, Portnova D, Radziejewska T, Raes M, Tchesunov MA, Vanaverbeke J, Van Gaever S, Venekey V, Bezerra TN, Flint H, Copley J, Pape E, Zeppilli D, Martinez PA, Galeron J (2010) The contribution of deep-sea macrohabitat heterogeneity to global nematode diversity. Mar Ecol 31:6–20

Venkataraman K, Wafar M (2005) Coastal and marine biodiversity of India. Indian J Mar Sci 34:57–75

Venkatesan V, Mohamed KS (2015) Bivalve classification and taxonomy. In: Director CMFRI (ed) Summer school on recent advances in marine biodiversity conservation and management. Cochin, CMFRI, pp 42–48

Warwick RM, Robinson J (2000) Sibling species in the marine pollution indicator genus *Pontonema* Leidy (Nematoda: Oncholaimidae), with a description of *P. mediterranea* sp. nov. J Nat Hist 34:641–662

Wehe T, Fiege D (2002) Annotated checklist of the polychaete species of the seas surrounding Arabian Peninsula: Red Sea, Gulf of Aden, Arabian Sea, Gulf of Oman, Arabian gulf. Fauna Arabia 19:7–238

Wetzel M, Weber A, Giere O (2002) Re-colonization of anoxic/sulfidic sediments by marine nematodes after experimental removal of macroalgal cover. Mar Biol 141:679–689

Zeppilli D, Sarrazin J, Leduc D, Arbizu PM, Fontaneto D, Fontanier C, Gooday AJ, Kristensen GM, Ivanenko VN, Sørensen MV, Vanreusel A, Thébault J, Mea M, Allio N, Andro T, Arvigo A, Castrec J, Danielo M, Foulon V, Fumeron R, Hermabessiere L, Hulot V, James T, Augen RL, Bot TL, Long M, Mahabror D, Morel Q, Pantalos M, Pouplard E, Raimondeau L, Rio-Cabello A, Seite S, Traisnel G, Urvoy K, Stegen TV, Weyand M, Fernandes D (2015) Is the meiofauna a good indicator for climate change and anthropogenic impacts? Mar Biodivers 45:505–535

Chapter 15
Microbial Ecology of Chilika Lagoon

Pratiksha Behera, Madhusmita Mohapatra, and Gurdeep Rastogi

Abstract Microbial communities in the coastal lagoons have been widely appreciated for their role in nutrient cycling, mineralization of organic matter, and detoxification of toxic contaminants. Chilika like other coastal lagoons remains an understudied ecosystem with respect to the microbes and microbial communities. In this chapter, we discuss different culture-dependent and culture-independent methods that have been applied in examining the microbial communities from the Chilika Lagoon. We discuss the original research studies which contribute to our existing knowledge and also identify the knowledge gaps that still exist in the microbial ecology of the Chilika Lagoon. The advent of high-throughput sequencing techniques will facilitate the characterization of microbial communities, their spatiotemporal variability in relation to the environmental factors or to link biogeochemical cycling to the specific microbial communities. Integrated approaches using 'omics' and culture methods would be useful in providing an in-depth knowledge about the structure and function of the microbial communities of the Chilika Lagoon.

Keywords Spatiotemporal · Brackish · Salinity · Microbial communities

15.1 Introduction

Coastal estuarine lagoons such as Chilika are one of the highly productive ecosystems that support a variety of flora and fauna due to their diverse salinity regime ranging from fully marine to brackish water and freshwater. These are ecologically and economically important habitats due to their diverse ecosystem services which directly or indirectly are linked to the nutrient biogeochemical cycling (Mitsch and Gosselink 2000). The ecosystem services of a coastal lagoon are positively correlated with the prokaryotic biodiversity and a loss of biodiversity due to climate

P. Behera · M. Mohapatra · G. Rastogi (✉)
Wetland Research and Training Centre, Chilika Development Authority, Balugaon, Odisha, India

C. M. Finlayson et al. (eds.), *Ecology, Conservation, and Restoration of Chilika Lagoon, India*, Wetlands: Ecology, Conservation and Management 6,
https://doi.org/10.1007/978-3-030-33424-6_15

change and anthropogenic pressure may also lead to loss of ecosystem services (Bellard et al. 2012; Rombouts et al. 2013). Coastal lagoons are also important in context to their contribution to global greenhouse gas emissions (Verma et al. 2002). From the ecological point of view, coastal lagoons act as a sink and accumulate nutrients and toxic compounds which are subsequently recycled by benthic and pelagic microbial communities and thereby preventing eutrophication of coastal and seawater (Hunter et al. 2006; Oni et al. 2015; Obi et al. 2016).

Coastal lagoon supports diverse microbial communities due to resource heterogeneity in their benthic and pelagic zone which is mostly due to the tidal influx, strong wind currents, and an influx of freshwater from Catchment Rivers. Generally, shallow coastal ecosystems that are dominated by macrophytes, the rate of primary production is much higher than what can be consumed by herbivores (Martin et al. 2015). Under these circumstances, a major fraction of organic matter tends to accumulate in the sediments in form of detritus which is eventually processed by benthic microbial communities making it available to higher trophic levels. The productivity of Chilika Lagoon is high mostly due to the penetration of light in the entire shallow water column that allows macrophytes and macroalgae to proliferate and attain higher biomass. Chilika is also a macrophyte dominated lagoon wetland that supports a luxuriant growth of *Potamogeton, Naja, Ulva, Hydrilla,* and seagrasses in different regions of the lagoon (Jaikumar et al. 2011). Benthic microbial communities of Chilika Lagoon are the key component in the recycling of nutrients which is crucial to sustaining high productivity of the lagoon. The remineralization of organic matter and nutrient cycling, especially nitrogen may play a key role in determining the productivity in the water column due to benthic-pelagic coupling (Knoppers 1994). Sediment microbial communities can rapidly metabolize and degrade autochthonous and allochthonous organic matter and play a pivotal role in carbon cycling. Benthic sediment communities are composed of diverse bacterial and archaeal communities containing specific physiological groups, such as nitrogen-fixers, nitrifiers, ammonia-oxidizers, methane-oxidizers, methanogens, and sulfate-reducing bacteria (Rusch et al. 2009; Oni et al. 2015; Behera et al. 2017b). These microbial communities mediate the carbon, nitrogen, and sulfur cycling in the wetlands and drive the energy flow in the aquatic food web. Compared to the pelagic zone, benthic sediment communities play much active role in the nutrient cycling as sediment surface provides a solid support and a rich availability of nutrients available for microbial growth (Hunter et al. 2006; Oni et al. 2015).

Understanding the spatial variability in sediment bacterial communities is a challenging task as large number of biotic and abiotic factors play role in it. Sediment microbial communities are highly variable due to spatial and temporal gradients in physicochemical characteristics namely salinity, pH, and nutrients (e.g. organic carbon, nitrogen, phosphorus) (Song et al. 2010; Wang et al. 2012; Liu et al. 2015). In addition, root exudates of aquatic macrophytes also act as one of the major determinants that govern the microbial community composition in sediments (Borruso et al. 2014; Behera et al. 2017b). Rhizosphere sediment communities are distinct from the bulk sediment communities not only with respect to their diversity and composition but also demonstrate higher rates of nitrification, denitrification, and methane production (Lamers et al. 2012; Borruso et al. 2014). These differences in community composition compared to bulk sediments are mostly due to root exudates which is a rich source of sugars, amino acids, etc. and enrich a specific rhizosphere microbial community that is often host-specific (Carvalhais et al. 2010). In addition, the partial aerobic conditions which often prevail in the rhizosphere sediments due to radial losses of oxygen from root surfaces of wetland plants, allow coupling of aerobic and anaerobic biogeochemical processes such as methane-oxidation and generation, sulfur-oxidation and reduction to occur simultaneously in sediments. Thus, a holistic understanding of microbial communities of coastal lagoons remain a fundamental research area to understand how biogeochemical cycles may respond to changing environmental conditions that eventually will drive the entire productivity of the lagoon.

15.2 Status of the Microbial Research on Chilika Lagoon

Chilika Lagoon has been historically evaluated from a 'macrobial' perspective as it is dominated by macrophytes and macrofauna. These macrobial flora and fauna such as fishes, macrophytes, algae, and benthic invertebrates have been assessed for their diversity and spatial and temporal changes in their community structure. However, the microbial component of the lagoon containing bacterial and archaeal communities of the benthic and pelagic zones remain mostly unexplored. A literature search reveals that studies on the microbiology of Chilika Lagoon are very limited and need a special focus. Microbial communities in an environmental sample can be studied using two broad methods (i) culture-dependent and (ii) culture-independent. Figure 15.1 summarizes the techniques which have been used so far in examining the microbes and microbial communities of Chilika Lagoon. This chapter aims to summarize the microbial research studies which have been conducted so far on Chilika Lagoon (Table 15.1). We discuss in detail the key findings from these existing studies and discuss the gaps therein to be considered in future research. In line with the lack of knowledge on the microbial ecology of Chilika Lagoon, most of the existing studies focus on culture-based isolation and characterization of novel bacterial species whereas studies on the characterization of microbial communities as a whole are much more limited.

15.2.1 Culture-Dependent Studies

Culture-dependent methods are highly important in understanding the ecological role of microbes in natural environments. In addition, these methods are also important in isolating and screening the pure cultures for various biological activities that may have great potential in the discovery of novel drugs and pharmaceutical

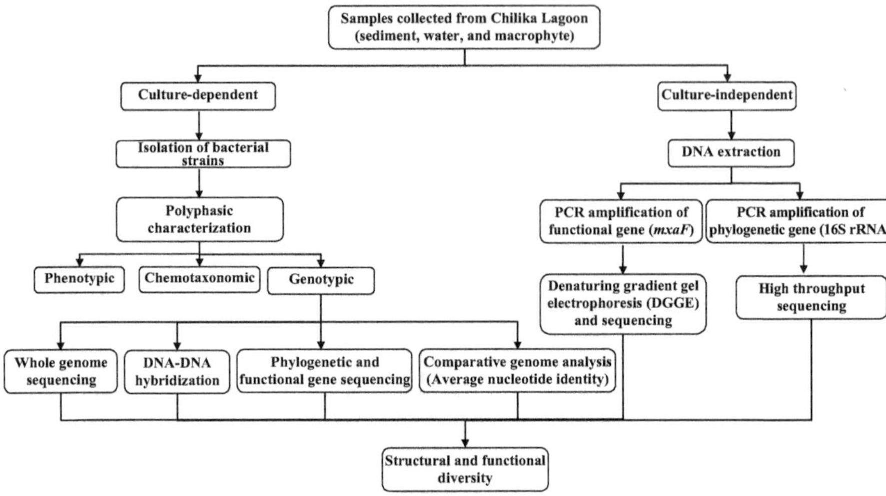

Fig. 15.1 Different methods used in the isolation, identification and analysis of structural and functional diversity of bacterial communities of the Chilika Lagoon

compounds from marine environments (Xiong et al. 2013). However, culture-dependent techniques on an average can culture only 0.1–1% of bacteria in any given sample and are not suitable to examine the whole microbial community structure in an environmental sample (Rastogi and Sani 2011). In sediment, it has been estimated that 0.25% of total cell counts can be cultured using traditional culture methods (Tamaki et al. 2005). It is evident that several studies have targeted isolation and characterization of individual bacterial species from sediments and water samples of the Chilika Lagoon. Novel species of bacteria affiliated to *Shewanella, Thiorhodococcus, Rhizobium, Mangrovibacter, Halobacillus, Pseudomonas,* and *Streptomyces* isolated from water, sediment, and plant tissue samples have been characterized using a polyphasic taxonomic approach. The cultured diversity of endolithic bacteria and their metabolites were examined from beach sand samples collected from the sea mouth of Chilika Lagoon (Parag et al. 2013). Based on 16S rRNA gene sequence phylogenetic analysis, cultured isolates showed similarities to *Acinetobacter johnsonii, Microbacterium esteraromaticum, Rhizobium yanglingense, Lysobacter soli, Kocuria palustris* and *Acinetobacter iwofii.*

The diversity of cultured methanol-oxidizing bacteria from sediments of Chilika Lagoon was assessed using PCR-restriction fragment length polymorphism (PCR-RFLP) of 16S rRNA and *mxaF* genes (Meena et al. 2015). *mxaF* gene codes for the α-subunit of methanol dehydrogenase enzyme that is a functional and phylogenetic marker for methanotrophs. Sequence analysis of 16S rRNA genes from cultured isolate revealed that *Methylobacterium radiotolerans, M. extorquens, M. hispanicum, M. organophilum, M. lusitanum, M. zatmanii, Hyphomicrobium facile, Methyloversatilis* sp., *Mycobacterium brisbanense* and *Pseudomonas* sp. were methanol-oxidizing bacteria in the sediments. Based on *mxaF* gene

Table 15.1 Summary of the microbial studies carried out on Chilika Lagoon

S. no.	Sampling location	Isolation source	Method	Major finding	References
Culture-dependent approach					
1	Southern part of Chilika Lagoon, near Barkul village	Sediment	Polyphasic approach for describing novel strain.	*Shewanella chilikensis* sp. nov., a moderately alkaliphilic bacterium was isolated and characterized.	Sucharita et al. (2009)
2	Bird's island coast in the southern sector	Sediment	Polyphasic approach for describing novel strain.	*Thiorhodococcus modestalkaliphilus* sp. nov. a novel purple sulfur bacterium was isolated and characterized.	Sucharita et al. (2010)
3	Central and Southern sectors and Outer Channel area of the lagoon	Water	Isolation of strains and studied their resistance properties.	*Shigella dysenteriae*, *Streptococcus lactis*, *Bacillus cereus* and *Klebsiella pneumoniae* with strong antigenic character were isolated.	Parida et al. (2012)
4	Boat dockyard near Barkul village jetty	Sediment	Polyphasic approach for describing novel strain.	A novel actinobacterial strain *Streptomyces chilikensis* sp. nov. was isolated and characterized.	Ray et al. (2013)
5	Fish dumping yard at Barkul village jetty	Sediment	Polyphasic approach for describing novel strain.	A novel actinobacterial strain *Streptomyces barkulensis* sp. nov. was isolated and characterized.	Ray et al. (2014)
6	No information on sampling location within the lagoon	Water	Isolation of strain and studied its plant growth properties.	A chitinolytic actinomycete *Streptomyces vinaceusdrappus* was isolated and characterized plant growth promoting and biocontrol activities.	Yandigeri et al. (2015)
7	Various sites from the periphery, middle area and near sea mouth	Water	Isolation of strain and purification of enzyme.	The extracellular halo tolerant, detergent and organic solvent stable, alkaline cholesterol oxidase produced by a newly isolated *Rhodococcus* sp. PKPD-CL was purified and characterized.	Kasabe et al. (2015)

(continued)

Table 15.1 (continued)

S. no.	Sampling location	Isolation source	Method	Major finding	References
8	Satapada near sea mouth	Sediment	Polyphasic approach for describing novel strain.	*Thiorhodococcus fuscus* sp. nov., a moderately halophilic bacterium was isolated and characterized.	Lakshmi et al. (2015)
9	Fish dumping yard at Barkul village jetty	Sediment	Polyphasic approach for describing novel strain.	A novel actinobacterial strain *Streptomyces chitinivorans* sp. nov., was isolated.	Ray et al. (2016)
10	Kalupadaghat	Water	Illumina MiSeq	The genome annotation of *Halobacillus* sp. showed various gene clusters for tolerance to stress, such as elevated pH, salt concentration, and toxic metals.	Panda et al. (2016)
11	Bhaseramundia	Sediment	Illumina MiSeq	The *Pseudomonas* sp. was capable of producing proteases and the genome analysis showed various genes related to plant growth promoting.	Mishra et al. (2016a)
12	Bhaseramundia	Rhizospheric region of *Phragmites karka*	Illumina MiSeq	The *Acinetobacter* sp. was capable of degrading cellulose and the genome analysis showed various genes related to plant growth promoting.	Mishra et al. (2016b)
13	Northern shoreline	Root of *Phragmites karka*	Illumina MiSeq	The genome sequence has provided information on putative genes that code for proteins involved in oxidative stress, uptake of nutrients, and nitrogen fixation that might offer niche specific ecological fitness and	Behera et al. (2016)

				explain the invasive success of *Phragmites karka* in Chilika Lagoon.	
14	Northern shoreline	Root of *Phragmites karka*	Polyphasic approach for describing novel strain.	Endophytic nitrogen fixing bacterium, *Mangrovibacter phragmitis* sp. nov., was isolated and characterized.	Behera et al. (2017a)
15	Kalupadaghat	Water sample	Polyphasic approach for describing novel strain.	A novel *Halobacillus* sp. was isolated from the water sample. The strain hydrolyzed casein and gelatin and can tolerate NaCl up to 25%.	Panda et al. (2018)
16	Northern shoreline	Rhizosphere sediment of *Phragmites karka*	Polyphasic approach for describing novel strain and Illumina MiSeq.	A novel *Pseudomonas* sp. was isolated from the rhizosphere sediment of *Phragmites karka*. The genome sequence has provided information on putative genes that code for proteins involved in iron acquisition, stress response, metabolism of amino acids, sulphur, potassium, siderophore production, and biofilm formation.	Behera et al. (2018b)
Integrated approach					
17	Sea mouth of Chilika lake near Satpada	Sand	Polyphasic approach for describing novel strain and 454 pyrosequencing of 16S rRNA genes.	Nitrogen fixing strain, *Rhizobium endolithicum* sp. nov. was isolated and characterized. The study indicated a rich bacterial diversity with majority of the OTUs belonging to the phylum *Proteobacteria*.	Parag et al. (2013)
18	Balugaon, Bhusandapur, Panaspada, Nalaban island, and Breakfast island	Sediment	Isolation of methylotrops. ARDRA and RFLP of 16S rRNA and *mxaF* genes. DGGE of *mxaF* genes of unculturable methylotrops.	In the study, the culture-dependent approach revealed the relatedness of α-*Proteobacteria* and *Methylobacterium*, *Hyphomicrobium* and	Meena et al. (2015)

(continued)

Table 15.1 (continued)

S. no.	Sampling location	Isolation source	Method	Major finding	References
				Ancyclobacter sp. And the culture-independent analyses revealed that up to 90% of the methylotrophs were unculturable.	
Culture-independent approach					
19	Rambhartia and Kalupadaghat	Sediment	454 pyrosequencing of 16S rRNA	The study observed stress produced due to high salinity resulting in less number of phyla from the sample of Rambhartia. *Proteobacteria* was the most dominating phyla from both the sites.	Pramanik et al. (2015)
20	Bulk sediment from 4 sectors, *Phragmites karka* from northern and southern shoreline, and seagrass from the outer channel	Sediment	Illumina MiSeq	The study observed that the bacterial communities were linked to salinity and 'the rhizosphere effect'. *Proteobacteria* was the most dominating phyla from all the three type of sediment.	Behera et al. (2017b)
21	*Phragmites karka* from northern and southern shoreline.	Rhizosphere sediment	Illumina MiSeq	The study demonstrated a high spatial variability in the community structure and carbon metabolic profiles of rhizosphere communities of *P. karka* suggesting that 'rhizosphere effect' is not the sole factor that shape microbial community.	Behera et al. (2018a)

sequencing of cultured isolates, it was found that *Methylobacterium* was the dominant genus followed by *Methylophilus* and *Hyphomicrobium* sp. Many of these isolates showed very low similarity to known methanotrophs in the NCBI database suggesting that about 90% of the methylotrophs were unculturable and these could be the potential novel methanol-oxidizing bacteria. Other studies have carried out a more detailed genetic characterization of cultured isolates using whole genome sequencing. For example, genomes of cultured bacteria affiliated to *Halobacillus, Pseudomonas, Acinetobacter,* and *Mangrovibacter* have been sequenced to get a molecular insight on their niche adaptation, nitrogen-fixation and plant-growth promoting properties.

15.2.2 Culture-Independent Studies

Culture-independent approaches for examining the microbial communities involve direct extraction of DNA from an environmental sample. The DNA is subsequently amplified using conserved primers targeting the phylogenetic or functional marker genes. PCR amplified products can be cloned in a vector and sequenced to analyze the taxonomic composition of microbial communities. PCR products amplified from environmental DNA can also be analyzed directly by genetic-fingerprinting techniques such as denaturing gradient gel electrophoresis (DGGE), temperature gradient gel electrophoresis (TGGE), single-strand conformation polymorphism (SSCP), random amplification of polymorphic DNA (RAPD), amplified ribosomal DNA restriction analysis (ARDRA), terminal restriction fragment length polymorphism (T-RFLP), length heterogeneity (LH)-PCR, and ribosomal RNA (rRNA) intergenic spacer analysis (RISA). The clone library method remained one of the most widely used methods in examining the microbial communities from a variety of environment till they have been almost replaced by high-throughput sequencing techniques. With the advent of high-throughput sequencing techniques such as 454 pyrosequencing and Illumina, it has now become possible to sequence the PCR products directly without cloning. In addition, hundreds of samples can be multiplexed in a single run making the cost per sequence much cheaper than traditional clone-library method. So far, only a few studies have applied culture-independent approaches for examining the bacterial communities of the Chilika Lagoon. The studies on spatial and temporal changes in microbial communities and their relationship with the functioning of the lagoon are still at its infancy but available preliminary studies suggest that sediment bacterial communities due to their diverse biogeochemical potential play a crucial role in the ecosystem functioning of the Chilika Lagoon.

Parag et al. (2013) applied 454-pyrosequencing of 16S rRNA genes to characterize the endolithic bacterial communities of two beach sand samples collected from Satpada region (near sea mouth) of Chilika. Majority of the bacterial sequences were represented by *Proteobacteria* followed by *Firmicutes* and phototrophic members belonging to genera *Rhodovulum, Rhodobacter, Chromatium, Marichromatium,*

Thiophaeococcus, and *Thiorhodococcus.* This study also noted marked species variation in bacterial communities between the stratified layers of golden and black sand samples. This study identified only 6 phyla and 16 genera from the golden layer whereas from the black layer, a total of 17 phyla and 286 genera were identified. The golden layer was composed of taxa affiliated to *Proteobacteria, Spirochaetes, Firmicutes, Chlamydiae, Tenericutes* and *Planctomycetes* whereas in the black layer *Proteobacteria, Firmicutes, Actinobacteria, Bacteriotedes, Acidobacteria, Cholofelxi, Verrumicrobia, Lentisphaerae, Spirochaetes, Plantomycetes, Deinococcus-Thermus, Gemmatimonadetes*, and *Cyanobacteria* were detected.

Meena et al. (2015) used DGGE analysis of *mxaF* PCR products amplified from sediment samples collected from Chilika Lagoon to examine the diversity of aerobic methanol-oxidizers. Sequencing analysis of DNA fragments recovered from DGGE gels revealed sequences related to *Methylobacterium, Ancyclobacter, Burkholderiales,* and *Hypomicrobium* in the methanol-oxidizing bacterial communities. The abundance of methylotrophs quantified using q-PCR of *mxaF* gene copies estimated $4.9 \times 10^6 - 1.25 \times 10^7$ gene copies g^{-1} of sediment. Comparison of q-PCR data with culturable bacteria counts (colony forming units) suggested that only 10% of the methanol-oxidizers were culturable.

Pramanik et al. (2015) have applied 454-pyrosequencing of V1–V3 region of 16S rRNA gene sequences to investigate the bacterial communities in the soil sediments of Chilika Lagoon. Although this study claims to use a metagenomic approach, the authors have used a PCR dependent-approach which in true sense does not qualify under the metagenomics. Metagenomics is a PCR independent approach in which sequencing of entire genomic DNA content of an environmental sample is carried out. The authors have collected two sediment samples representing high (Rambhartia region) and low (Kalupadaghat region) salinity zones of Chilika Lagoon. Analysis of 16S rRNA gene sequence dataset revealed 39 and 44 phyla in sediments collected from Rambhartia and Kalupadaghat region, respectively. Distinct differences in the relative abundances of phyla namely *Proteobacteria, Choloroflexi, Firmicutes, Acidobacteria, Actinobacteria,* and *Planctomycetes* were noted between high and low salinity zones suggesting that salinity could be an important factor in driving these changes.

Behera et al. (2017b) applied Illumina MiSeq sequencing on a large number of bulk and rhizosphere sediment samples to examine the bacterial communities and major drivers responsible for causing changes in community composition along spatial and temporal scales. A total of 100 bulk sediment samples across a network of 30 GPS fixed stations were sampled in three different seasons during the year 2014–2015. These stations were located across different sectors of the lagoon representing freshwater, brackish water, and marine salinity regimes. In addition to bulk sediments, rhizosphere sediment samples from *Phragmites karka* and *Halodule uninervis* (seagrass) were also collected. PCR amplification targeting the V6 region of 16S rRNA genes was carried out and detailed spatial analysis of community composition was conducted (Fig. 15.2). It was observed that *Proteobacteria* and lineages within it namely α, β, and γ-*Proteobacteria* were the most abundant bacteria

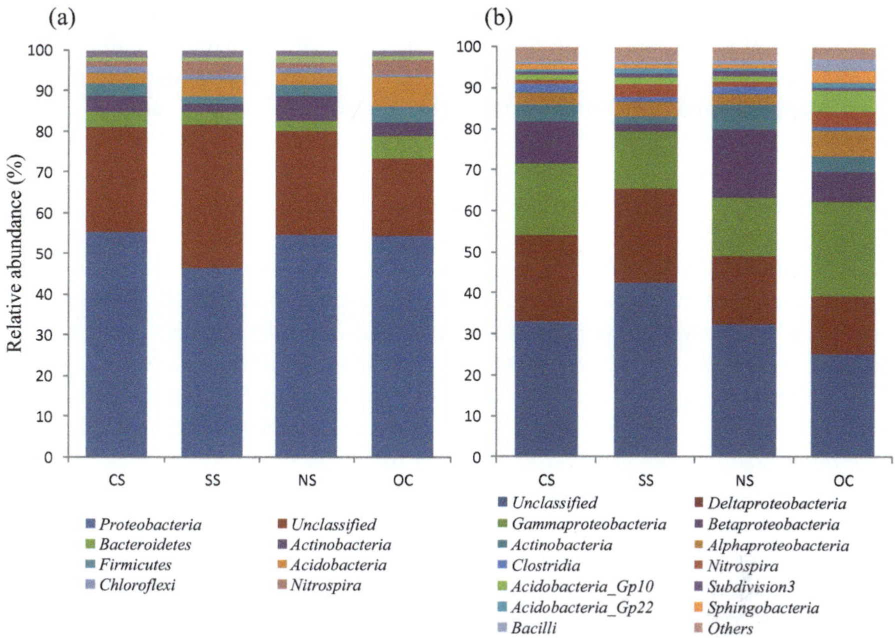

Fig. 15.2 Bacterial community composition of sediments ($n = 100$) collected from Chilika Lagoon at phylum (**a**) and class (**b**) levels obtained by Illumina MiSeq sequencing of 16S rRNA. The category 'others' accounted for phyla or classes which were composed of <1% sequences in the sample of all the four sectors. CS: Central sector, SS: Southern sector, NS: Northern sector, OC: Outer channel

across various sectors of the Chilika Lagoon. In addition, bacterial abundances in sediments were estimated using q-PCR of 16S rRNA genes which estimated a higher bacterial density in *P. karka* rhizosphere ($7.52 \pm 9.42 \times 10^8$ cells g^{-1}) than the bulk ($3.67 \pm 4.68 \times 10^8$ cells g^{-1}) sediments. The bacterial species richness reflected by OTUs was found higher in bulk (mean 789 $\pm$ 115) than the rhizosphere (mean 699 $\pm$ 237) sediments. Bulk sediment communities were also distinct in their community composition from the rhizosphere sediment communities of *P. karka* and seagrasses suggesting the influence of 'rhizosphere effect' in driving the community composition. In addition, rhizosphere sediment communities of *P. karka* and seagrasses were also differentiated based on their composition suggesting the plant species-specific nature of rhizosphere bacterial communities. This study has also assessed the relationship between environmental variables namely salinity, pH, total organic carbon, total phosphorus, and dissolved oxygen using redundancy analysis. This analysis showed that salinity was a major factor that drove the community composition of bulk sediment communities. General additive modeling of *Proteobacteria* against salinity data revealed differential responses in which abundance of γ and α-*Proteobacteria* increased with increase in salinity whereas β-*Proteobacteria* showed a sharp decline. Besides detailed analysis on sediment community composition and their biotic and abiotic drivers, the study by Behera et al. (2017b) also conducted a phenotype mapping of the taxonomic data which predicts the metabolic features in bacterial communities. The taxonomic-to-metabolic mapping predicted different metabolic phenotypes such as sulfate-reducer, ammonia-oxidizer, dehalogenation, nitrite-reducer and nitrogen-fixation in the sediment bacterial communities. These findings suggest that sediment bacterial communities could play an important role in carbon, nitrogen, and sulfur cycling in the lagoon.

In another detailed investigation, the structural and metabolic diversity of the rhizosphere bacterial and archaeal community structure was studied (Behera et al. 2018a). Illumina MiSeq sequencing was carried out targeting the bacterial and archaeal 16S rRNA genes of the rhizosphere sediment samples of *P. karka* collected from different locations. *Thiobacillus*, *Methylotenera*, *Bacillus*, *Steroidobacter* and *Escherichia/Shigella* were the most abundant bacterial genera whereas *Methanomassiliicoccus* was one of the most abundant genus in the archaeal communities. Analysis of community composition revealed a spatial variation in the bacterial and archaeal community structure suggesting that 'rhizosphere effect' was not the sole factor that shapes the microbial community structure and salinity also played an important role in it. BIOLOG based carbon substrate metabolic profiling showed that rhizosphere bacterial community were metabolically diverse and highest utilization of carbon substrate was recorded for tween 40, D-mannitol, L-asparagine, 4-hydroxybenzoic acid, putrescine, and D-galacturonic acid. Metabolic mapping of the bacterial communities suggested that *P. karka* rhizosphere microbial communities could play a critical role in maintaining the ecological health of coastal wetlands through nutrient cycling, pollutant biodegradation, and supporting reed growth. Although *Phragmites* is typically considered as an invasive weed, the ecosystem services provided by these reeds must be included in the management and conservation plan of wetlands.

15.3 Future Directions

The research studies compiled in this chapter clearly demonstrates that understanding of microbial communities residing in the benthic and pelagic zone of Chilika Lagoon is far from complete. Considering the ecosystem services of Chilika Lagoon and the role that microbial communities perform in mediating these services, research focussing on the microbial ecology should be a priority research area in coming years. These studies will be crucial to gauge changes in the nutrient cycling that are likely to happen due to changes in the hydrology of Chilika Lagoon predicted as a result of climate change and anthropogenic interventions. However, due to a high intrinsic variability of coastal lagoons coupled with high genetic diversity in microbial communities, a single approach or technique cannot provide enough resolution to examine the structure and function of microbial communities. An integrated approach combining culture efforts with molecular approaches along with detailed characterization of local and global environmental factors are required for a comprehensive understanding of microbial communities of Chilika Lagoon. Detailed metabolic analysis of microbial communities through techniques such as stable isotope probing and micro-autoradiography would be required to link the microbial community structure with function in the natural environment. A better understanding of environmental factors that determine the biogeographical patterns in microbial communities will allow us to predict changes in microbial communities and their functions under changing climate.

The development in high-throughput sequencing technique will facilitate the application of metagenomics and metatranscriptomics to examine the structure and functional potential of microbial communities including uncultured microorganisms. In addition, targeted metatranscriptomics based on sequencing of functional genes such as methyl-coenzyme A reductase (*mcrA*), ammonia monooxygenase (*amoA*), and particulate methane monooxygenase gene (*pmoA*) can reveal the microbial communities that are actively involved in carbon and nitrogen cycling in wetland sediments. Large-scale sequencing of microbial communities will improve our ability to examine the microbial communities and their relationship not only among themselves but also with their local environment. In addition, metagenomic analysis of microbial communities also has immense economic and ecological values. The bacterial communities residing in the lagoon are adapted to tolerate varying stress related to change in salinity and oxygen gradient and are like gold mines for novel genomes and metabolic pathways which could be harnessed for novel biocatalysts and drugs or degradation of various contaminants. Sequencing of the metagenome will also shed light on the ecology of benthic and pelagic microbial communities that is crucial for provisioning the ecosystems services of the lagoon.

Recent development in high-throughput sequencing techniques has put the aim of achieving water and sediment metagenome within our reach. Comprehensive sequencing efforts targeting water and sediment samples collected across varying spatial and temporal gradients will provide sufficient data for a deeper understanding of the microbial ecology of Chilika Lagoon. The information generated through the metagenomic projects will also serve as a starting platform for 'omics' approaches namely proteomics and metabolomics to provide a deeper understanding of microbial activities and their expression patterns in a variety of niches within the lagoon such as in the rhizosphere of different macrophytes where active biogeochemical cycling occurs. These 'omics' approaches will provide clues in understanding the microbial basis of the invasive success of certain macrophytes such as *P. karka* in the lagoon. In addition new approaches to culture the unculturable majority will also be highly essential to assign metabolic functions to a vast number of unknown or hypothetical genes that are recovered in metagenomic surveys.

References

Behera P, Vaishampayan P, Singh NK, Mishra SR, Raina V, Suar M, Pattnaik AK, Rastogi G (2016) The draft genome sequence of *Mangrovibacter* sp. MP23, an isolated from the roots of *Phragmites karka*. Genomics Data 9:128–129

Behera P, Ramana VV, Maharana B, Joseph N, Vaishampayan P, Singh NK, Shouche Y, Bhadury P, Mishra SR, Raina V, Suar M, Pattnaik AK, Rastogi G (2017a) *Mangrovibacter phragmitis* sp. nov., an endophyte isolated from the roots of *Phragmites karka*. Int J Syst Evol Microbiol 67:1228–1234

Behera P, Mahapatra S, Mohapatra M, Kim JY, Adhya TK, Raina V, Suar M, Pattnaik AK, Rastogi G (2017b) Salinity and macrophyte drive the biogeography of the sedimentary bacterial communities in a brackish water tropical coastal lagoon. Sci Total Environ 595: 472–485

Behera P, Mohapatra M, Adhya TK, Suar M, Pattnaik AK, Rastogi G (2018a) Structural and metabolic diversity of rhizosphere microbial communities of *Phragmites karka* in a tropical coastal lagoon. Appl Soil Ecol 125:202–212

Behera P, Mahapatra M, Seuylemezian A, Vaishampayan P, Ramana VV, Joseph N, Joshi A, Shouche Y, Suar M, Pattnaik AK, Rastogi G (2018b) Taxonomic description and draft genome of *Pseudomonas sediminis* sp., isolated from the sediment of *Phragmites karka*. J Microbiol 56:458–466

Bellard C, Bertelsmeier C, Leadley P, Thuiller W, Courchamp F (2012) Impacts of climate change the future of biodiversity. Ecol Lett 15:365–377

Borruso L, Bacci G, Mengoni A, Philippis RD, Brusetti L (2014) Rhizosphere effect and salinity competing to shape microbial communities in *Phragmites* (Cav.) Trin. ex-Steud. FEMS Microbiol Lett 359:193–200

Carvalhais LC, Dennis PG, Fedoseyenko D, Hajirezaei MR, Borriss R, Wirén N (2010) Root exudation of sugars, amino acids, and organic acids by maize as affected by nitrogen, phosphorus, potassium, and iron deficiency. J Plant Nutr Soil Sci 000:1–9

Hunter EM, Mills HJ, Kostka JE (2006) Microbial community diversity associated with carbon and nitrogen cycling in permeable shelf sediments. Appl Environ Microbiol 72:5689–5701

Jaikumar M, Chellaiyan D, Kanagu L, Senthil PK, Stella C (2011) Distribution and succession of aquatic macrophytes in Chilka Lake – India. J Ecol Nat Environ 3:499–508

Kasabe PJ, Mali GT, Dandge PB (2015) Assessment of alkaline cholesterol purified from *Rhodococcus* sp. PKPD-CL for its halo tolerance, detergent and organic solvent stability. Protein Expression Purif 116:30–41

Knoppers B (1994) Aquatic primary production in coastal lagoons. In: Kjerfve B (ed) Coastal lagoon processes, 60. Elsevier, Amsterdam, pp 243–286

Lakshmi KVNS, Divyasree B, Sucharita K, Sasikala C, Ramana CV (2015) *Thiorhodococcus fuscus* sp. nov., isolated from a lagoon. Int J Syst Evol Microbiol 65:3938–3943

Lamers LP, van Diggelen JM, Op den Camp HJ, Visser EJ, Lucassen EC, Vile MA, Jetten MS, Smolders AJ, Roelofs JG (2012) Microbial transformations of nitrogen, sulfur, and iron dictate vegetation composition in wetlands: a review. Front Microbiol 25:156

Liu S, Ren H, Shen L, Lou L, Tian G, Zheng P, Hu B (2015) pH levels, bacterial community structure in sediments of the Qiantang River as determined by 454 pyrosequencing. Front Microbiol 6:285

Martin P, Isabelle A, Daniele M, Gilles T, Amo DY, Franck D, Benedicte T (2015) Phytoplankton versus macrophyte contribution to primary production and biogeochemical cycles of a coastal mesotidal system. A modelling approach. Estuar Coast Shelf Sci 165:52–60

Meena KK, Kumar M, Mishra S, Ojha SK, Wakchaure GC, Sarkar B (2015) Phylogenetic study of methanol oxidizers from Chilika-Lake sediments using genomic and metagenomic approaches. Indian J Microbiol 55:151–162

Mishra SR, Panda AN, Ray L, Sahu N, Mishra G, Jadhao S, Suar M, Adhya TK, Rastogi G, Pattnaik AK, Raina V (2016a) Draft genome sequence of *Pseudomonas* sp. strain, BMS12, a plant growth-promoting and protease-producing bacterium, isolated from the rhizosphere sediment of *Phragmites karka* of Chilika Lake, India. Genome Announc 3:e00342–e00316

Mishra SR, Ray L, Panda AN, Sahu N, Xess SS, Jadhao S, Suar M, Adhya TK, Rastogi G, Pattnaik AK, Raina V (2016b) Draft genome sequence of *Acinetobacter* sp. Strain BMW17, a cellulolytic and plant growth-promoting bacterium isolated from the rhizospheric region of *Phragmites karka* of Chilika Lake, India. Genome Announc 4:e00395–e00316

Mitsch WJ, Gosselink JG (2000) The value of wetlands: importance of scale and landscape setting. Ecol Econ 35:25–33

Obi CC, Adebusoye SA, Ugoji EO, Ilori MO, Amund OO, Hickey WJ (2016) Microbial communities in sediments of Lagos Lagoon, Nigeria: elucidation of community structure and potential impacts of contamination by municipal and industrial wastes. Front Microbiol 7:1213

Oni OE, Schmidt F, Miyatake T, Kasten S, Witt M, Hinrichs K-U, Friedrich MW (2015) Microbial communities and organic matter composition in surface and subsurface sediments of the Helgoland mud area, North Sea. Front Microbiol 6:1290

Panda AN, Mishra SR, Ray L, Sahu N, Acharya A, Jadhao S, Suar M, Adhya TK, Rastogi G, Pattnaik AK, Raina V (2016) Draft genome sequence of *Halobacillus* sp. Strain KGW1, a moderately halophilic and alkaline protease-producing bacterium isolated from the rhizospheric region of *Phragmites karka* from Chilika Lake, Odisha, India. Genome Announc 4:e00361–e00316

Panda AN, Mishra SR, Ray L, Das S, Rastogi G, Pattanaik AK, Adhya TK, Suar M, Raina V (2018) Taxonomic description and genome sequence of *Halobacillus marinus* sp. nov., a novel strain isolated from Chilika Lake, India. J Microbiol 56:223–230

Parag B, Sasikala C, Ramana CB (2013) Molecular and culture dependent characterization of endolithic bacteria in two beach sand samples and description of *Rhizobium endolithicum* sp. nov. Antonie Van Leeuwenhoek 104:1235–1244

Parida S, Jena RC, Samal KC, Chand PK (2012) Isolation and identification of pathogenic bacteria from brackish waters of Chilika Lagoon, Odisha, India for pharmaceutical use. Malays J Microbiol 8:197–202

Pramanik A, Basak P, Banerjee S, Sengupta S, Chattopadhyay D, Bhattacharyya M (2015) Pyrosequencing based profiling of the bacterial community in the Chilika Lake, the largest lagoon of India. Genomics Data 4:112–114

Rastogi G, Sani RK (2011) Molecular techniques to assess microbial community structure, function, and dynamics in the environment. In: Ahmad et al (eds) Microbes and microbial technology: agricultural and environmental applications. Springer, New York, pp 29–57

Ray L, Suar M, Pattnaik AK, Raina V (2013) *Streptomyces chilikensis* sp. nov., a halophilic streptomycete isolated from brackish water sediment. Int J Syst Evol Microbiol 63:2757–2764

Ray L, Mishra SR, Panda AN, Rastogi G, Pattanaik AK, Adhya TK, Suar M, Raina V (2014) *Streptomyces barkulensis* sp. nov., isolated from an estuarine lake. Int J Syst Evol Microbiol 64:1365–1372

Ray L, Mishra SR, Panda AN, Das S, Rastogi G, Pattanaik AK, Adhya TK, Suar M, Raina V (2016) *Streptomyces chitinivorans* sp. nov., a chitinolytic strain isolated from estuarine lake sediment. Int J Syst Evol Microbiol 66:3241–3248

Rombouts I, Beaugrand G, Artigas LF, Dauvin J-C, Gevaert F, Goberville E, Kopp D, Lefebvre S, Luczak C, Spilmont N, Travers-Trolet M, Villanueva MC, Kirby RR (2013) Evaluating marine ecosystem health: case studies of indicators using direct observations and modelling methods. Ecol Indic 24:353–365

Rusch A, Hannides AK, Gaidos E (2009) Diverse communities of active bacteria and archaea along oxygen gradients in coral reef sediments. Coral Reefs 28:15–26

Song H, Du B, Zhang M, Fu W, Lu X, Li Z, Ding Y (2010) Effect of environmental factors on bacterial community in Lake Dongping sediment. Wei Sheng Wu Xue Bao 50:1065–1071

Sucharita K, Sasikala C, Park SC, Baik KS, Seong CN, Ramana CV (2009) *Shewanella chilikensis* sp. nov., a moderately alkaliphilic gammaproteobacterium isolated from a lagoon. Int J Syst Evol Microbiol 59:3111–3115

Sucharita K, Sasikala C, Ramana CV (2010) *Thiorhodococcus modestalkaliphilus* sp. nov. a phototrophic gammaproteobacterium from Chilika salt water lagoon, India. J Gen Appl Microbiol 56:93–99

Tamaki H, Sekiguchi Y, Hanada S, Nakamura K, Nomura N, Matsumura M, Kamagata Y (2005) Comparative analysis of bacterial diversity in freshwater sediment of a shallow eutrophic lake by molecular and improved cultivation-based techniques. Appl Environ Microbiol 71:2162–2169

Verma A, Subramanian V, Ramesh R (2002) Methane emissions from a coastal lagoon: Vembanad Lake, West Coast, India. Chemosphere 47:883–889

Wang Y, Sheng HF, He Y, Wu JY, Jiang YX, Tam NF, Zhou HW (2012) Comparison of the levels of bacterial diversity in freshwater, intertidal wetland, and marine sediments by using millions of illumina tags. Appl Environ Microbiol 78:8264–8271

Xiong ZQ, Wang JF, Hao YY, Wang Y (2013) Recent advances in the discovery and development of marine microbial natural products. Mar Drugs 11:700–717

Yandigeri MS, Malviya N, Solanki MK, Shrivastava P, Sivakumar G (2015) Chitinolytic *Streptomyces vinaceusdrappus* S5MW2 isolated from Chilika lake, India enhances plant growth and biocontrol efficacy through chitin supplementation against *Rhizoctonia solani*. World J Microbiol Biotechnol 31:1217–1225

Chapter 16
Survey, Characterization, Ecology, and Management of Macrophytes in Chilika Lagoon

Ajit K. Pattnaik, Pratap C. Panda, and Gurdeep Rastogi

Abstract Chilika lagoon is a macrophyte dominated ecosystem. The diversity and distribution of macrophytes provide a key to understand the ecological health of a wetland ecosystem. Chilika lagoon had been in a degraded condition and was included in the Montreux record in 1993 by Ramsar Convention, due to the change in its ecological characters. The restoration intervention by Chilika Development Authority resulted in the enhancement of fishery resources, the reappearance of native fish species, decrease in the spread of freshwater invasive species, expansion of seagrass meadows, and overall improvement of the ecosystem. To make an assessment of the changes in the phytodiversity of the lagoon after hydrological restoration, as a part of the doctoral thesis research work by the first author, phytodiversity of islands, shorelines, sandbar, and the littoral zone was surveyed for a period of 5 years (1998–2002). Subsequently, after a decade, a re-assessment was made for a period of 4 years (2012–2016) by the second author under the World Bank supported Integrated Coastal Zone Management Project, Odisha. Remote sensing and Geographic Information System tools were used for the assessment of the diversity, distribution, and seasonal changes in the aquatic angiosperms as well as for monitoring the spread of the invasive species. The extensive floristic survey after the hydrological intervention revealed a reduction in the area covered by water hyacinth and water fern, invasion of *Phragmites karka* in the northern sector and expansion of seagrass meadows. Four species of seagrasses i.e., *Halodule pinifolia, Halodule uninervis, Halophila ovata,* and *Cymodocea serrulata* were recorded for the first time from the lagoon. An occurrence of 748 species of angiosperm belonging to 486 genera under 127 families were recorded, identified, and preserved as herbarium specimens. Based on the outcome of this systematic survey, the strategies

A. K. Pattnaik (✉)
Chilika Development Authority, Bhubaneswar, Odisha, India

P. C. Panda
Regional Plant Resource Centre, Bhubaneswar, Odisha, India

G. Rastogi
Wetland Research and Training Centre, Chilika Development Authority, Balugaon, Odisha, India

C. M. Finlayson et al. (eds.), *Ecology, Conservation, and Restoration of Chilika Lagoon, India*, Wetlands: Ecology, Conservation and Management 6,
https://doi.org/10.1007/978-3-030-33424-6_16

for the management of macrophytes in Chilika lagoon and the scope of future studies are recommended.

Keywords Phytodiversity · Macrophytes · Invasive species · Remote sensing · Seagrass

16.1 Introduction

Aquatic macrophytes are photosynthetic plants, large enough to be seen with naked eye, that grow permanently or periodically submerged below, floating on, or growing up through the water surface (Chambers et al. 2008). Macrophytes play an important role in the balancing of lake ecosystems, vary in their biomass and capability to recycle nutrients through the release of oxygen and organic carbon in the sediments. The type and distribution of macrophytes, to a large extent, is determined by the nutrients available in the sediments and the water column. A clear understanding of the influence of the aquatic macrophytes on different physical, chemical, and biological processes of the wetland, especially with regard to the water quality and biodiversity is critical for the management of a wetland ecosystem. Emergent, free-floating, and submerged macrophytes grow in the shoreline of most wetlands and are influenced by the geomorphology, environmental factors, and biotic interactions (Carpenter and Lodge 1986; Engel 1988). Macrophyte also provides a substrate for colonization of algae and invertebrates (Dvořaki and Bestz 1982; Cattaneo 1983; Schramm et al. 1987). Macrophytes are also well recognized to affect water and sediment biogeochemistry, productivity, and biotic interactions (Crowder and Cooper 1982; Heck and Crowder 1991).

Chilika is a complex ecosystem with a spatiotemporal salinity gradient and complex hydrological regime due to simultaneous connectivity with rivers and sea. The lagoon is broadly divided into four ecological sectors based on the salinity regime, i.e. northern, southern, central, and outer channel. Salinity is found to conspicuously influence the composition and distribution of the macrophytic vegetation of the lagoon. The factors that affect growth and distribution of the macrophytes are sediment texture, nutrients, and other inputs from the drainage basin along with the salinity regime. Due to the high silt load, which is almost to the tune of 0.686 million tons per annum (CDA 2001), the northern sector of the lagoon is the shallowest and is covered by the emergent macrophytes mainly *Phragmites karka*.

A systematic and comprehensive survey of the macrophytes of Chilika has not been attempted earlier. Although the diatoms and other algal forms have been studied in detail, no concerted attempt has been made for higher plants of this region. Das and Samal (1988) have applied remote sensing technique for the survey of vegetation and land use pattern. Panda and Patnaik (1988) have made a significant contribution and reported the occurrence of 352 angiosperms belonging to 272 genera and 96 families from this region. Of these, *Macrotyloma ciliatum* (Panda et al. 1985) and *Heliotropium curassavicum* (Panda and Patnaik 1988) were new plants records from eastern India and Odisha respectively. Subsequently, Panda and

Pattnaik (2002) enumerated 546 species of angiospermic plants from Chilika and its immediate neighbourhood. A comprehensive and extensive survey of the flora of this region was carried out by Pattnaik et al. (2003) and Panda and Pattnaik (2002) and based on this, a checklist of 711 species of angiosperms belonging to 492 genera and 119 families has been reported. Pattnaik et al. (2008) studied the diversity, distribution, and taxonomy of five seagrass species of Chilika lagoon and reported the occurrence of *Halophila ovata*, *Halodule pinifolia,* and *Halodule uninervis* for the first time. From the review of literature, it is apparent that the studies carried out till 1998 on macrophyte diversity of Chilika are sparse and there is scope for a more detailed floristic survey from this region to make a complete inventory of the diversity of plant resources with particular emphasis on the flora of the uninhabited islands, shorelines, sandbar, and aquatic macrophytes.

16.2 Methods

16.2.1 Study of the Aquatic Angiosperms

The digital data of Indian Remote Sensing Satellite (IRS) 1D LISS III with a spatial resolution of 23.5 m was used for species-wise classification of the macrophytes in the Chilika (Pattnaik 2003). The digital data indented from the National Remote Sensing Agency (NRSA) Hyderabad was analyzed using Earth Resources Data Analysis System (ERDAS) imagine image processing software. The data were geo-rectified, mosaiced and then the aquatic angiosperms were classified. Before classifying the macrophytes, the landmass of islands was masked out. By visual interpretation, the image was enhanced using different types of enhancing procedures and the image was interpreted with reference to field data generated by way of rigorous ground truthing with the help of the Global Positioning System (GPS). The output was in the form of the classified vegetation map of the lagoon with 20 classes indicating the group-wise distribution and seasonal changes. The digital data was further analyzed for generating the vegetation map of the aquatic angiosperms.

16.2.2 Plant Collection, Preservation, Identification, and Nomenclature

Survey on the phytodiversity was carried out from four ecological sectors of the lagoon and their immediate neighborhood (Fig. 16.1). Each and every plant encountered during the survey, whether it was in flowering, fruiting, or vegetative stage was collected. Meta-data was recorded on the spot that included locality, habitat, collector's name, collection number, date, local name, and uses. For each species, 6 specimens were collected and a field number was tagged to each of them. The field and herbarium methods followed during the present work were as per the guidelines provided by Jain and Rao (1977) and Forman and Bridson (1989).

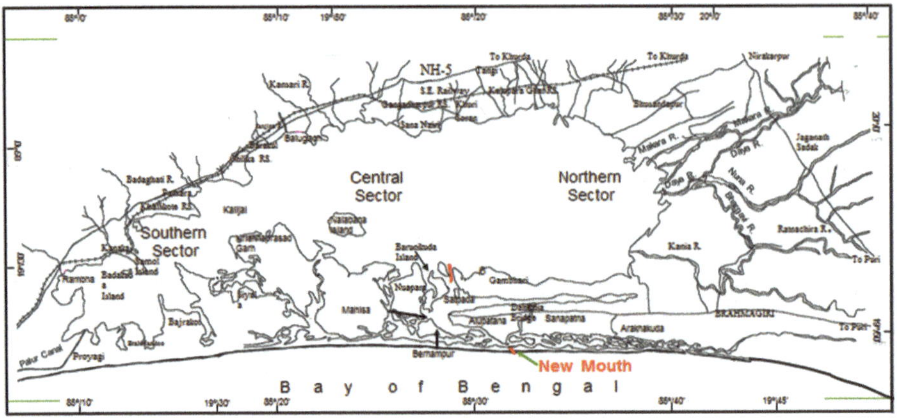

Fig. 16.1 Map of Chilika Lagoon showing macrophyte sampling locations

The identity of the plants were determined with the help of taxonomic keys provided by Haines (1921–1925), Mooney (1950), Saxena and Brahmam (1994–1996), and other flora, monographs and taxonomic revisions using the diagnostic characters recorded in the phytography of the plant. The correct botanical name was ascertained to each taxon as per the rules and provisions of the recent International Code of Botanical Nomenclature (ICBN). The important synonym(s) occurring in the Haines (1921–1925) and Mooney (1950) have been referred to assign a correct name to each plant.

16.2.3 Estimation of Biomass

Biomass of hydrophytes was assessed from four ecological sectors and sampling was done by forty randomly selected quadrates. For quantitative sampling, a quadrate sampler of 1 sq. m was used. The sites were sampled in summer, monsoon, and winter regularly for 1 year. Plant materials collected from each quadrate was sorted out and fresh weight was recorded.

16.2.4 Physicochemical Analysis of Water

The physicochemical parameters as reported by (Bhatta and Pattnaik 2002) and (Barik et al. 2017) from 30 monitoring stations by Chilika Development Authority (CDA) was used for interpretation of macrophyte diversity. The physicochemical parameters: air and water temperature, depth, transparency, specific conductivity, pH, total alkalinity, salinity, dissolved oxygen and concentrations of nitrate and phosphate were collected from each station.

16.3 Results and Discussion

16.3.1 Macrophytes of the Lagoon

The flora of Chilika was quite rich in terms of species content and vegetational diversity. A critical analysis of the vegetation revealed the occurrence of species that were rare, threatened, wild relatives of crop plants, economic and medicinal plants and those that were endemic to Chilika. The macrophytes encountered were classified into four broad groups, i.e. emergent, rooted floating-leaved, submerged, and free-floating. Emergent macrophytes are mostly perennial higher plants growing on periodically inundated or submerged soils with their basal portions submerged in water and tops above the water level. The rhizomes and the roots spread laterally into the deeper water to adjust with the changing water levels of the lagoon. The emergent species like *Phragmites karka, Schoenoplectus littoralis, Schoenoplectus articulatus, Typha angustata, Cyperus platystylis,* and *Cyperus compressus* were common in the eulittoral zone of the northern sector from adjoining Kalupadaghat to Mangalajodi village, extending up to the river confluence point of Daya, Bhargavi, and Luna. The western river confluence points of Kansari and Salia also supported the luxuriant growth of many emergent macrophytes.

Rooted floating-leaved macrophytes are the plants that are anchored by roots at the bottom with leaves floating on water surface. Floating leaves are attached to roots or rhizomes with a flexible, sturdy stem and in many cases by a leaf stalk. This group of aquatic plants were restricted to the northern sector of the lagoon, which remains predominantly freshwater for more than 10 months in a year. The characteristic species in the shallow sheltered zones of northern sector adjoining to village Mangalajodi to Kalupadaghat, where the wind action was minimum, were *Nymphaea pubescens, Nymphaea nouchali, Nymphoides hydrophylla,* and *Nymphoides indica.* The situation prevailed up to 4 km towards Sorono village where species of both *Nymphaea* and *Nymphoides* occurred, profusely. Species of *Nymphoides* were also abundant from Borkudi up to the river confluence points of Daya and Bhargavi.

Submerged macrophytic vegetation of the lagoon was much diversified and well distributed in all sectors. In the northern sector, the major submerged species encountered were *Hydrilla verticillata, Aponogeton natans, Potamogeton nodosus, Potamogeton octandrus, Potamogeton pectinatus, Utricularia aurea, Najas minor, Najas graminea, Najas indica,* and *Ceratophyllum demersum.*

Potamogeton pectinatus and *Najas graminea* along with the seagrass species like *Halophila beccarii, Halophila ovata, Halophila ovalis, Halodule uninervis,* and *Halodule pinifolia* are the dominant species of the central sector, and their occurrence has been recorded from Nalabana bird sanctuary and the creeks of Krushna Prasad Island with the sandy substratum. The seagrass meadows provided an excellent nursery ground for the various species of fish, shrimp, and crab. These meadows also harboured many invertebrates that constituted the food for the fishes and birds and provided a surface for the growth of epiphytic algae. The prominent

Fig. 16.2 Dominant macrophytes of the lagoon. (**a**) *Gracilaria verrucosa* and (**b**) *Enteromorpha intestinalis* from southern sector. Occurrence of *Eichhornia crassipes* (**c**, **d**) before and (**e**) after the hydrological intervention. (**f**) *Phragmites karka* and (**g**) *Schoenoplectus littoralis* from northern shoreline. (**h**) *Halophila beccarii*, (**i**) *Halophila ovalis*, (**j**) *Halophila ovata*, (**k**) *Halodule uninervis*, (**l**) *Halodule pinifolia*, (**m**) *Ruppia maritima* (**n**) bed of *Ruppia maritima* from the southern sector. (**o**) *Basella alba*, (**p**) *Sesuvium portulacastrum*, (**q**) *Salicornia brachiata* from islands of the lagoon. Macrophytes from (**r**) Somolo Island, (**s**) Chheliakuda Island. (**t**) *Hydrophylax maritima*, (**u**) *Sesuvium portulacastrum*, (**v**) *Ipomoea pes-caprae*, (**w**) *Ipomoea pes-caprae*'s flower from the sand spit. (**x**) Shifting of vegetation in summer season in the sandy areas

macroalgae in this zone were *Gracilaria verrucosa, Enteromorpha intestinalis, Enteromorpha compressa,* and *Chaetomorpha linum.*

The photic zone of the southern sector extended to an average of 1.0 m and supported a luxuriant growth of seagrasses. The area adjoining Birds Island up to the Palur Canal with sandy substratum supported a luxuriant growth of the seagrasses. *Gracilaria verrucosa* was the dominant red algal species along the shoreline adjoining village Pathara up to Somolo Island. The dominant macroalgae encountered in this sector were *Gracilaria verrucosa* (Fig. 16.2a), *Polysiphonia subtilissima, Polysiphonia sertularioides, Ceramium elegans, Grateloupia filicina, Ulva lactuca, Enteromorpha intestinalis* (Fig. 16.2b), *Chaetomorpha linum, Cladophora glomerata,* and *Lyngbya aestuarii.*

The outer channel from Barunikuda Island, including the entire stretch adjoining village Khirisahi up to the end of the Morei jana (end of the Brahampura Island), supported good seagrass meadows. Interestingly, *Gracilaria verrucosa* was found near the jetties along the outer channel in discontinuous patches that were apparently eutrophic. The other dominant macroalgae found were *Ulva lactuca, Enteromorpha intestinalis, Chaetomorpha linum,* and *Cladophora glomerata.*

Free-floating macrophyte float on or just below the water surface with their roots were observed in the northern sector and the river confluence points of the central sector. *Eichhornia crassipes* were found at the river confluence points. *Eichhornia crassipes* (water hyacinth) that used to occur abundantly in the northern sector, perished due to increased salinity level of lagoon caused by the hydrological intervention by CDA (Pattnaik 2000) (Fig. 16.2c, d, e). Patches of *Eichhornia crassipes* were observed only at the river confluence points. *Pistia stratiotes, Salvinia cucullata,* and small free-floating ones like *Azolla pinnata, Spirodela polyrhiza,* and *Lemna minor* were the dominant species of this group. Interestingly, *Azolla pinnata* that finds its way into the lagoon from the nearby paddy fields, grow and form thick mat near Kalupadaghat and Mangalajodi during post-monsoon months. Azolla mats gradually disappear as the salinity level of these area improved from April onwards (Pattnaik 2001b).

16.3.2 Vegetation of Shoreline

The shoreline is an ecotone, *i.e.* a transition interface between two radically different ecosystems containing very different environmental conditions and communities. The composition of the vegetation on the shoreline was determined by the nature of slope, nature of substratum, edaphic factors, the period of inundation, and the physicochemical parameters of the lagoon water.

The major shoreline plants of the outer channel along the village Satapada up to Manikpatna, where the soil was mainly sandy were *Pandanus fascicularis, Crotalaria pallida, Croton bonplandianus, Opuntia stricta var. dillenii, Boerhavia diffusa, Fimbristylis acuminata, Cassia tora, Solanum trilobatum, Ocimum sanctum, Ziziphus oenoplia, Cyperus rotundus, Cyperus compressus, Paspalum distichum, Jatropha gossypifolia, Mollugo pentaphylla, Grangea maderaspatana, Cressa cretica, Mimosa pudica, Emilia sonchifolia, Achyranthes aspera, Alternanthera sessilis, Heliotropium curassavicum, Ipomoea carnea,* and *Ipomoea pes-caprae.* The dominant algal forms along the shoreline were *Enteromorpha intestinalis* and *Gracilaria verrucosa* that showed luxuriant growth being attached to the stones and the concrete structures of the jetties. The sandy substratum and continuous aeration due to the movement of the boats favoured the luxuriant growth of the *Gracilaria sp.,* in this zone.

The species diversity along the shoreline of the northern sector was quite rich and dominated by freshwater aquatics, grasses and sedges like, *Cyperus arenarius, Cyperus castaneus, Cyperus distans, Cyperus haspan, Cyperus iria, Schoenoplectus articulatus, Phragmites karka* (Fig. 16.2f), *Schoenoplectus littoralis* (Fig. 16.2g), *Paspalum distichum, Panicum repens, Alternanthera philoxeroides, Alternanthera sessilis, Ludwigia adscendens, Eleocharis dulcis, Mariscus paniceus,*

Aeschynomene indica, Ammannia baccifera, Eclipta alba, Polygonum barbatum, Brachiaria ramosa, Murdannia nudiflora, Centrostachys aquatica, Centella asiatica, Grangea maderaspatana, Coldenia procumbens, Centipeda minima, Oplismenus compositus, Commelina benghalensis, Heliotropium indicum, Heliotropium curassavicum, Hemarthria compressa, Marsilea quadrifolia, Ipomoea aquatica, Echinochloa stagnina, Diplachne fusca, and *Typha angustata.*

The shoreline plants of the central sector were a combination of the freshwater and brackish water species. The freshwater elements encountered in the river confluence points and the shorelines were; *Lindernia antipoda, Lindernia viscosa, Fimbristylis aestivalis, Fimbristylis miliacea, Fimbristylis schoenoides, Paspalum scrobiculatum, Brachiaria ramosa, Diplachne fusca, Hemarthria compressa, Lindernia viscosa, Cyperus castaneus, Murdannia spirata, Oplismenus burmannii, Mukia maderaspatana, Alloteropsis cimicina, Mariscus paniceus, Panicum psilopodium, Echinochloa colona,* and *Echinochloa stagnina* in association with the emergents like *Typha angustata, Schoenoplectus articulatus, Schoenoplectus littoralis, Eichhornia crassipes* and *Phragmites karka.*

The species diversity on the shoreline of the southern sector was poor in comparison to the northern and central sectors. *Pandanus fascicularis, Crotalaria pallida, Croton bonplandianus, Opuntia stricta var. dillenii, Cressa cretica, Heliotropium curassavium, Sesuvium portulacastrum, Salicornia brachiata, Dichrostachys cinerea, Diplachne fusca, Schoenoplectus littoralis, Cyperus rotundus, Amaranthus spinosus, Solanum virginianum, Cynodon dactylon, Mollugo pentaphylla, Ocimum basilicum, Tribulus terrestris, Aristolochia indica, Evolvulus alsinoides, Allmania nodiflora, Polycarpon prostratum, Parkinsonia aculeata, Heliotropium indicum, Parthenium hysterophorus, Ocimum sanctum, Sauropus bacciforme, Dichanthium pertusum, Phyllanthus virgatus, Andrographis echioides,* and *Eclipta alba* were some conspicuous shoreline plants of this sector.

16.3.3 Seagrass Meadows

Five species of seagrasses i.e., *Halodule uninervis, Cymodocea serrulata, Halodule pinifolia, Halophila ovalis, Halophila ovata,* and *Halophila beccarii* were recorded for the first time from the lagoon (Fig. 16.2h, i, j, k, l). Out of this, the occurrence of *Halodule uninervis, Halodule pinifolia, Halophila ovata,* and *Cymodocea serrulata* from the lagoon was a new distributional record. After the hydrological intervention, a proliferation of the seagrass meadows into the deep-water zone and appearance of species like *Halodule uninervis, Halodule pinifolia,* and *Halophila ovalis* was recorded (Pattnaik et al. 2008). Well-established seagrass meadows were recorded from the central, southern and outer channel after the opening of the new mouth. The seagrass meadows were found at their best in calm sheltered areas with sandy substratum starting from the southeastern part of the southern sector extending through the creeks of the central sector up to the village Arakhakuda in the outer channel. In the central sector, rich seagrass meadows composed of *Halophila ovata,*

Halophila ovalis, and *Halodule uninervis* in association with *Ruppia maritima* were recorded (Fig. 16.2m, n). In Nalabana Island and the creeks of Krushna Prasad Island with shallow water, soft bottom and less fluctuation of salinity supported excellent seagrass meadows. The epiphytic macroalgae associated with the seagrasses were *Enteromorpha intestinalis, Enteromorpha compressa, Chaetomorpha linum, Cladophora glomerata, Gracilaria verrucosa,* and *Polysiphonia sertularioides.*

The shoreline of the southern sector from village Pathara up to Somolo Island supported excellent seagrass beds of *Halodule uninervis* and *Halophila ovalis* in association with the red algae *Gracilaria verrucosa. Cymodocea serrulata* was recorded from the shoreline of Odialpur from the southern sector. The sheltered bay behind the Ghantasila hill up to the point of origin of Palur canal also supported meadows of *Halophila ovata* associated with *Ruppia maritima.*

The seagrass beds along the outer channel disappeared during August/September, when there was a fall in salinity and poor transparency due to the unidirectional flow of water from the lagoon towards the sea discharging the turbid floodwater. However, seagrass bed re-appeared along the entire shoreline of the outer channel once salinity increased during post-monsoon. In the outer channel, all along the shoreline from Barunikuda up to the end of Morei jana, which was relatively sheltered, luxuriant growth of seagrasses composed of species like, *Halophila beccarii, Halophila ovata, Halophila ovalis, Halodule uninervis,* and *Halodule pinifolia* in association with the *Ruppia maritima* was observed. Excellent meadows of *Halophila beccarii* were encountered in the outer channel near the village Arakhakuda.

16.3.4 Vegetation of Islands

The dominant species of plants recorded from the islands were *Gyrocarpus americanus, Thespesia populnea, Ficus rumphii, Ficus retusa, Helicteres isora, Trichosanthes tricuspidata, Canavalia virosa, Tinospora cordifolia, Dalbergia rubiginosa, Basella alba, Cissus repens, Ipomoea alba, Sarcostemma acidum, Biophytum sensitivum, Porteresia coarctata, Dichanthium bladhii,* and *Fimbristylis bisumbellata* (Fig. 16.2o, p, q). Interestingly, during the survey from Birds Island, the occurrence of a single plant of *Aegiceras corniculatum*, a mangrove associate growing on exposed rocks with minimal soil cover was recorded. *Cassipourea ceylanica* apparentely an endemic species was recorded from Barkuda and Sanakuda Islands in a severely degraded state. In India, it has been recorded earlier from Tamil Nadu (Nair and Henry 1983), Kerala (Beddome 1872; Nayar et al. 2006), Orissa (Brahmam et al. 2001), and Goa (Prabhugaonkar et al. 2007). The uninhabited islands within the lagoon supported many rare, threatened and endangered plants like *Cassipourea ceylanica, Capparis roxburghii, Gyrocarpus americanus,* and *Dimorphocalyx glabellus.* An *in situ* conservation of these island habitats is essential to allow natural regeneration of these important species.

Narayanswami and Carter (1922) enumerated 142 species of flowering plants and ferns from Barkuda Island. These authors have reported large trees of *Pongamia pinnata* along the periphery of the island and as many as eight species of *Ficus* in the island. In our survey, 27 species listed by them could not be traced and instead, additional taxa were documented. In our survey, no *Pongamia pinnata* in tree form was encountered from the island. *Ficus gibbosa, Ficus arnottiana,* and *Ficus geniculata* were conspicuously missing from the island. The population of *Capparis roxburghii* was also observed to be sparse. The threatened species *Cassipourea ceylanica* was in a severely degraded condition and the existing plants were only in the form of coppiced shoots without proper flowering and fruiting. *Calotropis acia* and *Caralluma adscendens*, reported by Narayanswami and Carter (1922) have been eliminated from the island. Invasive exotic species such as *Lantana camara, Chromolaena odorata, Mikania micrantha, Parthenium hysterophorus,* and *Alternanthera spp*. were spreading rapidly in the island preventing the establishment of other plant species. Due to heavy biotic pressure, virtually no other arboreus member could be found except a few scattered trees of *Morinda coreia*. Besides *Cassipourea ceylanica*, other species that occurred in Sanakuda Island were *Ipomoea pilosa, Crateva religiosa, Meyna pubescens, Colubrina asiatica, Derris scandens, Pergularia daemia, Kalanchoe pinnata, Capparis sepiaria,* and *Capparis brevispina*.

Somolo, another island with a dense human settlement, *Pandanus fascicularis* formed a dense thicket along the southeastern shoreline. *Anacardium occidentale, Mangifera indica,* and *Artocarpus heterophyllus* were common fruit-bearing trees planted by the villagers besides raising many kinds of cereal and pulses on the island (Fig. 16.2r). The natural vegetation of this island was comprised of *Lepisanthes tetraphylla, Grewia disperma, Mimosa rubicaulis, Vitex negundo, Derris trifoliata, Heliotropium curassavicum, Canavalia virosa, Ipomoea carnea, Lippia javanica,* and *Opuntia dillenii var stricta*. Close to Somolo Island, there lies Chheliakuda Island with a good number of trees, shrubs, and climbers (Fig. 16.2s). Of these, *Lepisanthes tetraphylla, Strychnos nux-vomica, Salvadora persica, Ficus auriculata, Phoenix sylvestris, Holarrhena antidysenterica, Clerodendrum inerme, Canavalia virosa, Dioscorea wallichii, Olax scandens,* and *Luffa acutangula* deserved special mention. The vegetation of this island was in a degraded condition because of severe human interference.

Clumps of *Bambusa arundinacea* represented the natural vegetation on Ghantasila Island and other species like *Cassia fistula, Erythrina suberosa, Diospyros chloroxylon, Helicteres isora, Tarenna asiatica, Caesalpinia digyna, Croton caudatus, Flacourtia indica, Hyptis suaveolens, Blepharis boerhavifolia, Rhynchosia capitata, Bacopa monnieri, Malachra capitata, Gloriosa superba, Tylophora indica, Commelina paludosa, Commelina erecta, Chrysopogon aciculatus, Heteropogon contortus, Parthenium hysterophorus, Centella asiatica,* and *Dichanthium bladhii* were recorded during the survey. The northeastern part of the island with very poor soil cover has disturbed natural vegetation due to grazing by buffaloes, which swim across from the nearby villages.

Gopakuda Island which is situated close to the Chheliakuda Island, the natural vegetation has almost been eliminated by the villagers from the adjacent Tentuliapadar and Keshpur village for raising crops and plantations of *Anacardium occidentale, Mangifera indica,* and *Artocarpus heterophyllus.* The natural vegetation still existed in few discrete patches and blocks with higher abundance of *Crateva magna, Azadirachta indica, Catunaregam spinosa, Cressa cretica, Croton bonplandianus, Capparis brevispina, Calotropis gigantea, Chromolaena odorata, Crotalaria verrucosa, Plumbago zeylanica, Mimosa rubicaulis, Capparis roxburghii, Ipomoea sepiaria, Derris scandens, Grangea maderaspatana, Lippia geminata, Cassia occidentalis, Sida cordifolia, Aristolochia indica, Barleria prionitis, Opuntia stricta var. dillenii, Vernonia cinerea, Ziziphus mauritiana, Celosia argentea,* and *Canavalia virosa.*

Malatikuda is another small uninhabited island, which gets submerged during monsoon. As water level recedes, plants characteristic to salt marshes like *Salicornia brachiata, Suaeda nudiflora, Paspalum paspalodes,* and *Sesbania procumbens* were observed to appear as immediate colonizers.

The vegetation on the Islands of the central sector was mostly in a degraded condition due to biotic pressures of several form and magnitudes. The floral diversity of the Islands in the central sector was poorer as compared to the islands of the southern sector. Vasaramunda Island of the central sector had been planted with *Tectona grandis, Acacia auriculiformis, Acacia nilotica,* and *Anogeissus accuminata* by the Forest Department. The dominant plant species recorded from this island were *Barringtonia acutangula, Azadirachta indica, Pongamia pinnata, Crateva magna, Thevetia peruviana, Strychnos nux-vomica, Bambusa arundinacea, Morinda tomentosa, Ziziphus mauritiana, Andrographis paniculata, Chromolaena odorata, Ecbolium viride, Calotropis gigantea, Ziziphus oenoplia, Croton bonplandianus, Achyranthes aspera, Passiflora foetida, Emilia sonchifolia, Abutilon hirtum, Cissus quadrangularis, Breynia rhamnoides, Vernonia cinerea, Jatropha gossypifolia, Asparagus racemosus, Tridax prucumbens, Cardiospermum halicacabum, Mimosa pudica,* and *Paspalum paspaloides.* Plantations of *Acacia auriculiformis* and *Acacia nilotica* have been raised in parts of Kalijugeswar Island, but the western part was entirely devoid of soil and vegetation. Natural vegetation comprising of few species like *Morinda tinctoria, Ziziphus mauritiana, Abutilon hirtum, Urena lobata, Achyranthes aspera, Aerva lanata, Pupalia lappacea, Croton bonplandianus, Pavonia odorata, Tiliacora acuminata,* and *Rottboellia cochinchinensis* were found to occur in this Island. Species like *Catunaregam spinosa, Lepisanthes tetraphylla, Alangium salvifolium, Cipadessa baccifera, Cissus quadrangularis, Aristolochia indica, Mimosa rubicaulis, Toddalia asiatica, Eclipta alba, Clerodendrum inerme,* and *Anisomeles indica* form the major component of the vegetation of Tampara Island. Ambokona Island, which was quite close to Tampara Island was also in a degraded condition and covered with planted species like *Acacia auriculiformis.* The natural vegetation of this island was composed of *Cipadessa baccifera, Strychnos nux-vomica, Toddalia asiatica, Urena lobata, Lygodium flexuosum, Plumbago zeylanica, Merremia emarginata, Mucuna pruriens, Cocculus hirsutus, Derris scandens, Commelina longifolia, Eragrostis unioloides,* and *Alysicarpus monilifer.*

The Nalabana Island is a wintering ground for many migratory birds and gets completely submerged during monsoon. As the water recedes from the month of November, many species characteristic to salt marshes like *Paspalum distichum, Cressa cretica, Salicornia brachiata, Croton bonplandianus, Sesuvium portulacastrum,* and *Heliotropium curassavium* make their appearance. The grasses like *Cynodon dactylon, Diplachne fusca,* and *Paspalum paspalodes* started appearing as soon as the land surface became dry. The bar-headed goose (*Anser indicus*) forage on *Cynodon dactylon* and the grassy land provides an ideal habitat for nesting of Gull-billed Tern (*Gelochelidon nilotica*), River Tern (*Sterna aurantia*), Little Tern (*Sterna albifrons*), and waders like Black-winged Stilt (*Himantopus himantopus*) and Collared Pratincole (*Glareola pratincola*).

The Kalijai Island was under great biotic pressure due to grazing from the animals like goats. The plant species recorded from this island were *Crateva magna, Cressa cretica, Bulbostylis barbata, Barleria prionitis, Strychnos nux-vomica, Breynia rhamnoides, Morinda tomentosa, Ficus benghalensis, Cissus quadrangularis, Datura stramonium, Amaranthus spinosus,* etc. Besides, *Myriostachya wightiana* and *Porteresia coarctata* were occasionally found along the island margin close to the water level.

16.3.5 Vegetation on the Sandbar

The sand spit and the adjoining barren sandy areas covered an area of ~ 46 sq. km and formed dynamic coastal landforms subjected to erosion and accretion. *Hydrophylax maritima, Ipomoea pes-caprae, Sesuvium portulacastrum,* and *Launaea sarmentosa* were the common broad-leaved runners on sand dunes (Fig. 16.2t, u, v, w). *Hydrophylax maritima, Sesuvium portulacastrum,* and *Launaea sarmentosa*, growing on the neo-dunes bind the sand with the help of their extensive root systems. These species were fast succeeded by *Ipomoea pes-caprae* and other dominant herbaceous plants like *Cyperus arenarius, Spinifex littoreus, Bulbostylis barbata, Geniosporum tenuiflorum, Hedyotis graminifolia sp. arenaria, Brachiaria remota, Macrotyloma ciliatum, Acalypha lanceolata, Bacopa monnieri, Vigna sublobata, Crotalaria nana, Gisekia pharnaceoides, Alternanthera sessilis, Cyperus compressus, Cyperus rotundus, Mollugo pentaphylla, Fimbristylis acuminata, Croton bonplandianus, Tephrosia villosa, Cassia tora, Opuntia stricta var. dillenii, Oxalis corniculata, Heliotropium curassavicum, Allmannia nudiflora, Grangea maderaspatana, Crotalaria pallida,* and *Solanum trilobatum* appeared in the next stage of successional vegetation development on sand dunes followed by *Pandanus fascicularis, Morinda tinctoria, Calotropis gigantea,* and *Jatropha gossypifolia* (Fig. 16.2x). The most abundant species were *Ipomoea pes-caprae* and *Spinifex littoreus* that bind sand with their extensive running stems and nodal roots. *Pandanus odoratissimus* attained excellent growth due to congenial edaphic factor along the sandbar forming discontinuous patches. The common associates of screw pine were *Euphorbia tirucalli, Pentatropis capensis, Asystasia gangetica, Ipomoea obscura,* and *Tinospora cordifolia.* Shrubs like *Abutilon hirtum* and *Abelmoschus ficulneus* also occur in profusion.

The sandbar from Pitisal up to the Sanapatna was planted with *Casuarina equisetifolia* by the Forest Department exhibited a luxuriant growth and natural regeneration. Plants of *Azadirachta indica*, were observed in between the plantation of *Casuarina equisetifolia.* Sporadic nesting of Olive Riddley turtles takes place towards the sea side of the sandbar during nesting season. The wildlife experts are of the view that *Casuarina* plantation should be avoided along the sandbar as it hinders the turtle nesting.

16.3.6 Mangroves

Gupta and Khandelwal (1990, 1992) through palynostratigraphic studies reconstructed the mangrove flora of Chilika lagoon. They predicted the occurrence of some core as well as associate mangroves such as *Rhizophora, Bruguiera, Excoecaria, Sonneratia, Avicennia,* and *Nypa.* However, only a few mangrove associates like *Aegiceras corniculatum, Salvadora persica, Pongamia pinnata,* and *Cassipourea ceylanica* could be recorded in a degraded condition in the form of shrubs during the present survey. The confluence point of the rivers, Daya, Bhargavi, and Luna, which used to harbour mangrove vegetation, was now utterly devoid of such species. This was due to accelerated siltation and change in the land-use pattern in the lagoon basin. The lone mangrove species *Aegiceras corniculatum* was encountered on Bird Island. The mangrove associates recorded during the survey were *Pongamia pinnata, Clerodendrum inerme, Porteresia coarctata,* and *Myriostachya wightiana.* The mangrove species namely, *Rhizophora mucronata, Aegiceras corniculatum, Ceriops decandra,* and *Kandelia candel* were planted during 2005 on the mud flats along the outer channel of the lagoon were found to be growing luxuriantly.

16.3.7 Rare and Threatened Species

The species of rare or threatened plants recorded were *Cassipourea ceylanica* (Rhizophoraceae), *Colubrina asiatica* (Rhamnaceae), *Capparis roxburghii* (Capparaceae), *Maerua oblongifolia* (Capparaceae), *Macrotyloma ciliatum* (Fabaceae), *Indigofera aspalathoides* (Fabaceae), *Dimorphocalyx glabellus* (Euphorbiaceae), *Heliotropium curassavicum* (Boraginaceae), and *Halophila beccarii* (Hydrocharitaceae).

16.4 Plants of Special Habit and Habitat

16.4.1 Insectivorous Plants

Sundews and bladderworts were the common insectivorous plants encountered during the survey. *Drosera burmannii* and *Drosera indica* were the two sundews occurring along the sandbar and dunes. Bladderworts like *Utricularia aurea* and *Utricularia stellaris* occurred as submerged hydrophytes in the northern sector.

16.4.2 Parasites

Although a few parasitic angiosperms were found to occur in and around the lagoon, they are an interesting group of plants from the ecological standpoint. The parasites belong to two distinct types viz. stem parasites and root parasites and may be total or partial parasites. *Cuscuta reflexa* represents the stem parasites, *Cassytha filiformis, Dendrophthoe falcata,* and *Viscum articulatum*, the first two being leafless total parasites. *Cuscuta reflexa* occurred as a parasite on *Ziziphus mauritiana, Dichrostachys cinerea, Acacia nilotica ssp. indica*, *Morinda coreia,* and *Syzygium cumini*.

16.4.3 Epiphytes

During the present survey, no epiphytic species could be collected from the islands of the lagoon. However, epiphytic plants adjoining the lagoon were represented by three species of orchids namely *Acampe praemorsa, Cymbidium aloifolium,* and *Vanda tessellata* that were observed frequently on the trees like *Mangifera indica* and *Diospyros melanoxylon* growing along the roads in the periphery of the lagoon and adjacent villages. *Scindapsus officinalis* grew as an epiphyte on many host plants and rock surfaces.

16.4.4 Lithophytes

Caralluma adscendens, Sarcostemma acidum, *Tephrosia maxima,* and *Anisochilus carnosus* were the lithophytes growing on the barren rock surfaces of the central sector islands.

16.5 Ecology of Aquatic Plants

Transparency is one of the most important factors which affects the abundance and distribution of aquatic macrophytes in shallow coastal lagoons. Generally, submerged macrophytes grow up to a depth of 2–3 times the Secchi depth (Canfield et al. 1985; Chambers and Kalff 1985). In Chilika, the Secchi depth transparency varied between 28.0 and 202.0 cm which was reasonably good for the growth of macrophytes. The water transparency in the northern and central sectors was relatively poor due to the inflow of sediments through the rivers. The bottom sediment was re-suspended during summer due to churning action induced by the wind which limited the light penetration and thereby affected the growth of macrophytes.

The biomass productivity of the northern sector was highest. The dominant emergent species like *Phragmites karka* occurred almost in pure communities in this zone and produced high biomass of 19,040 g fresh wt/m^2. The biomass recorded from the central sector and the southern sector was 10,600 g fresh wt/m^2 and 7560 g fresh wt/m^2, respectively. The outer channel with the low nutrient in the water column and sediment, the biomass recorded was 576 g fresh wt/m^2.

Salinity played a distinct role with regard to occurrence, distribution, and abundance of the macrophytes. Maximum species diversity was noted in the northern sector, which remained predominantly freshwater during most part of the year. With the increase in the salinity level, the emergent species *Schoenoplectus littoralis* and the submerged aquatic taxa gradually decomposed and the blue-green algae *Lyngbya aestuarii* grew at its luxuriance on the decomposed mass. With the increase in the salinity during May 2001 post-hydrological intervention, the most conspicuous change observed was a decrease in the area covered by the water hyacinth (Pattnaik 2003).

The salinity level in the central sector was observed to fluctuate between freshwater to brackish (Pattnaik 2003). The dominant submerged aquatic plants of this sector were *Potamogeton pectinatus* and *Najas minor*. During winter, Nalabana Island was inhabited by submerged brackish water aquatics like *Potamogeton pectinatus, Halophila beccarii, Halophila ovata, Ruppia maritima*, and *Najas graminea* and few of the emergent species like *Schoenoplectus littoralis*. From January onward, as the water level receded, the island becomes exposed and the grasses, sedges, and herbs characteristic of salt marshes colonised and spread to the entire island by the end of March. *Paspalidium flavidum, Paspalum distichum, Cyperus rotundus, Panicum montanum, Sesuvium portulacastrum,* and *Salicornia brachiata* were the dominant species in the Nalabana Island.

Macrophytes are key elements in providing a suitable feeding and breeding ground to the birds of Nalabana Island. As a habitat management activity by the State Forest Department, narrow navigational channels were created. Due to evaporation of retained water during summer, these water bodies turned hypersaline resulting proliferation of *Salicornia brachiata* in the areas previously dominated by *Cyprus* species. This has altered the habitat used by the waders and their population was observed to be sharply declining due to shrinkage of habitat, which is posing a serious management problem.

The salinity level of the southern sector was brackishly ranging between 7.29 and 13.2 ppt (Barik et al. 2017). The shoreline vegetation encountered in this sector were emergent species like *Phragmites karka, Schoenoplectus littoralis, Paspalidium flavidum, Paspalum distichum,* and *Salicornia brachiata.* The submerged species like *Potamogeton pectinatus, Ruppia maritima,* and *Najas marina* were recorded in this sector. Throughout the year very less fluctuation of salinity was recorded in this sector. This was due to very less freshwater inputs and exchange of water from Rushikulya estuary through Palur canal. This supported healthy meadows of seagrass i.e., *Halophila ovata, Halophila beccarii, Halodule uninervis, Halodule pinifolia,* and *Cymodocea serrulata* in this zone. The macroalgae *Gracilaria verrucosa, Enteromorpha intestinalis,* and *Chaetomorpha linum* grew luxuriantly in this sector.

The salinity level of the outer channel happened to be similar to seawater except during monsoon. Due to favourable salinity, the seagrasses like *Halophila beccarii, Halophila ovata,* and *Halophila ovalis* were found growing in abundance in this sector. Extremely coarse-textured sediment (sand) along the shoreline of the outer channel, being nutritionally poor, did not support the growth of macrophytes. However, in the northern sector, the fine clay and sediment brought by the river and rivulets provided a favourable habitat for the luxuriant growth of the emergent macrophytes. The areas adjoining the jetties, the creeks of Nalabana Bird Island are the shallow zones from Kalupadaghat to Bhusandapur were eutrophic in nature due to high nutrient inputs. Interestingly, the high dissolved oxygen level was recorded from the creeks of Nalabana Island, which become eutrophic during winter because of the guanos deposited by the birds.

Aquatic plants provide feeding and sheltering habitat to a wide variety of organisms. The physical and chemical changes that macrophytes bring about in the littoral zone have a profound impact on other living organisms. There is a close relationship between the macrophytes and the epiphytes that grow on them. The epiphytes and the algal mats on the macrophytes are observed to be grazed by gastropoda, amphipods, isopods, and mycids. The dominant detritus feeders of economic importance in Chilika are shrimp and crab species, which mostly colonise in the littoral zone. A number of invertebrates and few species of fish and waterfowl, however, feed directly on aquatic macrophytes.

The Nalabana bird sanctuary is known to be an excellent spawning nursery ground for finfish and shellfish. The sanctuary is known to serve as feeding and nursery for 45 species of fish and spawning of 11 resident species, contributing significantly to the lake fishery (Pattnaik et al. 2018). The creeks of Krushna Prasad Island, the littoral zone of the southern sector and the outer channel with luxuriant seagrass meadows serve as a nursery ground for many shrimp and crab species. Similarly, *Potamogeton pectinatus* provides very good shelter for the sea bass in the central sector. The beds of *Enteromorpha compressa* growing luxuriantly attached to the rocky surface along the shoreline of islands serve as a feeding ground for the mullets.

The waterfowls like wigeon, pintail, shovellor, common poachard, tufted poachard, and curlew sandpiper were observed to feed on macrophytes like *Najas foveolata, Ruppia maritima, Halophila ovata,* and *Potamogeton pectinatus* in Nalabana Island. It was further observed that luxuriant growth of seagrasses like, *Halophila beccarii, Halophila ovata, Halodule uninervis,* and *Halodule pinifolia,* provide an excellent platform for the epiphytic and epizoic species to grow. Thus they provide an excellent habitat for the dabbling waterfowl that forage on the epiphytic algal mats in Nalabana Island. Invertebrates associated with the macrophyte beds are also an important component of the lagoon ecosystem. They produce the protein that is vital to many waterfowls and allied waterbirds (Swanson et al. 1979). The northern sector with the rich growth of emergents like *Phragmites karka* and *Schoenoplectus littoralis*, provide shelter to the avian fauna. The fruits of *Nymphaea pubescens, Nymphaea nouchali,* and *Nymphoides sp.* were good sources of food for the waterfowl. Many parts of the lagoon with submerged species like *Potamogeton crispus, Potamogeton nodosus, Potamogeton octandrus, Aponogeton natans, Ceratophyllum demersum, Hydrilla verticillata, Utricularia aurea, Najas graminea,* and *Najas minor*, were flocked by waterfowl. Wigeons were observed to forage on *Ruppia maritima*. The nutrient import and recycling by the migratory birds in a wetland ecosystem is also very significant (Anderson and Polis 1999).

16.6 Strategies for Management of Macrophytes

A host of ecosystem services flow from macrophytes, they provide refuge from predators to many aquatic animal species, roosting and nesting sites for the birds and also spawning and nursery grounds for a variety of finfish and shellfish and host of aquatic animal species. The initial step in the management will be to determine the types and quantities of macrophytes present, whether the macrophyte populations are expanding or declining, and the relationship(s) between the vegetation and the lake uses (Johnstone 1986). In a wetland, when a particular plant species grow in such a large proportion that they reduce or eliminate the growth of other desirable plants, or affect other biota, then they are assumed as weed. Aquatic weeds can be broadly defined as "unwanted and undesirable plants which grow and reproduce in an aquatic environment" (Lawrence 1966). This holds good for the macrophytes of the northern sector of Chilika. The dense growth of macrophytes like *Phragmites, Eichhornia,* and *Salvinia* cause hindrance in the movement of the waterfowls and ducks (e.g. in diving and dabbling ducks), and reduction in availability of food as the preferred species of macrophytes are eliminated due to shading effect.

Large volumes of water hyacinth enter into the Chilika lagoon through the distributaries of Mahanadi River and cover a sizeable water surface during monsoon. The abundance, spread and distribution of such floating and emergent species can be monitored through remote sensing techniques (Pattnaik 2003). The area covered by water hyacinth and water ferns was estimated to be 21.34 sq. km in October 2000 in Chilika lagoon (Fig. 16.3), based on the remote sensing tool (Pattnaik et al. 2002).

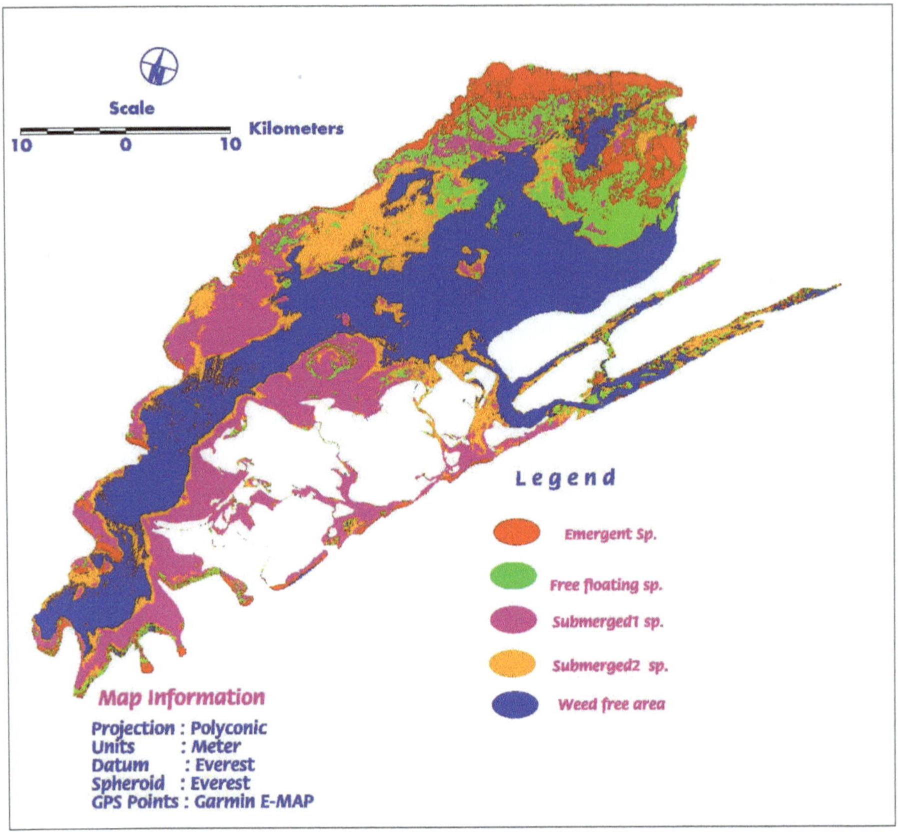

Fig. 16.3 Classified aquatic vegetation map of Chilika Lagoon. (Data Source (IRS 1D, LISS III). Date of Pass: 23/10/2000)

After the opening of the new mouth, the salinity regime of the northern sector improved which resulted in decomposition of water hyacinth and the water fern in the northern sector (Pattnaik 2003) (Fig. 16.4). The ecosystem approach adopted by the CDA for the restoration of the lagoon was effective in controlling the proliferation of invasive aquatic weeds (Pattnaik 2002).

Most of the emergent survive the dry periods with the help of their underground rhizomes, as they are adapted to grow in a wide range of water depths. It has been observed that *Phragmites karka* is fast spreading in the northern sector of the lagoon in large patches, particularly adjoining village of Mangalajodi and Kalupadaghat, which was previously infested with water hyacinth. *Phragmites karka* is known to spread quite rapidly by its rhizomes and can form pure stands in favourable conditions. *Phragmites* is a perinneal grass that can produce dense monoculture patches resulting into reduction of floral diversity (Hudon et al. 2005). The plant can survive for several years through underground rhizomes but shoots only survive for an year (Roberts 2013). *Phragmites* stands have potential to impact on ecological processes

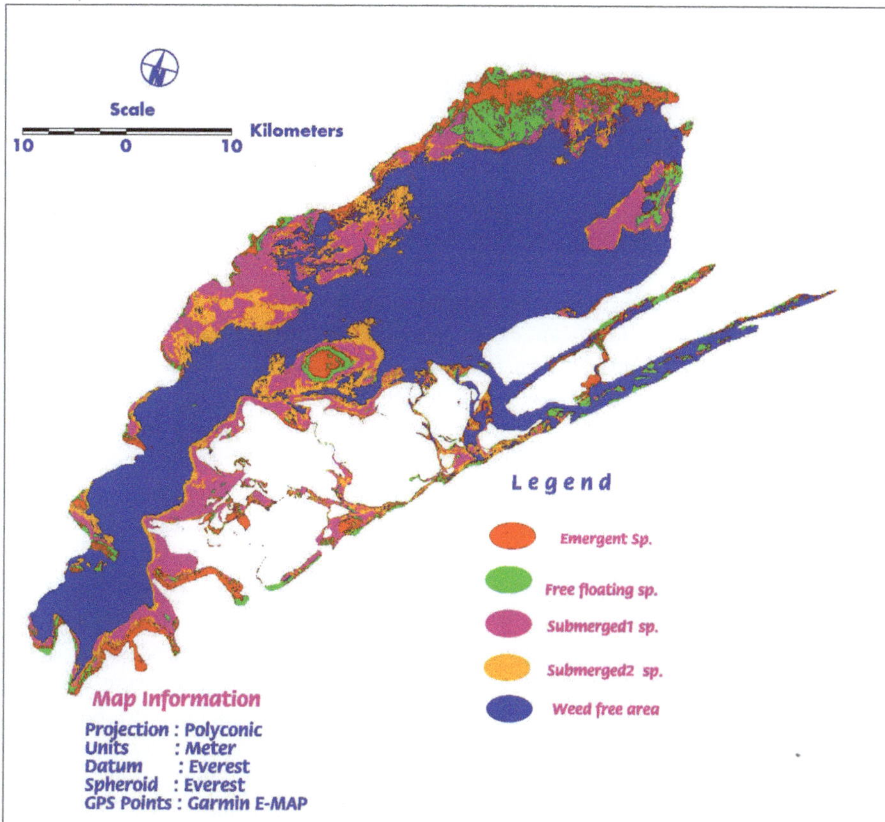

Fig. 16.4 Classified aquatic vegetation map of Chilika Lagoon. (Data Source (IRS 1D, LISS III). Date of Pass: 11/05/2001)

leading to habitat deterioration for some faunal groups (Hudon et al. 2005; Kodric-Brown et al. 2007; (Sun et al. 2007).

The area occupied by *Phragmites* increased from 27.82 sq. km in October 2000 to 50.00 sq. km in 2011 as assessed by use of the image-processing tool. In Chilika the dense monospecific stand attaining a mean tallest height of 334 cm with 121 stem has resulted in the reduction of available open water space and bringing about an alteration in the structure and function of the diverse aquatic eco-system including reduced species richness. The monospecific stand of *Phragmites* in the northern sector of the lagoon is accelerating sedimentation, causing hindrance to the movement of the boats, fishery activities and in the movement of the waterfowl.

Chemical and biological control of macrophytes is a complex process. Many chemicals are used to kill the aquatic plants selectively. In a large wetland like Chilika, it has got a limited scope for application, as no herbicide is safe to any aquatic ecosystem. Apart from this, the use of herbicides to kill the targeted species of macrophyte, may result in release of a large quantum of nutrients to the water leading to an eutrophic condition. Since most of the herbicides are species specific, when applied to a mixed population, tolerant species grow at the cost of susceptible ones. An attempt to address *Phragmites* invasion in the lagoon using chemical methods was attempted by CDA in 2007–2008, in collaboration with Orissa University of Agriculture and Technology (OUAT), on a pilot basis. Herbicide Glyphosate (15 ml per liter) of water was applied to foliage over 5.4 hectares of area infested by *Phragmites.* This, however, was of limited success and raised several questions related to the short and long-term implications of the herbicide on the ecology of the wetland as well as the biota living therein. The treated area regenerated after 2 years, and the *Phragmites* was back. In addition, controlling *Phragmites* with herbicides can cause negative impacts on other flora species (Güsewell 2003). One observation made during the survey is that *Phragmites* is not expanding to deeper zone, beyond the water depth of 0.5 m. It is being observed that after 2011 there has been no further expansion of *Phragmites* in the northern sector beyond 50.00 sq. km. The strategy adopted by CDA to reduce the silt flow in to the lagoon by adopting integrated management of drainage basin, would be helpful in controlling the spread of *Phragmites* in the northern sector in long-term.

Seagrass is an essential indicator of the health of the ecosystem in which they grow, and provide a host of ecosystem services. During the last 18 years after the hydrological intervention, extensive seagrass meadows have flourished along the creeks of the central sector and the shorelines of the southern sector. Prior to the opening of the new mouth the seagrass meadows of the lagoon were 24.8 sq. km. However, after restoration, the seagrass meadows expanded to 86.84 sq. km. in 2004 and 102 sq. km. in 2012 and to 152 sq. km. in November 2018 (CDA). The present survey also indicates the increase in the biomass of the *Halophila ovalis* after the opening of the new mouth. The biomass of *Halophila ovalis* reported from the southern sector and the outer channel by Patnaik (1973) and during the present study has been estimated at 250.5 g fresh wt./m^2 and 2880 g fresh wt./m^2 respectively, during winter season. During the survey period four species of seagrasses, i.e. *Halodule pinifolia, Halodule uninervis, Halophila ovata* and *Cymodocea serrulata* were recorded for the first time from the lagoon. Therefore, existing seagrass beds need to be preserved and sustainably managed to facilitate their growth and spread. Restoration of seagrass beds will help in the improvement of the water quality and provide habitat and food for both birds and fishes. It was found during the survey that most of the seagrass meadows in the creeks of Krushna Prasad Island are degrading due to unauthorized shrimp culture by raising earthen embankments across the creeks. The seagrasses are also being damaged by the propellers of the boats and due to fixing of the prawn traps by the fishermen. The expansion of the seagrass meadows and the species diversity are good signs of improvement of the ecosystem, which need to be monitored regularly.

In Chilika lagoon, the coverage of the macrophytes is more than 50% of the total water spread area and is associated with an intricate and complex food web. Chilika is a macrophyte dominated ecosystem, so care should be taken before launching any massive de-weeding programme, as it may lead to appearance and spread of algal blooms, as experienced in case of Dal lake (Ticku and Justshi 1993). Macrophytes are a vital component of the aquatic ecosystem, and they assume more significance in case of Chilika because of its shallow nature and high level of productivity. The river basin approach adopted by CDA for management of drainage basin of the lagoon, appears to be appropriate to regulate the flow of nutrient into the lagoon (Pattnaik 2001a). Based on the present survey, the measures recommended for conservation and management of macrophytes in the lagoon are as follows;

(i) The integrated management of the drainage basin, adopted by CDA to reduce the silt flow into the lagoon needs to be continued till all the degraded micro watersheds are treated. (ii) To check the nutrients, flow from the drainage basin to the lagoon, rational use of chemical fertilizer, insecticides, and pesticides and organic farming need to be promoted. (iii) Monitoring of nutrient load and pollutants due to the land based activities in the drainage basin both from the point and nonpoint sources need to be meticulously carried out. (iv) The shoreline of the lagoon is an important ecotone that supports important bio-geochemical cycles, luxuriant growth of macroalgae and seagrasses, which needs to be kept unaltered and encroachment-free. (v) The seagrass meadows of central sector along the Krushna Prasad Island are adversely affected due to encroachment for shrimp culture, and these sensitive ecosystems need to be kept entirely free from encroachment. (vi) Artificial regeneration of seagrass in the areas freed from encroachment along the shorelines may be attempted on a pilot basis following standard protocols. (vii) Some uninhabited islands are excellent sites for occurrence of interesting species and for the process of speciation. The occurrence of an apparent endemic species namely, *Cassipourea ceylanica* was recorded from the Barkuda and Sanakuda Islands which are now in severely degraded conditions due to biotic pressure, warrants special attention. (viii) For management of the invasive species in a wetland like Chilika with complex food webs, a clear understanding is essential before taking any decision for removal/eradication of a particular species or component. The eco-system approach adopted by CDA for the restoration of the lagoon seems appropriate for curbing the growth of invasive species like water hyacinth. (ix) Monitoring of the macrophytes should be carried out at a regular interval by use of remote sensing technique which was found to be very effective and useful. (x) Steps need to be taken to reintroduce the mangrove in potential areas like the river confluence points and along the outer channel and the shorelines of the Islands, which used to harbour mangrove vegetation earlier.

Acknowledgement The authors would like to sincerely acknowledge the World Bank supported Integrated Coastal Zone Management Project Odisha (Credit No. 4765-IN) to Chilika Development Authority through which project entitled "Qualitative and quantitative assessment of biological diversity and distribution pattern of macrophytes (angiospermic plants) of Chilika Lagoon and its adjoining regions" was implemented by the Regional Plant Resource Centre, Bhubaneswar, Odisha, India. The support received from the project field staff Dr. D. Sahoo, Dr. P. K. Acharya, Sri D. P. Barik, Sri S. K. Kar, Sri P. K. Tripathy, Sri S. Mohanty who worked as Research Fellow under the project is sincerely acknowledged.

References

Anderson WB, Polis GA (1999) Nutrient fluxes from water to land: seabirds affect plant nutrient status on Gulf of California islands. Oecologia 118:324–332

Barik SK, Muduli PR, Mohanty B, Behera AT, Mallick S, Das A, Samal RN, Rastogi G, Pattnaik AK (2017) Spatio-temporal variability and the impact of Phailin on water quality of Chilika lagoon. Cont Shelf Res 136:39–56

Beddome RH (1872) The flora sylvatica for Southern India, vol 194. Gantz Brothers, Madras

Bhatta KS, Pattnaik AK (2002) Environmental monitoring of Chilika Lagoon. In: Proceedings of the international workshop on sustainable development of Chilika Lagoon, Chilika Development Authority, pp 95–113

Brahmam M, Dhal NK, Rout NC (2001) Floristic diversity of Orissa-a critical assessment of new, rare, endemic and threatened taxa for in-situ and ex-situ conservation. In: Sahoo S, Ramesh DB, Rao YR, Debata BK, Mishra VN (eds) Conservation and utilization of medicinal and aromatic plants. Orissa Botanical Society, Bhubaneswar, pp 14–21

Canfield DE, Langeland KA Jr, Linda SB, Haller WT (1985) Relations between water transparency and maximum depth of macrophyte colonization in lakes. J Aquat Plant Manag 23:25–28

Carpenter SR, Lodge DM (1986) Effects of submerged macrophytes on ecosystem processes. In: Sand-Jensen K (ed) Submerged macrophytes: carbon metabolism, growth regulation and role in macrophyte-dominated ecosystems. Aquatic botany, vol 26. Elsevier, Amsterdam, pp 341–370

Cattaneo A (1983) Grazing on epiphytes. Limnol Oceanogr 28:124–132

CDA (2001) Report on Chilika: environmental monitoring, vol 103. Chilika Development Authority, Bhubaneswar

Chambers PA, Kalff J (1985) Depth distribution and biomass of submersed macrophyte communities in relation to secchi depth. Can J Fish Aquat Sci 42:701–709

Chambers PA, Lacoul P, Murphy KJ, Thomaz SM (2008) Global diversity of aquatic macrophytes in freshwater. In: Freshwater animal diversity assessment. Hydrobiologia, vol 595. Springer, Dordrecht, pp 9–26

Crowder LB, Cooper WE (1982) Habitat structural complexity and interaction between bluegills and their prey. Ecology 63:1802

Das NK, Samal RC (1988) Environmental survey of Chilika. In: Chilika, the pride of our wetland heritage. Orissa Environment Society, Bhubaneswar, pp 96–103

Dvořaki J, Bestz EPH (1982) Macro-invertebrate communities associated with the macrophytes of Lake Vechten: structural and functional relationships. Hydrobiologia 95:115–126

Engel S (1988) The role and interactions of submerged macrophytes in a shallow Wisconsin lake. J Freshw Ecol 4:329–341

Forman LL, Bridson DM (1989) The herbarium hand-book. Royal Botanic Gardens, Kew

Gupta HP, Khandelwal A (1990) Mangroves of India: history and palynostraigraphy of Chilika lake, Orissa. In: Jain KP, Tiwari RS (eds) Proceedings of the symposium "Vistas in Indian paleobotany". Paleobotanist, vol 38. Birbal Sahni Institute of Palaeobtoany, Lucknow, pp 379–393

Gupta HP, Khandelwal A (1992) Mangrove development at Nalban Island, Chilika lake: a palynological interpretation. In: Venkatchela BS, Jain KP, Awasthi N (eds) Proceedings of the Birbal Sahni birth centenary palaeobotanical conference. Geophytology, vol 22. Palaeobotanical Society, Lucknow, pp 235–238

Güsewell S (2003) Management of *Phragmites australis* in Swiss fen meadows by mowing in early summer. Wetl Ecol Manag 11:433–445

Haines HH (1921–1925) The botany of Bihar and Orissa. In: Singh BSMP (eds) Botanical survey of India, Dehradun, vol 3, pp 825–1372

Heck KL, Crowder LB (1991) Habitat structure and predator-prey interactions in vegetated aquatic systems. In: Bell SS, McCoy ED, Mushinsky HR (eds) Habitat structure, the physical arrangement of objects in space. Chapman & Hall, London, pp 281–299

Hudon C, Gagnon P, Jean M (2005) Hydrological factors controlling the spread of the common reed (*Phragmites australis*) in the St. Lawrence River (Quebec, Canada). Ecoscience 12:347–357
Jain SK, Rao RR (1977) A handbook of field and herbarium methods. Today & Tomorrow's Printers and Publishers, New Delhi
Johnstone IM (1986) Macrophyte management: an integrated perspective. N Z J Mar Freshw Res 20:599–614
Kodric-Brown A, Wilcox C, Bragg J, Brown J (2007) Dynamics of fish in Australian desert springs: role of large-mammal disturbance. Divers Distrib 13:789–798
Lawrence JM (1966) Aquatic weed control in fish pond. FAO Fish Rep 44:76–91
Mooney HO (1950) Supplement to the botany of Bihar and Orissa. Catholic Press, Ranchi
Nair NC, Henry AN (1983) Flora of Tamil Nadu, India, vol 1. Botanical Survey of India, Calcutta, p 148
Narayanswami V, Carter HG (1922) Systematic list of the plants of Barakuda. Mem Asiat Soc Beng 7:289–319
Nayar TS, Begum AR, Mohanan N, Ragakumar G (2006) Flowering plants of Kerala. Tropical Botanical Garden and Research Institute, Palode, Tiruvanathapuram, pp 501
Panda PC, Patnaik SN (1988) A contribution to the flora and vegetation of Chilika wild life Sanctuary, Orissa. In: Chilka (ed) The pride of our wetland heritage. Orissa Environmental Society, Bhubaneswar, pp 81–85
Panda PC, Pattnaik SN (2002) An enumeration of the flowering plants of Chilika Lake and its immediate neighbourhood. In: Proceedings of the international workshop on sustainable development of Chilika lagoon, Chilika Development Authority, Bhubaneswar, pp 121–138
Panda PC, Choudhury BP, Patnaik SN (1985) The genus *Macrotyloma* (Wt. & Arn.). Verdc. (Fabacea) in Orissa. J Econ Taxon Bot 7:631–633
Pattnaik AK (2000) Desiltation of outer channel of Chilika lagoon: a step to restore its ecosystem. 6th INTECOL, Quebec. Millennium wetland event. Canada, 176 (Abstract)
Pattnaik AK (2001a) Participatory watershed management of Chilika lagoon – a Ramsar site of India. 9th international conference on the conservation & management of lakes, Japan, pp 454–457
Pattnaik AK (2001b) Opening of a new mouth – a step to restore the ecosystem of Chilika lagoon – a Ramsar site of India. In: Proceedings of the workshop on environment manual & wise use of Brackish water lakes, lagoons & shallow bays. 9th international conference on conservation and management of lakes, Otsu, Japan, pp 14–18
Pattnaik AK (2002) Rejuvenation of Chilika lagoon, India. An example of ecological restoration with community participation. Workshop on environmental governance and sustainable development: the contribution of The Ramsar Convention; GBF 17 – Valencia, pp 1–4
Pattnaik AK (2003) Survey and characterisation of phytodiversity of Chilika Lake. PhD thesis, Utkal University, Bhubaneswar
Pattnaik AK, Samal RN, Bhatta KS (2002) Aquatic vegetation monitoring of Chilika Lagoon through remote sensing technique (Abstract) internat. Workshop on restoration of Chilika Lagoon. Chilika Development Authority, Bhubaneswar, pp 15
Pattnaik AK, Panda PC, Patnaik SN (2003) Flora and vegetation of Chilika lagoon, Orissa: a synoptic analysis. Plant Sci Res 25(1&2):1–12
Pattnaik AK, Panda PC, Patnaik SN (2008) The diversity, distribution and taxonomy of sea grasses of Chilika lake, Orissa- the largest brackish water lake of India. J Econ Taxon Bot 32:381–392
Pattnaik AK, Mohanty SK, Parida S, Patnaik S (2018) Can scientific evidence to depict the ecosystem services effectively resolve the conflict between the protected area mangers and the local dependent community? Nalabana Bird Sanctuary; Chilika Lake, A wetland of international importance, India; A case study (Abstract). 17th world lake conference, Ibaraki, Japan, pp 131
Prabhugaonkar A, Dessai JRN, Janarthanam MK (2007) Extended distribution of *Cassipourea ceylanica*, an inland Rhizophoraceae. Indian J For 30:81–82

Roberts J (2013) *Phragmites australis*: knowledge to support its management on GAB springs. In: Gotch T (ed) Groundwater-dependent ecosystems of the Western Great Artesian Basin. National Water Commission Canberra, Canberra

Saxena HO, Brahmam M (1994–96) Flora of Orissa, vol 1–4. OFDC & RRL, Bhubaneswar

Schramm HL, Jirka KJ, Hoyer MV (1987) Epiphytic macroinvertebfrates on dominant macrophytes in two Central Florida lakes. J Freshw Ecol 4:151–161

Sun H, Brown A, Coppen J, Steblein P (2007) Response of *Phragmites* to environmental parameters associated with treatments. Wetl Ecol Manag 15:63–79

Swanson GA, Krapu GL, Serie JR (1979) Foods of laying female dabbling ducks on the breeding grounds. In: Bookhout TA (ed) Waterfowl and wetlands—an integrated review. North- Central Section, The Wildlife Society, Madison, pp 47–57

Ticku A, Zutshi DP (1993) Effects of weed cutting on species composition and abundance of plankton population. Int J Ecol Environ Sci 19:197–204

Correction to: Fish and Fisheries of Chilika: Post-Restoration Scenario

Surya K. Mohanty and Debabrata Panda

Correction to:
Chapter12 in: C. M. Finlayson et al. (eds.), *Ecology, Conservation, and Restoration of Chilika Lagoon, India*, Wetlands: Ecology, Conservation and Management 6, https://doi.org/10.1007/978-3-030-33424-6_12

The reference section of this chapter was inadvertently left out of the original publication. The references are listed below.

References

Annandale N, Kemp S (1915) Fauna of the Chilika Lake: introduction. Mem Indian Mus 5(1):1–20

Barman RP, Mishra SS, Kar S, Mukherjee P, Saren SC (2007) Marine and estuarine fish fauna of Orissa. Rec Zool Surv India, Occ Paper 260:1–186

Chaudhuri BL (1916a) Description of two new fishes from Chilika Lake. Rec Indian Mus 12 (3):105–108

Chaudhuri BL (1916b) Fauna of the Chilika Lake: fish part I. Mem Indian Mus 5(4):403–440

Chaudhuri BL (1917) Fauna of the Chilika Lake: fish part III. Mem Indian Mus 5(6):491–508

Chaudhuri BL (1923). Fauna of the Chilika Lake. Fish, part IV. Mem Indian Mus 5(11):711–736

CIFRI (2015) Annual report (2014–15). ICAR-Central Inland Fisheries Research Institute, Barrackpore, Kolkata, 164p

Dujovny E (2009) The deepest cut: political ecology in the dredging of a new sea mouth in Chilika Lake, Orissa, India. Conserv Soc 7(3):192–204

Eschmeyer WN, Fong JD (2014) Species of fishes by family/subfamily. Online version dated. 26.03.2014. http://research.calacademy.org/research/ichthyology/catalog/SpeciesByFamily.asp

Froese R, Pauly D (eds) (2015) FishBase. World Wide Web electronic publication, version 08/ 2015. Accessed at http://www.fishbase.org, 13 Sept 2015

Hora SL (1923) Fauna of the Chilika Lake: fish part-V. Mem Indian Mus 5(11):737–769

The online version of the chapter can be found at
https://doi.org/10.1007/978-3-030-33424-6_12

C. M. Finlayson et al. (eds.), *Ecology, Conservation, and Restoration of Chilika Lagoon, India*, Wetlands: Ecology, Conservation and Management 6,
https://doi.org/10.1007/978-3-030-33424-6_17

https://en.wikipedia.org/wiki/List_of_Ramsar_sites_in_India. Accessed on 20.04.2016
https://en.wikipedia.org/wiki/List_of_largest_fish. Accessed on 15 Apr 2019
IUCN (2014) IUCN red list of threatened species, Version 2014.2. Electronic Database accessible at http://www.iucnredlist.org/. Captured on 13 Oct 2014
Jeyabaskaran R, Wafar S, Wafar M (2002) CD on Brachyuran Crabs of West Coast, India. National Institute of Oceanography, Dona Paula, Goa, India
Jha BC, Mukhopadhaya MK, Nath D, Bhaumick U, Tyagi RK, Mishra RN, Samanta S (2005) Assessment of fish yield potential of Chilika Lake (Final report), 115p
Jhingran VG, Natarajan AV (1966) Final report on the fisheries of the Chilika Lake (1957–1965). Central Inland Fisheries Research Institute, Bulletin No 8:1–12
Jhingran VG, Natarajan AV (1969) A study of the fisheries and fish populations of the Chilika Lake during the period 1957–1965 India. J Inland Fish Soc 1:49–126
Jhingran VG, Natarajan AV (1973) Fishery resources of the Chilika Lake and its bearing on fisheries in adjacent areas of Bay of Bengal. In: Proceedings of the symposium on living resources of the seas around India. Special Publication, Central Marine Fisheries Research Institute, Cochin, pp 365–372
Jones S, Sujansingani KH (1952) Notes on the crab fishery of the Chilika Lake. J Bombay Nat Hist Soc 51(1):128–134
Jones S, Sujansingani KH (1954) Fish and fisheries of the Chilika Lake with statistics of the catches for the years 1948–1950. Indian J Fish 1(1&2):256–344
Lakra WS, Sarkar UK, Gopalakrishnan A, Kathirvelpandian A (2010) Threatened freshwater fishes of India. National Bureau of Fish Genetic Resources (ICAR), Lucknow, 20p
Mangla B (1989) Chilika Lake: desilting Asia's largest brackish water lagoon. AMBIO J Hum Environ 18(5):298
Maya Deb 1995. Crustacea: Brachyura. Wetland ecosystem series 1: Fauna of Chilika Lake. Zool Surv India:345–366.
Menon AGK (2004) Threatened fishes of India and their conservation. Zoological Survey of India Publication, 170p
Mohanty SK (1975) Further observations on the crab landings in the Chilika Lake during 1973. Bull Dept Mar Sci, University of Cochin 6(3):645–652
Mohanty RK, Mohapatra A, Mohanty SK (2009) Assessment of the impacts of a new artificial lake mouth on the hydrobiology and fisheries of the Chilika Lake, India. Lakes Reserv Res Manag 14:231–245
Mohapatra A, Mohanty SK, Mishra SS (2015) Fish and shellfish fauna of Chilika lagoon: an updated checklist. In: Venkataraman K, Sivaperuman C (eds) Marine faunal diversity in India, taxonomy, ecology & conservation. Academic Press
Mohanty SK, Mohapatra A, Mohanty RK, Bhatta KS, Pattnaik AK (2006) Occurrence and biological outlines of two species of Scylla (De Haan) in Chilika lagoon, India. Indian J Fish 53(2):191–202
Mohapatra A, Mohanty RK, Mohanty SK, Bhatta KS, Das, NR (2007) Fisheries enhancement and biodiversity assessment of fish, prawn and mud crab in Chilika lagoon through hydrological intervention. Wetlands Ecol Manag 15(3):229–251
Mohanty SK, Bhatta KS, Mohanty RK, Mishra S, Mohapatra A, Pattnaik AK (2008) Eco-restoration impact on fishery biodiversity and population structure in Chilika Lake. In: Monitoring and modeling lakes and coastal environments. Springer, Dordrecht, pp 1–21
Mohanty SK, Mishra SS, Khan M, Mohanty RK, Mohapatra A, Pattnaik AK (2015) Ichthyofaunal diversity of Chilika Lake, Odisha, India: an inventory, assessment of biodiversity status and comprehensive systematic checklist (1916–2014). Check List 11(6):1–19
Molur S, Walker S (eds) (1998) Report of the workshop "Conservation assessment and management plan for freshwater fishes of India". In: Zoo outreach organization, conservation breeding specialist group, India, Coimbatore, India, 156p
Nelson JS (2006) Fishes of the world, 4th ed. Wiley, Hoboken, 601p

Ng PKL, Guinot D, Davie PJF (2008) Systema Brachyurorum: Part I. An annotated checklist of extant Brachyuran Crabs of the world. Raffles Bull Zool (National University of Singapore) 17:1–286

Panda D (2013) Final report on fisheries and fish yield potential of Chilika lagoon during the post-engineering intervention. Central Inland Fisheries Research Institute (ICAR), Barrackpore, Kolkata, 104p

Pattnaik AK, Kumar R (2016) Lake Chilika (India): ecological restoration and adaptive management for conservation and wise use. In: Finlayson CM, Milton GR, Prentice RC, Davidson NC (eds) The wetland book: distribution, description and conservation. Springer, Dordrecht, pp 1–14. https://doi.org/10.1007/978-94-007-6173-5

Ponniah AG (1993) Categorisation of Indian threatened fishes. In: Dehadrai P-I, Das SR, Verma P, Natcon SR (eds) Threatened fishes of India, Pub No. 4, pp 375–387

Radhakrishnan EV, Deshmukh VD, Maheswarudu G, Josileen J, Dineshbabu AP, Philipose KK, Sarada PT, Lakshmi Pillai S, Saleela KN, Chakraborty R, Dash G, Sajeev CK, Thirumilu P, Sridhara B, Muniyappa Y, Sawant AD, Vaidya NG, Johny RD, Verma JB, Baby PK, Unnikrishnan C, Ramachandran NP, Vairamani A, Palanichamy A, Radhakrishnan M, Raju B (2012) Prawn fauna (Crustacea: Decapoda) of India – An annotated checklist of the Penaeoid Sergestoid, Stenopodid and Caridean prawns. J Mar Biol Ass India 54(1):50–72

Rajan S, Pattnaik S, Basu NC (1968) New records of fishes from the Chilika Lake. J Zool Soc India 20(1&2):80–93

Ramarao KV (1995) Pisces. In: Fauna of Chilika Lake. Wetland ecosystem series 1: ZSI, Calcutta, India, pp 483–506

Reddy KN (1995) Crustacea: Decapoda prawns & shrimps. Wetland ecosystem Series 1: Fauna of Chilika Lake. Zool Sur India:367–389

Remesan MP, Pravin P, Pradhan BK, Meenakumari B (2011) Fishing methods of Chilika Lagoon, CIFT Information Series, Central Institute of Fisheries Technology, Cochin, 24p

Sarkar SK, Bhattacharya A, Bhattacharya AK, Satapathy KK, Mohanty AK, Panigrahi S (2012) Chilika Lake. In: Bengtsson L, Herschy RW, Fairbridge RW (eds) Encyclopedia of lakes and reservoirs. Springer, pp 148–155

Sekhar NU (2007). Social capital and fisheries management: the case of Chilika Lake in India. Environ Manag 39(4):497–505. https://doi.org/10.1007/s00267-006-0183-0

Suresh VR, Manna RK, Panda D, Manas HM, Samanta S, Nag SK (2015) Post-restoration assessment of the ecology and fisheries diversity of Chilika Lake. In: 7th Bi-annual report of CIFRI/CDA-ICZMP Consultancy Research Project, 2015, 61p

FSC
www.fsc.org
MIX
Papier | Fördert
gute Waldnutzung
FSC® C083411

Kolibri 360 GmbH
Ferdinand-Jühlke-Straße 7
99095 Erfurt, Deutschland
produktsicherheit@kolibri360.de